CURRENT TRENDS IN ORGANIC SYNTHESIS

CURRENT TRENDS IN ORGANIC SYNTHESIS

Edited by

Carlo Scolastico and Francecso Nicotra

University of Milan
Milan, Italy

Springer Science+Business Media, LLC

Library of Congress Cataloging-in-Publication Data

Current trends in organic synthesis / edited by Carlo Scolastico and
    Francesco Nicotra.
        p.   cm.
    "Proceedings of the 12th International Conference on Organic
    Synthesis, held June 28-July 2, 1998, in Venezia, Italy"--T.p.
    verso.
    Includes bibliographical references and index.
    ISBN 978-1-4613-7175-5
    1. Organic compounds--Synthesis Congresses.   I. Scolastico,
    Carlo. II. Nicotra, Francesco. III. IUPAC Symposium on Organic
    Synthesis (12th : 1998 : Friuli-Venezia Giulia, Italy)
    QD262.C87 1999
    547'.2--dc21   99-32449
                                                                CIP

Proceedings of the 12th International Conference on Organic Synthesis, held June 28–July 2, 1998, in Venezia, Italy

ISBN 978-1-4613-7175-5   ISBN 978-1-4615-4801-0 (eBook)
DOI 10.1007/978-1-4615-4801-0

© 1999 Springer Science+Business Media New York
Originally published by Kluwer Academic/Plenum Publishers in 1999
Softcover reprint of the hardcover 1st edition 1999

10 9 8 7 6 5 4 3 2 1

A C.I.P. record for this book is available from the Library of Congress.

PREFACE

The last two decades have seen a rapid growth in the synthetic processing of both simple and complex molecules, aimed at meeting the needs of society in all aspects of life. Many efforts have been devoted to the development of new biologically active compounds, new materials with innovative properties such as bio-compatibility, new catalysts that allow highly selective transformations, and technologies that facilitate the synthetic processes.

This book is a compendium of recent progress in all these aspects of synthetic chemistry. It collects the lectures of the XII International Conference on Organic Synthesis, held in Venice from June 28 to July 2, 1998, in which the present state of art of this discipline has been reported.

The topics covered include: combinatorial chemistry, new synthetic methods, stereo-selective synthesis, metal-mediated synthesis, and target oriented synthesis. The book collects the contributions, in the mentioned topics, of 43 scientists from 19 different countries. The contributions presented in the Conference as plenary lectures are reported in the first section of the book. Particular attention has been dedicated to combinatorial chemistry, a new and promising methodology for the synthesis of libraries of pharmaco-logically interesting compounds in order to allow the automatic pharmacological screening of thousands of compounds. The Conference has dedicated to combinatorial chemistry a mini-symposium in which scientists from academy and companies have described the current trends of this very new technology.

CONTENTS

Plenary Lectures

Minisymposium on Combinatorial Chemistry

viii

New Synthetic Methods

Stereoselective Synthesis

Metal-Mediated Synthesis

Target Oriented Synthesis

CHIRAL RELAY AUXILIARIES

Steven D. Bull, Stephen G. Davies,* David J. Fox, A. Christopher Garner, and Thomas G. R. Sellers

Dyson Perrins Laboratory
University of Oxford
South Parks Road
Oxford
UK OX1 3QY

INTRODUCTION

Chiral auxiliaries and templates are effective tools for the asymmetric synthesis of homochiral molecules.[1] Most chiral auxiliaries are small heterocyclic compounds which rely on sterically demanding functional groups to control the conformation of their ring systems. Under ideal circumstances, the conformation of an auxiliary should be constrained to ensure that its prochiral centre reacts with a reagent *via* diastereoisomeric transition states which are sufficiently different in energy to ensure that only a single diastereoisomer is formed as product. In order to maximise the diastereoselectivity observed for an auxiliary, it would appear reasonable that the stereocontrolling functional group adopts a position in space as close as possible to the newly forming stereogenic centre. Structural considerations dictate that realisation of this ideal is not always attainable and numerous examples of chiral auxiliaries that rely on relatively remote stereogenic centres to control diastereoselectivity are known. Alkylation of the enolates of Seebach's imidazolidinone (**1**),[2] or Schöllkopf's bis-lactim ether auxiliary (**2**),[3] for example, are controlled *via* 1,3- and 1,4- asymmetric induction respectively (scheme 1).

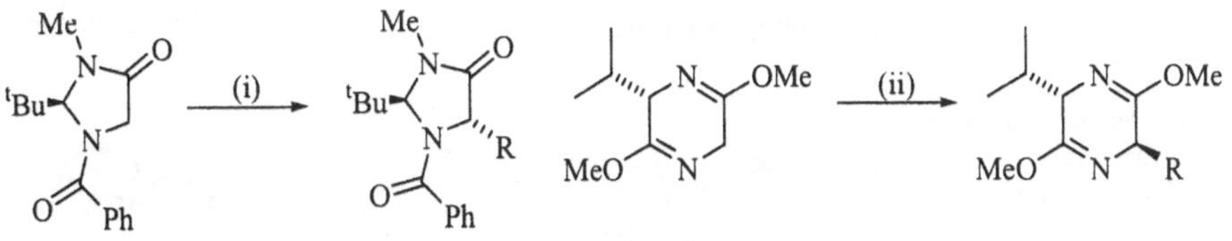

1 2

Reagents: (i) LDA, RX; (ii) ⁿBuLi, RX

Scheme 1

Current Trends in Organic Synthesis
Edited by Scolastico and Nicotra, Kluwer Academic/Plenum Publishers, 1999

PRINCIPLES OF CHIRAL RELAY AUXILIARIES

A vast body of work has demonstrated that the factors which control the diastereoselectivities of chemical reactions are a complex function of both steric and electronic interactions, which combine to transfer, or 'relay', stereochemical information from the stereogenic centre to the site of reaction. The complex interplay of these interactions is often very subtle, and numerous examples exist where small changes in bond angles, or heteroatom hybridisation can result in large changes in diastereoselectivities. Seebach *et al*, for example, have reported that modifying the *N*-protecting group of the imidazolidinone derived auxiliaries (**3**) results in large improvements in the observed diastereoselectivities during enolate alkylation (scheme 2).[2]

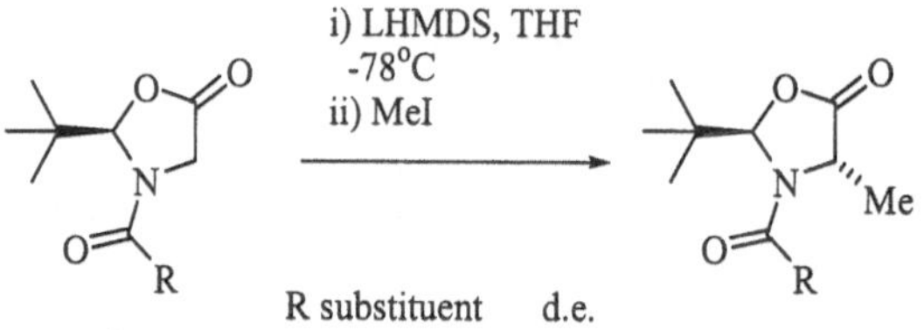

R substituent	d.e.
^tBu	50%
Ph	88%
OPh	96%

Changing protecting group improves diastereoselectivity

Scheme 2

Since it is increasingly apparent that protecting groups do not always act as passive spectators, an alternative strategy for auxiliary design can be proposed in which an achiral conformationally flexible group is inserted between the stereogenic centre and the prochiral reactive centre (scheme 3). Steric interactions between the functional group **R** of the stereogenic centre and the conformationally mobile group **Y** serve to fix the relative 1,2-stereochemistry antiperiplanar, thus directing any incoming reactant at the prochiral centre *anti* to the conformationally mobile group. In ideal circumstances, the conformationally flexible group should serve to both relay, and amplify the stereochemical information of the existing stereogenic centre, thus enabling efficient control of diastereoselectivity.

Conventional auxiliary design   *trans*- product

R=stereocontrolling group
X=heteroatom
Y=relay group

Chiral relay auxiliary design   *cis*- product

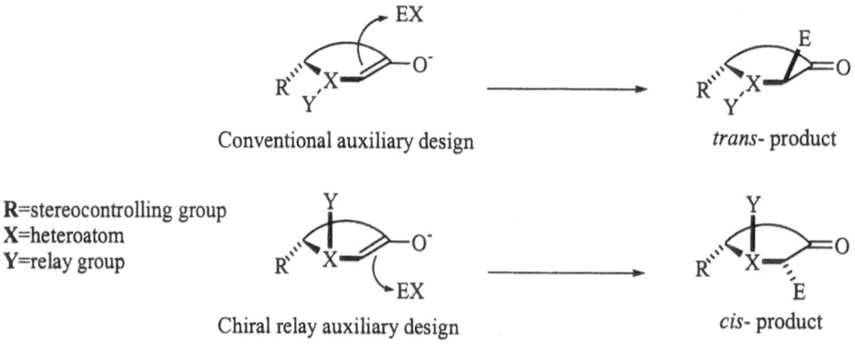

Scheme 3

CHIRAL AUXILIARIES EMPLOYING A SINGLE RELAY UNIT

It is important to note that the presence of a single chiral relay within a chiral auxiliary results in inversion of the stereodirecting information of the stereogenic centre, thus providing a simple diagnostic tool to determine when this type of effect is operating. Careful examination of the structures of known chiral auxiliaries revealed a morpholin-2-one based system where changing *N*-protecting group results in dramatic changes in facial selectivity. The enolate of *N*-Boc morpholinone **4** was benzylated in 99% d.e. with *trans*-selectivity, while the enolate of the corresponding *N*-benzyl morpholinone **5** was benzylated to afford the *cis*- product as the major diastereoisomer in 94% d.e. (scheme 4).[4]

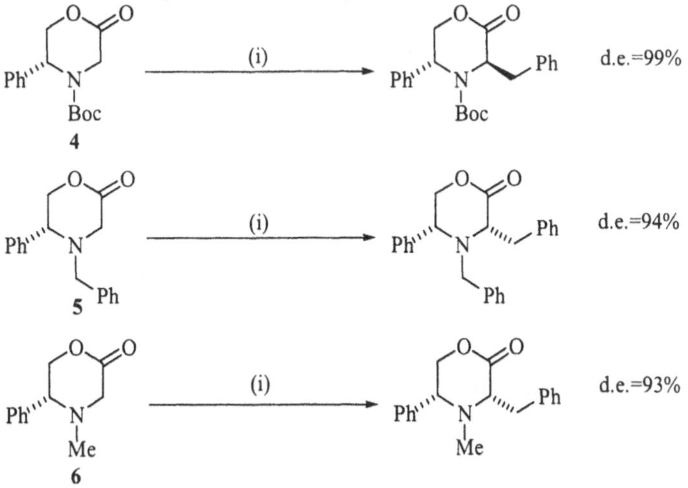

Ph
N
Boc
4
(i)
d.e.=99%
5
d.e.=94%
Me
6
d.e.=93%

Reagents and conditions: (i) LiHMDS, $PhCH_2Br$, THF, -78°C

Scheme 4

While the selectivity observed for the *N*-Boc morpholin-2-one **4** was easily explained by invoking a conventional 1,3-asymmetric induction argument where the *N*-Boc protecting group plays no role in controlling facial selectivity, the reversal in selectivity observed for *N*-benzyl **5** was explained by invoking a chiral relay network. Studies on the corresponding *N*-methyl morpholinone **6** (*cis*-benzyl, 93% d.e.), revealed that the high *cis*-diastereoselectivity observed for **5** was a consequence of the conformation adopted by the enolate of the auxiliary, which was controlled by two cooperative effects:[5]

(i) A stereoelectronic effect that places the nitrogen lone-pair in a *pseudo*-equatorial environment to avoid interaction with the π-system of the enolate.

(ii) A chiral relay effect where the *N*-benzyl group occupies a *pseudo*-axial position *anti* to the C_5 phenyl group thus directing the incoming electrophile to the *Si* face of the enolate to afford *cis*-substituted product (figure 1).

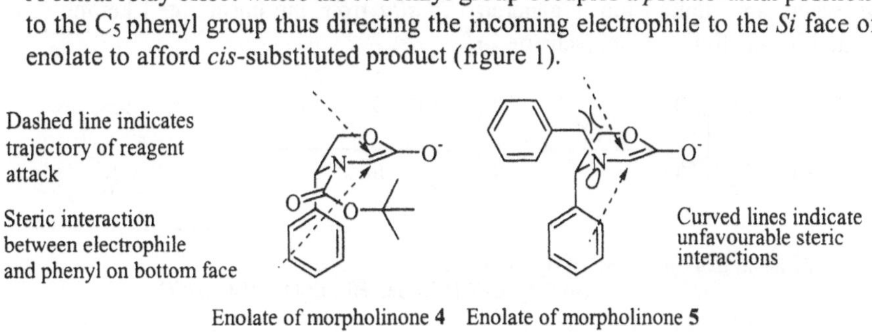

Dashed line indicates trajectory of reagent attack
Steric interaction between electrophile and phenyl on bottom face
Curved lines indicate unfavourable steric interactions
Enolate of morpholinone **4**
Enolate of morpholinone **5**

Figure 1

The apparently anomalous stereoselectivities reported for other chiral auxiliaries reported in the literature may also be rationalised by invoking a single chiral relay concept. Rapoport has reported a five-membered ring system **7**, where an exocyclic enolate is *cis*-alkylated in 78% d.e. It is likely that this selectivity is due to a chiral relay effect, similar to that operating in the morpholinone **5**, where the *N*-benzyl group is oriented *anti* to the heptyl side-chain (scheme 5).[6]

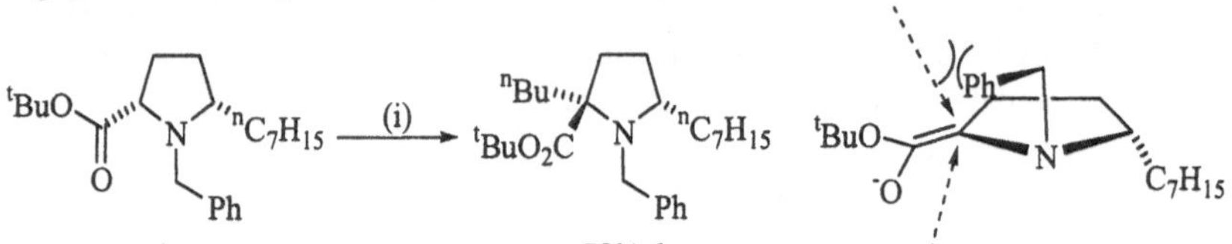

7 78% d.e.

Reagents and conditions: (i) LDA, THF, -78°C, ^nBuBr Transition state

Scheme 5

Craig *et al.* have recently reported on a chiral auxiliary **8** based on the stereoselective alkylation of a cyclic iminium species. The iminium species is generated *via* an S_N1' mechanism, since it was shown that the reaction proceeds with identical stereoselectivity regardless of the stereochemistry of the C_4 tosyl group. This iminium species is attacked by a nucleophilic species with high diastereoselectivity, to give the *syn* isomer as the exclusive product. A simple 1,3- asymmetric induction argument would predict that the reaction should proceed with *anti* selectivity, but instead, the tosyl group must clearly block the *Si*-face of the iminium species, thus inverting the stereochemical information of the C_2 substituent to afford *cis*-alkylated product (scheme 6).[7]

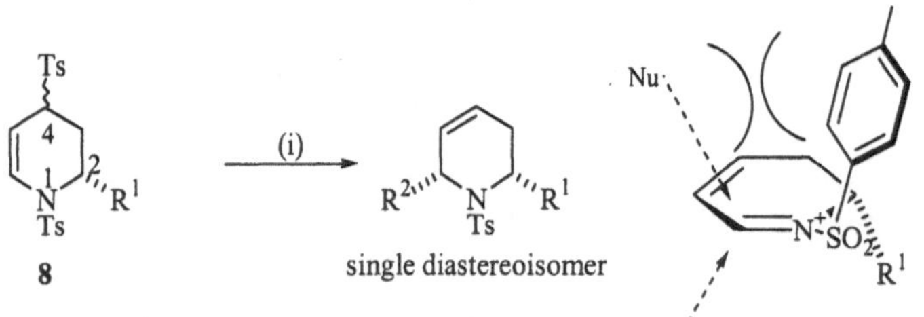

8 single diastereoisomer

Reagents and conditions: (i) Me_3Al or Et_2AlCl or $CH_2{=}CHCH_2TMS/SnCl_4$ or $^tBuCOCH_2OTBS/SnCl_4$ Nu Transition state

Scheme 6

A similar mechanism is likely to be operating for the system described by Hopman *et al.* where an iminium ion is generated *in situ* from the unsaturated ketone **9**, and *cis*-alkylated by an allylic silane, (scheme 7).[8]

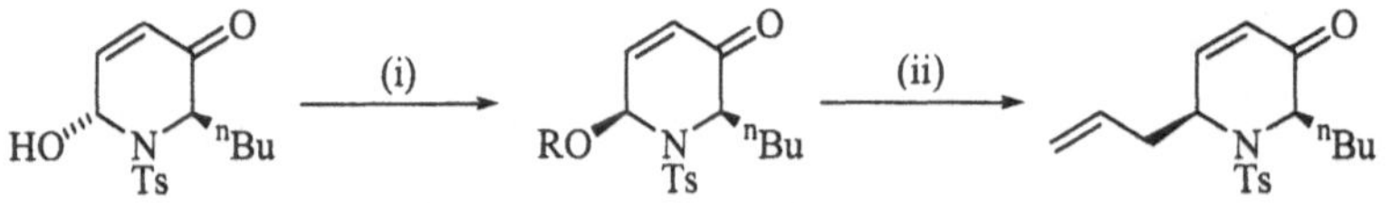

9 single diastereoisomer

Reagents and conditions: (i) H_2SO_4, ROH
(ii) $CH_2{=}CHCH_2SiMe_3$, $BF_3.Et_2O$, CH_2Cl_2, 0°C

Scheme 7

CHIRAL AUXILIARIES EMPLOYING TWO RELAY UNITS

The chiral relay approach is not confined to simple systems employing single relays. Clayden *et al.* have reported on a novel bis-amide aryl based system **10** which uses two tertiary amide units to transfer chiral information in a double relay system. The stereochemical information of the C_1 benzylic position fixes the conformation of the proximal amide unit at C_2 orthogonal to the aryl ring. This conformation ensures that the bulky diisopropyl unit is fixed *anti* to the phenyldimethylsilyl group, which in turn directs the diisopropyl unit of the C_3 amide unit *syn* to the phenyldimethylsilyl group at C_1. This chiral relay network may be deployed for asymmetric synthesis by deprotonation of the C_4 benzylic position to afford a carbanion which is silylated *anti* to the bulky diisopropyl unit of the C_3 amide group (scheme 8).

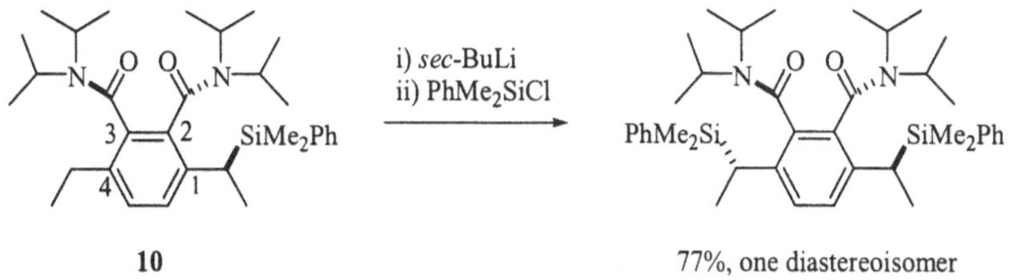

i) *sec*-BuLi
ii) PhMe$_2$SiCl

10

77%, one diastereoisomer

Scheme 8

This example serves to illustrate the general principle that while a single chiral relay unit results in net inversion of the original stereochemical information, two chiral relay units result in net retention of stereochemical information. In a similar vein, we have shown that a diketopiperazine (DKP) derived auxiliary **11** can employ a chiral relay network to control the diastereoselectivity of alkylation at its C_6 enolate. X-Ray crystallographic analysis of DKP **11** revealed that the isopropyl group at C_3 fixed the proximal N_4 *para*-methoxybenzyl group *anti*, which fixed the conformation of the distal N_1 *para*-methoxybenzyl group *syn* to the isopropyl group. The close proximity of the N_1 benzyl group to the point of enolate alkylation at C_6 results in very high d.e.s for this system (> 90% d.e. for all electrophiles). Evidence that a chiral relay network was in fact responsible for the very high d.e.s observed for this class of enolate was obtained from studies involving the *N,N'*-dimethyl DKP enolate **12**, which was alkylated with a poor 33% d.e. (scheme 9).[10]

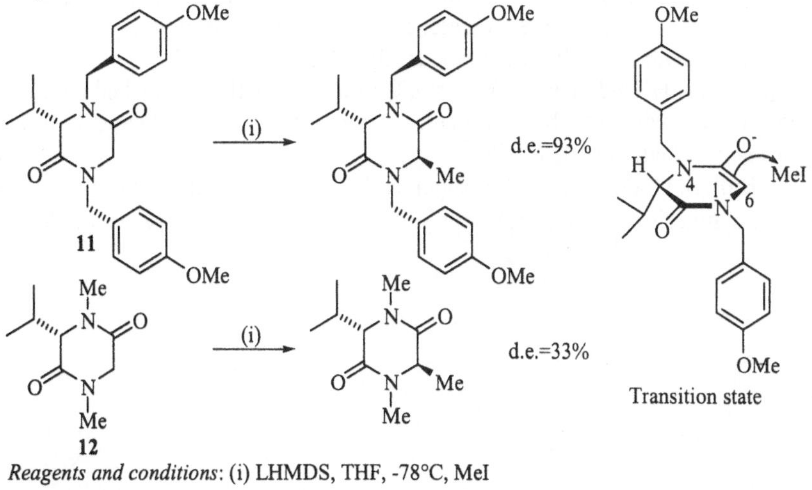

(i)

d.e.=93%

11

(i)

d.e.=33%

12

Transition state

Reagents and conditions: (i) LHMDS, THF, -78°C, MeI

Scheme 9

5

CONCLUSION

The principle of using chiral relay networks to enhance asymmetric induction has been shown to be an effective mechanism for the preparation of homochiral molecules. This approach demonstrates that it is not always necessary for the stereodirecting group to be in close proximity to the newly forming stereogenic centre provided that the orientation of any intervening conformationally mobile groups can be efficiently controlled. Incorporation of these powerful ideas into the design of chiral auxiliaries ensures that efficient amplification and inversion of stereochemical information can occur. It is highly likely therefore that the further application of this powerful tool to other scenarios will lead to a new generation of highly efficient chiral auxiliaries.

REFERENCES

1   G. Procter, *Asymmetric Synthesis*, Oxford University Press, Oxford (1996).
2.   D. Seebach, A. R. Sting, and M. Hoffmann, Self-regeneration of stereocentres (SRS)-applications, limitations, and abandonment of a synthetic principle, *Angew. Chem., Int. Ed. Engl.*, 35:2708 (1996).
3.   U. Schöllkopf, U. Groth, and C. Deng, Asymmetric syntheses *via* heterocyclic intermediates. Enantioselective synthesis of (*R*)-amino acids using L-valine as chiral agent, *Angew. Chem., Int. Ed. Engl.*, 20:798 (1981).
4.   J. F. Dellaria Jr., and B. D. Santarsiero, Enantioselective synthesis of α-amino acid derivatives *via* the stereoselective alkylation of a homochiral glycine enolate synthon, *J. Org. Chem.*, 54:3916 (1989).
5.   S. D. Bull, S. G. Davies, D. J. Fox, and T. G. R. Sellers, Chiral relay effects influence the facial selectivity of *N*-alkylated 5-phenyl-morpholin-2-one enolates, *Tetrahedron: Asymmetry*, 9:1483 (1998).
6.   K. Shiosaki, and H. Rapoport, α-Amino acids as chiral educts for asymmetric products. Chirospecific syntheses of the 5-butyl-2-heptylpyrrolidines from glutamic acid, *J. Org. Chem.*, 50:1229 (1985).
7.   D. Craig, R. McCague, G. A. Potter, and M. R. V. Williams, 1,4-Bis(arylsulfonyl)1,2,3,4-tetrahydropyridines in synthesis. Highly regio- and stereoselective S_N1' and alkylation reactions, *Synlett*, 55 (1998).
8.   J. C. P. Hopman, E. van den Berg, L. Ollero Ollero, H. Hiemstra, and W. N. Speckamp, Stereoselective carbon-carbon bond formation *via* allylic *N*-sulfonyliminium ions, *Tetrahedron Lett.*, 36:4315 (1995).
9.   J. Clayden, J. H. Pink, and S. A. Yasin, Conformationally interlocked amides: remote asymmetric induction by mechanical transfer of stereochemical information, *Tetrahedron Lett.*, 39:105 (1998).
10. S. D. Bull, S. G. Davies, S. W. Epstein, and J. V. A. Ouzman, A chiral relay auxiliary for the synthesis of homochiral α-amino acids. *J.C.S., Chem. Commun.*, 659 (1998).

ASYMMETRIC CATALYSIS WITH CHIRAL LEWIS BASES[¥]

Scott E. Denmark*, Robert A. Stavenger, Xiping Su, Ken-Tsung Wong and Yutaka Nishigaichi

Department of Chemistry, University of Illinois at Urbana-Champaign
Urbana, Illinois, 61801, USA

Catalytic, asymmetric aldol additions have come to the forefront of synthetic methodology due to both the synthetic utility of the products and the challenge of designing such a transformation.[1] There are now a number of reports concerning the addition of silyl enol ethers or silyl ketene acetals to aldehydes using chiral Lewis acids that proceed with good-to-high enantioselectivity.[2] Despite the advantages of chiral Lewis acid catalysis, an asymmetric catalytic aldol reaction with the generality and selectivity of the well known stoichiometric methods still remains to be developed. We envisioned the possibility of devising a new reaction that uses chiral Lewis base catalysis, that is, activation of the nucleophile, Scheme 1.[3,4]

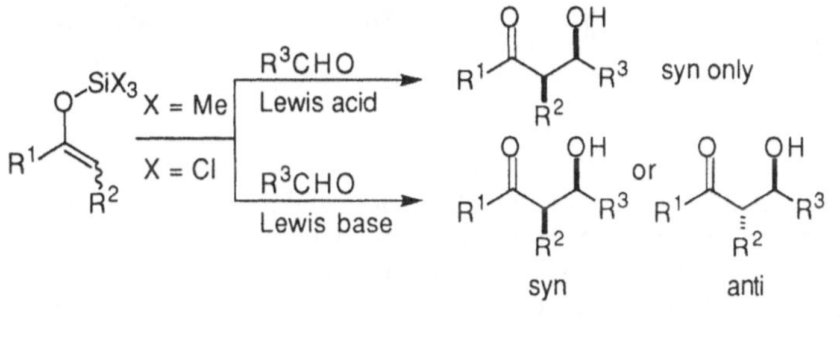

Scheme 1

The challenges that face the development of nucleophilic catalysis of the aldol addition are outlined in Scheme 2. In this case the enoxymetal derivative is activated by pre-association with a chiral Lewis basic group. The ate complex must be more reactive than the free enolate for the ligand accelerated catalysis to be observed. Next, association of this still Lewis acidic ate complex with the Lewis basic carbonyl oxygen of the aldehyde produces a hyper-reactive complex in which the metal has expanded its valence by two. It is expected that this complex between enolate, aldehyde and the chiral Lewis basic group reacts through a closed type transition structure with a high degree of information transfer to produce the metal aldolate product. This represents a single turnover event and for catalysis to be observed, the complex must undergo the expulsion of the G* group with the formation of the chelated metal aldolate product.

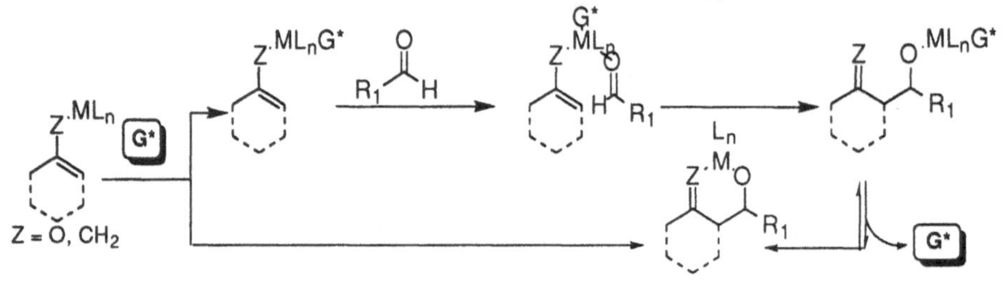

Scheme 2

To formulate the criteria that are necessary to invent such a process we must consider the design elements that go into the enoxy metal and the G* group. For the metal, the ML_n subunit must be able expand its valence by two and balance nucleophilicity of enolate with electrophilicity to coordinate both the Lewis basic aldehyde and the chiral G* group. Such metals that would satisfy these criteria are those that can expand their valence such as silicon, tin, titanium, zirconium and aluminum. To impart sufficient Lewis acidity to the metal group, the ligands (L) should be strongly electron withdrawing such as halogen or carboxyl groups. The criteria necessary for the chiral Lewis base G* group are that it must be able to activate the addition without cleaving OML_n linkage and provide an effective asymmetric environment with single point attachment. Candidates for the G* Lewis basic group would include phosphine oxides and derivatives such as phosphoramides, N-oxides, sulfoxides but not negatively charged alkoxides or amines and carboxylates. Thus, to reduce this concept to practice, we envisioned the use of a new class of aldol reagents, trichlorosilyl enolates, in conjunction with the Lewis basic phosphoramides. These agents can be seen as chiral analogs of HMPA the Lewis basicity of which is well documented.

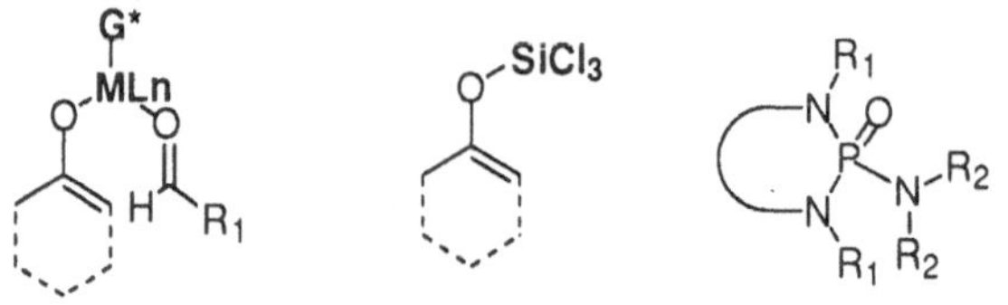

A ready and efficient preparation of trichlorosilyl enol ethers was needed. A number of different approaches have been devised initially based on observations by Baukov who first described trichlorosilyl ketene acetals of esters.[5] Trichlorosilyl enolates of ketones have been prepared beginning with enol acetates of the starting methyl ketones for example in this case pinacolone. The α–tri-butylstannyl ketone undergoes a metathesis with neat silicon tetrachloride at 0 °C. Fractional distillation at reduced pressure afforded the O-bound trichlorosilyl enol ether, Scheme 3.

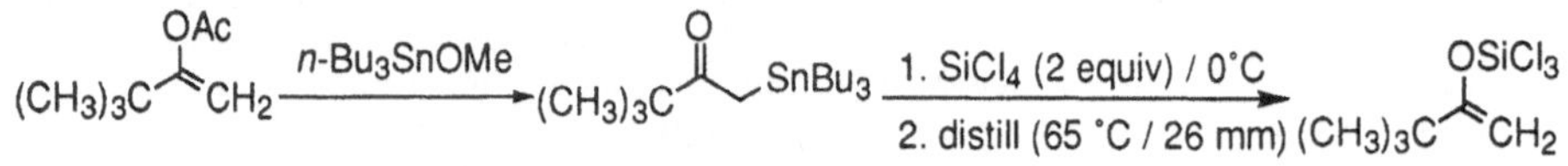

Scheme 3

A much more practical approach involves the metathesis of trimethylsilyl enol ethers easily prepared directly from ketones. One process involves metathesis with trichlorosilyl triflate to produce the trichlorosilyl enolate. An alternative and milder processes uses silicon tetrachloride itself in the presence of a catalytic amount of mercuric acetate, Scheme 4.[6]

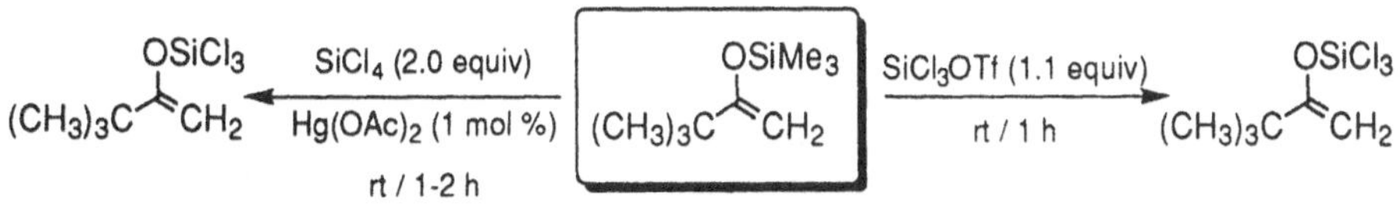

Scheme 4

The generality of the approach for the synthesis of trichlorosilyl enol ethers using the exchange of trimethylsilyl for trichlorosilyl with silicon tetrachloride in the presence of mercuric acetate is shown in Table 1. The enoxytrichlorosilanes of simple aliphatic methyl ketones as well cyclic ketones are formed in good yield and high efficiency.

Table 1. Preparation of Enoxytrichlorosilanes from $SiCl_4$

ketones	enoltrimethylsilanes	yield (%)	enoxytrichlorosilanes	yield (%)
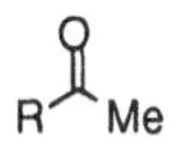	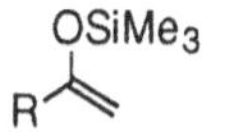	R = Me: 92 R = i-Pr: 90 R = n-Bu: 84 R = i-Bu: 80 R = t-Bu: 79 R = TBSOCH$_2$: 27 R = Ph: 88	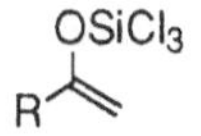	R = Me: 60 R = i-Pr: 83 R = n-Bu: 76 R = i-Bu: 74 R = t-Bu: 81 R = TBSOCH$_2$: 65 R = Ph: 69
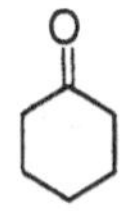	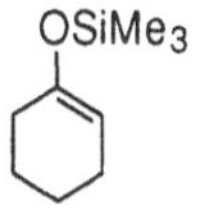	90	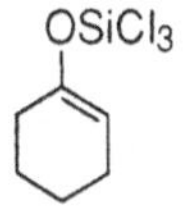	60

To survey the reactivity of these reagents, the trichlorosilyl enol ether of a variety of acyclic ketones were combined with aldehydes at room temperature in 0.5 M methylene chloride solution. In reactions with benzaldehyde as the acceptor a variety of substituted methyl ketones gave excellent yields between 91 and 96% of the aldol addition products.

Furthermore, reaction of α,β-unsaturated aldehydes with a number of different methyl ketone derivatives gave exclusively 1,2-addition with no trace of Michael addition being detected. Highly enolizable aldehydes such as phenylacetaldehyde undergo smooth aldol addition reaction with the pinacolone trichlorosilyl enolate. Aliphatic aldehydes, linear branched and highly substituted types all gave good yields in combination with methyl ketone enol ethers, Scheme 5.

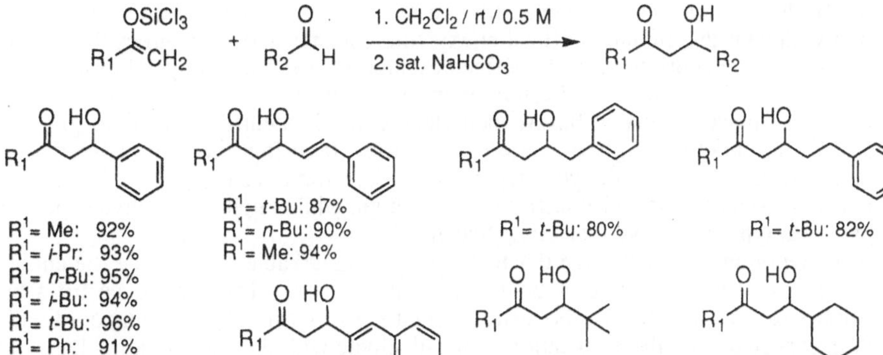

Scheme 5

While these reagents are shown to be synthetically useful, the most important objective is to demonstrate asymmetric catalysis of the aldol addition. A number of structurally diverse chiral phosphoramides was surveyed for the reaction of trichlorosilyl enol ethers and benzaldehyde at -78°C in 0.1 M methylene chloride solution. Remarkably, only a catalytic amount of the chiral phosphoramide was required. After an extensive survey, the best selectivities were obtained with the N,N-dimethylphospholidine derived from stilbenediamine in which the phosphorus bears a piperidino group.[7] In this reaction the acetone aldol addition product is isolated in 93% yield and 85% enantiomeric excess.

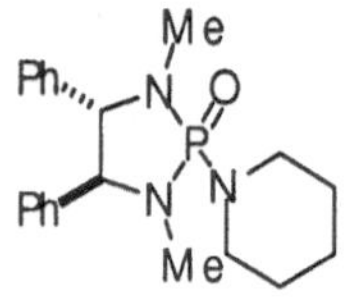

With the stilbenediamine phosphoramide as catalyst at 5 mole % loading a variety of chlorosilyl enol ethers and aldehydes was surveyed to evaluate the generality of the reaction. The yields of analytically pure materials are shown underneath the aldol products and the enantiomeric excesses are in parenthesis, Scheme 6. The trend is clearly seen that the enantioselectivity is dependent upon the bulk of the spectator substituent of the ketone: the enantioselectivity increases with decreasing size of the spectator group such that the n-butyl methyl ketone and acetone aldol addition products proceed with the highest selectivities roughly in the middle 80% ee's.

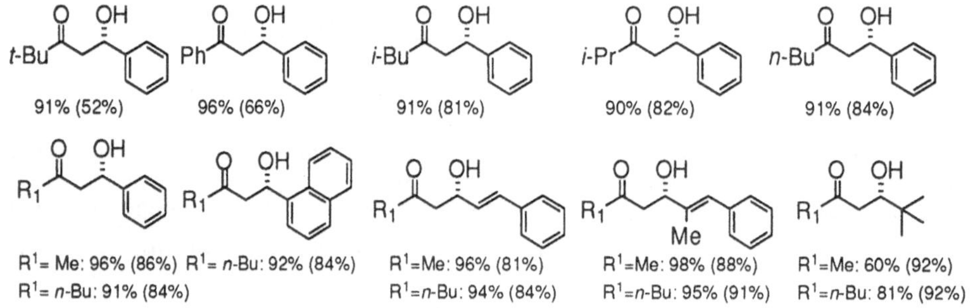

Scheme 6

To further explore reaction generality we next examined the diastereoselectivity of reaction with configurationally defined enoxy chlorosilanes derived from cyclic ketones. The reaction of the configurationally defined trichlorosilyl enolate of cyclohexanone with aldehydes can give rise to two diastereomers, syn and anti as clearly defined. These reactions take place readily in 0.5 M methylene chloride solution at 0° C to give high yields and remarkably high diastereoselectivities favoring the syn isomer. The diastereoselectivities are aldehyde dependent. For aromatic, olefinic and acetylenic aldehydes the syn diastereoselectivities are quite high, while aliphatic aldehydes of linear and branched type are less syn selective, Scheme 7.

The predominance of the syn diastereomer is a remarkable result since the enolates are E-configured which implies a boat transition structure. The preference for a boat transition structure in this case is easily understood in terms of the pentacoordination which is expected from the association of the aldehyde with the enoxychlorosilane. The apical chlorine avoids severe non-bonded interactions with the aldehydic R group by proceeding through a boat transition structure.[8]

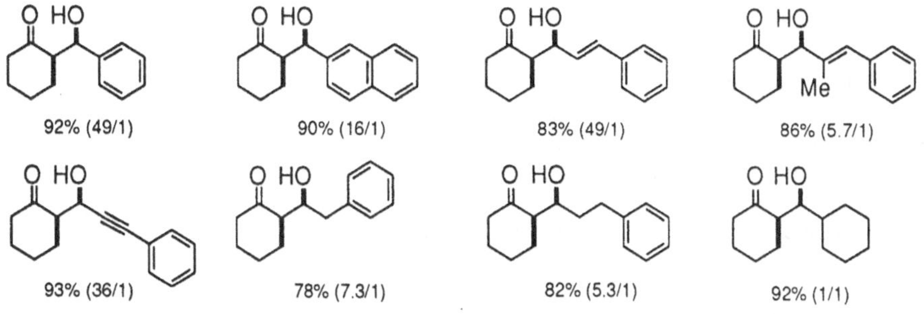

92% (49/1)	90% (16/1)	83% (49/1)	86% (5.7/1)
93% (36/1)	78% (7.3/1)	82% (5.3/1)	92% (1/1)

Scheme 7

The ability of catalytic quantities of chiral phosphoramides to promote the addition of the cyclohexanone trichlorosilyl enolate with aldehydes was then assayed, Scheme 8. The diastereoselectivity of the addition was found to be remarkably sensitive to the structure of the phosphoramide promoter. The results of the uncatalyzed reaction are reiterated to recall a 49/1 selectivity favoring the syn diastereomer. When the reactions are promoted by a catalytic amount of HMPA the syn selectivity drops to 3/1. Other phosphoramide structures bring about a striking change in the diastereoselectivity to the point where with the *N,N*-dimethylstilbenediamine-derived phosphoramide, the selectivity completely reverses now favoring the anti diastereomer to the extent of 49/1. Moreover, the enantiomeric excess of the anti diastereomer is 93%. Overall that product is obtained in 94% yield.

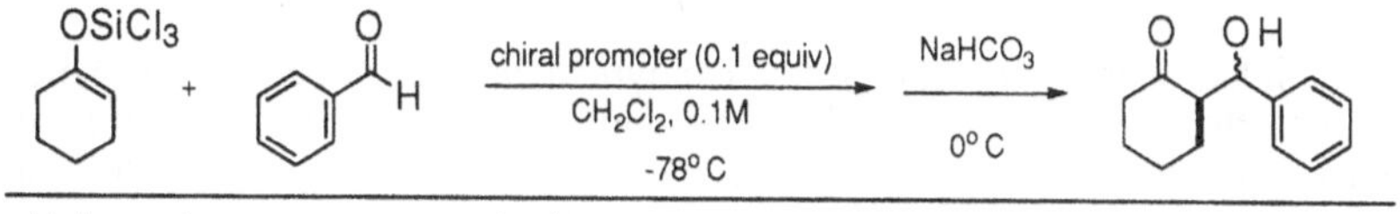

chiral promoter	syn / anti	syn ee (%)	anti ee (%)	yield (%)
none	49 / 1	-	-	92
HMPA	3 / 1	-	-	90
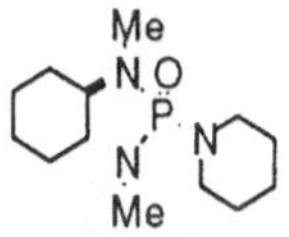 (phosphoramide structure)	1 / 2	0	50	93
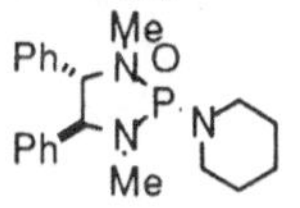 (phosphoramide structure)	1 / 49	-	93	94

Scheme 8

The generality of this catalytic, asymmetric reaction was illustrated for a variety of different aldehyde types. It was found that aromatic and olefinic as well as acetylenic aldehydes all react with good diastereoselectivity in some cases exclusive anti selectivity and with respectable enantiomeric excess ranging from 82 - 97% ee. The yields in all cases of analytically pure materials are quite high.

To illustrate that these reactions are indeed proceeding through highly-organized, closed type transition structures, it was important for us to demonstrate the response of the aldol addition reaction to a change in enolate geometry. Combining the trichlorosilyl enolate derived from propiophenone (which was demonstrated to be exclusively Z configured) with a variety of aldehydes at 0 °C in 0.5 M methylene chloride solution gave rise, in good yield, to the aldol addition products. In this case the anti diastereomers weakly predominated. It is not surprising that the anti-selectivity is not high considering again the preferred boat-like arrangement of the groups in the trigonal bipyramidal transition structures. In the boat, the apical chlorine atom now experiences non-bonded

11

interactions with one of the substituents on the Z-enolate, in this case a methyl group. Thus, obtaining anti diastereomers from a Z-enoxysilane implies a boat transition state albeit with much lesser preference due to pseudoaxial interactions.

Nevertheless, the reactions promoted by the *N,N*-dimethylstilbenediamine-derived phosphoramide proceeded rapidly at -78° C to afford the aldol products with good syn-diastereoselectivity and in many cases with excellent enantiomeric excess, Scheme 9. A variety of aromatic and olefinic aldehydes reacted with syn-diastereoselectivity, and the syn-diastereomer was highly enantiomerically enriched.

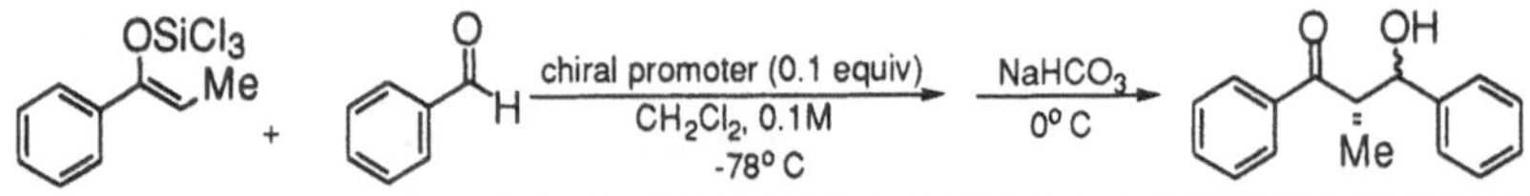

chiral promoter	syn / anti	syn ee (%)	anti ee (%)	yield (%)
none	1 / 2	-	-	74
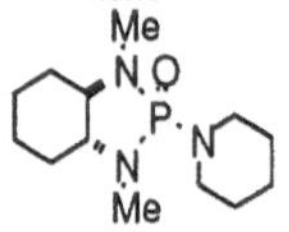 (cyclohexyl phosphoramide)	1 / 2.7	15	5	72
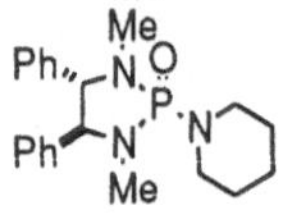 (stilbene phosphoramide)	11.5 / 1	94	11	77

Scheme 9

Thus, it has been clearly shown that catalytic quantities of chiral phosphoramides are capable of effectively promoting the aldol addition reactions of trichlorosilyl enolates derived from methyl ketones as well as *E*- and *Z*- configured substituted enolates of ketones.

While the synthetic utility of trichlorosilyl enolates is now documented, the origin of their remarkable reactivity is not clearly understood. The most important feature of this new reaction is the susceptibility of the trichlorosilyl enolates to catalysis by trace amounts of the phosphoramide. The most important effects attributable to the chiral promoter are summarized in Scheme 10.

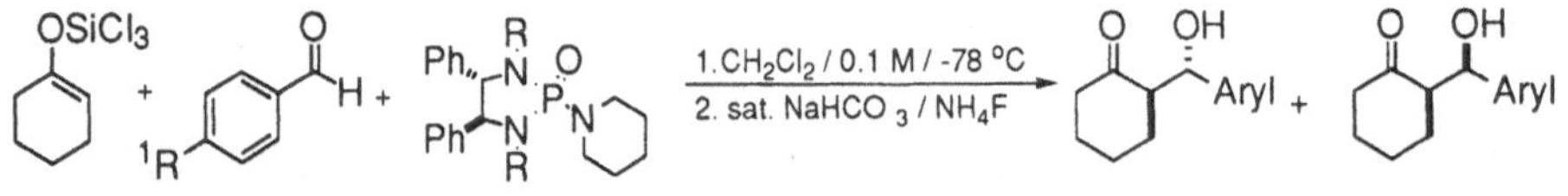

Rate of Reaction

R=Me	0.0 equiv	8 min/-78°C	2% yield (96% recovery)
R=Me	0.1 equiv	8 min/-78°C	99% yield

Diastereoselectivity

			anti		syn
	0.0 equiv	2 h/0°C	1	/	49
R=Me	0.1 equiv	8 min/-78°C	60	/	1
R=Ph	0.1 equiv	3 h/-78°C	1	/	97

Enantioselectivity

			anti	syn
R=Me	0.1 equiv	8 min/-78°C	92% ee	
R=Ph	0.1 equiv	3 h/-78°C		51% ee

Scheme 10

To understand the origin of phosphoramide catalysis, it is first necessary to consider the origin of activation of the chlorosilyl enolates in their reactions with aldehydes alone.

12

The nature of this process is similar to the behavior of boron enolates in their reactions with aldehydes.[9] Both boron enolates and trichlorosilyl enolates are non-nucleophilic species. The boron enolate bears an empty orbital which withdraws electron density from the enolate. Complexation with the Lewis basic oxygen of an aldehyde now generates a boron ate complex which is doubly activated. The enolate is nucleophilically activated as a result of B-O weakening and interruption of the oxygen lone pair backbonding. The aldehyde is electrophilically activated by polarization of the carbonyl oxygen. Through this ate complex the dual nucleophilic and electrophilic activation leads to rapid reaction rates and high diastereoselectivities characteristic of the chair-like arrangement of groups; namely Z-enolate to syn-aldol product and E-enolate to anti-aldol product, Scheme 11.

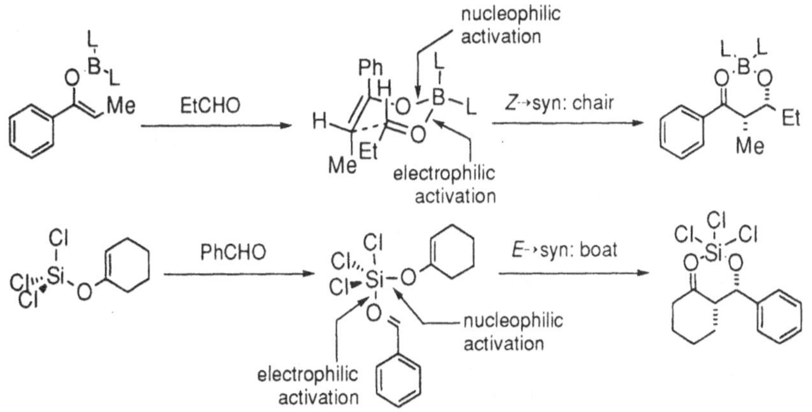

Scheme 11

Trichlorosilyl enolates are also non-nucleophilic. Upon complexation of the Lewis basic oxygen of an aldehyde, the resulting trigonal bipyramidal ate complexes enjoy a similar dual activation. Electrophilic activation of the aldehyde occurs by polarization of electron density towards the silicon and nucleophilic activation of the enolate occurs by formation of the negatively charged ate complex. Reactions of these species also illustrate the characteristics of closed transition structures proceeding through boat like arrangements in the correlation of E-enolate to syn aldol product and Z-enolate to anti aldol product. Aldehyde coordination of these species activates both nucleophilic and electrophilic components and reaction takes place only in the coordination sphere of the boron or silicon.

This analysis is relatively straightforward and finds good analogies in other areas of chemistry. However to understand the origin of the remarkable catalysis due to the chiral phosphoramides, we must understand the changes in bonding that take place around the coordination sphere of silicon as the Lewis bases coordinate, Scheme 12. One of the most powerful and universal indices of bonding changes around a central atom is the re-distribution of s-character in the bonds connected to a central atom.[10]

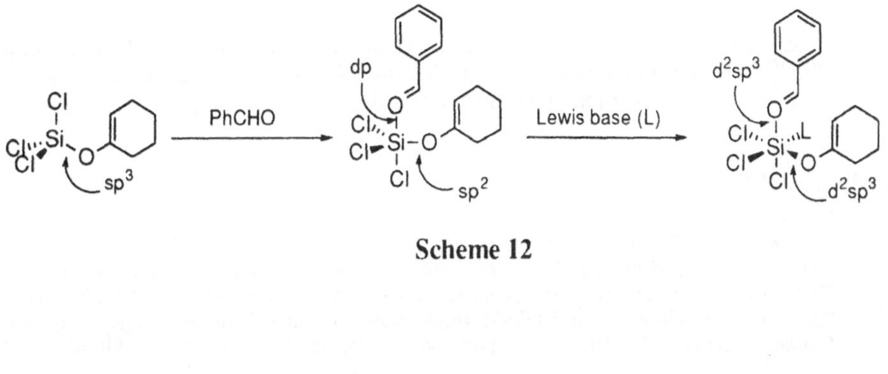

Scheme 12

13

An interesting insight is revealed if one considers the redistribution of s-character upon bonding the fifth and sixth ligands to the trichlorosilyl enolate reagents. In the tetracoordinate enoxytrichlorosilanes the silicon is formally sp^3 hybridized. Coordination of the Lewis basic oxygen of the aldehyde results in a trigonal bipyramidal complex in which the electronegative positively charged oxygen of the aldehyde presumably takes up an apical position. If we analyze the redistribution of s-character in a trigonal bipyramid we note that the apical bond to the aldehyde is devoid of s-character being composed of a dp-type hybrid. Furthermore, the equatorial or basal bond to the enolate oxygen is now sp^2 hybridized.

It is interesting to consider the consequences of s-character redistribution upon coordination of the sixth Lewis basic ligand which activates the aldol addition over the background unpromoted reaction. Upon coordination of another Lewis basic group to form an octahedral hexacoordinate silicon ate complex, the apical bond to the aldehyde changes from a dp to a d^2sp^3 hybrid (assuming equal s-character distribution along the directions of the six equivalent bonds in the octahedron). This constitutes an increase s-character from dp to d^2sp^3, i.e. from 0 to 16% s-character in the change from a trigonal bipyramid to an octahedron. Thus, the bond strength to the aldehyde *increases* thereby increasing the polarization and activation of the aldehyde towards nucleophilic attack. Upon coordination of a sixth Lewis basic ligand the change in s-character of the basal enolate bond (formerly sp^2 that is 33% s-character) changes again to a d^2sp^3 hybrid now 16% s-character. Thus, the bond from the silicon to the enolate *decreases* in s-character upon changing from a trigonal bipyramid to an octahedron. Consequently the bond strength decreases, thus increasing the activation of the enolate by polarizing electron density away from the silicon towards the nucleophilic end of the enolate.

Interestingly, the overall of the consequence of the coordination of the sixth ligand is that it generates a highly reactive species which experiences a simultaneous activation of both reacting partners in the coordination sphere of the silicon. This process constitutes a unique kind of simultaneous activation both of electrophilic and nucleophilic nature by a redistribution of the bonding character to the two participating species.

In summary, the key features that have been highlighted in this lecture are that: (1) trichlorosilyl enolates of ketones are easily prepared from trimethylsilyl enol ethers or stannyl ketones, (2) trichlorosilyl enolates of ketones react rapidly at room temperature with aldehydes, (3) the uncatalyzed aldol reactions of enoxychlorosilanes provide syn products from *E*-enolates and thus proceed via boat-like transition structures, (4) the reactions of trichlorosilyl enolates of ketones are susceptible to nucleophilic catalysis with chiral phosphoramides in high de and excellent ee, (5) the reactions appear to proceed via closed transition structures, with a phosphoramide in a hexacoordinate silicon species.

We believe that this aldol addition represents a new, emerging class of reactions characterized by the ability to simultaneously employ Lewis acid and Lewis base activation. These reactions proceed around a defined organizational center and include ligands that can craft an effective asymmetric environment for the reaction. Finally, the invention of new catalytic systems and the design of effective catalytic agents constitutes an important challenge for the future and involves the interplay of physical-organic, structural, computational and synthetic chemistry.

ACKNOWLEDGMENT

We are grateful to the National Science Foundation for generous financial support (CHE 9500397), R.A.S. thanks the Eastman Chemical Co. for a graduate fellowship. Y.N. thanks the Ministry of Education (Japan) for a postdoctoral fellowship.

REFERENCES

¥   The Chemistry of Trichlorosilyl Enolates. 7.

1. For reviews on catalytic asymmetric aldol additions, see (a) T. Bach, *Angew. Chem., Int. Ed. Engl.* 34:417 (1994). (b) A.S. Franklin and I. Paterson *Contemp. Org. Synth.* 1:317 (1994). (c) M. Braun and H. Sacha, *J. Prakt. Chem.* 33: 653 (1993). (d) M. Sawamura and Y. Ito in *Catalytic Asymmetric Synthesis*, I.Ojima, Ed.; VCH: New York, 367 (1993). (e) H. Yamamoto, K. Maruoka and K.

Ishihara, *J. Synth. Org. Jpn.* 52:912 (1994). (f) M. Braun in *Stereoselective Synthesis, Methods of Organic Chemistry (Houben-Weyl)*; Edition E21; G. Helmchen, R. Hoffman, J. Mulzer and Schaumann, E. Eds.; Thieme: Stuttgart 3: 1730 (1996).

2. For leading references on catalytic asymmetric aldol additions see: (a) M. Sodeoka, R. Tokunoh, F. Miyazaki, E. Hagiwara and M. Shibasaki, *Synlett* 463 (1997). (b) Y.M.A. Yamada, N. Yoshikawa, H. Sasai, and M. Shibasaki. *Angew. Chem., Int. Ed. Engl.* 36:1871(1997). (c) S.G. Nelson *Tetrahedron : Asymmetry* 9:37 (1998).

3. For examples of primarily anti-selective catalytic asymmetric aldol additions, see: (a) D.A. Evans, D.W.C. MacMillan, K.R. Campos, *J. Am. Chem. Soc.* 119:10859 (1997). (b) S. Kobayashi, M. Horibe, I. Hachiya, *Tetrahedron Lett.* 36:3173 (1995).

4. (a) S.E. Denmark, S.B.D.Winter, X.Su and K.-T. Wong *J. Am. Chem. Soc.* 118:7404 (1996). (b) S.E. Denmark, K.-T. Wong and R.A. Stavenger, *J. Am. Chem. Soc.* 119:2333 (1997). (c) S.E. Denmark and S.B.D.Winter, *Synlett* 1087 (1997). (d) S.E. Denmark, R.A. Stavenger and K.-T.Wong, *J. Org. Chem.* 63:918(1998). (e) S.E. Denmark, R.A. Stavenger and K.-T. Wong, *Tetrahedron* 0000 (1998).

5. (a) G.S. Burlachenko, B.N. Khasapov, L.I. Petrovskaya, Baukov , I. Yu. and I.F. Lutsenko, *J. Gen. Chem. USSR (Engl. Transl.)* 36:532 (1996). (b) I.F. Lutsenko, Baukov, I. Yu, G. S. Burlachenko, B.N. Khasapov, *J. Organomet. Chem.* 5:20 (1996). (c) G.S .Burlachenko, Baukov, I.Yu, T.G. Dzherayan and I.F. Lutsenko, *J. Gen. Chem. USSR (Engl. Transl.)* 45: 73 (1975). (d) Baukov, I.Yu, I.F. Lutsenko, *Moscow Univ. Chem. Bull. (Engl. Transl.)* 25:72 (1970). (e) S.V. Ponomarev, Baukov, I. Yu, O.V. Dudukina, I.V. Petrosyan and L.I. Petrovskaya, *J. Gen. Chem. USSR (Engl. Transl.)* 37:2092 (1967). (f) R.A. Benkeser and W.E. Smith, *J. Am. Chem. Soc.* 90:5307 (1968). (g) G.S. Burlachenko, Baukov, I.Yu. and I.F. Lutsenko, *J. Gen. Chem. USSR (Engl. Transl.)* 40:88 (1970).

6. We view this formal silicon-silicon metathesis as involving initial formation of an α-mercurio ketone followed by O-complexation to $SiCl_4$ and loss of HgX_2 as an electrofugal group. For examples of the synthesis of α-mercurio ketones from silyl enol ethers see: (a) H.O. House, R.A. Auerbach, M. Gall and N.P. Peet, *J. Org. Chem.* 38:514.1973). (b) Y.Yamamoto and K. Maruyama, *J. Am. Chem. Soc.* 104:232 (1982). (c) N.Bluthe, M. Malacria and J. Gore, *Tetrahedron* 40:3277 (1984). (d) J. Drouin, M.-A. Boaventura and J.-M. Conia, *J. Am. Chem. Soc.* 107:1726 (1985).

7. S.E. Denmark, D.M. Coe, N.E. Pratt and B.D. Griedel, *J. Org. Chem.* 59:6161 (1994).

8. We have previously demonstrated that the E-enolate to syn adduct correlation is characteristic of uncatalyzed aldol reactions of enoxysilacyclobutanes. The putative pentacoordinate siliconate transition structures were shown computationally to prefer boat-like arrangements in this array. S.E. Denmark, B.D. Griedel, D.M. Coe and M.E. Schnute *J. Am. Chem. Soc.* 116:7026 (1994).

9. For reviews of the aldol reaction of boron enolates see: (a) B.M. Kim; S.F. Williams, S. Masamune in *Comprehesive Organic Synthesis: Additions to C-X π-Bonds Part 2*; C.H. Heathcock, Ed. Pergamon Press: Oxford, 1991; Chapt. 1.7. (b) C.J. Cowden and I. Paterson, *Org. React.* 51:1 (1997).

10. H.A. Bent, *Chem. Rev.* 61:275 (1961).

ABOUT THE USE OF MOLECULAR WORKBENCHES AND PLATFORMS IN ORGANIC SYNTHESIS

Johann Mulzer[*][1,] Karin Schein[2], Ingo Böhm[2] and Dirk Trauner[1]

[1]Institut für Organische Chemie der Universität Wien
 Währinger Strasse 38, A-1090 Wien, Austria
[2]Institut für Organische Chemie der Johann Wolfgang Goethe Universität
 Marie Curie Strasse 11, D-60439 Frankfurt, Germany

INTRODUCTION

The concept of molecular workbenches[1] is introduced and outlined in Figure 1. According to the distinction between disconnectible and non-disconnectible workbenches this account falls into two sections, the first one describing the application of auxiliary like disconnectible workbenches in the functionalization of double bonds, and the second one the incorporation of a non-disconnectible workbench into the ultimate target structure (dihydrocodeinone) at a very early stage of the total synthesis.

Definition of Molecular Workbenches

-Disconnectible
 and nondisconnectible workbenches:
-Arrays of rigid substructures
 (e.g. aromatic or heterocyclic systems)
-chemically relatively inert
-high force constants regarding
 vibration and deformation

Figure 1. Definition of molecular workbenches

DISCONNECTIBLE WORKBENCHES

The idea to construct such workbenches[1] occurred to us in connection with the dihydroxilation or epoxidation of an acyclic *(E)*-olefinic polyol such as **1** which gave unselective reaction to **2/3**-mixtures in the unmodified version (Figure 2).

The synthetic problem to solve was the stereoselective dihydroxylation of

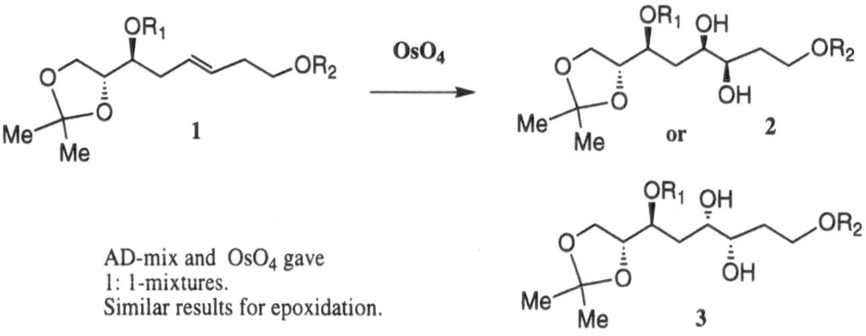

AD-mix and OsO$_4$ gave
1: 1-mixtures.
Similar results for epoxidation.

Figure 2. Unselective additions to acyclic (E) olefins

Therefore, it was decided to incorporate the acyclic olefin into a cyclic template (= workbench) by using the terminal functions as anchoring groups. The next poblem to solve was the helicameric inversion which was to be expected from cyclic *(E)*-olefins with ring sizes of >8 carbons (Figure 3). A possible solution to this problem was envisaged in form of the ansa macrolide template which should slow down helicameric mobility and, at the same time, exert diastereofacial shielding of the double bound by virtue of the benzenoid ring (Figure 4). Thus, the ansa chain **4** was prepared by routine methods and connected to the commercially available benzenoid platform **5**. After chromatographic separation of the anomers **6** the ansa macrolides **7** and **8** were obtained by macrolactonization and analyzed by single crystal diffraction (Figures 8a/b). It turned out that the configuration of the acetal center determines the helicity of the ansa olefin (Figure 6) and that the attack of the reagents (osmium tetroxide or m-chloroperbenzoic acid) occurred from the outside with varying selectivities. The stereoselectivity was dramatically increased for the 12-membered ring and the *trans*-acetal, whereas it was nonexistent for the 13-membered macrolide (Figure 6). This means that there must be a very close fit between ansa chain and platform, similar to the one in enzyme and substrate. Even an increase by one carbon atom in the ansa chain is intolerable for a stereoselective addition! Another issue was the conformational rigidity of the workbench during the addition. To this end the crystal structures of olefin **7** (R=Bn, n = 1) and its epoxide were superimposed to show only minimal discrepancies (Figure 9). This indicates a significantly high geometrical stability of the molecular workbench as postulated in Figure 1.
To disconnct the workbench after the functionalization of the double bond hydrogenolysis of the benzylic CO-bonds was employed to furnish polyol **10** from precursor **9**. In the case of the epoxide **11** hydrogenolysis was combined with an S$_N$2-type cyclization to generate the tetrahydrofuran derivative **12** stereoselectively. The p-methylbenzoate served as a protective group for one of the primary OH-functions which could easily be differentiated from the second primary OH-group (Figure 10).

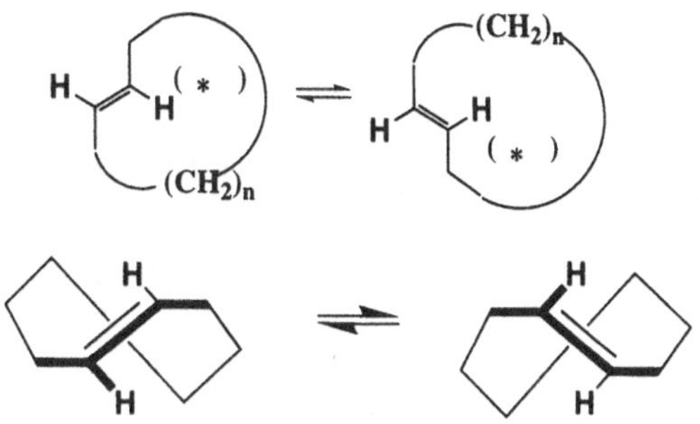

1. The smallest possible ring is of size 8 (too small for ansa compound ?)
2. in the 8-membered ring the helicamer inversion is slow
   (required temperature ca. 150 °C)
3. larger rings show rapid helicamer inversions
   (9-membered ring already at O°C!)

How to slow down inversion of larger rings ?

Figure 3. Helicameric mobility in cyclic (E) olefins

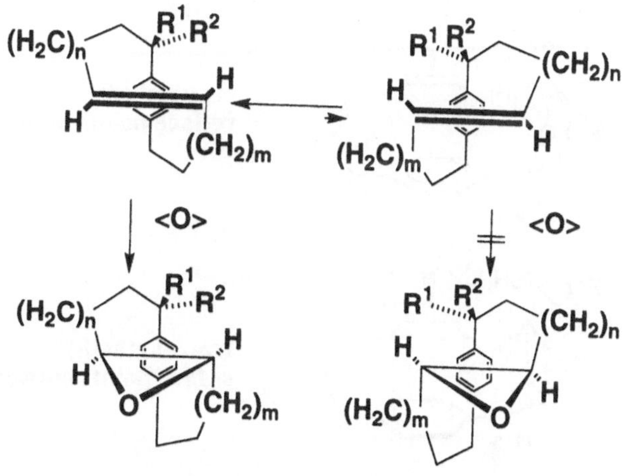

high selectivity

Figure 4. Concept of ansamacrolides as chiral molecular workbenches

19

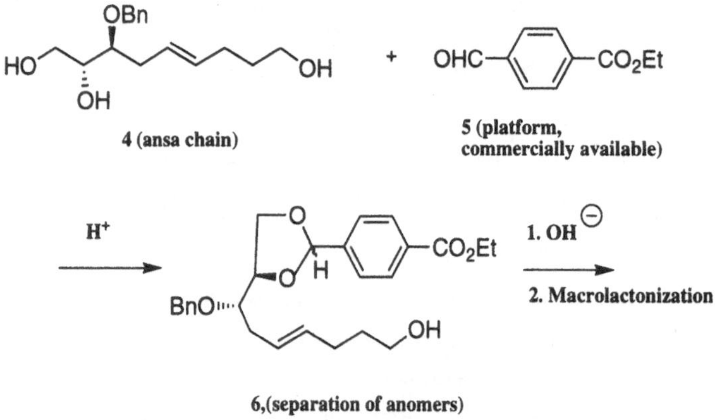

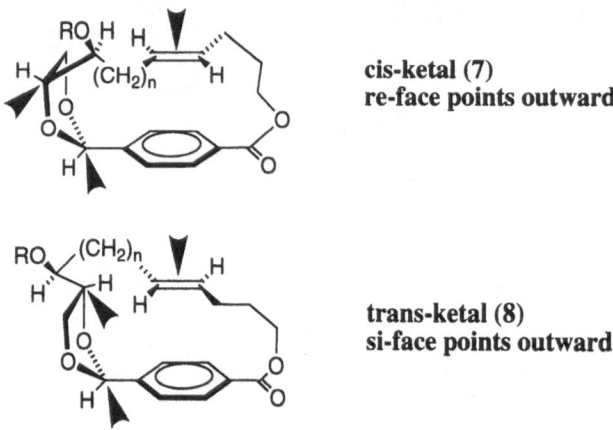

Figure 5. Synthesis of the chiral workbench

Ansamacrolides : n =1 and 2, R = Bn and Me (60-70% Yield)
The acetal center determines the helicity of the olefin

cis-ketal (7)
re-face points outward

trans-ketal (8)
si-face points outward

Figure 6. Helicity of the ansamacrolides **7** and **8**

Selectivities for Epoxidation (mCPBA) (A)
and Dihydroxilation (OsO4/NMO) (B)

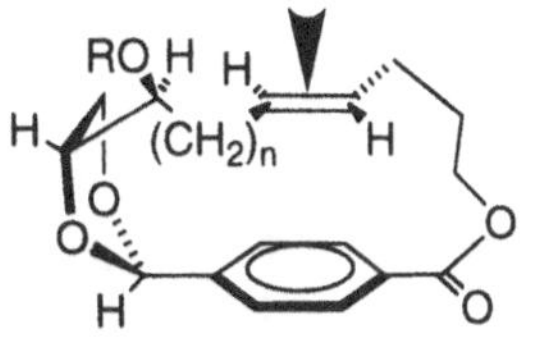

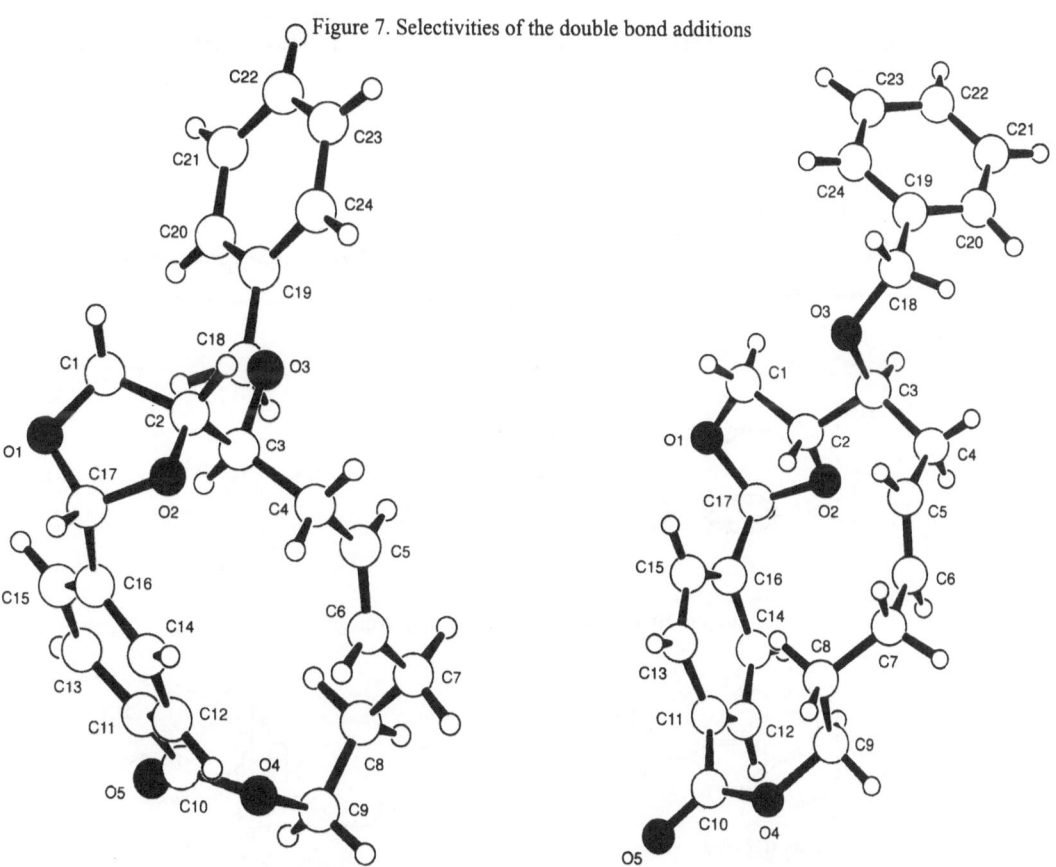

ring size = 12
A 90: 10   B > 95 : 5

ring size = 13
A 2:1   B 1.5 : 1

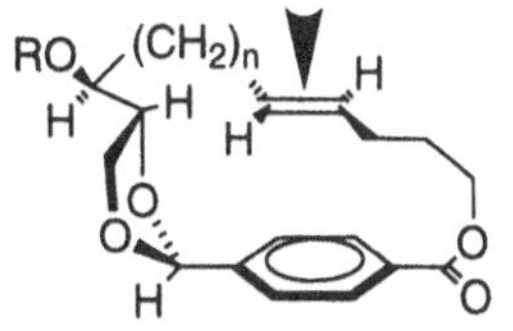

ring size = 12
A > 98: 2 B >99 : 1

ring size = 13
A 1.5 : 1   B 1.5 : 1

Platform and ansa chain must be adjusted to one another (like enzyme and substrate)

Figure 7. Selectivities of the double bond additions

Figure 8a. Crystal structure of **7** (n = 1, R = Bn)

Figure 8b. Crystal structure of **8** (n = 1, R = Bn)

Contrary to this approach which uses chiral ansa chains and an achiral platform, a second approach was tested with achiral ansa chains and chiral platforms (Figure 11). In fact similar selectivities were observed as before, the best results being obtained for 12-membered rings and trans-acetals (Figure 12). For instance the epoxidation of olefin **14** proceeded with >98% dr. The removal of the chiral platform was again achieved by catalytic hydrogenolysis.

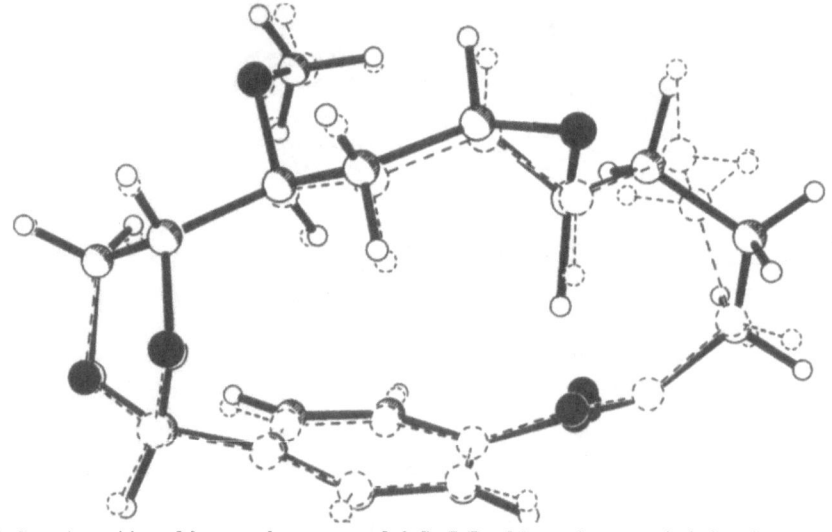

Figure 9. Superimposition of the crystal structures of olefin **7** (R = Me, n =1, empty circles) and its epoxide (black circles)

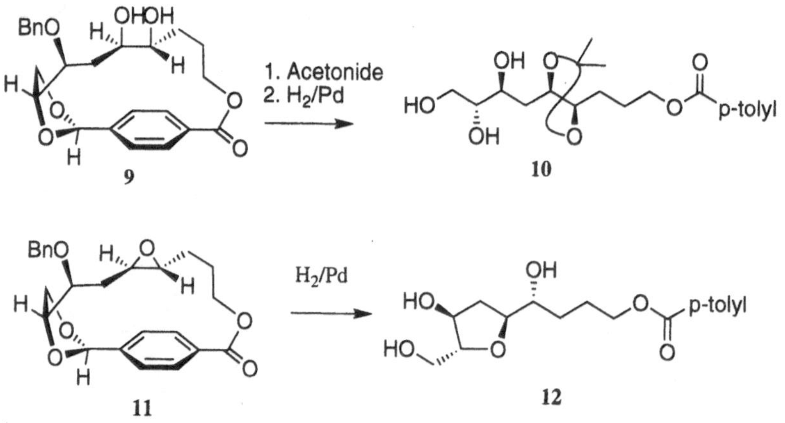

Figure 10. Removal of the platform by hydrogenolysis

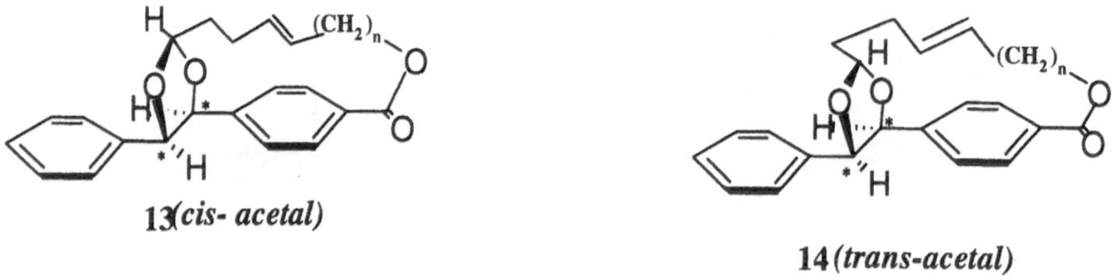

13(*cis- acetal*)

14(*trans-acetal*)

Helicity is independent of the acetal geometry

Figure 11. Chiral workbenches from chiral platform and achiral ansa chain

22

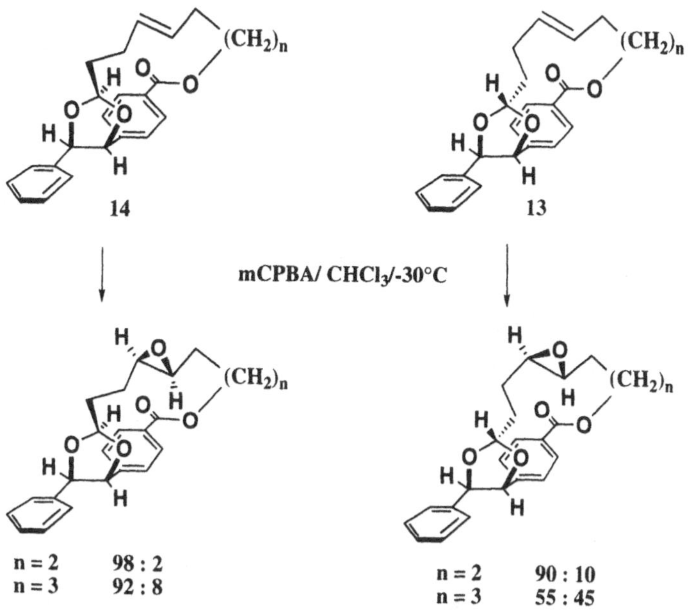

Figure 12. Selectivities for chiral workbenches with achiral ansa chains

NON-DISCONNECTIBLE CHIRAL WORKBENCHES

This approach is illustrated by the synthesis of morphinane alkaloids (e.g. dihydrocodeinone, **22**) which uses phenanthrenone **15** as a chiral workbench on which the remaining two rings are mounted successively (Figure 13)[2,3]. In this way all stereoproblems are easily solved, as the rigid workbench serves quite well as a template for the further construction of the molecular framework. One of the key steps is the diastereocontrolled 1,4-addition of an anionic synthon to **15** to establish the benzylic quaternary center and to construct the characteristic piperidine ring in **22**. (Scheme 1).

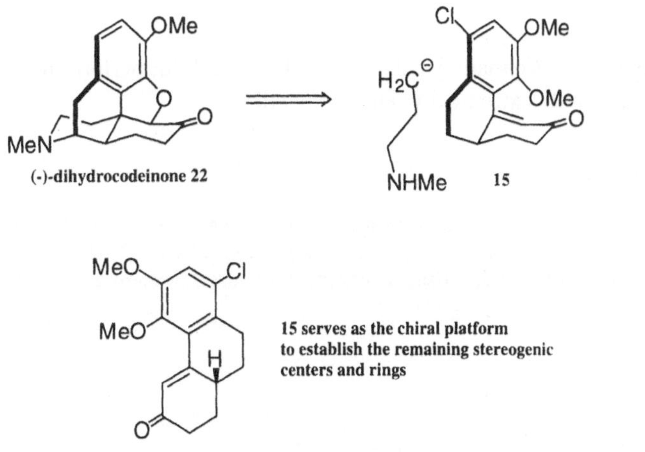

Figure 13. Phenanthrenone **15** as a non-disconnectible workbench for the synthesis of **22**.

23

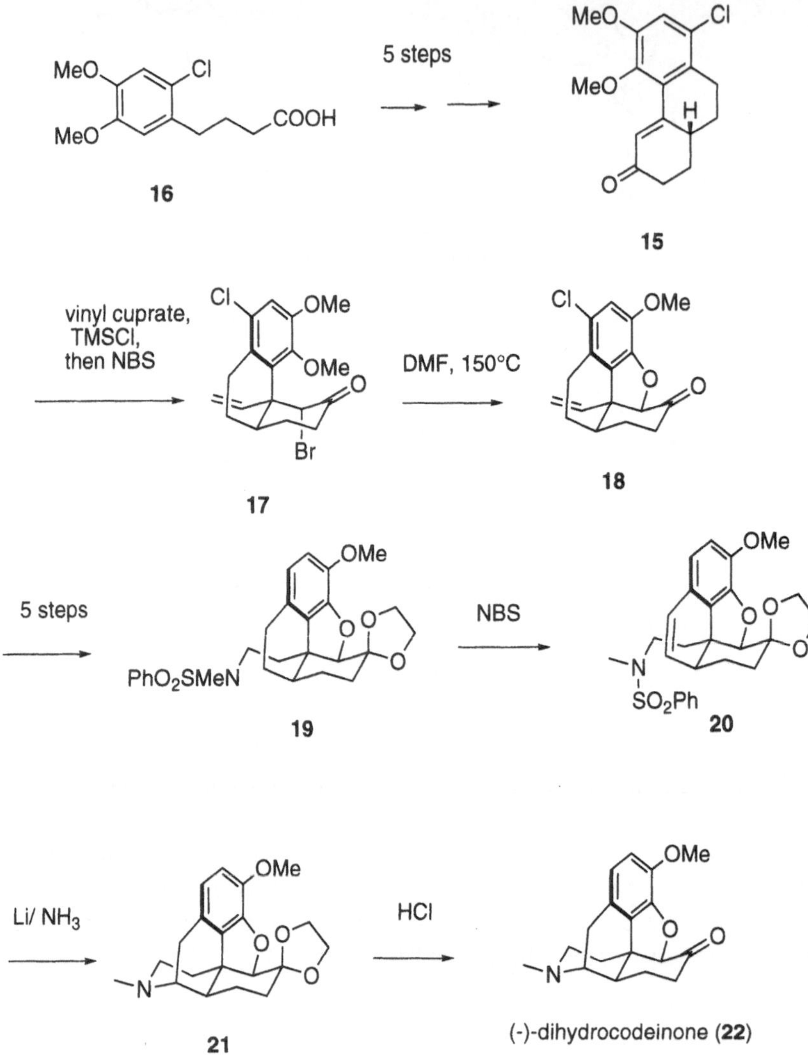

Scheme 1. Synthesis of dihydrocodeinone (22) via chiral workbench 15.

Acknowledgement. We thank Dr. J.-W.Bats, Frankfurt, for performing numerous X-ray crystal structures with extreme diligence and skill.

REFERENCES

1. J.Mulzer, K.Schein, J.W.Bats, J.Buschmann and P.Luger, *Angew.Chem.***110**: 1625-1628(1998).
2. J.Mulzer, G.Dürner and D.Trauner, Angew.*Chem.int.Ed.Eng.* **35**:2830-2832(1996).
3. D.Trauner, S.Porth, T.Opatz, J.W.Bats, G.Giester and J.Mulzer, *Synthesis*:653-664(1998).

FROM *D*-CAMPHOR TO THE TAXANES. HIGHLY CONCISE REARRANGEMENT-BASED APPROACHES TO TAXUSIN AND TAXOL

Leo A. Paquette, Mangzhu Zhao, Francis Montgomery, Qingbei Zeng, Ting Zhong Wang, Steven Elmore, Keith Combrink, Hui-Ling Wang, Simon Bailey, and Zhuang Su

Evans Chemical Laboratories
The Ohio State University
Columbus, Ohio 43210 USA

PURPOSE OF THE INVESTIGATION

Our involvement with the de novo acquisition of taxusin (**1**) and taxol (**2**) has been fueled in large part by the enchanting structural features of these molecules and the much heralded antitumor efficacy of **2**, particularly in patients beset with refractory ovarian, breast, and lung cancers.[1] In combination with the unusual mode of action of **2**, which exhibits a remarkable capacity for stabilizing microtubule assembly and deterring cell division,[2] the intrinsically exciting promise shown by taxol for benefiting human health has commanded unrivaled attention.

The studies to be described herein were formulated not only to reach these targets, but do so in as correlated a fashion as possible, and with the utmost brevity. At the initiation of these efforts, only a single pioneering synthesis of *ent*-taxusin had been realized.[3] The intervening time has been witness to a significantly more expedient approach to **1**[4] as well as to five elegant syntheses of **2**.[5-9]

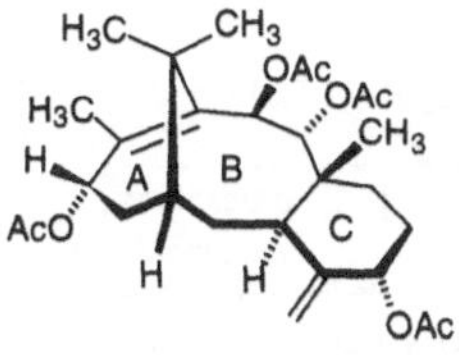

1

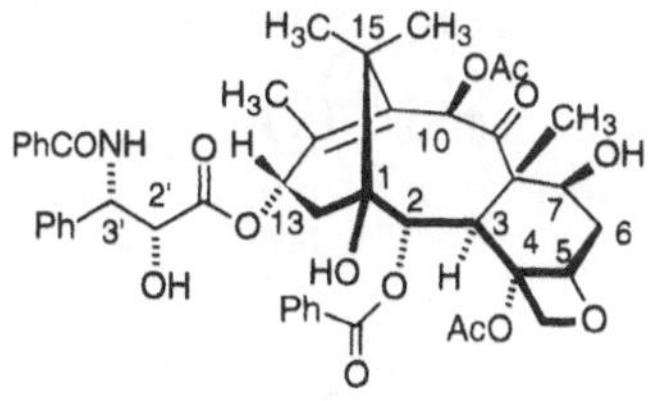

2

A second inducement was a projected unrivaled opportunity to deploy the anionic oxy-Cope rearrangement[10] in the specific context of bridgehead olefin construction.[11] Indeed, Nature has produced a far greater number of this generic type of novel unsaturated product than is generally appreciated. Therefore, a real need exists for improvising direct methods for

Current Trends in Organic Synthesis
Edited by Scolastico and Nicotra, Kluwer Academic/Plenum Publishers, 1999

elaborating in a global sense the relatively abundant and diverse members of this structural class.

SYNTHETIC PLANNING

While taxusin and taxol share the identical tricyclo[9.3.1.0^{3,8}]pentadecane framework, it is obvious at a glance that **2** is significantly more oxygenated than **1**. One of the more obvious points of contrast is bridgehead carbon C-1, which carries hydroxyl functionality in **2**. Also distinctive is the added oxygen functionality at C-2, C-4, and particularly C-7. Here, taxol is recognized to have aldol characteristics, a feature that holds special significance in our synthetic design and harmonizes the construction of these sectors in notably direct ways.

The retrosynthetic analysis illustrated in Scheme 1 defines an implicitly workable approach to taxusin. In order to reach **5**, endo coupling of either **8** or its enantiomer (as the respective lithio derivative) to the readily available enone **7** (100% ee)[12] was envisioned to set the stage for the targeted anionically-accelerated [3,3] sigmatropic event. The prospect for C-3 epimerization and dihydroxylation of **5** was intended to deliberately address the stereoelectronic requirements of the pinacol-like 1,2 Wagner-Meerwein shift required for proper bridge migration and generation of **3**. The availability of this triketone would leave only issues surrounding functional group manipulation to be dealt with.

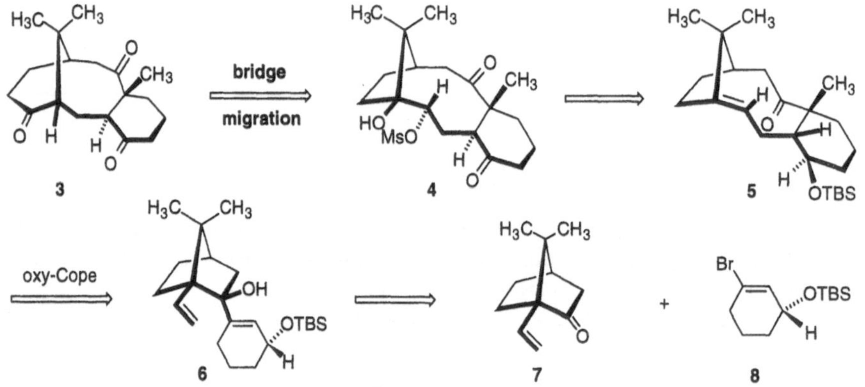

Scheme 1

The prospectus for a related approach to taxol, shown in Scheme 2, deliberately leaves open many crucial particulars. The most notable of these is the unspecified nature (loosely

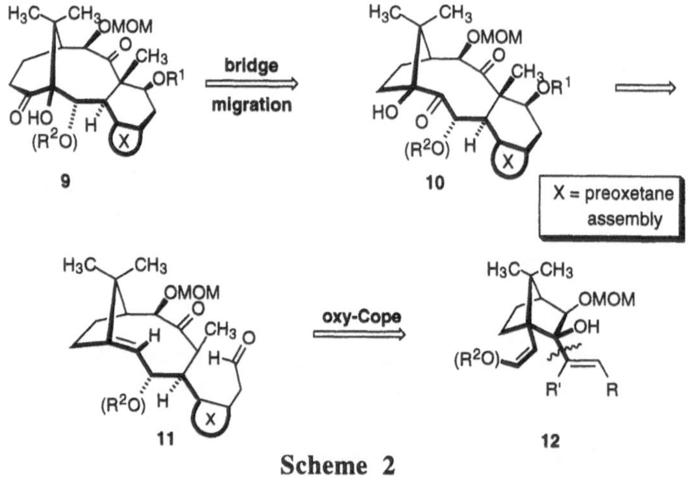

Scheme 2

defined as "X") of the functional group array that will ultimately serve to generate the oxetane ring. Several choices of relatively different merit are available. Also unaddressed are the manner and timing to be used for introduction of the C-2 oxygen center. Again, several protocols are possible including carrying the R^2O substituent through from **12** and α-oxygenating a suitably protected form of **10**. It is clear, however, that this matter requires attention prior to the α-ketol rearrangement that forms the basis for obtaining **9**. Note that construction of the C-ring with proper installation of the C-7,8,9 triad is to be accomplished via intramolecular aldolization.

THE PATHWAY TO TAXUSIN

The earliest experiments aimed at **1** showed that (+)- and (-)-**8** were most effectively obtained by enantioselective hydrolysis of the choroacetate with lipase P-30,[13] that arrival at **6** required use of the cerate in order to curtail enolization, and that the oxy-Cope rearrangement of **6** to give **5** was highly atroposelective[14] because of strict adherence to a so-called "endo-chair" transition state[15]. In effect, only five laboratory steps were required to advance from **7** to **13** (Scheme 3). Furthermore, the conversion of **13** to **3** could be readily implemented (Scheme 3). High efficiency can be routinely achieved if the bridge migration is promoted with diethylaluminum chloride.[16] Mangzhu Zhao subsequently recognized that the three ketone carbonyls in **3** could be readily distinguished. The transformation to **14**, which aptly demonstrates the steric crowding present in the area of C-9 and C-10, made possible the convenient stereocontrolled introduction of the necessary hydroxyl substituent at C-5.[17]

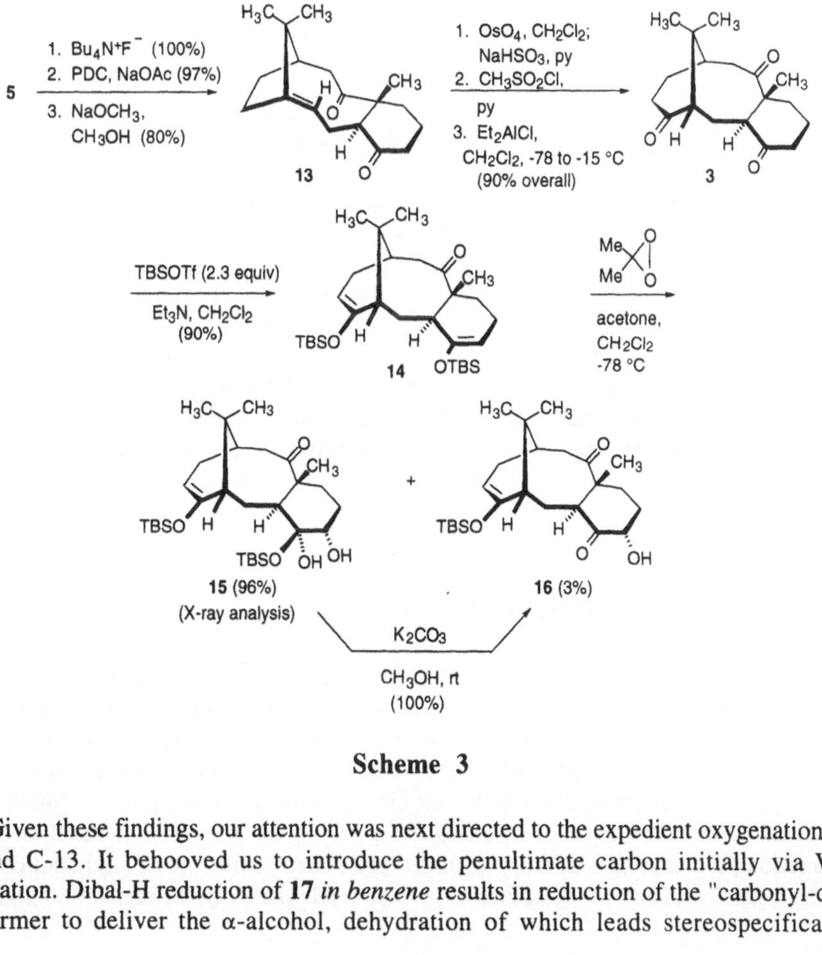

Scheme 3

Given these findings, our attention was next directed to the expedient oxygenation of C-10 and C-13. It behooved us to introduce the penultimate carbon initially via Wittig olefination. Dibal-H reduction of **17** *in benzene* results in reduction of the "carbonyl-down" conformer to deliver the α-alcohol, dehydration of which leads stereospecifically to cyclononene **18**, the less thermodynamically stable of the two possible trans diastereomers (Scheme 4).[17] Since **18** undergoes osmylation preferably at this site and only from the direction external to the ring, the oxygenation pattern found in **19** evolves as required for taxusin. The A-ring enolate anion generated from **19** is particularly amenable to oxidation, as are many related compounds.[18] Arrival at α-diketone **20** is consequently greatly facilitated. Reduction of **20** with lithium aluminum hydride is preceded by deprotonation of the enol, thereby transiently protecting that substituent from nucleophilic attack. Subsequent benzoylation afforded the very desirable product **21**, which capped completed construction of the B and C rings that had already materialized in **19**. Most significantly, only twenty laboratory steps are needed to arrive at this highly advanced stage.

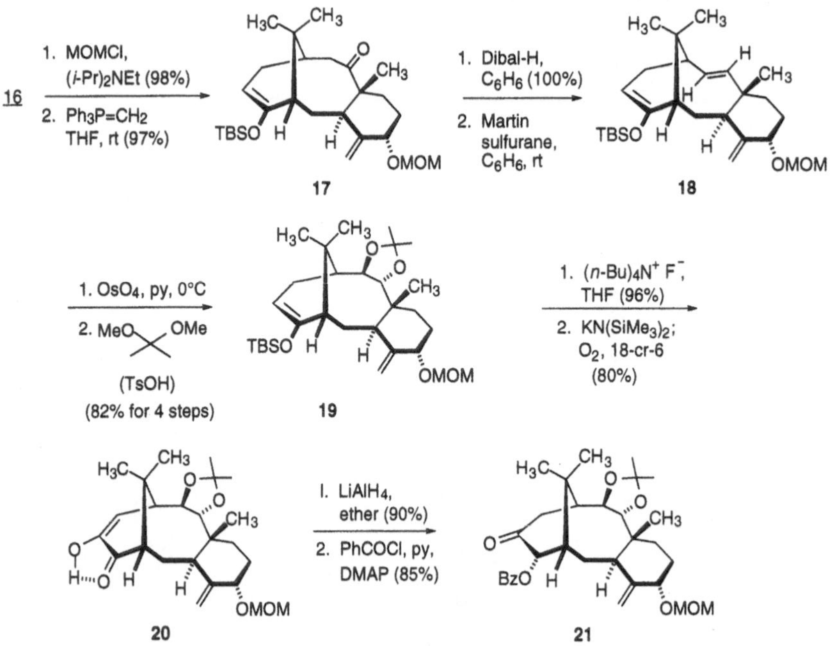

Scheme 4

The oxygen atom at C-14, originally introduced into **13** for the ultimate purpose of enabling the Wagner-Meerwein shift, had not yet outworn its usefulness. As shown in Scheme 5, its presence allows for kinetically favored deprotonation at C-12 where a methyl group and bridgehead olefinic center must be introduced. While an exocyclic double bond (as in **22**) or its intracyclic variant (as in **25**) can be readily established, and subsequent reductive cleavage of the α-benzoyloxy group is highly chemoselective, the need exists to merge these two tactics. The lecture will address appropriate means for accomplishing this task.

AN ABBREVIATED ROUTE TO TAXOL

We explored, in the first instance, the possibility of adapting synthetic building blocks of type **8** in a similarly direct way to taxol construction.[19,20] When it came to be recognized that introduction of the C-7 oxygen in this manner was problematical,[21] the retrosynthetic model embodied in Scheme 2 was adopted. The first objective was to examine the interplay of heightened functionalization of **7** with the efficiency of the oxy-Cope process. Much to our amazement, these camphor-based ketones undergo α-oxygenation predominantly from the exo direction (Scheme 6). For example, dimethyldioxirane oxidation of silyl enol ether **30** gives the α-ketol with an exo preference of 94-100%. Alternatively, the enolates of **28** and

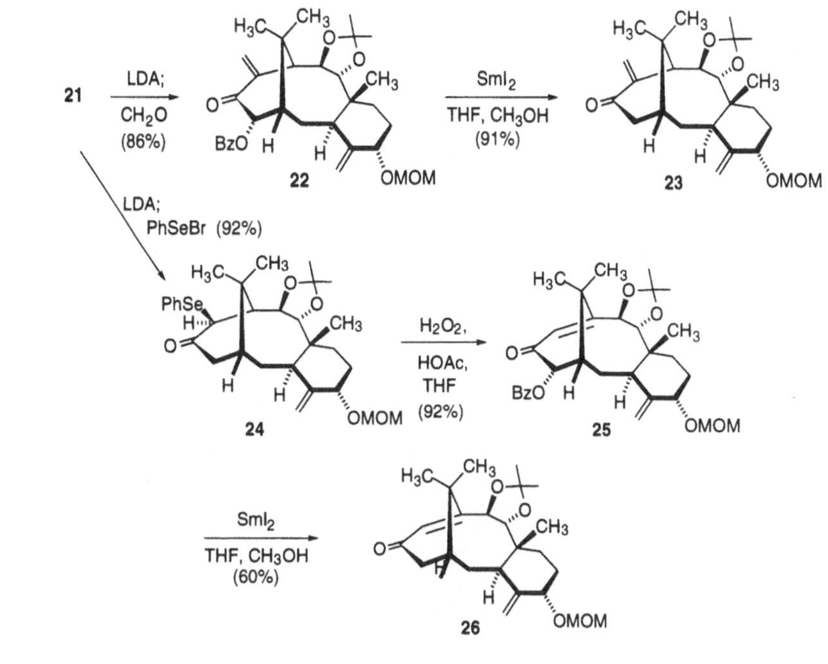

Scheme 5

29 enter into reaction with the Davis oxaziridine to give mixtures rich in **32** (5:1) and **33** (7:1).[22] It will be recognized that the exo-oriented OMOM substituent in **31-33** is destined to become the C-10 oxygenated center in taxol. Likewise, the -OPMP substituent will find itself ultimately positioned at C-2.

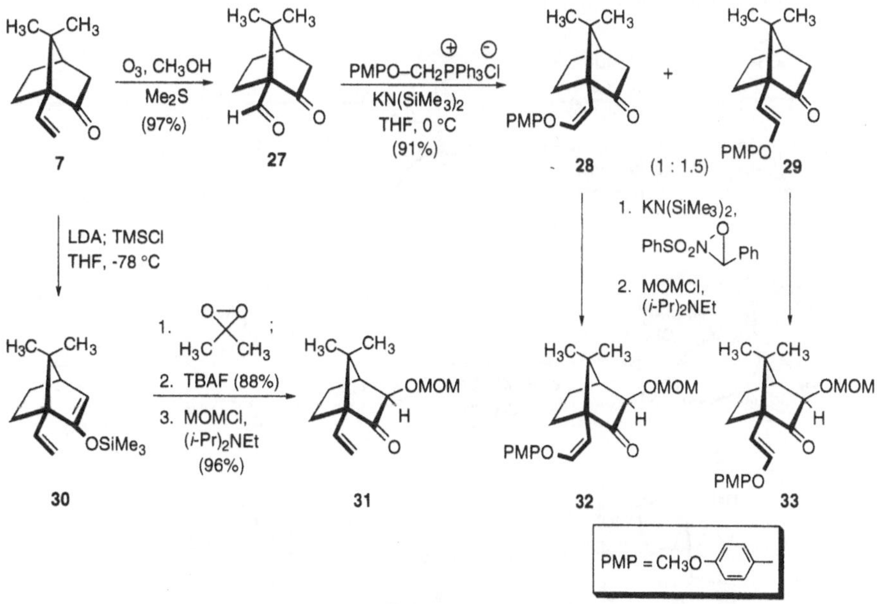

PMP = CH₃O─⟨⟩─

Scheme 6

This array of functionality proved to be most accommodating and especially conducive to our goals. Thus, the steric bulk of the OMOM group further enhanced endo addition to the adjacent carbonyl. Also, its inductive effect sufficiently reduced the acidity of the α-carbonyl proton that direct condensation with alkenyllithium reagents was now possible. In addition, the subsequent charge-accelerated sigmatropic rearrangement of the resulting exo carbinols delivered tricyclic products that feature the characteristic B-ring oxygenation pattern resident in **2** (Scheme 7). The process is stereospecific for **34** and **35** as a consequence of continued adherence to the "exo-chair" transition state geometry.[22]

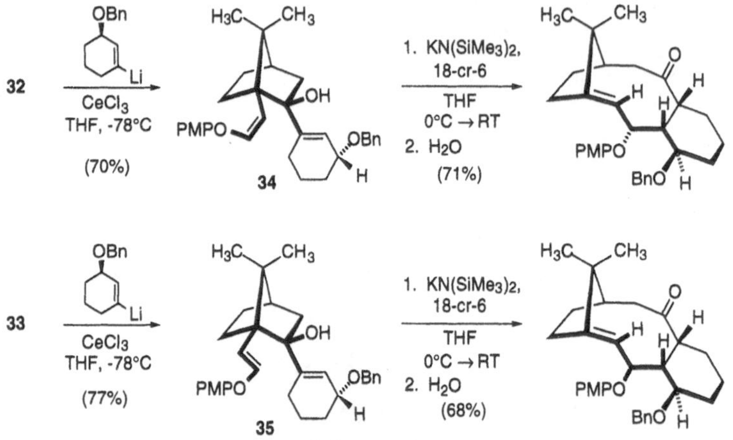

Scheme 7

Although the thermodynamics associated with the pending α-ketol equilibration had been evaluated early and shown to involve strain release in the desired direction,[23,24] this key transformation had yet to be performed when the nearby C-2 center was oxygen-substituted (as in **10 → 9**). The urge to show workability was followed up by the rapid assembly of prototype **40** from the readily available carbinol **36** as shown in Scheme 8.[25] Subsequent exposure of **40** to aluminum tri-*tert*-butoxide led efficiently to **41**. Another important dimension of this scheme was the fashion in which the benzoate group was introduced - *viz.*, by oxygenation of the enolate anion of **38**.

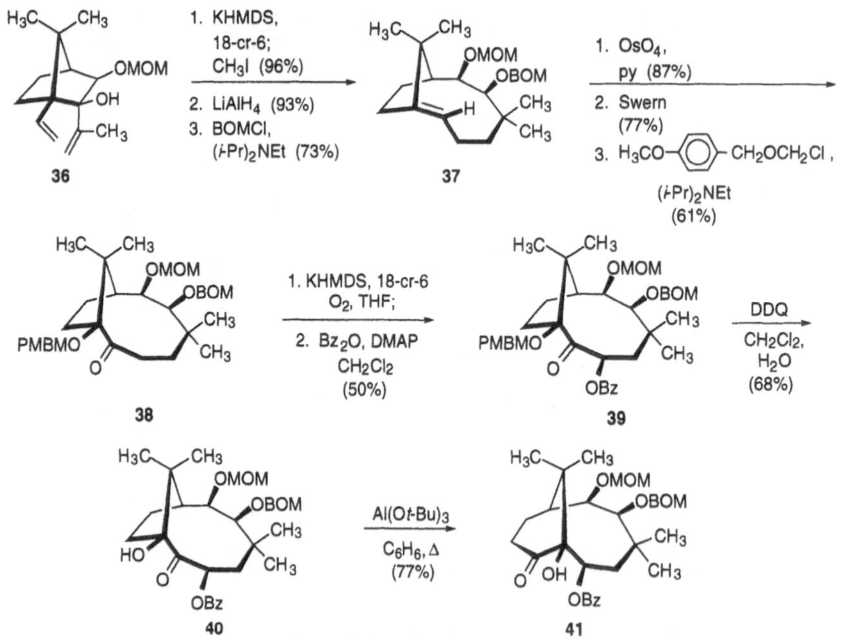

Scheme 8

The task of devising the proper C-ring synthon now had to be addressed. The enantiomerically pure (*Z*)-vinyl iodide **42**, conveniently available from D-ribose functioned admirably well up to the point of aldol cyclization (Scheme 9). The aldehyde obtained from **43** enters into ring closure, but only after β-elimination (see **44**) and recapture of methanol.[26] This was not acceptable.

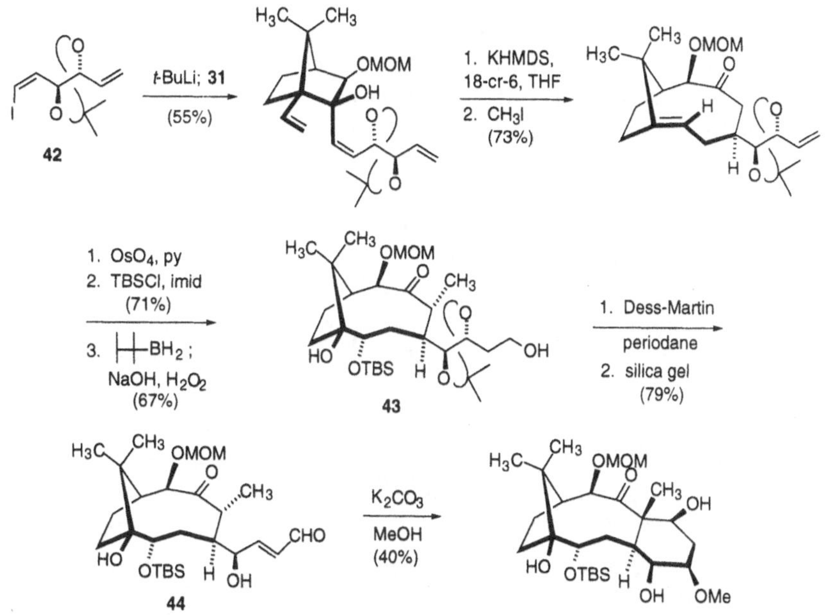

Scheme 9

Maneuvers of this type can be completely bypassed if a (*Z*)-iodide of type **45** is utilized instead.[27] The latter is produced from (*R*)-glyceraldehyde acetonide and likewise sets the stereocenters resident at C-7 and C-8 in their proper absolute configuration (Scheme 10). The convergent coupling of **45** to **31** provides a forum for the rapid elaboration of aldehyde **46**, the aldol cyclization of which proceeds without β-elimination. The outcome is the acquisition

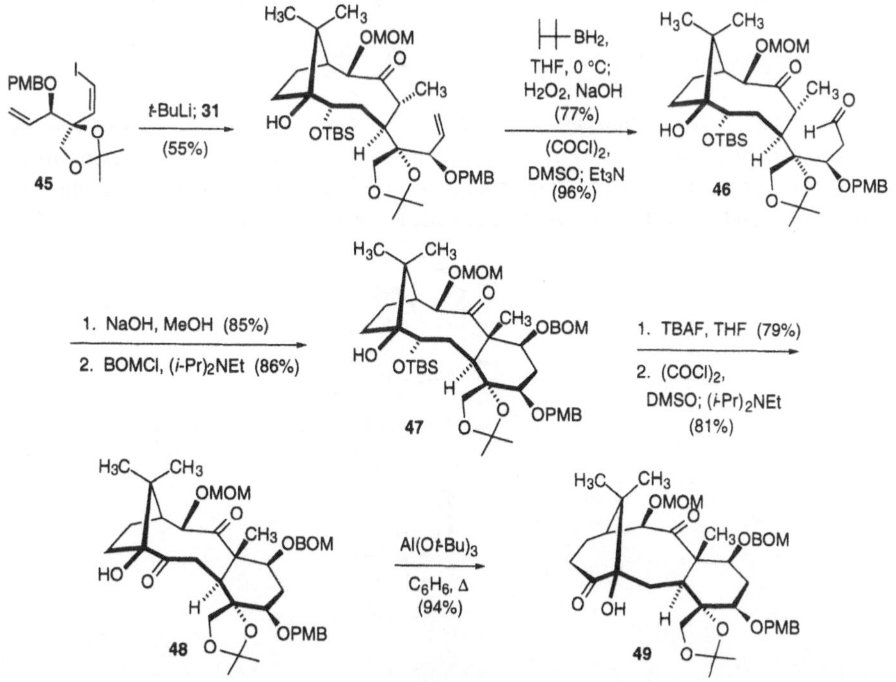

Scheme 10

of **47**, Swern oxidation of which affords **48** and ultimately its α-ketol tautomer **49**. When exploration of the α-oxygenation of **48** and protected forms thereof commenced, it quickly became apparent that the steric shielding provided by the acetonide subunit precluded all possibility of engaging these enolates in reaction. Recognition of this fact was adequate inducement to install the oxetane ring first. Thus, preference was given to selective hydrolysis of **47** and generation of tribenzoate **50** (Scheme 11). Although the cyclization of ring D proceeded without event and **53** could now be crafted from **51** according to precedent,[25] this protocol is certainly too lengthy and awkward for our purposes.

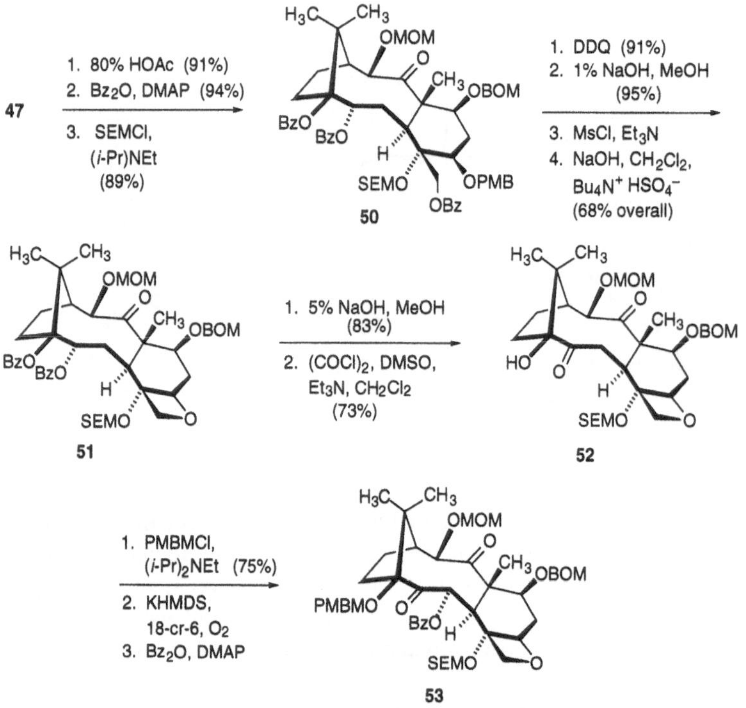

Scheme 11

The extensive array of steps associated with the oxetane assembly process in Scheme 11 can be dramatically reduced to only two in molecules related to **54**. In addition, no steric blockade is experienced at C-2 as long as an acetonide is absent. The manner in which these considerations can be effectively united will be made clear in our lecture.

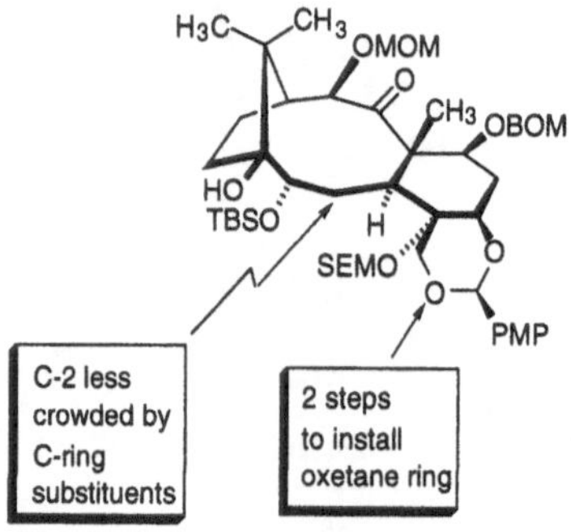

ACKNOWLEDGMENTS

This research was supported at its outset by the Bristol-Myers Squibb Company and more recently by the National Cancer Institute of the U.S. Public Health Service.

REFERENCES

1. E. K. Rowinsky and R. C. Donehower *Pharmacol. Ther.* **52**, 35 (1991).
2. P. B. Schiff, J. Fant, and S. B. Horwitz *Nature* **277**, 665 (1979).
3. R. A. Holton, R. R. Juo, H. B. Kim, A. D. Willliams, S. Harusawa, R. E. Lowenthal, and S. Yogai *J. Am. Chem. Soc.* **110**, 6558 (1988).
4. R. Hara, T. Furukawa, Y Horiguchi, and I. Kuwajima *J. Am. Chem. Soc.* **118**, 9186 (1996).
5. R. A. Holton, C. Somoza, H. B. Kim, F. Liang, R. J. Biediger, P. D. Boatman, M. Shindo, C. C. Smith, S. Kim, Y. Suzuki, C. Tao, P. Vu, S. Tang, P. Zhang, K. K. Murthi, L. N. Gentile, and J. H. Liu. *J. Am. Chem. Soc.* **116**, 1597 (1994). R. A. Holton, H. B. Kim, C. Somoza, F. Liang, R. J. Biediger, P. D. Boatman, M. Shindo, C. C. Smith, S. Kim, H. Nadizadeh, Y. Suzuki, C. Tao, P. Vu, S. Tang, P. Zhang, K. K. Murthi, L. N. Gentile, and J. H. Liu *J. Am. Chem. Soc.* **116**, 1599 (1994).
6. K. C. Nicolaou, P. G. Nantermet, H. Ueno, R. K. Guy, E. A. Couladouros, E. J. Sorensen *J. Am. Chem. Soc.* **117**, 624 (1995). K. C. Nicolaou, J.-J. Liu, Z. Yang, H. Ueno, E. J. Sorensen, C. F. Claiborne, R. K. Guy, C.-K. Hwang, M. Nakada, and P. G. Nantermet *J. Am. Chem. Soc.* **117**, 634 (1995). K. C. Nicolaou, Z. Yang, J.-J. Liu, P. G. Nantermet, C. F. Claiborne, J. Renaud, R. K. Guy, and K. Shibayama *J. Am. Chem. Soc.* **117**, 645 (1995). K. C. Nicolaou, H. Ueno, J.-J. Liu, P. G. Nantermet, Z. Yang, J. Renaud, K. Paulvannan, and R. Chadha *J. Am. Chem. Soc.* **117**, 653 (1995).
7. S. J. Danishefsky, J. J. Masters, W. B. Young, J. T. Link, L. B. Snyder, T. V. Magee, D. K. Jung, R. C. A. Isaacs, W. G. Bornmann, C. A. Alaimo, C. A. Coburn, and M. J. DiGrandi *J. Am. Chem. Soc.* **118**, 2843 (1996).
8. P. A. Wender, N. F. Badham, S. P. Conway, P. E. Floreancig, T. E. Glass, C. Gränicher, J. B. Houze, J. Jänichen, D. Lee, D. G. Marquess, P. L. McGrane, W. Meng, T. O. Mucciaro, M. Mühlebach, M. G. Natchus, H. Paulsen, D. B. Rawlins, J. Satkofsky, A. J. Shuker, J. C. Sutton, R. E. Taylor, and K. Tomooka *J. Am. Chem. Soc.* **119**, 2755 (1997). P. A. Wender, N. F. Badham, S. P. Conway, P. E. Floreancig, T. E. Glass, J. B. Houze, N. E. Krauss, D. Lee, D. G. Marquess, P. L. McGrane, W. Meng, M. G. Natchus, A. J. Shuker, J. C. Sutton, R. E. Taylor *J. Am. Chem. Soc.* **119**, 2757 (1997).
9. T. Mukaiyama, I. Shiina, H. Iwadare, H. Sakoh, Y. Tani, M. Hasegawa, K. Saitoh *Proc. Japan Acad., Ser. B.* **73**, 95 (1997).
10. L. A. Paquette *Angew. Chem., Int. Ed. Engl.* **29**, 609 (1990). L. A. Paquette *Tetrahedron* **53**, 13971 (1997).
11. L. A. Paquette *Chem. Soc. Rev.* **24**, 9 (1995).
12. N. Fischer and G. Opitz *Organic Syntheses*; Wiley, New York, 1973; Collect. Vol. V, p. 877.
13. L. A. Paquette, S. W. Elmore, K. D. Combrink, E. R. Hickey, and R. D. Rogers *Helv. Chim. Acta* **75**, 1755 (1992).
14. L. A. Paquette, K. D. Combrink, S. W. Elmore, and R. D. Rogers *J. Am. Chem. Soc.* **113**, 1335 (1991).
15. L. A. Paquette, N. A. Pegg, D. Toops, G. D. Maynard, and R. D. Rogers *J. Am. Chem. Soc.* **112**, 277 (1990).

16. L. A. Paquette, M. Zhao, and D. Friedrich *Tetrahedron Lett.* **33**, 7311 (1992).
17. L. A. Paquette and M. Zhao *J. Am. Chem. Soc.* **115**, 354 (1993).
18. S. W. Elmore and L. A. Paquette *J. Org. Chem.* **58**, 4963 (1993).
19. L. A. Paquette and R. C. Thompson *J. Org. Chem.* **58**, 4952 (1993).
20. Z. Su and L. A. Paquette *J. Org. Chem.* **60**, 764 (1995).
21. L. A. Paquette, Z. Su, S. Bailey, and F. J. Montgomery *J. Org. Chem.* **60**, 897 (1995).
22. S. W. Elmore and L. A. Paquette *J. Org. Chem.* **60**, 889 (1995).
23. L. A. Paquette, K. D. Combrink, S. W. Elmore, and M. Zhao *Helv. Chim. Acta* **75**, 1772 (1992).
24. S. W. Elmore, K. D. Combrink, and L. A. Paquette *Tetrahedron Lett.* **32**, 6679 (1991).
25. Q. Zeng, S. Bailey, T.-Z. Wang, and L. A. Paquette *J. Org. Chem.* **63**, 137 (1998).
26. L. A. Paquette and S. Bailey *J. Org. Chem.* **60**, 7849 (1995).
27. L. A. Paquette, F. J. Montgomery, and T.-Z. Wang *J. Org. Chem.* **60**, 7857 (1995).

NON-COVALENT SYNTHESIS OF ORGANIC NANOSTRUCTURES

Leonard J. Prins, Peter Timmerman and David N. Reinhoudt[*]

Supramolecular Chemistry and Technology
University of Twente
P.O.Box 217, 7500 AE Enschede
The Netherlands

INTRODUCTION

During the last decade self-assembly has emerged as a tool that enables relatively easy access to molecular assemblies of high molecular weight and nanoscopic dimensions[1]. The elegance of self-assembly lies in the fact that the extent of *covalent* synthesis is reduced to the level of the individual components or modules, that contain information necessary for the formation of the assembly. Like Nature, chemists employ noncovalent, reversible interactions for the recognition of the modules, for instance hydrogen bonding[2] or metal-ligand interactions[3]. The advantage of using this kind of reversible interactions over irreversible covalent interactions is that it allows for 'self-correction' since the assembly is at thermodynamic equilibrium. A large variety of assemblies based on noncovalent interactions has been reported over the past decade illustrating how the concept of self-assembly has triggered the creativity of chemists[4].

One of our recent activities concerns the development of artificial receptors using a modular approach. Hitherto we have focused on covalent structures, based on a combination of modules like calix[4]arenes, resorcin[4]arenes and cyclodextrins[5]. With the synthesis of the holand, comprising two calix[4]arenes and two resorcin[4]arenes, we have reached the boundaries of synthetically accessible receptor molecules[6]. To eliminate the elaborate syntheses and to create more flexibility in the receptor molecules we recently started a research program that should lead us towards *functional self-assembled aggregates*. Here we describe our recent achievements so far in this area.

SYNTHESIS AND CHARACTERIZATION OF HYDROGEN BONDED BOX-LIKE ASSEMBLIES

Complementary hydrogen bonding between cyanuric acid and melamine has received considerable attention as a structural motif for self-assembly. Work from the groups of Whitesides[7] and Lehn[8] has shown that the recognition between these components can be used to form different types of aggregates, either lineair or crinkled *tapes*, or cyclic *rosettes*. Which type of aggregate is formed critically depends on the extent of preorganization of the complementary binding units and on steric interactions between their substituents. Due to the facile synthesis and the high degree of controllability we have decided to use this recognition motif for the self-assembly of hydrogen bonded box-like assemblies[9, 10]

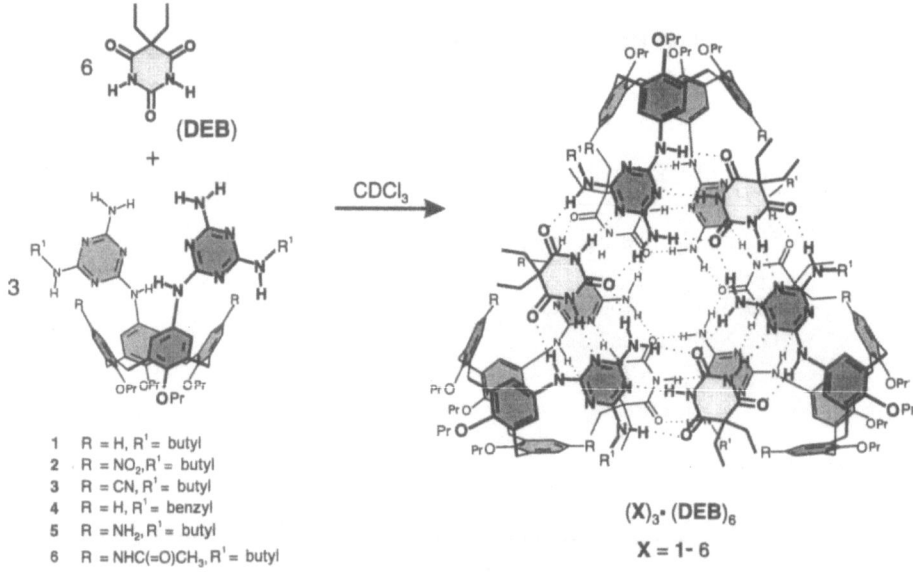

1 R = H, R' = butyl
2 R = NO$_2$, R' = butyl
3 R = CN, R' = butyl
4 R = H, R' = benzyl
5 R = NH$_2$, R' = butyl
6 R = NHC(=O)CH$_3$, R' = butyl

(X)$_3$· (DEB)$_6$

X = 1- 6

Figure 1 Bismelamine calix[4]arenes and **DEB** form a self-assembled aggregate

Calix[4]arene **1**, diametrically substituted at the upper rim with two melamine units, spontaneously forms a well-defined box-like assembly in the presence of two equivalents of 5,5-diethylbarbiturate (**DEB**) (Figure 1). The assembly comprises three calix[4]arenes and six diethylbarbiturates and is held together by a total number of 36 hydrogen bonds. The top and bottom of this box-like assembly consist of a cyclic hydrogen bonded platform whereas the calix[4]arene units act as side walls. The assembly is stable in apolar solvents like chloroform and toluene even at 10^{-4} M, but the stability significantly decreases when significant amounts (10-20%) of polar solvents (like DMSO and methanol). Characterization of assemblies of this type is not straightforward and structural proof is in general provided by a combination of several different techniques. Here we will discuss the use of ^{1}H NMR, X-ray spectroscopy and MALDI-TOF mass spectrometry for the structure determination of the assembly (**1**)$_3$·(**DEB**)$_6$.

^{1}H NMR Spectroscopy

Titration of bismelamine **1** with **DEB** in CDCl$_3$ shows several characteristic features (Figure 2). Already at low concentrations of **DEB** two signals become evident at very low field. These resonances, which remain at $\delta = 14.10$ and 13.32 regardless of the **1**:**DEB** ratio, are assigned to the hydrogen bonded NH protons of **DEB** in the complex. These protons are observed at different chemical shifts as a result of the unsymmetrical substitution of the melamine units, which gives both protons a chemically different environment in the complex. When the amount of **DEB** is increased, the signal corresponding to the NH_2 protons of free **1** decreases in intensity but remains at the same position. At the same time, two signals are observed around $\delta = 6.9$ (c) and 6.7 (d) for both the NH_2 protons. Two additional signals appear at $\delta = 8.37$ (e) and 7.43 (f), which correspond to the two secondary amine protons of **1** in the hydrogen bonded complex. The aromatic protons ortho to the melamine substitutents of calix[4]arene **1** give rise to signals at $\delta = 7.15$ (g) and 6.03 (h). In free **1** these protons display broad signals at $\delta = 6.65$-6.05. The resonance at $\delta = 6.03$ is in accordance with a *pinched cone* structure in which the two melamines approach each other.

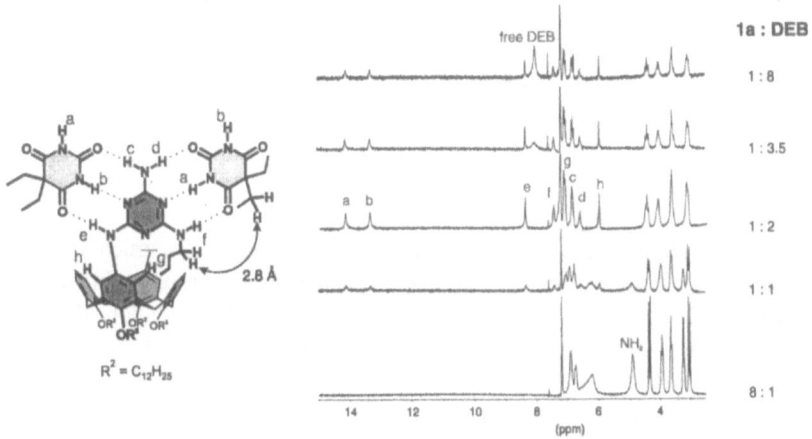

Figure 2 Titration of bismelamine **1** with **DEB**

At a ratio of **1**:**DEB** of 1:2 the spectrum is sharp, which indicates the absence of free **1**. The 1:2 ratio is consistent with the box-like assembly that is represented in Figure 1. When more than two equivalents of **DEB** are added, the signal for the NH protons of free **DEB** is observed alongside the two signals for the hydrogen bonded **DEB**. This indicates that exchange between hydrogen bonded and free **DEB** is slow on the NMR time scale.

Single Crystal X-Ray Diffraction Studies

Upon recrystallization from toluene the assembly $(2)_3 \bullet (DEB)_6$ (R = NO_2) forms large single crystals. The X-ray structure, which is shown in Figure 3, provides the first crystallographic information for this type of molecular boxes. The structure shows that the calix[4]arene units are fixed in a *pinched cone* conformation, which is the only conformation that allows simultaneous participation of the calix[4]arene units in both rosette motifs. The two rosettes tightly stack on top of each other with an interatomic separation of 3.5 Å at the edges to 3.2 Å in the centre of the box, which leaves little space for guest molecules. Interestingly, the structure reveals that the two rosette motifs are oriented in an *anti-parallel* fashion (see also Figure 1), which means that the assembly is *stereogenic*.[11]

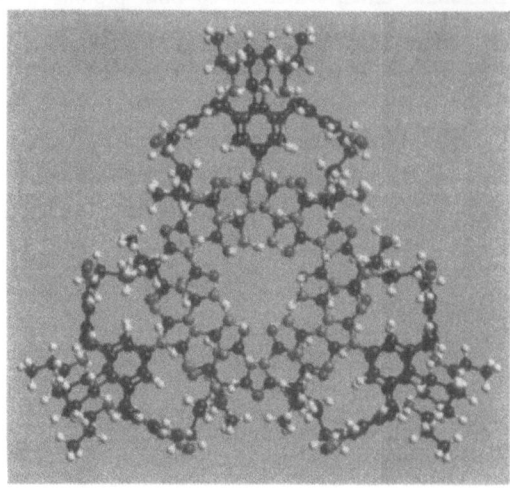

Figure 3 X-Ray crystal structure of assembly $(2)_3\bullet(DEB)_6$ (topview)

2D ^{1}H NMR Spectroscopy

Evidence for the fact that the solid state structure resembles the structure in solution was obtained by measuring the distances between the relevant protons in the box-like assembly with 2D NOESY experiments using the initial rate approximation. The distances are slightly larger than those observed in the crystal structure, but still provide sufficient evidence for the proposed assembly structure. A particularly strong NOE connectivity was observed between the NCH_2 protons of **2** (R = NO_2) and one of the ethyl-CH_2 groups of **DEB**. The interatomic distance of these protons *within one single* rosette is too large (*ca.* 6.4 Å) to cause the observed NOE connectivity. Therefore, it must arise from a proximity effect between the two rosette motifs, which consequently puts a limit to their mutual orientation. In the *anti-parallel* orientation (also observed in the solid state) the two protons are 2.8 Å apart, which is in perfect agreement with the strong NOE observed. The presence of this NOE connectivity provides strong evidence for the fact that the structure in solution closely resembles the solid state structure as determined by X-ray crystallography.

MALDI-TOF Mass Spectrometry

Additional evidence for the self-assembly of these structures was obtained by using a novel MALDI-TOF mass spectrometry technique using Ag^+ labelling[12]. The technique makes use of the high affinity of silver cations for a variety of aromatic π-donors and cyano groups. It provides for a nondestructive way to generate positively charged hydrogen bonded assemblies that can easily be detected by MALDI-TOF mass spectrometry. A sample prepared by stirring assembly $(3)_3.(DEB)_6$ with 1.5-2.0 equivalents of silver trifluoroacetate shows an intense signal at m/z=4220.0 (calcd. for $^{13}C_2{}^{12}C_{214}H_{282}N_{54}O_{30}.Ag^+$-complex: 4221.9) in the MALDI-TOF mass spectrum. Signals corresponding to partially formed aggregates or higher oligomers are not observed in the mass spectrum. Formation of the silver complex of assembly $(3)_3.(DEB)_6$ must result from coordination of a silver cation to one of the cyano groups, since neither $(1)_3 \cdot (DEB)_6$ nor $(2)_3 \cdot (DEB)_6$ shows any significant signal between m/z 1500 and 8000 in the corresponding MALDI-TOF mass spectrum.

π-Donors can be used as well for the complexation of Ag^+ which is illustrated by the fact that samples prepared by stirring assembly $(4)_3 \cdot (DEB)_6$ with $AgCF_3COO$ give a strong signal at $m/z=4278.3$ (calcd. $^{13}C_2{}^{12}C_{226}H_{276}N_{48}O_{30} \cdot Ag^+$-complex: 4276.1) in the MALDI-TOF spectrum. Presumably the silver cation is complexed between the benzyl substituent and one of the phenyl rings of the calix[4]arene (Figure 4). It is quite remarkable that assemblies $(1)_3 \cdot (DEB)_6$ and $(2)_3 \cdot (DEB)_6$ do not form stable Ag^+ complexes by themselves, since calix[4]arenes are known to interact strongly with Ag^+ ions through the aromatic π faces at the upper rim of the cavity. This is confirmed by the presence of intense signals for the Ag^+ complexes of calix[4]arenes 2 and 3 in the MALDI-TOF mass spectra. Apparently, the calix[4]arene skeleton loses its affinity for Ag^+ ions upon formation of the hydrogen bonded assembly. This is most probably a consequence of the extreme conformational change that the calix[4]arene skeleton undergoes when the hydrogen bonded assembly is formed. The X-ray crystal structure of assembly $(2)_3 \cdot (DEB)_6$ reveals that the melamine-substituted aromatic ring carbon atoms are 4.05 Å apart, which is 0.75 Å less than the optimal distance measured from the crystal structure of the calix[4]arene-Ag^+ complex. Formation of the hydrogen bonded assembly therefore leaves too little space for complexation of the Ag^+ ions in between the parallel aromatic rings of the calix[4]arene fragment.

From these results it can be concluded that this MALDI-TOF-MS technique after Ag^+ labeling is a convenient new tool for the mass spectrometric characterization of hydrogen bonded assemblies. The absence of any signal corresponding to fragments of assemblies in the mass spectrometer illustrates the unprecedented mildness of the technique. The method requires only a binding site for the soft Ag^+ ion in order to charge the noncovalent assembly in a nondestructive way. Aromatic π-donors, which can sandwich a Ag^+ ion, or cyano groups are adequate for this purpose, but in principle many other functionalities such as acetylenes, ethylenes, amines, and sulfur groups may interact strongly with Ag^+ ions.

CONFORMATIONAL ISOMERISM OF HYDROGEN BONDED ASSEMBLIES

In assembly $(1)_3 \cdot (DEB)_6$ the two melamines on each calix[4]arene are in an anti-parallel orientation with respect to each other and this conformational isomer is therefore referred to as the staggered conformer. Two other conformers of the assembly are possible as well in which the melamines on each calix[4]arene unit are facing the

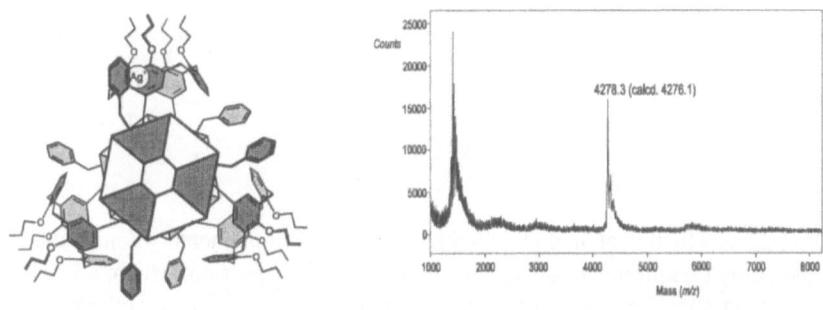

Figure 4 Ag^+ complexation between π-donors and resulting MALDI-TOF spectrum

same side of the calix[4]arene (Figure 5), *i.e.* the symmetrical eclipsed and the unsymmetrical eclipsed (Figure 6). In both assemblies the two rosette motifs are aligned on top of each other. The difference between these two eclipsed conformers arises from the different positions of the calix[4]arene components in the assembly. In the symmetrical eclipsed conformer all calix[4]arenes 'face' the same direction, yielding a C_3 symmetrical assembly. In the unsymmetrical eclipsed conformer, one of the three calix[4]arenes is oriented in an opposite direction compared to the other two, resulting in a C_1 symmetrical assembly. For the C_3 symmetrical conformers (staggered or symmetrical eclipsed) two ^{1}H NMR signals are expected for the imide protons of the cyanurate, for reasons given before. In the C_1 symmetrical conformer (unsymmetrical eclipsed) all imide protons of the cyanurates within a single rosette are in a different chemical environment and the assembly should therefore exhibit six different signals in the 13-15 ppm region.

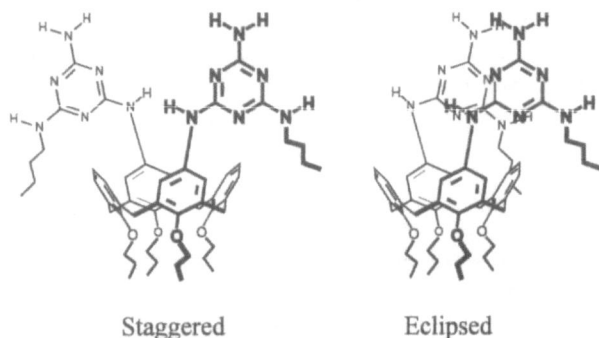

Staggered   Eclipsed

Figure 5 Staggered and eclipsed orientation of melamine units

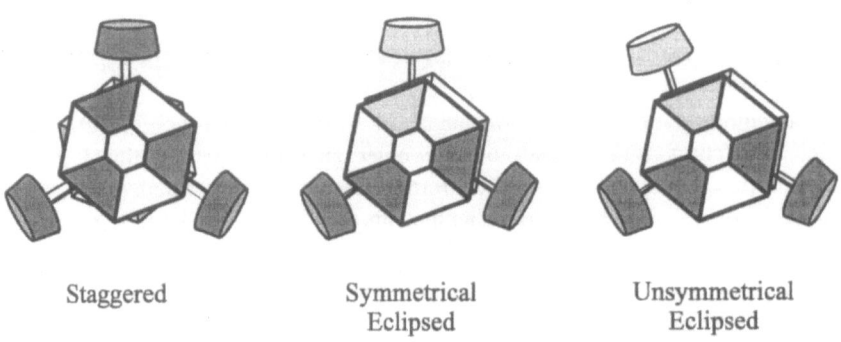

Staggered   Symmetrical   Unsymmetrical
                         Eclipsed   Eclipsed

Assemblies which contain diethylbarbiturate as the complementary compound show in all cases only two imide proton signals in the ^{1}H NMR spectrum, which, according to 2D NMR techniques, belong to the staggered conformer (*vide supra*). However, the ^{1}H

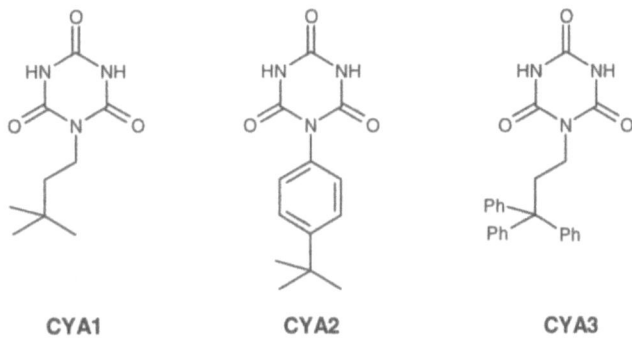

CYA1 **CYA2** **CYA3**

Figure 7 Cyanurates

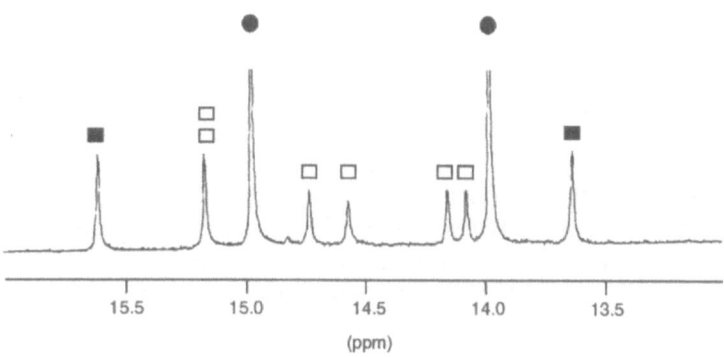

Figure 8 Part of the ¹H NMR spectrum of assembly (1)₃(CYA1)₆. Ten signals are observed originating from the staggered (●), symmetrical eclipsed (■) and unsymmetrical eclipsed (□) conformer (two signals coincide)

NMR spectrum of assembly **(1)₃.(CYA1)₆** (Figure 7) appears to be much more complicated[13,14]. In the diagnostic region at 13-15 ten peaks are observed for **(1)₃.(CYA1)₆** instead of the two signals observed for **(1)₃.(DEB)₆** (Figure 8). A detailed 2D ¹H NMR study revealed that these signals originate from the staggered (●), symmetrical eclipsed (■) and unsymmetrical eclipsed (□) conformers.

Apparently the assembly process is not so well-defined when neohexylcyanurate is used instead of diethylbarbiturate. The reason for the exclusive formation of the staggered conformer with diethylbarbiturate is likely to be steric, but the influence of electronic factors cannot be excluded. In the eclipsed conformers the barbiturates in upper and lower rosette would be aligned on top of each other causing a steric hindrance of the ethyl chains. Since in neohexylcyanurate the sp³ hybridized carbon is replaced by a sp² hybridized nitrogen the neohexyl substituent is oriented in the plane of the rosettes thus imposing a different steric effect on the assembly. In this case it is sterically possible to form the eclipsed conformers as well. Furthermore, the higher electronegativity of nitrogen compared to carbon causes the cyanurates to form stronger hydrogen bonded complexes with melamines[15] and might also favor the formation of the eclipsed conformers.

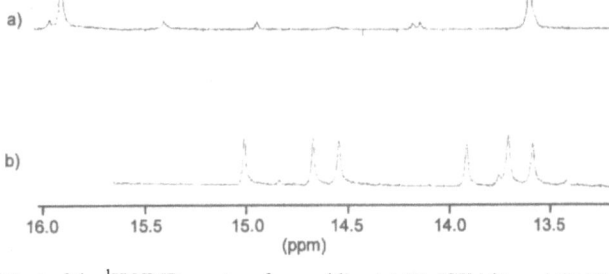

Figure 9 Part of the ^{1}H NMR spectra of assemblies (a) $(1)_3(CYA2)_6$ and (b) $(2)_3(CYA3)_6$

Further studies were carried out to determine if this behavior is a general phenomenon for *N*-substituted cyanurates. The results showed that the influence of the cyanurate substituents on the conformer distribution is difficult to predict. For instance, the ^{1}H NMR spectrum of assembly $(1)_3.(CYA2)$ shows two major peaks in the 13-16 ppm region together with 6 smaller signals, the latter accounting for about 10% of the total aggregate present (Figure 9a). The 6 signals are assigned to the unsymmetrical eclipsed conformer, but the two major signals can in principle originate from either the staggered or symmetrical eclipsed conformer.

To illustrate that the assembly process sometimes gives surprising results, part of the ^{1}H NMR spectrum of the assembly $(2)_3.(CYA3)_6$ is depicted in Figure 9. In this case only the 6 signals belonging to the unsymmetrical eclipsed conformer are observed. At this moment rationalization of this result has not been successful, demonstrating that the assembly process has not yet been comprehended to every extent. Most likely a combination of steric and electronic factors accounts for the difference in assembly behavior of cyanurates compared to diethylbarbiturate.

FUNCTIONAL GROUP DIVERSITY IN HYDROGEN BONDED ASSEMBLIES

In the hydrogen bonded assembly *six* functionalizable sites R (see Figure 1) are gathered around the box-like cavity. In view of our objective to generate functional group diversity in or around non-covalently assembled cavities, we studied how the nature of the functionality at position R affects the stability of such calix[4]arene box-like assemblies in solution[10].

The introduction of polar substituents at positions R, such as nitro or cyano, does not influence the stability of the cyclic hydrogen bonded assembly in solution to any significant extent. Compounds **2** (R= NO$_2$) and **3** (R= CN) readily assemble in chloroform in the presence of **DEB**, the ^{1}H NMR spectra of which are virtually identical to that of $(1)_3•(DEB)_6$ (R= H). Titration of **3** (R= CN) with **DEB** clearly proved the 3:6 stoichiometry of the box-like assembly formed.

Surprisingly, the stability of assembly $(5)_3•(DEB)_6$ (R= NH$_2$) is highly solvent dependent. The ^{1}H NMR spectrum in toluene-d_8 is very well-defined and shows two singlets at 14.55 and 13.85 ppm for the magnetically nonequivalent barbiturate NH protons. In CDCl$_3$, however, the spectrum is very broad and does not show any resonance in the region between 13 and 15 ppm, which indicates a preference for non-specific oligo- or polymer formation in this solvent. The decreased stability of $(5)_3•(DEB)_6$ in the more polar solvent CDCl$_3$ is most probably related to the hydrogen bond donating ability of the two NH$_2$ groups.

Compound **6** [R= NHC(O)CH₃] does not form the box-like assembly in either CDCl₃ or toluene-d_8. Both ¹H NMR spectra exhibit broad resonances indicating non-specific aggregation. Also in this case the formation of intramolecular hydrogen bonds, *i.e.* between the amide carbonyl and the melamine NH proton, seems to be primarily responsible for inhibiting the assembling ability of **5**.

Towards Dynamic Combinatorial Libraries

Combinatorial chemistry allows for the simultaneous synthesis of large libraries of structurally well-defined molecules. These libraries are prepared by statistical combination of reactive molecular fragments *via* the irreversible formation of covalent bonds. In principle, one could also use noncovalent interactions, like hydrogen bonding, to build in a reversible way libraries of small complementary molecules[16]. In the previous section it was demonstrated that a large variety of functional groups can be accumulated in a hydrogen bonded assembly.

Mixing homomeric assemblies N, carrying different functional groups should theoretically result in the formation of a mixture of heteromeric assemblies, M. This provides a rapid method to create a large number of assemblies at thermodynamic equilibrium and can be regarded as the build up of a dynamic combinatorial library.[17] The total number of assemblies P (*i.e.* N + M) present in such a library rapidly increases with increasing N following equation (1).

$$P = N + N(N-1)+[N(N-1)(N-2)]/6 \qquad (1)$$

For the smallest possible library (N=2, P=4) we have studied the combination of assemblies **(1)₃.(DEB)₆** and **(2)₃.(DEB)₆**. The four possible assemblies are schematically depicted in Figure 10.

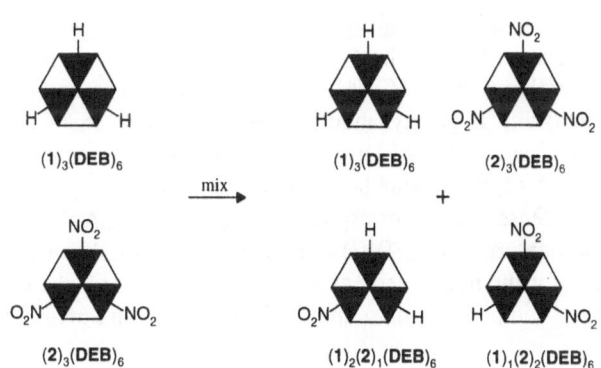

Figure 10 Self-assembled combinatorial library

Mixing of 5 mM solutions of **(1)₃.(DEB)₆** and **(2)₃.(DEB)₆** at 0 °C (ratio 2:1) in toluene-d_8 gives only a mixture of the two separate homomeric assemblies. Exchange of the bismelamine calix[4]arenes is extremely slow at this temperature and, as a consequence, formation of the heteromeric assemblies is not observed. When the mixture is warmed to temperatures over 15 °C the heteromeric assemblies start to form. After 2.5 h at 25 °C the system reaches the thermodynamic equilibrium. (Figure 11).

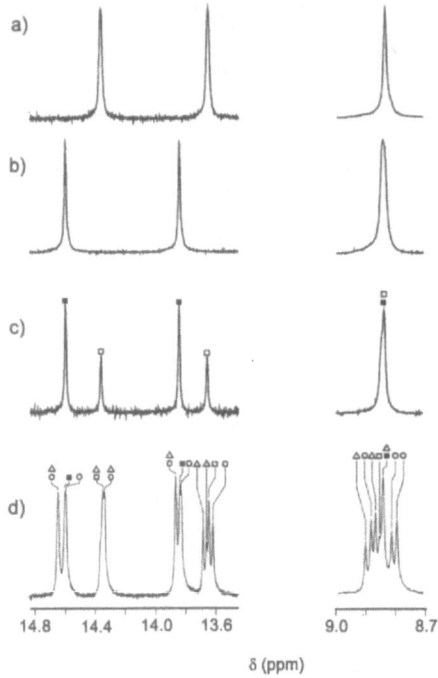

Figure 11 Temperature dependent formation of heteromeric assemblies as observed with ^{1}H NMR. Shown are the imide proton signals and one NH signal of the bismelamine of a) homomeric assembly **(1)$_3$.(DEB)$_6$**, b) homomeric assembly **(2)$_3$.(DEB)$_6$**, c) 2:1 mixture of **(1)$_3$.(DEB)$_6$** and **(2)$_3$.(DEB)$_6$** at 0 °C and d) 2.5 h after mixing **(1)$_3$.(DEB)$_6$** and **(2)$_3$.(DEB)$_6$** at 25 °C.

The unsymmetrical heteromeric rosettes **(1)$_2$(2)$_1$.(DEB)$_6$** and **(1)$_1$(2)$_2$.(DEB)$_6$** cause multiple signals in the characteristic 13-15 ppm region.Signals of all four assemblies could be assigned with 2D NMR measurements. The relative concentrations were 1:3:3:1, in agreement with a statistical distribution. The complex ^{1}H NMR spectrum for this small library shows that ^{1}H NMR spectroscopy might not be the appropriate technique for analyzing larger dynamic libraries. Therefore characterization was also performed with Ag$^+$ assisted MALDI-TOF mass spectrometry as described previously. Individual solutions of the assemblies, pretreated with 1.5 equiv. of CF$_3$CO$_2$Ag in chloroform, give intense signals corresponding to **(1)$_3$.(DEB)$_6$.Ag$^+$** and **(2)$_3$.(DEB)$_6$.Ag$^+$** respectively. The mixture of **(1)$_3$.(DEB)$_6$** and **(2)$_3$.(DEB)$_6$** on the other hand shows two additional signals for the heteromeric assemblies. Especially for the characterization of libraries with more components, mass spectrometry will be an important tool.

CONCLUSIONS

Here we have presented our achievements in creating functional hydrogen bonded assemblies. Characterization of these dynamic assemblies was performed by ^{1}H NMR spectroscopy, single crystal X-ray diffraction and MALDI-TOF mass spectrometry technique with Ag$^+$ labelling. From these studies it appeared that the **DEB**-containing assemblies are stereogenic, due to an anti-parallel orientation of the rosette motifs. However, other conformers are observed as well when N-substituted cyanurates were used to form rosettes instead of **DEB**. Furthermore, the possibility of bringing different functional groups together in a dynamic assembly was demonstrated. The possibility of creating heteromeric assemblies by mixing homomeric assemblies provides a new strategy that enables the noncovalent synthesis of a dynamic library of potential receptor molecules. Currently, we are actively pursuing the use of modified assemblies as receptor molecules for guests.

ACKNOWLEDGMENTS

The work described in this paper is the result of the efforts of several coworkers whose names appear in the references. We thank R. Fokkens and N. Nibbering (University of Amsterdam) for the MALDI-TOF MS measurements. We are grateful for financial support from the Netherlands Organization for Scientific Research (NWO), and the European Community (M.C.C.no.ERBFMBICT 961445).

REFERENCES

1. J.-M. Lehn. *Supramolecular Chemistry: Concepts and Perspectives*, VCH, Weinheim (1995)
2. Some examples of assemblies based on hydrogen bonding: M.M.Conn, J.Rebek, Jr., *Chem.Rev.* 97:1647-1668 (1997); S.C.Zimmerman, F.Zeng, D.E.C.Reichert, S.V.Kolotuchin, *Science*, 271:1095-1098 (1996); M.R.Ghadiri, *Adv.Mater.* 7:675-677 (1995); J.Yang, S.J.Geib, A.D.Hamilton, *J.Am.Chem.Soc.* 115:5314-5315 (1993)
3. Some examples of assemblies based on metal-ligand interaction: P.J.Stang, B.Olenyuk, *Acc.Chem.Res.* 30:502-518 (1997); M.Fujita, K.Ogura, *Bull.Chem.Soc.Jpn.* 69:1471-1482 (1996); W.T.S.Huck, L.J.Prins, R.H.Fokkens, N.M.M.Nibbering, F.C.J.M. van Veggel, D.N.Reinhoudt, *J.Am.Chem.Soc.* 120, in press (1998) ; W.T.S.Huck, F.C.J.M.van Veggel, B.L.Kropman, D.H.A.Blank, E.G.Keim, M.M.A.Smithers, D.N.Reinhoudt, *J.Am.Chem.Soc.* 117: 8293-8294 (1995); J.-P.Sauvage, *Acc.Chem.Res.* 23:319-327 (1990)
4. For reviews see: D.Philp, J.F.Stoddart, *Angew.Chem.Int.Ed.Engl.*, 35:1154-1196 (1996); D.S.Lawrence, T.Jiang, M.Levett, *Chem.Rev.* 95:2229-2260 (1995)
5. For some examples see: I.Higler, P.Timmerman, W.Verboom, D.N.Reinhoudt, *J.Org.Chem.* 61:5920-5931 (1997); E.Van Dienst, B.H.M.Snellink, I.van Piekartz, J.F.J.Engbersen, D.N.Reinhoudt, *J.Org.Chem.* 60:6537 (1995); P.Timmerman, W.Verboom, F.C.J.M. van Veggel, J.P.M. van Duynhoven, D.N.Reinhoudt, *Angew.Chem.Int.Ed.Engl.* 33:2345 (1994)
6. P.Timmerman, W.Verboom, F.C.J.M. van Veggel, W.P. van Hoorn, D.N.Reinhoudt, *Angew.Chem.Int.Ed.Engl.* 33:1292-1295 (1994)
7. For a review see: G.M.Whitesides, E.E.Simanek, J.P.Mathias, C.T.Seto, D.N.Chin, M.Mammen, D.M.Gordon *Acc.Chem.Res.* 28:37 (1995)
8. J.-M.Lehn, M.Mascal, A.DeCian, J.Fischer *J.Chem.Soc., Chem.Commun.* 479-481 (1990)
9. R.H.Vreekamp, J.P.M. van Duynhoven, M.Hubert, W.Verboom, D.N.Reinhoudt, *Angew.Chem.Int.Ed.Engl.* 35:1215-1218 (1996)
10. P.Timmerman, R.Vreekamp, R.Hulst, W.Verboom, D.N.Reinhoudt, K.Rissanen, K.A.Udachin, J.Ripmeester *Chem.Eur.J.* 3:1823-1832 (1997)
11. Recent results show that it is possible to selectively form one of the two enantiomers: L.J.Prins, P.Timmerman, D.N.Reinhoudt, unpublished results

12. K.A.Jolliffe, M.Crego Calama, R.Fokkens, N.M.M.Nibbering, P.Timmerman, D.N.Reinhoudt, *Angew.Chem.Int.Ed.Engl.* 37:1247-1251 (1998)
13. K.A.Jolliffe, P.Timmerman, D.N.Reinhoudt, unpublished results
14. For a discussion on conformational isomerism in hydrogen bonded assemblies see: D.N.Chin, E.E.Simanek, X.Li, M.I.M.Wazeer, G.M.Whitesides, *J.Org.Chem.* 62:1892-1895 (1997)
15. M.Mascal, P.S.Fallon, A.S.Batsanov, B.R.Heywood, S.Champ, M.Colclough, *J.Chem.Soc., Chem.Commun.* 805-806 (1995)
16. I.Huc, J.-M.Lehn, *Proc.Natl.Acad.Sci.USA.* 94:2106 (1997); S.J.Rowan, J.K.M. Sanders, *Chem.Commun.* 1407 (1997)
17. M.Crego Calama, R.Hulst, R.Fokkens, N.M.M.Nibbering, P.Timmerman, D.N.Reinhoudt *Chem.Commun.,* 1021-1022 (1998)

RECENT ADVANCES IN THE SYNTHESIS OF CARBOHYDRATE MIMICS

Pierre Sinaÿ

Ecole Normale Supérieure, Département de Chimie, Associé au CNRS,
24 rue Lhomond, 75231 Paris Cedex 05, France

INTRODUCTION

A C-disaccharide is a close mimic of a regular disaccharide in which the interglycosidic oxygen atom has been replaced by a methylene group (ref. 1). For example, C-maltose is a close mimic of maltose :

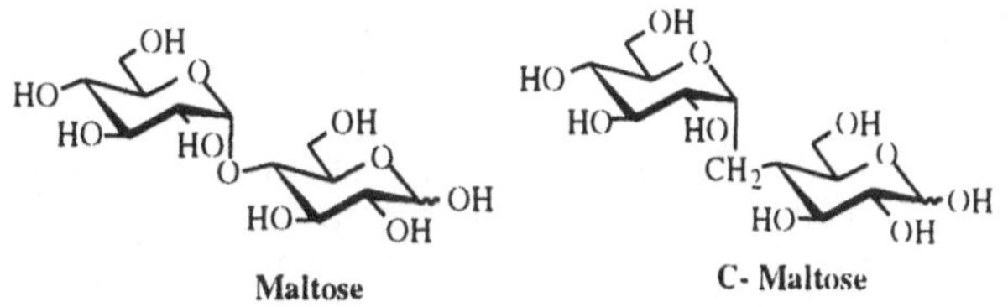

Maltose   **C- Maltose**

In the last few years, we have developed a general strategy for the stereoselective synthesis of these molecules, based on a 8 or 9 *endo-trig* radical cyclisation from two monosaccharides temporarily connected through a tether (ref. 2). A typical example of this strategy is shown in scheme 1.

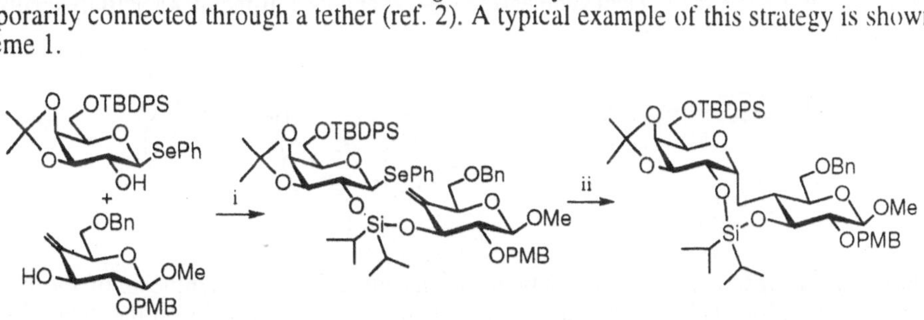

Scheme 1: Reagents and conditions: i) BuLi, THF, 0° C, iPr$_2$SiCl$_2$, 4 equiv., -78° C to r.t. DMAP. THF. r.t.(70%); ii) Bu$_3$SnH, AIBN, toluene. reflux (80%).

In doing so, three main goals are pursued :
(1) To build a molecular object which is fully resistant to glycosidases, a feature which could be of importance for the development of an oligosaccharide-based drug. On the other hand, such compounds might be competitive inhibitors of glycosidases.
(2) To get information on the potential direct intervention and importance of an interglycosidic oxygen atom on the hydrogen bonding network with a receptor protein.
(3) To evaluate the contribution of the *exo*-anomeric effect on the conformation. and then on the biological activity, of a di- or oligosaccharide.
Concerning the first goal, recent studies have indeed demonstrated that oligosaccharides are involved in a steadily increasing number of biological processes and constitute potential active substances for drug development. Several structural modifications have been developed to

increase the resistance of oligosaccharides to enzymatic hydrolysis : introduction of S-interglycosidic bonds (ref. 3), and the alkylation (ref. 4) or acylation (ref. 5) of non critical hydroxyl functions. The currently studied introduction of a C-interglycosidic bond is another option.

Concerning the third goal, a critical feature with any mimic is that the biological activity must be preserved or increased. The question is thus to evaluate to what extend the replacement of the oxygen atom by a methylene group - which removes the *exo* anomeric effect - affects the conformational properties of the molecule, then the biological activity. In the first part of this lecture, we would like to describe the preparation of a biologically active mixed C,O pentasaccharide.

Another way to prepare hydrolytically stable close mimics is to build up a so called pseudo sugar, in which the *endo* cyclic oxygen atom of a glycoside has been replaced by a methylene group.

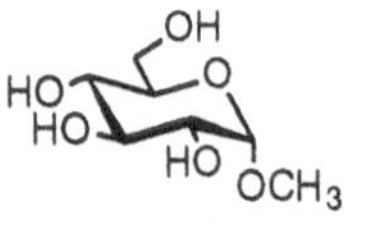

methyl α-D-glucopyranoside

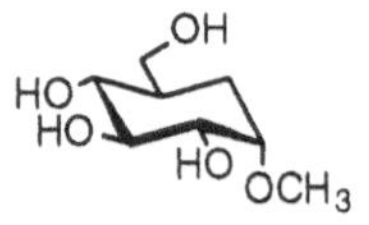

pseudosugar mimic

Such a strategy will be described in the second part of this lecture.

INTRODUCING A C-INTERGLYCOSIDIC BOND IN A BIOLOGICALLY ACTIVE PENTASACCHARIDE HARDLY AFFECTS ITS BIOLOGICAL PROPERTIES

Among the restricted number of synthetic oligosaccharides clinically investigated is a pentasaccharide that reproduces the exact sequence required in heparin for binding and activation of antithrombin III (AT III) (ref.6), and which is a good candidate as an antithrombotic drug. The availability of a well-established series of biologically active heparin pentasaccharides allowed us to investigate how the introduction of a C-disaccharide bond would influence the interaction with the target protein receptor, AT III. The methylated pentasaccharide **1** has already been synthesized (ref. 7) and its anti-factor Xa activity evaluated (1180 units/mg) :

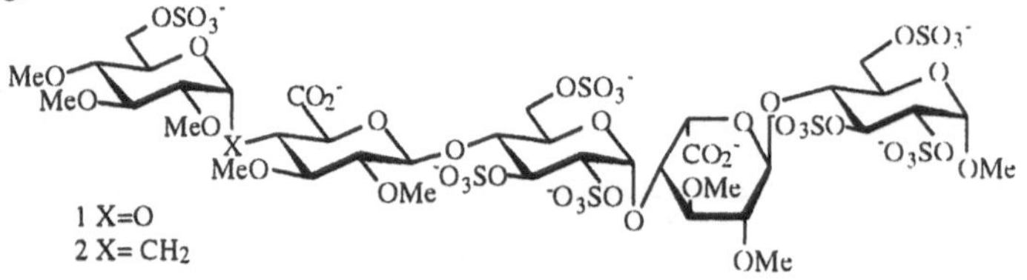

1 X=O
2 X= CH₂

We have synthesized the mixed C, O-pentasaccharide **2**.The retrosynthesis of the key C-disaccharide trichloroacetimidate is shown in scheme 2 (ref. 8). It follows the general route previously delineated.

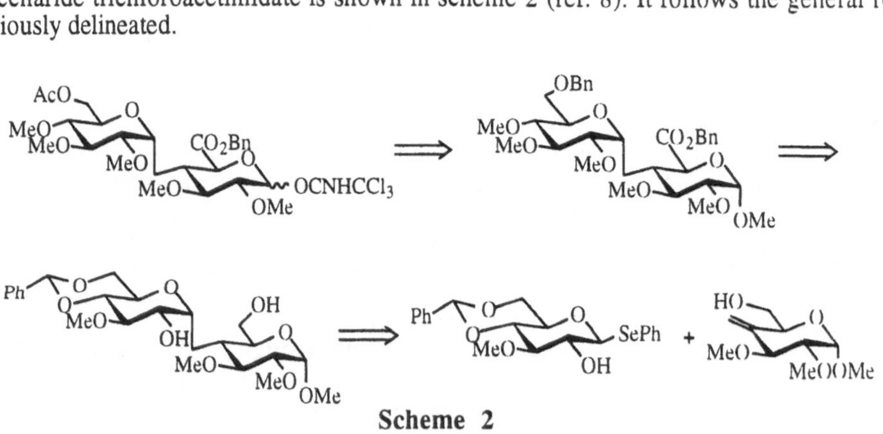

Scheme 2

48

This compound has then been transformed into the target molecule through a sequence presented in scheme 3 (ref. 9).

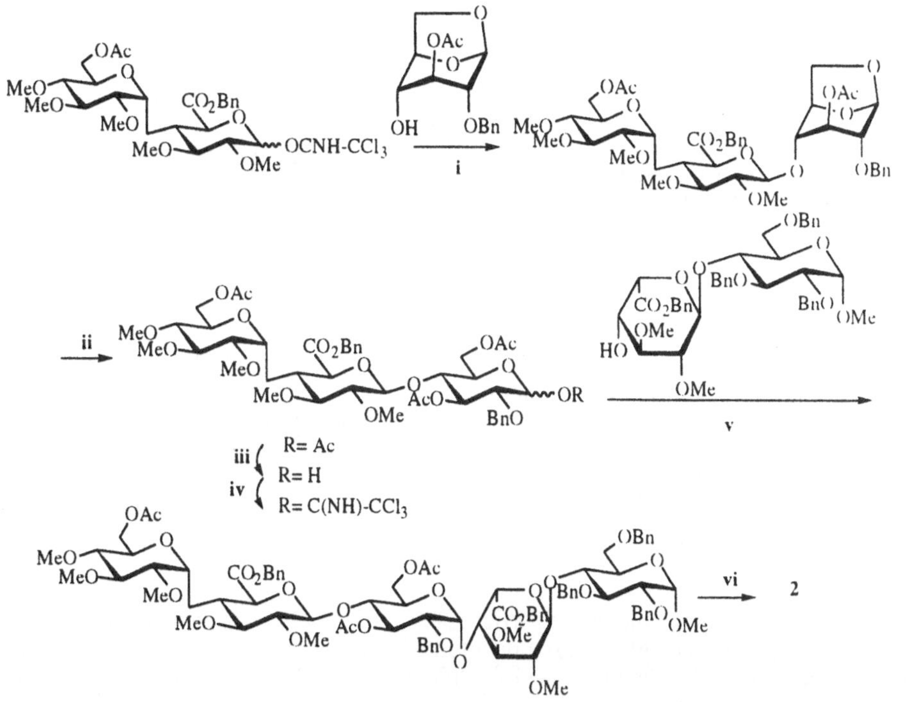

Scheme 3: Reagents and conditions: i) TMSOTf, CH_3CN, -37°C, 88% ii) H_2SO_4, AcOH. Ac_2O. -20°C. 100%; iii) NH_3NH_2OAc, DMF, 90%; iv) CCl_3CN, DBU, CH_2Cl_2, 94%; v) TMSOTf, CH_2Cl_2. -20°C. 79%; vi) 1. H_2, Pd/C, CH_2Cl_2, MeOH, 99%, 2. NaOH, H_2O, MeOH, 0°C, 77%, 3. SO_3. Et_3N, DMF, 55°C. 85%

It has been found that such a structural modification only slightly affected the affinity of the compound for AT III, as well as the active factor Xa activity (880 anti Xa/mg). This compound represents the first example of a new class of anti-factor Xa pentasaccharides containing a C-interglycosidic bond.

ONE-STEP STEREOSELECTIVE CONVERSION OF A SUGAR DERIVATIVE INTO A PSEUDO-SUGAR.

It is appealing to invent reactions to transform a carbohydrate derivative into a substituted cyclohexane derivative. We have discovered (ref. 10) that reaction of a carbohydrate vinyl acetal (6-deoxy-hex-5-exopyranoside) with an excess of triisobutylaluminum at 40 °C resulted in the transposition of an oxygen atom on the ring with the exocyclic carbon atom (scheme 4)

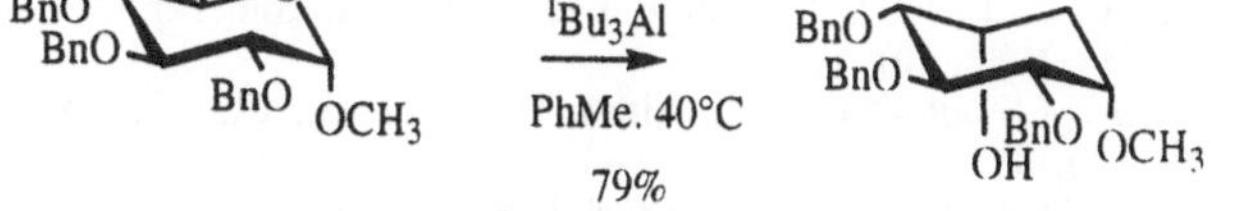

Scheme 4

A postulated mechanism of this transposition is presented in scheme 5.

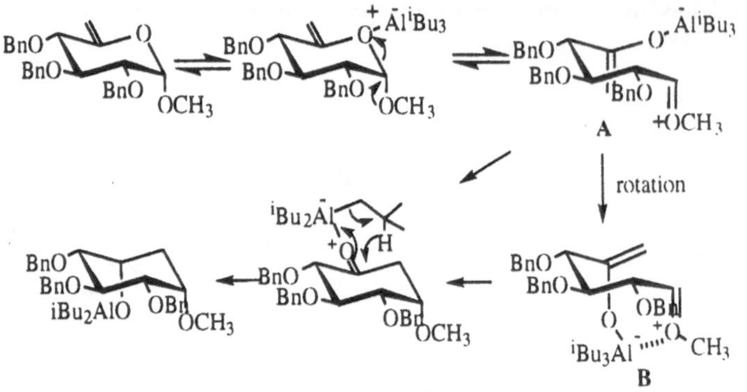

Scheme 5.

When the reaction was applied to a β-methyl glucoside, the stereochemical information of the anomeric centre was again retained, as shown in scheme 6.

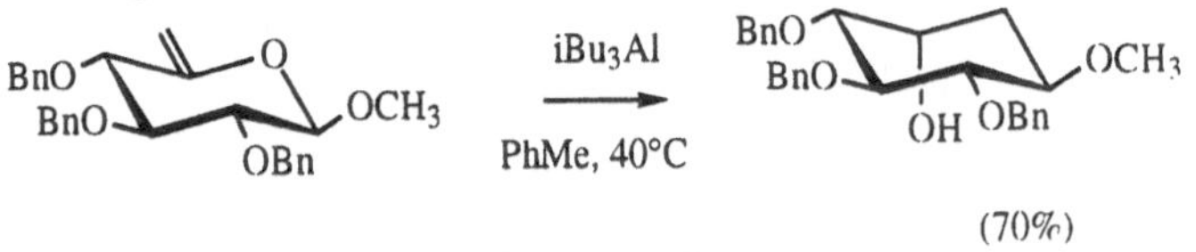

(70%)

Scheme 6

This novel reaction complements the Ferrier-II reaction (ref. 11), which inherently requires an *exo* cleavage to eject the aglycon. We then turned our attention to the case of titanium (IV) derivatives and found that the Lewis acid Ti(OiPr$_2$)Cl$_3$ (ref. 12) resulted in an almost quantative rearrangement, as shown in scheme 7.

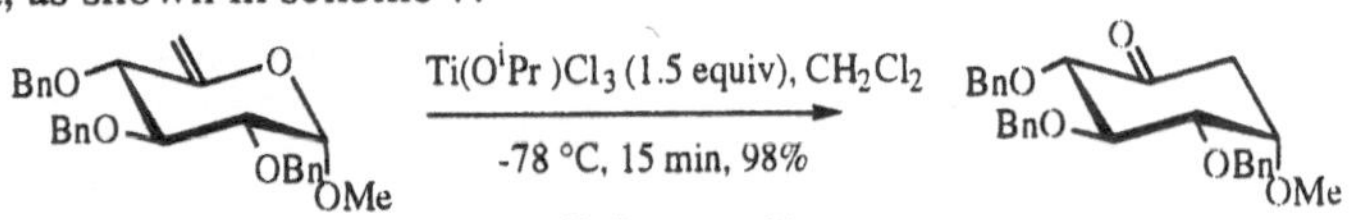

Scheme 7

In comparison with the previously disclosed triisobutylaluminum mediated rearrangement, this Ti (IV) version involves much milder reaction conditions (-78°C) and does not result in the reduction of the keto function.

REFERENCES

1   Rouzaud, D.; Sinaÿ, P.*J. Chem. Soc., Chem. Commun.*, **1983**, 1353.
2   a) Xin, Y. C.; Mallet, J. -M.; Sinaÿ, P. *J. Chem. Soc., Chem. Commun.* **1993**, 864; b) Vauzeilles. B.; Cravo, D.; Mallet, J. -M.; Sinaÿ, P. *Synlett* **1993**, 522; c) Chénedé, A.; Perrin, E.; Rekaï, E. D.; Sinaÿ, P. *Synlett* **1994**, 420; d) Mallet, A.; Mallet, J. -M.; Sinaÿ, P. *Tetrahedron: Asymmetry* **1994**, 2593. d) Rubinstenn, G.; Mallet, J. -M.; Sinaÿ, P. *Tetrahedron Lett.* 39, **1998**, 3697
3   Driguez, H. *Top. Curr. Chem.* **1997**, *187*, 85.
4   Jaurand, G.; Basten, J.; Lederman, I.; van Boeckel, C. A. A.; Petitou, M. *Bioorg. Med. Chem. Lett.* **1992**, 2, 897.
5   Petitou, M.; Coudert, C.; Level, M.; Lormeau, J.-C.; Zuber, M.; Simenel, C.; Fournier, J.-P.; Choay, J. *Carbohydr. Res.* 236, **1992**, 107.
6   Choay, J.; Petitou, M.; Lormeau, J.-C.; Sinaÿ, P.; Casu, B.; Gatti, G. *Biochem. Biophys. Res. Commun.* **1983**, *116*, 492.
7   Westerduin, P.; van Boeckel, C. A. A.; Basten, J. E. M.; Broekhoven, M. A.; Lucas, H.; Rood, A.; van Der Heijden, H.; van Amsterdam, R. G. M.; van Dinther, T. G.; Meuleman, D. G.; Visser, A.; Vogel, G. M. T.; Damm, J. B. L.; Overklift, G. T. *Bioorg. Med. Chem.* **1994**, 2, 1267.
8   Helmboldt, A.; Mallet, J.-M. Petitou, M.; Sinaÿ, P. *Bull. Soc. Chim. Fr.* **1997**, 134, 1057
9   Helmboldt, A.; Petitou, M.; Mallet, J.-M.; Hérault, J.-P.; Lormeau, J.-C.; Driguez, P.-A.; Herbert, J.-M.; Sinaÿ, P. *Bioorg. Med. Chem. Lett.* **1997**, 7, 1507.
10 Das, S. K.; Mallet, J.-M.; Sinaÿ, P. *Angew. Chem., Int. Ed. Engl.* **1997**, *36*, 493.
11 Ferrier, R. J. *J. Chem. Soc., Perkin Trans. 1* 1979, 1455.
12 Sollogoub, M. Mallet, J.-M. Sinaÿ.P. *Tetrahedron Lett.*, 39 , **1998**, 3471.

SOME NEW ASPECTS OF ASYMMETRIC CATALYSIS
WITH CHIRAL FERROCENYL LIGANDS

Antonio Togni*, Romano Dorta, Christoph Köllner, and Giorgio Pioda

Laboratory of Inorganic Chemistry, Swiss Federal Institute of Technology
ETH-Zentrum, Universitätstrasse 6, CH-8092 Zürich, Switzerland

INTRODUCTION

It has been previously demonstrated that chiral ligands, derived from a 1,2-disubstituted ferrocene constitute one of the most successful class of auxiliaries used in asymmetric catalysis.[1] Thus, ligands of the *Josiphos*-type[2] have recently reached the stage of industrial applications. As shown in Scheme 1, a new synthesis of biotin (vitamine H) was developed at LONZA Ltd., involving the use of such ligands in a Rh-catalyzed asymmetric hydrogenation reaction of a fully substituted C-C double bond.[3] Moreover, what appears to be the largest-scale industrial application of asymmetric catalysis so far, is an Ir-catalyzed imine reduction that makes use of a ferrocenyl diphosphine for the synthesis of the herbicide (S)-Metolachlor at Novartis Ltd.[4]

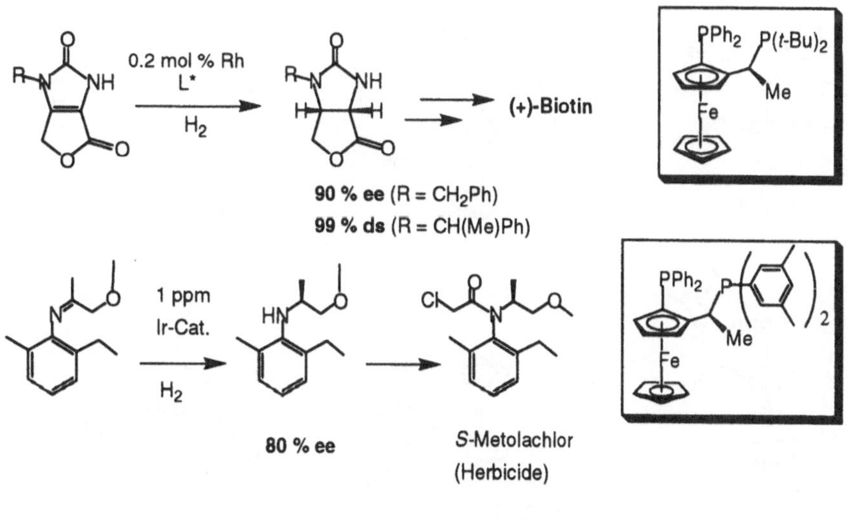

Scheme 1. Examples of production-scale applications of ferrocenyl diphosphine

Current Trends in Organic Synthesis
Edited by Scolastico and Nicotra, Kluwer Academic/Plenum Publishers, 1999

The modular synthetic approach to this kind of ligands represents one of their most important and unique features. Indeed, it is possible to readily vary the nature of the two ligating fragments L^1 and L^2 attached to the 1-ferrocenylethyl backbone by a two-step procedure, starting from a common precursor (see Scheme 2). L^1 and L^2 are typically phosphino and/or pyrazolyl fragments. Thus, the opportunity is given to expeditiously fine-tune both the steric and electronic properties of the ligands, according to the needs of a particular reaction, with a particular substrate.

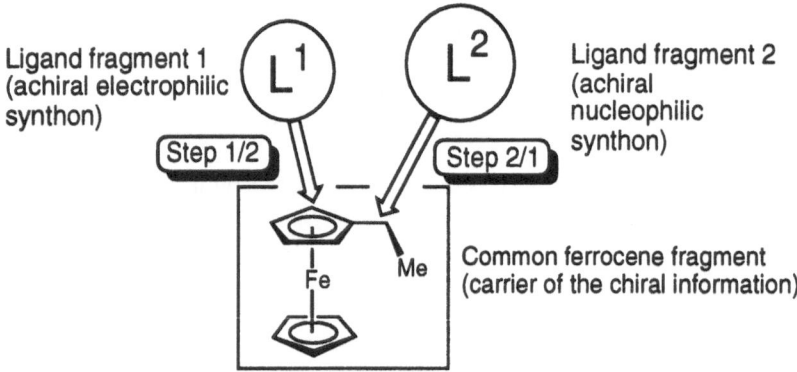

Scheme 2. General strategy for the synthesis of new ferrocenyl ligands bearing two different ligand fragments L^1 and L^2.

Examples of ligands that have been recently prepared following this strategy and that were successfully applied in asymmetric reactions, such as Rh-catalyzed hydroboration of olefins,[5] Pd-catalyzed allylic amination,[6] hydrosilylation,[7] Heck-reactions,[8] Ir-catalyzed olefin hydroamination,[9] etc., are shown in Scheme 3.

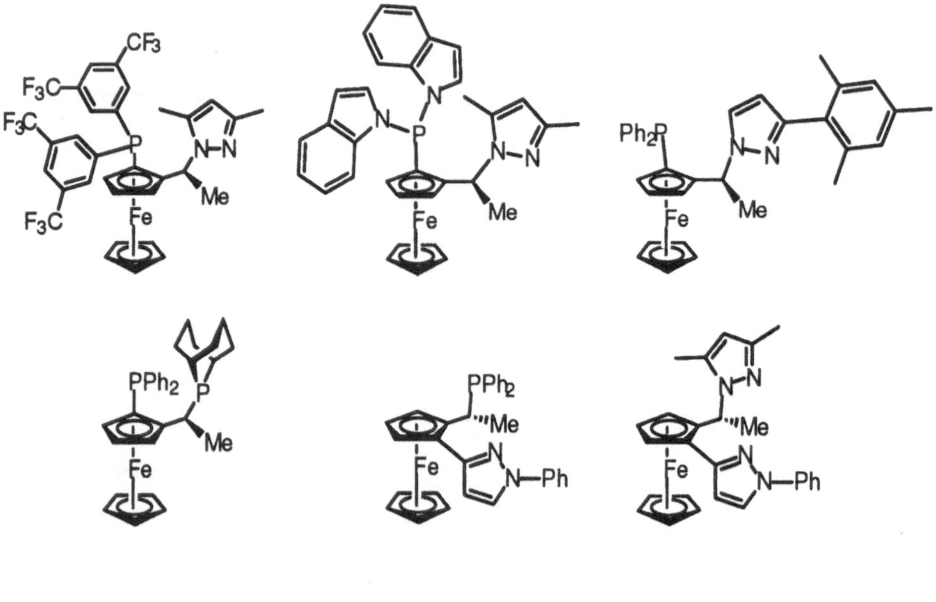

Scheme 3. Examples of ferrocenyl ligands bearing a variety of phosphino- and pyrazolyl fragments.

Pd-CATALYZED HYDROSILYLATION OF OLEFINS

We have previously shown that P,N-ferrocenyl ligands may be easily tuned in both a steric and electronic manner, in order to achieve optimum stereoselectivities. We recently found that such ligands are suitable auxiliaries in the Pd-catalyzed hydrosilylation of olefins with trichlorosilane.[10] As shown in Scheme 4, a relatively brief, empirical screening of ligand structures allowed us to achieve an enantioselectivity of 99% ee for the model olefin norbornene.

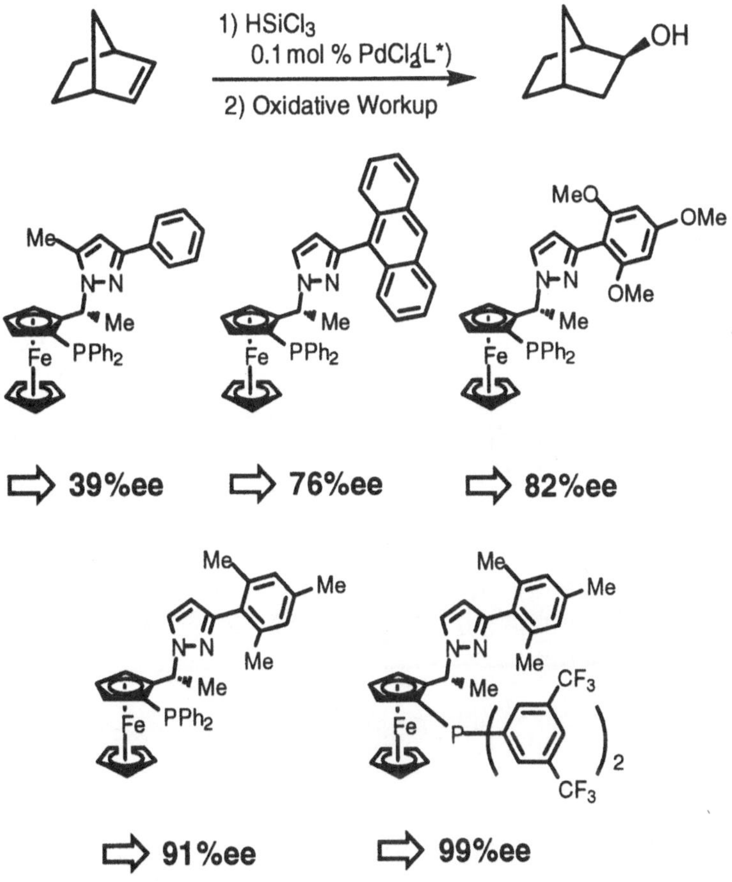

Scheme 4. Steric and electronic tuning of P,N-ferrocenyl ligands for the Pd-catalyzed hydrosilylation of norbornene with trichlorosilane.

The representative ligand series of Scheme 4, clearly shows that this particular reaction requires a relatively large substituent at position 3 of the pyrazolyl fragment. In the present case, the best substituent appears to be 2,4,6-trimethylphenyl. Furthermore, the replacement of the unsubstituted phenyls at the phosphorus atom by 3,5-bis(trifluoromethyl)phenyl groups, i.e., by increasing the π-acidity of the phosphine, led to the ligand affording the highest known enantioselectivity for this hydrosilylation. This is an illustration that the steric optimization of the ligand should be followed by a corresponding electronic fine-tuning, best achieved by the introduction and/or a variation of peripheral substituents, in terms of their electron-withdrawing or donating properties.

For this Pd-catalyzed hydrosilylation important electronic effects exerted by the substrate olefin have also been observed. As shown in Scheme 5, the *para*-substituents in styrenes were found to affect the enantioselectivity of the reaction in a very important manner. Thus, electron-withdrawing and electron-donating groups, respectively, lead to the formation of opposite product enantiomers. Furthermore, the correlation of $\log[(S)/(R)]$ with $\sigma_p{}^+$ indicates the development of a positive charge in the transition state of the enantioselectivity-determining step, thus suggesting that after hydride insertion and before the formation of the carbon-silicon bond, the benzyl ligand is coordinated in an η^3 mode (allylic). This is an important piece of information for a reaction for which a detailed mechasnistic understanding is still lacking.

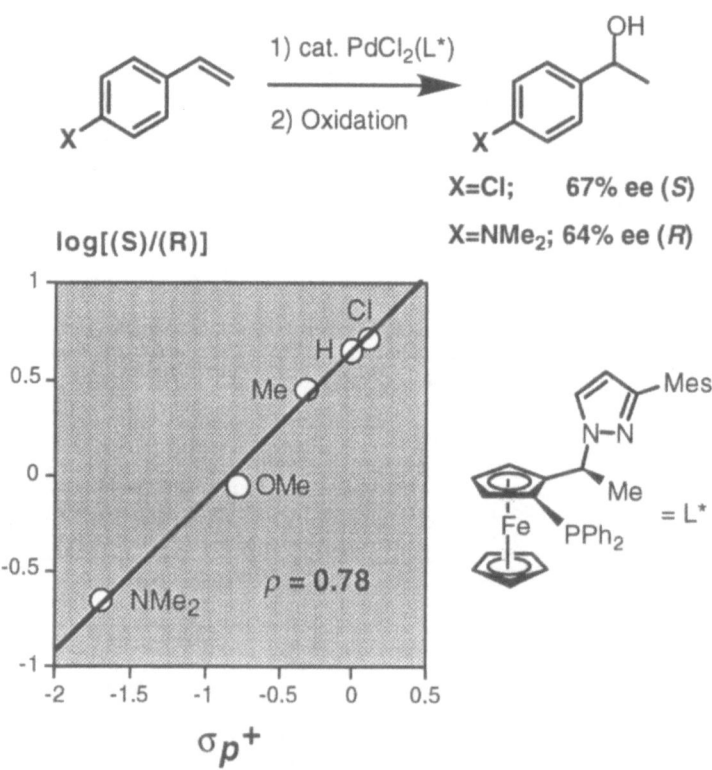

Scheme 5. Linear free-energy relationship for the hydrosilylation of *p*-substituted styrenes

THE DEVELOPMENT OF DENDRITIC CATALYST SYSTEMS.

Because of the proven efficacy of chiral ferrocenyl ligands in a number of asymmetric reactions, we set up a study aimed at creating dendrimers which carry catalytically active units at their periphery.[11] Thus, combining several, otherwise independent catalysts into one single nanoscopic particle, several advantages related to the properties of such macromolecules could be envisaged. Catalytically active dendrimers, in particular if they contain cationic transition metal complexes should be less soluble than their monomeric counterpart. Recovery by selective precipitation may thus be an opportunity. Furthermore, due to their size, these materials are amenable to ultrafiltration methods (membrane reactor).

For the incorporation of ferrocenyl ligands into dendritic structures a functionalization is required. The choice of the point of attachment of the spacer on the ferrocenyl ligand molecule is of great potential importance. From a synthetic point of view, different anchoring points may entail different, more or less complex synthetic routes. From a structural point of view, location and length of the spacer may influence catalyst performance, by virtue of intramolecular interactions. Scheme 6 illustrates this general idea and gives the structure of the currently used intermediates according to their anchoring strategy.

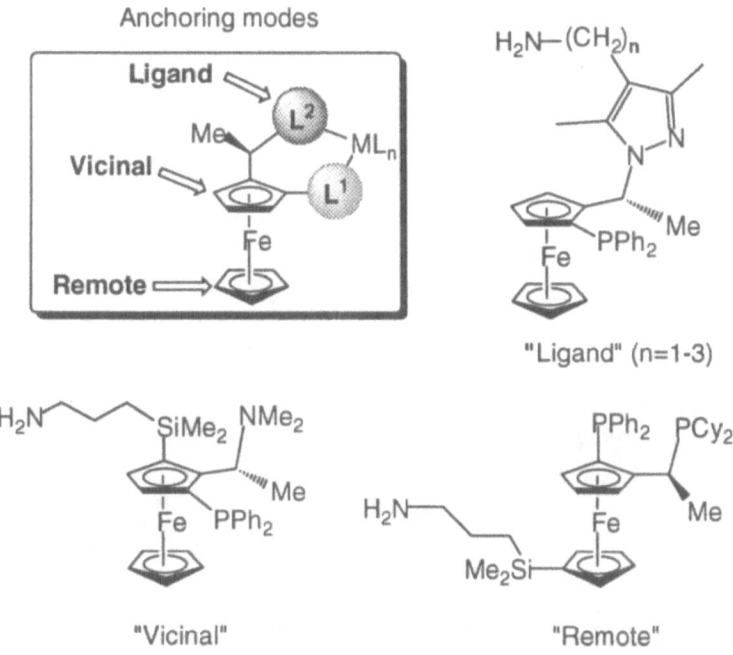

Scheme 6. Illustration of the different anchoring modes for ferrocenyl ligands to dendrimers. The three specific compounds shown are currently being used for the synthesis of G0, G1, and G2 dendrimers.

The derivatives depicted in Scheme 6 are accessible in three, four, and six steps, respectively, from the common, enantiomerically pure starting material (*R* or *S*)-N,N-dimethyl-1-ferrocenyl ethylamine in moderate to good overall yields. The remote anchoring mode intuitively appears to be the best suited because of the minimized intramolecular steric interactions between ligand fragments and linker. Thus, using the corresponding intermediate the so far largest dendrimers of this type containing six or eight Josiphos units has been prepared, as illustrated in Scheme 7, in good yields and in gram quantities. The core of the dendrimers is constituted in these two cases by a 1,3,5-trisubstituted benzene ring and by a 1,3,5,7-tetrasubstituted adamantane, respectively. By way of example, the discussion shall be focussed on these two derivatives only. As one would expect, in the ^{31}P NMR spectra only one pair of doublets is observed, with the typical long range $^4J_{PP}$ coupling constant of ca. 34 to 37 Hz. This indicates the equivalence of the ferrocenyl units. However, broad signals in the ^{1}H NMR 300 MHz spectra are observed at room temperature, possibly indicating slow conformational equilibria in the sterically rather crowded inner core of the dendrimers.

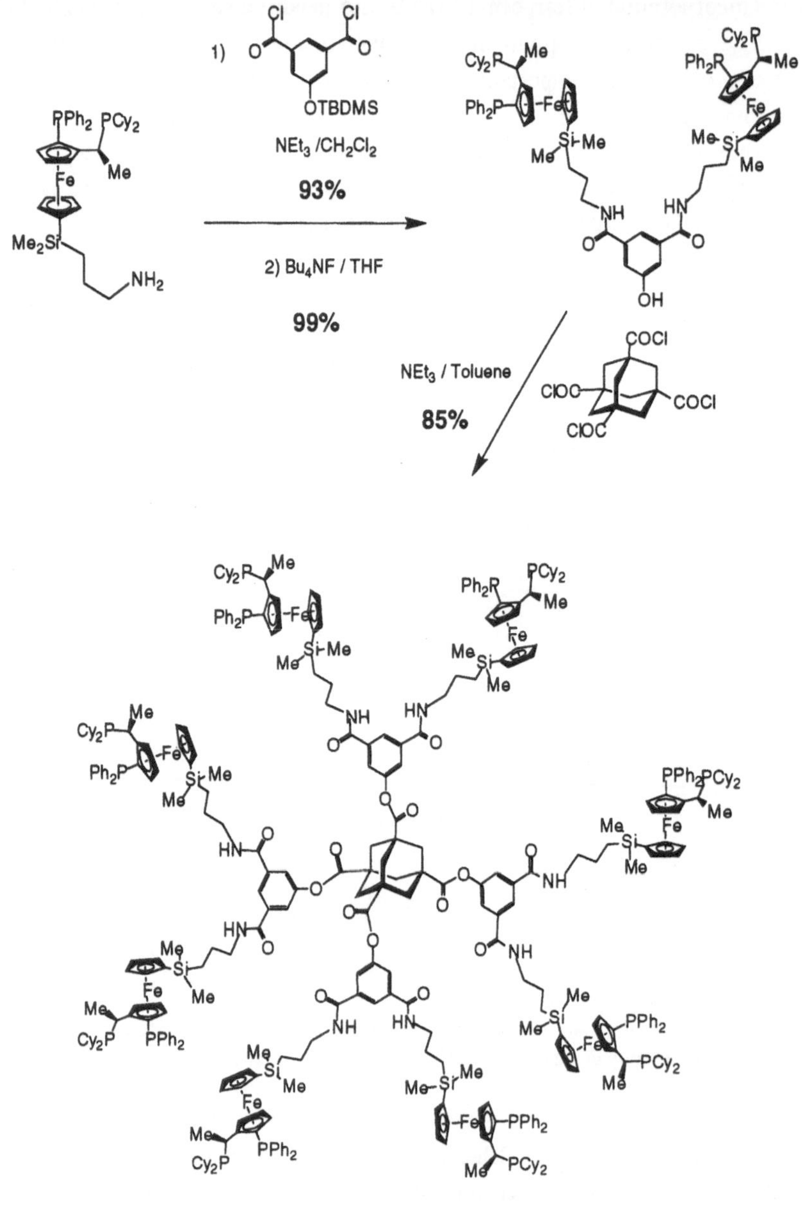

Scheme 7. Typical convergent synthesis of Josiphos-containing dendrimer.

Nonetheless, well resolved spectra for these two compounds are obtained when corresponding DMSO-d_6 solutions are measured at 80°C and 120°C, respectively. The new high-molecular-weight compounds were also characterized by MALDI-TOF mass spectrometry. Molecular peaks at m/z values in very good agreement with the calculated ones, as well as in-source fragmentation[12] patterns very similar to those of the „monomeric" counterpart Josiphos under EI conditions were observed. Sections of such MALDI-TOF spectra are illustrated in Figure 1.

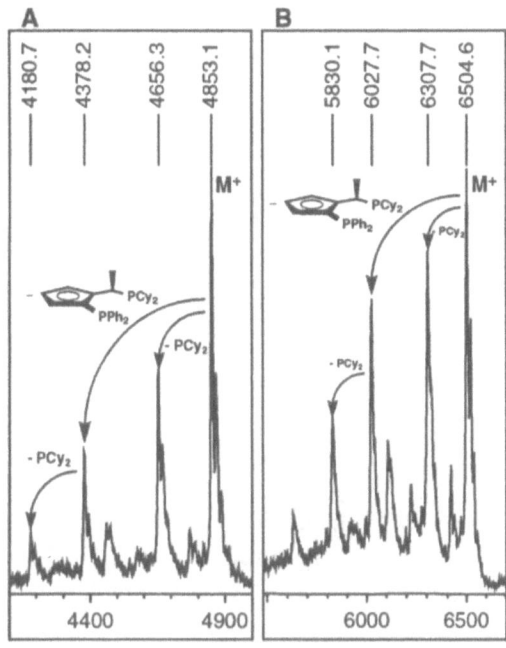

Figure 1. Sections of the MALDI-TOF mass spectra of dendrimers containing six (A) and eight (B) Josiphios units, showing typical in-source fragmentation patterns (calculated molecular mass = 4853.15 and 6502.95, respectively).

In order to gain insight as to the behavior of the new dendritic ligands in catalytic reactions, the simple Rh-catalyzed asymmetric hydrogenation of dimethyl itaconate in MeOH was chosen as a standard reaction for comparing their performances. In situ catalyst preparation was attained by mixing the ligand (1 eq / number of ferrocenyl units) with 1 eq of [Rh(COD)$_2$]BF$_4$ in CH$_2$Cl$_2$ and stirring for 15 min under Ar. The orange solid obtained after evaporation of the solvent was used as a catalyst precursor. ^{31}P NMR spectra of these materials showed a single AMX spin-system, as expected when all ferrocenyl sites are bonded to Rh. Hydrogenation experiments performed using 1 mol% Rh under 1 bar of hydrogen pressure, as previously reported,[2a] gave very similar results. In all cases hydrogen take-up ceased after ca. 20 min, and substrate was no longer detectable. The enantioselectivities afforded by the dendritic catalysts (between 98.0% ee and 98.6% ee) are only slightly lower than that obtained using the corresponding mononuclear Josiphos catalyst (99.0% ee). However, a weak trend to lower enantioselectivities seems to parallel the increasing size of the dendritic ligands.

The synthetic strategy illustrated in Scheme 7 implies an intrinsic limitation, since it starts from the complete functionalized ligand Josiphos. For a wider scope of application, one would like to modify the ligand fragments according to the needs of a specific reaction. From this point of view it is desirable to complete the ligand synthesis *after* assembling the dendrimer. Thus, a PPFA derivative, still containing a dimethylamino group amenable to nucleophilic substitution reactions (Scheme 6) , has been prepared and incorporated into a G0 dendrimer. In order to test the possibility of a multiple functionalization, the preparation of a P,N-ligand system has been achieved in the last syntetic step, as (Scheme 8).

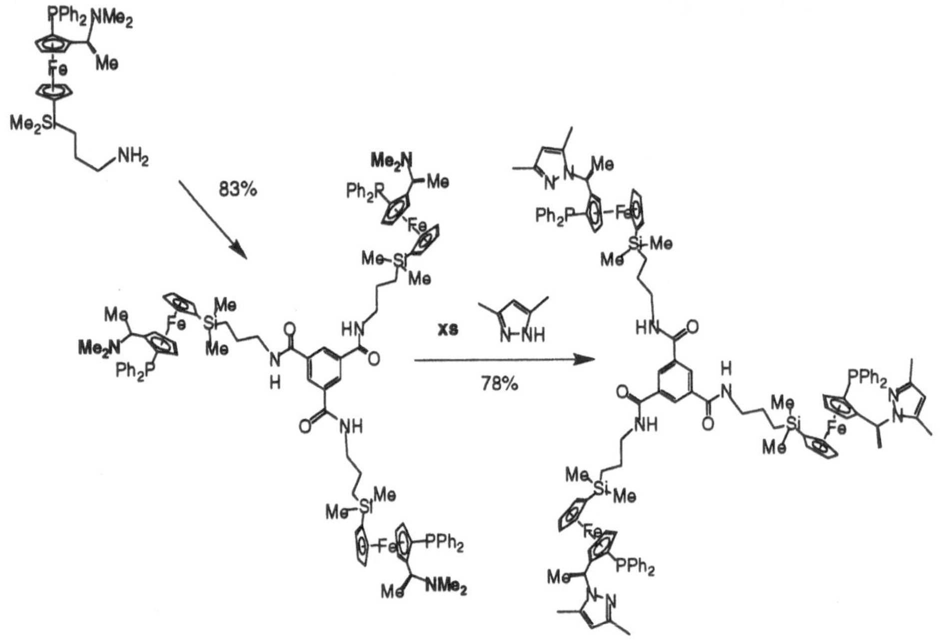

Scheme 8. Multiple functionalization of a G0 dendrimer as the final step in the preparation of a P,N-ligand.

We are currently exploiting compounds of the type shown in Scheme 8, however of higher dendrimer generation for the preparation of ligands having different combinations of donor fragments (both P,P and P,N).

THE DEVELOPMENT OF CATALYTIC ASYMMETRIC HYDROAMINATION OF OLEFINS

Despite its synthetic potential and practical relevance, the asymmetric catalytic olefin hydroamination is still an undeveloped reaction. Indeed, besides the lanthanide-catalyzed enantioselective olefin hydroamination / cyclization reported by Marks and co-workers,[13] to the best of our knowledge no intermolecular version of this reaction is known. So far, transition metal catalysts for the intermolecular hydroamination of unactivated olefins hardly afford turnover numbers of more than ca. 0.08 h^{-1} at 1 atm and medium temperature.[14] Organo-f-element complexes are able to catalyze the same reaction of unactivated olefins, such as 1-pentene, at a rate of up to N$_t$ 0.4 h^{-1} at 60°C.[13b]

The feasibility of catalytic hydroamination via N-H activation has been demonstrated previously.[14a] In particular, it was shown that the electron-rich Ir complex $[Ir(PEt_3)_2(C_2H_4)Cl]$ cleanly oxidatively adds aniline. The Ir(III) hydrido amido species thus formed undergoes insertion of norbornene leading to a well characterized complex containing a chelating alkylamino ligand, from which the amination product is released upon reductive elimination. However, this system did not afford more than up to six turnovers in catalytic experiments, in the presence of apparently beneficial Lewis acids such as $ZnCl_2$.

We found that $[IrCl(C_2H_4)_4]$, or $[IrCl(COE)_2]_2$ smoothly reacts with Josiphos-type ligands and with the axial chiral auxiliaries BINAP and Biphemp, respectively, in toluene solution, affording in good yields the new dinuclear derivatives shown in Scheme 9. The ferrocenyl derivatives were isolated as an inseparable mixture of cis/trans isomers.[9]

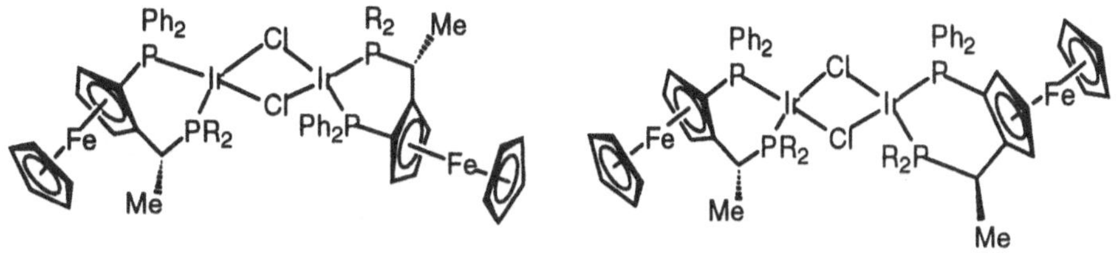

(R=Cy, [IrCl((R)-(S)-Josiphos)]₂)
(R=Ph)
(R=3,5-Me₂Ph)

(R=Cy)
(R=Ph)
(R=3,5-Me₂Ph)

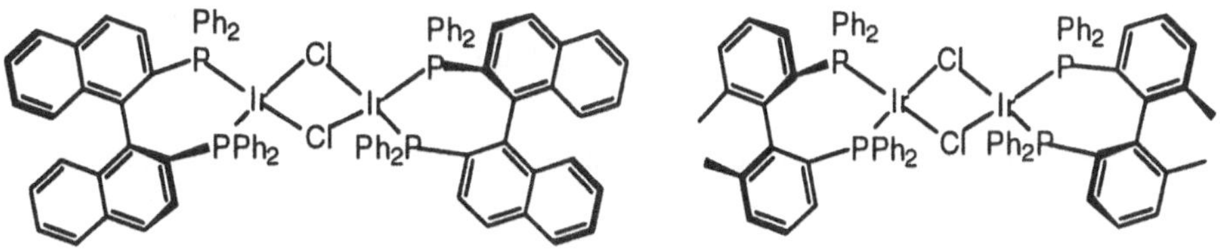

([IrCl((S)-BINAP)]₂)

([IrCl((R)-Biphemp)]₂)

Scheme 9. The new dinuclear Ir(I) complexes used as catalyst precursors in the asymmetric addition of aniline to norbornene.

The complexes shown in Scheme 9 were used as catalyst precursors in the model hydroamination of norbornene with aniline. The reactions were performed without solvent and with a catalyst concentration of 1 or 2 mol% Ir. First experiments, performed with the (R)-(S)-Josiphos complexes at 50°C, have shown the clean but slow formation of exo-(2-phenylamino)norbornane in 51% ee (2S), accompanied by traces of the corresponding endo-isomer (ca. 15% conversion in 3d). We reasoned that the good π-donating properties of fluoride as a ligand, together with its tendency to form hydrogen bridges, could enhance the propensity towards N-H oxidative addition. However, we were not able so far to prepare the fluoro analogues of any of the complexes used as catalysts precursors. Nevertheless, both activity and enantioselectivity benefited from the addition of co-catalytic amounts of „naked" fluoride, added as a benzene solution of phosphazenium-fluoride-P2. For all catalysts used the addition of fluoride (0.5 to 4 eq per Ir) led to both an increase of activity and significantly higher enantioselectivity. Under the cocatalytic activity of fluoride it was possible to obtain the so far highest enantioselectivities for this kind of transformation, using BINAP as chiral ligand. A selection of results is shown in Scheme 10.

Due to the high basicity of the „naked" fluoride anion, anilide could be produced by deprotonation of aniline. It has been shown that addition of LiNHPh is necessary to promote the condensation of aniline with norbornene when a rhodium catalyst is used.[14b,c] However, when one eq of LiNHPh per Ir was added to the catalyst, instead of fluoride, all the activity was lost. Furthermore, anilido-bridged Ir(I) complexes were prepared and characterized and shown not to posses any catalytic activity.[15] From these experiments we conclude that fluoride is likely to behave as a ligand, replacing chloride in the coordination sphere of Ir.

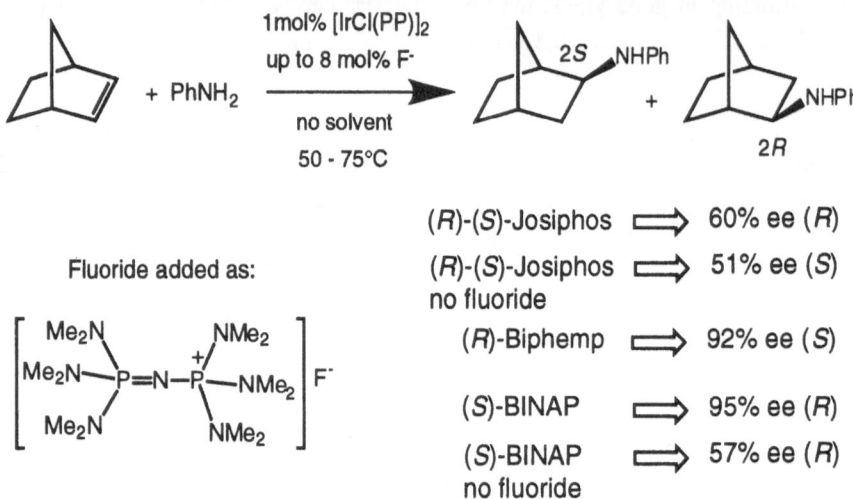

Scheme 10. Some examples of asymmetric Ir-catalyzed addition of aniline to norbornene using different chiral ligands, and with and without cocatalytic fluoride.

The precise role of fluoride in enhancing both activity and selectivity is so far still a matter of speculation. Given that this anion is acting as a ligand, the formation of a hydrogen bridge to aniline would contribute in activating the N-H bond and at the same time in predetermining the relative orientation of the aniline molecule in the coordination sphere of Iridium. A schematic representation of the possible elementary processes leading to the formation of Ir(III) complexes containing fluoro, hydrido, and amido ligands is shown in Scheme 11. At the Ir(III) stage and after olefin insertion into the Ir-N bond, the lability of the fluoro ligand and hence the rapid dissociation of F⁻ could accelerate the final step in the catalytic cycle, i.e., the reductive elimination of the product.

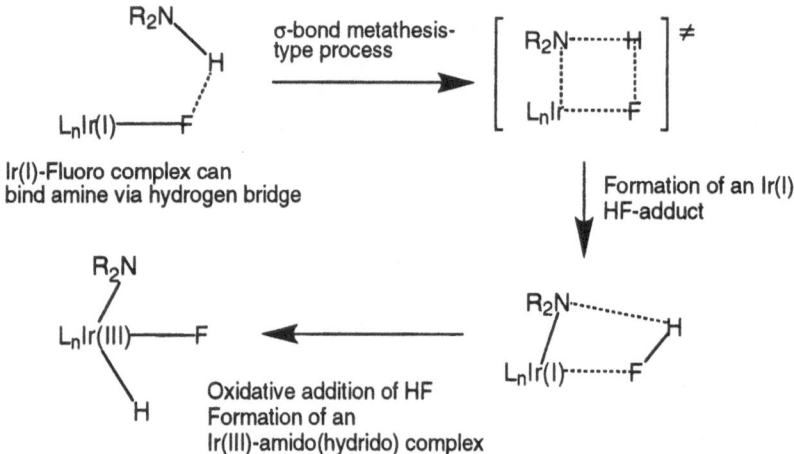

Scheme 11. A speculative role of fluoride in the Ir-catalyzed hydroamination of olefins, postulating the formation of reactive Ir fluoride complexes.

CONCLUSION

We have shown that chiral ferrocenyl ligands derived from 1,2-disubstituted ferrocenens, by virtue of their modular synthetic approach, may be adapted successfully to several types of transition-metal catalyzed asymmetric reactions, thus affording very high stereoselectivities. Moreover, the possibility to further functionalize such ligands offers the opportunity to incorporate them into dendritic structures.

The development of the asymmetric hydroamination of olefins, still very much in its infancy, has shown how important the effect of additives may be. In particular, fluoride is an anion whose possible role in catalytic processes has rather been neglected so far.

REFERENCES

1.   For a review, see: a) T. Hayashi In *FERROCENES: Homogeneous Catalysis, Organic Synthesis, Materials Science*, A. Togni, T. Hayashi, Eds., VCH, Weinheim (New York), **1995**, 105-142, and references cited therein. First report: b) T. Hayashi, K. Yamamoto, K. Kumada, *Tetrahedron Lett.* **1974**, 4405. For recent reviews , see: c) A. Togni, *Chimia* **1996**, *50*, 86. d) A. Togni, *Angew. Chem.* **1996**, *108*, 1581 (*Angew. Chem. Int. Ed. Engl.* **1996**, *35*, 1475).

2.   a) A. Togni, C. Breutel, A. Schnyder, F. Spindler, H. Landert, A. Tijani, *J. Am. Chem. Soc.* **1994**, *116*, 4062. b) U. Burckhardt, L. Hintermann, A. Schnyder, A. Togni, *Organometallics* **1995**, *14*, 5415.

3.   J. McGarrity, F. Spindler, R. Fuchs, M. Eyer, Eur. Pat. Appl. EP 624 587 A2, (*LONZA AG*). *Chem. Abstr.* **1995**, *122*, P81369q.

4.   For an account about the successful development of this process, see: F. Spindler, B. Pugin, H.-P. Jalett, H.-P. Buser, U. Pittelkow, H.-U. Blaser, In *"Catalysis of Organic Reactions"*, R.E. Malz, Jr., Ed. (Chem. Ind. Vol. *68*), Dekker, New York, **1996**, pp. 153-166.

5.   a) A. Schnyder, L. Hintermann, A. Togni, *Angew. Chem.* **1995**, *107*, 996 (*Angew. Chem. Int. Ed. Engl.* **1995**, *34*, 931). b) A. Schnyder, A. Togni, U. Wiesli, *Organometallics*, **1997**, *16*, 255.

6.   A. Togni, U. Burckhardt, V. Gramlich, P. S. Pregosin, R. Salzmann, *J. Am. Chem. Soc.* **1996**, *118*, 1031.

7.   G. Pioda, A. Togni, to be published

8.   M. Baumann, A. Togni, to be published.

9.   R. Dorta, P. Egli, F. Zürcher, A. Togni *J. Am. Chem. Soc.* **1997**,*119*, 10857.

10.   For other successful applications of asymmetric hydrosilylation, see: a) T. Hayashi, *Asymmetric Reactions Catalyzed by Palladium-MOP Complexes* in *Organic Synthesis via Organometallics OSM 5*, G. Helmchen, J. Dibo, D. Flubacher, B. Wiese, (Eds.), Vieweg, Braunschweig/Wiesbaden, **1997**. b) Y. Uozomi, S.Y. Lee, T. Hayashi, *Tetrahedron Lett.* **1992**, *33*, 7185. c) Y. Uozomi, K. Kitayama, T. Hayashi *Tetrahedron: Asymmetry* **1993**, *4*, 2419. d) Y. Uozomi, T. Hayashi *Tetrahedron Lett.* **1993**, *34*, 2335. e) T. Hayashi *Acta Chem. Scand.* **1996**, *50*, 259.

11.   For previous reports, see: a) J.W.J. Knapen, A.W. van der Made, J.C. de Wilde, P.W.N.M. Leeuwen, P. Wijkens, D.M. Grove, G. van Koten, *Nature* **1994**, *372*, 659. b) D. Seebach, R.E. Marti, T. Hintermann, *Helv. Chim. Acta* **1996**, *79*, 1710. b) P.B. Rheiner, H. Sellner, D. Seebach, *Helv. Chim. Acta* **1997**, *80*, 2027. c) C. Bolm, N. Derrien, A. Seger, *Synlett* *1996*, 387. d) H. Brunner, *J. Organomet. Chem.* **1995**, *500*, 39.

12.   For examples of mass spectrometric characterization of dendrimers, see, e.g.: a) J.C. Hummelen, J.L.J. van Dongen, E.W. Meijer, *Chem. Eur. J.* **1997**, *3*, 1489. b) K.L. Walker, M.S. Kahr, C.L. Wilkins, Z. Xu, J.S. Moore, *J. Am. Soc. Mass Spectrom.* **1994**, *5*, 731. c) For a discussion of in-source fragmentations in MALDI MS, see: D.C. Reiber, T.A. Grover, R.S. Brown, *Anal. Chem.* **1998**, *70*, 673, and references cited therein.

13.   See, e.g.; a) M.A. Giardello, V.P. Conticello, L. Brard, M.R. Gagné, T.J. Marks, *J. Am. Chem. Soc.* **1994**, *116*, 10241. b) Y. Li, T.J. Marks, *Organometallics* **1996**, *15*, 3770. c) Y. Li, T.J. Marks, *J. Am. Chem. Soc.* **1996**, *118*, 9295, and further references cited therein.

14.   a) A.L. Casalnuovo, J.C. Calabrese, D. Milstein, *J. Am. Chem. Soc.* **1988**, *110*, 6738. b) J.-J. Brunet, D. Neibecker, K. Philippot, *J. Chem. Soc. Chem. Commun.* **1992**, 1215. c) J.-J. Brunet, G. Commenges, D. Neibecker, K. Philippot, *J. Organomet. Chem.* **1994**, *469*, 221. For recent reviews, see, e.g.: d) J.-J. Brunet, *Gazz. Chim. Ital.* **1997**, *127*, 111. e) D.M. Roundhill, *Catalysis Today* **1997**, *37*, 155.

15.   R.Dorta, A. Togni, to be published.

DESIGNER LEWIS ACIDS FOR SELECTIVE ORGANIC SYNTHESIS

Hisashi Yamamoto, Akira Yanagisawa, Kazuaki Ishihara, Susumu Saito

Graduate School of Engineering, Nagoya University, CREST, Japan Science and Technology Corporation (JST), Chikusa, Nagoya 464-8603, Japan

INTRODUCTION

An excellent candidate as a proton substitute in man-made organic reactions is a Lewis acid. The goal of the research was to engineer an artificial proton of a special shape, which could be utilized as an effective tool for chemical reactions, by harnessing the high reactivity of the metal atom towards a variety of functional groups. Such a concept was initially researched by examining the influence of a specially designed organometallic reagent on a typical organic reaction. Michael addition of simple organolithium and magnesium reagent to aromatic aldehydes in the presence of a bulky organoaluminum reagent is described as an example of the concept. The successful discrimination observed led to examine the more intricate question of enantioface differentiation. A variety of Lewis acid reagents were utilized for Diels-Alder, aldol, and ene reactions with high enantioselectivities. The origins of the selectivity of these reactions are discussed. The review describes these points with a variety of designer Lewis acids selected to illustrate the utility of the concept.

SELECTIVE CARBON-CARBON BOND FORMATION USING ALUMINUM TRIS(2,6-DIPHENYLPHENOXIDE) (ATPH)

The selective functionalization of an aromatic nucleus has become increasingly important in synthetic organic chemistry. However, little is known about nucleophilic addition to aromatic nucleus covalently attached by a carbonyl functionality which serves as an electron-withdrawing group with weak activating capability. Accordingly, aromatic carbonyl compounds have long been believed as rather inactivated aromatics which do not allow the aromatic functonalization by the attack of nucleophiles but usually give addition at their carbonyl carbons. We recently discovered that organolithiums undergo conjugate addition to aromatic carbonyl compounds by complexation with aluminum tris(2,6-diphenylphenoxide) (ATPH) (ref. 1). This novel strategy represents a recent application to a number of nucleophiles including silyl- (ref. 2), aryl-, allyl-, and vinyllithiums (ref. 3) as well as lithium enolates (ref. 4), leading to a powerful and general functionalization—dearomatization sequence for a wide range of aromatic carbonyl compounds.

ATPH

CHO, OLi, OBu^t, BuLi, Ph$_2$MeSiLi, MePh$_2$Si, CO$_2$Bu^t, Me$_3$Si, OBu^t, TBAF/MeOH, CO$_2$Bu^t, ATPH, Li, OLi, OMe, Li, CO$_2$Me, CHO

Using ATPH as a key catalyst, a new directed aldol condensation was realized (ref. 5). Presented below is a new different strategy for combining two different carbonyl compounds using lithium diisopropyl amide (LDA), in which both of the substrates are complexed with ATPH (ref. 6). Several key issues have been documented here: (1) the two different carbonyl reactants and ATPH should be mixed together prior to treatment with a base to give effective cross-coupling, (2) conjugated carbonyl compounds including aldehydes, ketones, and esters (ref. 7) all demostrated to work as effective nucleophiles (Fig. 2), (3) neither the α-carbon of aromatic ketones nor the α'-carbon of α,β-unsaturated ketones were directed site for deprotonation. Thus, (4) deprotonation and ensuing alkylation are quite regioselective at an allylic terminus of given nucleophiles which serve as extended dienolates. Of particular note is the regioselective aldolization of highly conjugated esters which have several possible site for functionalization, (5) this transformation displayed high *E*-selectivity with respect to the γ-aldolization.

LEWIS ACID ASSISTED CHIRAL BRØNSTED ACID (LBA)

Enantioselective protonation of prochiral silyl enol ethers is a very simple and attractive route for preparing optically active carbonyl compounds (ref. 8). However, it is difficult to achieve high enantioselectivity using simple chiral Brønsted acids because of the conformational flexibility in the neighborhood of the proton. We expected that the coordination of a Lewis acid to a Brønsted acid would restrict the direction of the proton and increase its acidity. In 1994, we found that the Lewis acid assisted chiral Brønsted acid (LBA) is a highly effective chiral proton donor for the enantioselective protonation (ref. 9). LBA **1** is generated in situ from optically pure BINOL and tin tetrachloride in toluene, and is stable in the solution even at room temperature. In the presence of a stoichiometric amount of (*R*)-**1**, the protonation of the TMS enol ether **2** derived from 2-phenylcyclohexanone (**3**) proceeded at -78 °C to give the (*S*)-**3** with 97% ee. The observed absolute stereopreference can be understood in terms of the proposed transition state assembly. The trialkylsiloxy group is directed opposite to the binaphthyl moiety in order to avoid any steric interaction, and the aryl group stacks on this naphthyl group.

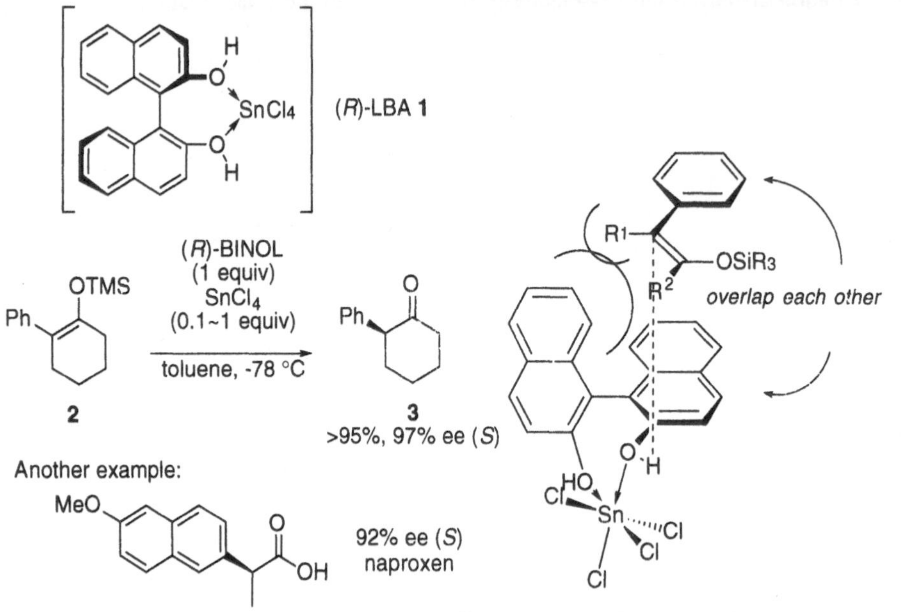
H
O
SnCl4
O
H
(R)-LBA 1
(R)-BINOL
(1 equiv)
SnCl4
(0.1~1 equiv)
OTMS
Ph
toluene, -78 °C
O
Ph
2
3
>95%, 97% ee (S)
Another example:
MeO
O
OH
92% ee (S)
naproxen
R1
OSiR3
R2
overlap each other
O
H
HO
Cl
Sn
Cl
Cl
Cl

The Proposed Transition State Assembly 4

In further studies, we succeeded in the enantioselective protonation using a stoichiometric amount of an achiral proton source and a catalytic amount of LBA **5** in place of LBA **1** (ref. 10). In the presence of 8 mol% of tin tetrachloride, 10 mol% of the monomethyl ether of (*R*)-BINOL, and stoichiometric amounts of 2,6-dimethylphenol as an achiral proton source, the protonation of the ketene bis(trimethylsilyl)acetal **6** derived from 2-phenylpropanoic acid (**7**) proceeded at –80 °C to give the (*S*)-**7** with 94% ee. The catalytic system was applied to the enantioselective synthesis of various α-arylcarbonyl compounds such as 2-phenylcycloheptanone, 2-(naphthyl)cyclohexanone and ibuprofen.

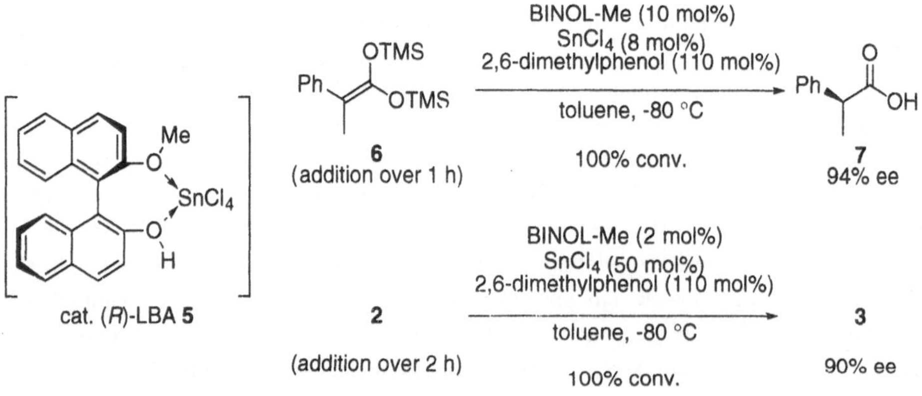
Me
O
SnCl4
O
H
cat. (R)-LBA 5
OTMS
Ph
OTMS
6
(addition over 1 h)
BINOL-Me (10 mol%)
SnCl4 (8 mol%)
2,6-dimethylphenol (110 mol%)
toluene, -80 °C
100% conv.
O
Ph
OH
7
94% ee
2
(addition over 2 h)
BINOL-Me (2 mol%)
SnCl4 (50 mol%)
2,6-dimethylphenol (110 mol%)
toluene, -80 °C
100% conv.
3
90% ee

The mechanism of the catalytic cycle was investigated by ^{1}H NMR analysis of the 1 to 1 reaction mixtures of the silyl enol ether and chiral LBAs **1** and **5** at –78 °C. In this case, two singlets for the TMS groups of TMSCl and the mono TMS ether of (*R*)-BINOL were observed at a molar ratio of 15 to 85. In another case, only one singlet for TMSCl was observed. The presence of TMSCl leads us to anticipate the generation of these tin aryloxide intermediates. The catalytic cycle can be reasonably explained by assuming that the tin aryloxide intermediate is reconverted to the chiral LBA by receiving a proton and a chloride from 2,6-dimethylphenol and TMSCl or tin tetrachloride, respectively.

chiral LBA **5**

[R3 = alkyl] [R3 = OSiMe₃]

BINAP·SILVER(I)-CATALYZED ASYMMETRIC REACTIONS

Enantioselective addition of an allyl group to carbonyl compounds to provide optically active secondary homoallylic alcohols is a valuable synthetic method since the products are readily transformed into β-hydroxy carbonyl compounds and various other chiral compounds (ref. 11). Although numerous important works on the reaction using a stoichiometric amount of chiral Lewis acids have been reported, there are few methods available for a catalytic process including a chiral (acyloxy)borane (CAB) complex (ref. 12) or a binaphthol-derived chiral titanium complex (ref. 13) as a catalyst. In contrast, we found that a BINAP·silver(I) complex also catalyzes the asymmetric allylation of aldehydes with allylic stannanes, and high γ-, anti-, and enantioselectivities are obtained by this method. The chiral phosphine-silver(I) catalyst can be prepared simply by stirring an equimolar mixture of chiral phosphine and silver(I) compound in THF at room temperature. Treatment of benzaldehyde with allyltributyltin under the influence of 5 mol % of (S)-BINAP·silver(I) triflate in THF at -20 °C provides the corresponding (S)-enriched homoallylic alcohol in 88% yield with 96% ee (ref. 14). The reaction furnishes high yields and remarkable enantioselectivities not only with aromatic aldehydes but also with α,β-unsaturated aldehydes and aliphatic aldehydes. Enantioselective addition of methallyltributylstannane to aldehydes can also be achieved using this method (ref. 15).

Condensation of γ-substituted allylmetals with aldehydes is a fascinating subject with respect to regioselectivity (α/γ) and stereoselectivity (E/Z or anti/syn). Addition of (E)-crotyltributyltin (E/Z = 95/5) to benzaldehyde in the presence of 20 mol % of (R)-BINAP·AgOTf in THF at -20 °C ~ r.t. exclusively gives the γ-adducts with an anti/syn ratio of 85/15 (ref. 15). The anti-isomer indicates 94% ee with a 1R,2R configuration. Use of (Z)-crotyltributyltin (E/Z = 2/98) or a nearly 1:1 mixture of the (E)- and (Z)-crotyltributyltin also results in a similar anti/syn ratio and enantioselectivity.

E/Z ratio of crotyltin	yield, %	anti (% ee)/syn (% ee)
95/5	56	85 (94)/15 (64)
2/98	72	85 (91)/15 (50)
53/47	45	85 (94)/15 (57)

Reaction of aldehydes with 2,4-pentadienylstannanes is also catalyzed by BINAP·silver(I) complex, and the corresponding γ-pentadienylated optically active alcohols are obtained with high enantioselectivity (ref. 16). When benzaldehyde is reacted with 1 equiv of pentadienyltributyltin (E/Z = 97/3) and 0.1 equiv of (S)-BINAP·AgOTf at -20 °C, the γ-product is obtained in 61% yield with 90% ee. Pentadienyltrimethyltin offers a chemical yield and enantioselectivity comparable to those of pentadienyltributyltin. Ketones are inert under the standard reaction conditions.

The asymmetric Mukaiyama aldol reaction is a popular method for preparing optically active β-hydroxy carbonyl compounds, and has been widely applied to the synthesis of natural products. A variety of chiral Lewis acid catalysts have been developed for the enantioselective reaction of silyl enol ethers or ketene silyl acetals with carbonyl compounds (ref. 11c and 17). In contrast, the use of organotin(IV) enolates for aldol reactions has so far received little attention. Recently, we found that the aldol reaction of tributyltin enolates with aldehydes is catalyzed by a BINAP·silver(I) complex with high diastereo- and enantioselectivities (ref. 18). The tributyltin enolate is easily prepared from the corresponding enol acetate and tributyltin methoxide in the absence of solvent. The tin enolate thus obtained exists in O–Sn form and/or C–Sn form; however, both species can be used for the aldol reaction of the present system. Several different solvents were tested for the reaction and THF was found to provide the best result. The catalytic aldol reaction of a variety of tributyltin enolates with typical aromatic, α,β-unsaturated, and aliphatic aldehydes was investigated and the highest ee (95% ee) was obtained when the tin enolate prepared from pinacolone was added to benzaldehyde. Addition of substituted enol stannanes to aldehydes also proceeds to furnish high diastereo- and enantioselectivities using this chiral catalyst. For example, treatment of the tributyltin enolate of cyclohexanone (1 equiv) with benzaldehyde (1 equiv) under the influence of 10 mol % of (R)-BINAP·AgOTf complex in dry THF at -20 °C gives the optically active anti aldol product preferentially with an anti/syn ratio of 92/8. The anti-isomer indicates 93% ee with a 2S,1′R configuration. In contrast, the Z-enolate derived from tert-butyl ethyl ketone provides the syn aldol adduct nearly exclusively with 95% ee. These results show that the diastereoselectivity depends on the geometry of enol stannane and that six-membered cyclic transition-state structures A and B are probable models.

Probable cyclic transition-state structures.

References

1. K. Maruoka, M. Ito and H. Yamamoto. *J. Am. Chem. Soc.* **117**, 9091 (1995).
2. S. Saito, K. Shimada, H. Yamamoto, E. Martínez de Marigorta and I. Fleming. *Chem Commun.* 1299 (1997).
3. S. Saito, K. Shimada, M. Ito, K. Maruoka and H. Yamamoto. in preparation.
4. S. Saito, T. Sone, K. Shimada and H. Yamamoto. in preparation.
5. S. Saito, M. Shiozawa, M. Ito and H. Yamamoto. *J. Am. Chem. Soc.* **120**, 813 (1998).
6. T. Mukaiyama. In Organic Reaction Vol. 28 (G. A. Boswell, Jr, R. F. Hirshmann, S. Danishefsky, A. S. Kende, H. W. Gschwend, L. A. Paquette, R. F. Heck, G. H. Posner, B. M. Trost and B. Weinstein, eds.), pp. 203. John Wiley & Sons, New York (1982).
7. S. Saito, M. Shiozawa and H. Yamamoto. in preparation.
8. C. Fehr, *Angew. Chem. Int. Ed. Engl.*, *35*, 2566 (1996).
9. (a) K. Ishihara, M. Kaneeda and H. Yamamoto, *J. Am. Chem. Soc.*, **116**, 11179 (1994).
   (b) K. Ishihara, S. Nakamura and H. Yamamoto, *Croat. Chem. Acta*, **69**, 513 (1996).
   (c) For the enantioselective protonation of prochiral allyltrimethyltins, see: K. Ishihara, Y. Ishida, S. Nakamura and H. Yamamoto, *Synlett*, 758 (1997).
10. (a) K. Ishihara, S. Nakamura, M. Kaneeda and H. Yamamoto, *J. Am. Chem. Soc.*, **118**, 12854 (1996).
   (b) A. Yanagisawa, K. Ishihara and H. Yamamoto, Synlett, 411 (1997).
11. Reviews: (a) W. R. Roush. In Comprehensive Organic Synthesis (B. M. Trost, I. Fleming and C. H. Heathcock, eds.), vol. 2, pp. 1-53. Pergamon Press, Oxford (1991). (b) Y. Yamamoto and N. Asao. *Chem. Rev.* **93**, 2207 (1993). (c) T. Bach. *Angew. Chem. Int. Ed. Engl.* **33**, 417 (1994). (d) A. H. Hoveyda and J. P. Morken. *Angew. Chem. Int. Ed. Engl.* **35**, 1262 (1996).
12. (a) K. Furuta, M. Mouri and H. Yamamoto. *Synlett*, 561 (1991). (b) K. Ishihara, M. Mouri, Q. Gao, T. Maruyama, K. Furuta and H. Yamamoto. *J. Am. Chem. Soc.* **115**, 11490 (1993).
13. (a) S. Aoki, K. Mikami, M. Terada and T. Nakai. *Tetrahedron* **49**, 1783 (1993). (b) A. L. Costa. M. G. Piazza, E. Tagliavini, C. Trombini, A. Umani-Ronchi. *J. Am. Chem. Soc.* **115**, 7001 (1993). (c) G. E. Keck, K. H. Tarbet, L. S. Geraci. *J. Am. Chem. Soc.* **115**, 8467 (1993).
14. A. Yanagisawa, H. Nakashima, A. Ishiba and H. Yamamoto. *J. Am. Chem. Soc.* **118**, 4723 (1996).
15. A. Yanagisawa, A. Ishiba, H. Nakashima and H. Yamamoto. *Synlett*, 88 (1997).
16. A. Yanagisawa, Y. Nakatsuka, H. Nakashima and H. Yamamoto. *Synlett*. 933 (1997).
17. Reviews: (a) T. K. Hollis and B. Bosnich. J. Am. Chem. Soc. **117**, 4570 (1995). (b) M. Braun. In Houben-Weyl: Methods of Organic Chemistry (G. Helmchen, R. W. Hoffmann, J. Mulzer and E. Schaumann, eds.), vol. E 21, pp. 1730-1735. Georg Thieme Verlag, Stuttgart (1995). (c) S. G. Nelson. *Tetrahedron: Asymmetry* **9**, 357 (1998).
18. A. Yanagisawa, Y. Matsumoto, H. Nakashima, K. Asakawa and H. Yamamoto. *J. Am. Chem. Soc.* **119**, 9319 (1997).

COMBINATORIAL CHEMISTRY:
PART OF THE EVOLUTION IN DRUG DISCOVERY

John J. Baldwin, Jonathan J. Burbaum, Nolan Sigal

Pharmacopeia, Inc.
3000 Eastpark Boulevard
Cranbury, NJ 08512

Combinatorial chemistry, automated synthesis, high-throughput screening, the explosion of molecular targets from genomic research and advances in informatics have come together in an interdependent evolution to provide an enormous opportunity not only for the life sciences but also for the materials sciences as well. Drug discovery has been the initial focus of this revolution because high-speed chemistry has rapidly evolved to meet the compound demands of high throughput, low volume screening which had been put in place to support the mechanistic based targets from molecular biology.

Combinatorial chemistry was not only integrated successfully into this cluster of new technologies, but now serves as the engine which drives the lead discovery process. This synthesis method is perhaps the most powerful of the various automated chemistries since it uses a true combinatorial process to prepare a set of compounds from sets of building blocks. Combinatorial chemistry has its roots in solid phase peptide synthesis pioneered in the 1960's by Merrifield but assumed its present role as a method for multiple compound synthesis in the peptide directed work of Geysen[1], Houghton[2], Furka[3] and Lam[4]. It was not until the work of DeWitt, Ellman and others that the early focus on amide chemistry expanded to include the synthesis of small, drug-like molecules on solid support.

The solid phase synthesis of small molecules fall into two general categories, either spacially addressable parallel synthesis[5,6] or split synthesis. Although small combinatorial libraries can be prepared by either parallel or split synthesis, for large libraries, the split synthesis technique is clearly preferred. The classical "mix and split" method[3,4] involves an interative process of repeated reacting, pooling and dividing sets of solid supports and therefore gives rise to a statistical distribution of all possible individual library members with some being over represented and others underrepresented. The "direct divide" method was developed[7] to provide a more even distribution of library members. In this technique, the solid supports in each reaction vessel from step one are evenly apportioned into the next set of reactors. After the second reaction is complete, the solid supports are again individually reapportioned from each vessel into the next set of reactors and so on. The number of unique library members (N) produced by either split synthesis or the direct divide method is equal to $n_1 \times n_2 \times \cdots n_y$ where $n_1, n_2 \cdots n_y$ are the number of synthons at each combinatorial step.

The initial problem in applying the split synthesis approach to nonsequencable small molecules was in determining the structure of the compound built on the solid support since each bead or each micro-reactor is anonymous. Both direct method for structural assignment, primarily using spectroscopy[8,9] and indirect approaches including deconvolution and encoding methods[10,11,12,13,14] have been developed. Encoding, which records the reaction history experienced by each particular reaction, can be as simple as labeling a "tea bag"[2], placing readable optical characters on a polyethylene tubes[15], etching bar codes on polystyrene grafted ceramic plates[16] or utilizing a radio frequency chip encased in glass then either polymer coated or placed in a "tea bag" like reaction can filled with resin[17, 18].

Chemical encoding can be performed on the solid support beads to record the reaction history that each bead has experienced and hence the structure of the compound built on a particular bead can be inferred. Orthogonal encoding methods include oligonucleotides[10,11], peptides[12,13], GC separable halophenyl ethers [14], GC separable amines released from amides on acid hydrolysis and captured as dansylamides[19], isotopically varied tags[20], chromophores and fluorophores[21,22]. When sufficient compound is available directly from the bead or a "tea bag" type reactor, mass spectrometry can be successful especially if each member of the library has a different molecular weight and the compound can be ionized in tact in the gas phase to observe the molecular ion.

The first large encoded library used the electrophoric halophenylethers and defined a 6,727 membered library[23]. A similarly encoded 1,200 membered dihydrobenzopyran library yielded inhibitors of carbonic anhydrase II as low as 15nM in potency[24]. A library of 56,000 members, also encoded with electrophoric tags, was evaluated against two related G protein coupled receptors; twenty four compounds were found to bind to one of these and 86 to the other with potencies as low as 53nM[25]. Four chemically encoded combinatorial libraries have been built around the statine and related classes of transition state analogs designed for aspartic acid proteases. In this case, the design was biased toward inhibitors of plasmepsin II, a digestive enzyme of *Plasmodium falciparum*. The design included the more lipophilic leucyl statine and phenylalanyl statine as the spanning elements for the scissile bond and included alanyl statine derivatives as the control set. Selective inhibitors of plasmepsin II as low as 50 nM were found with the related enzyme cathepsin D serving as the specificity control[26]. At Pharmacopeia, using the Still electrophoric tags, over 90 ECLiPS™ (Encoded Combinatorial Library on Polymeric Support) libraries have been prepared to date containing a total of nearly four million individual, non-peptidic, small molecules and encompassing well over 300 structural motifs.

ECLiPS™ libraries have been used for both lead identification and lead optimization, and experience with these libraries is beginning to establish some general concepts. For non-biased collections, those libraries which contain the higher number of active compounds, as defined by decoded structures, also contain the more potent compounds.

Targeted libraries, that is those based on some level of structural or mechanistic information, tend to have larger numbers of more potent compounds relative to discovery libraries. Such libraries have been uniformly productive across a range of target classes in generating bioactive compounds. In our laboratories the G protein class of receptor targets has been the most intensely screened with novel antagonists being identified in seven libraries against a panel of 11 GPCRs.

Split synthesis libraries have proved to be valuable in lead discovery and optimization efforts, in part, because of the significant advantages of solid phase chemistry. Once reaction conditions are optimized, compounds upon detachment from the resin can be obtained in high purity and in reasonable yield. However, optimization often requires excess of the reactant and/or multiple cycles to drive the reaction to completion. This leads directly to a significant disadvantage of solid phase chemistry, i.e., large quantities of reactants are needed hence, implementation often depends on commercially available material. This reagent requirement can limit the overall size and diversity of a library especially in the parallel approach where several milligrams are produced and multistep custom synthons often needed. Split synthesis can overcome this obstacle to achieve high diversity within a large, optimally designed collection. For example, in a 48,050 (31 x 5 x 31 x 10) membered library on TentaGel beads (loaded with lysine to double the functionality to 0.5mmol/g), only 16g of resin would be required to prepare 100 copies of each library member. In this example, only 116 mg of a 150 MW synthon would be required for each of the 31 reactions vessels to provide a two equivalent excess and only 720mg of a 150 MW synthon for each of the 5 reaction vessels in Step 2 (again assuming 3 equivalents of the reactant) (Figure 1). Each 180 Micron TentaGel bead in the library will release between 300pmol to 1,000pmol of product depending upon chemistry, linker and extractability from the solid phase. Thus, in a standard 100µl assay up to three assays at 1µM concentration can be performed.

Figure 1.

Example Library Numbers

31 X 5 X 31 X 10 = 48,050 member library

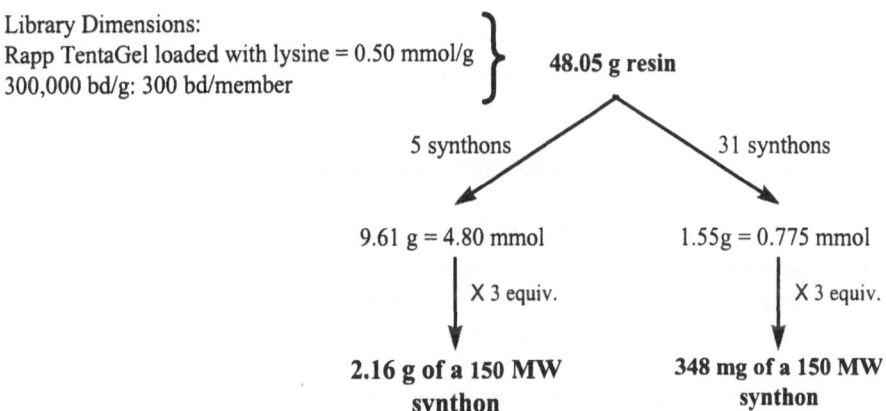

Library Dimensions:
Rapp TentaGel loaded with lysine = 0.50 mmol/g
300,000 bd/g: 300 bd/member
} 48.05 g resin

5 synthons   31 synthons

9.61 g = 4.80 mmol   1.55g = 0.775 mmol

X 3 equiv.   X 3 equiv.

2.16 g of a 150 MW synthon   **348 mg of a 150 MW synthon**

To achieve higher compound concentration lower assay volumes can be used. For example, if one could reduce the volume of the assay to 1µl, 50 to 100 assays could easily be run from the compound available from a single bead and in higher concentrations. With such miniaturization of biological testing, a library would not need to be prepared in 100 fold redundancy. Decreasing the redundancy would decrease the synthon requirement and still produce enough compound for thousands of test points. This is the potential of miniaturization chemistry being combined with miniaturized bioformats.

The origin of miniaturized bioassays can be traced back to the 96 well plate format developed by Sever for measuring viral titers[27]. Analytical use of the 96 well plate expanded to include various clinical analyses, such as the enzyme linked immunosorbent assay (ELISA) along with concomitant advances in automated plate readers and pipettes. This format has been the mainstay for today's higher throughput screening and recently this same footprint has been adapted to 96 well reaction vessel blocks for automated synthesis.

As indicated previously, miniaturization is a rapidly evolving technology in both biology and chemistry as efficiency improvements in both cost and throughput became increasingly important. The ability to rapidly prepare thousands of compounds for lead discovery requires miniaturization to control cost per compound, and as the number of assays expand compound cost per assay becomes increasingly important. To extend the life of a sample collection, whether it be classical or one in library format, the amount of compound used per assay becomes increasingly important. Similarly, miniaturization of bioassays will not only save on the cost of biological reagents per assay but will increase throughout, a critical factor as more targets come on streams from genomics.

In the effort to achieve the benefits of miniaturization, two general directions are evolving for screening formats, an open container approach and a closed system which depends on microfluidics. In the open system, a one μl test volume is the smallest practical one considering the limitations imposed by evaporation. A 96 well plate footprint containing 1536 2μl wells has been developed by Pharmacopeia and Corning. This high-density well plate provides the foundation for an entire new bioassay system which includes reformatting capability, precise liquid handling and highly sensitive bioreadout technology.

For assay volumes less than 1μl, evaporation becomes a significant problem and closed containers using microfluidics are needed. Instruments providing evaporation control and microfluidic capabilities are under development by Caliper Technologies[28] and Orchid BioComputer. For example, the latter is combining microsynthesis with miniaturized bioassays using microfabricated silicon wafers with nanoliter reaction and analysis vessels to integrate a common synthesis and analysis system.

The difficulties associated with bioassay miniaturization, as indicated above, include the need to develop highly specialized equipment, often at the cutting edge of technology, frequently linked to new assay methods and reagents. For example, alternatives are required for today's classic radioassay methods since counting a 100 fold smaller sample to the same confidence limits would require an increase of 10,000 fold counting time per sample.

Fluorescence is a particularly appealing approach because of its high-sensitivity and its adaptability to microvolumes. Evotec, for example, is using fluorescence correlation spectroscopy[29] in its evolving HTS instruments, and Wallac uses its Delfia system (dissociation enhanced lanthanide fluorescence immunyassay system[30]) to provide broad platforms for biological assays. To accommodate these advances in instrumentation, new highly sensitive fluorescent reagents are being developed for bioassays. Thus, reagent development is becoming a critical component in the evolution of mircovolume high-throughput screening. The development of sensitive readout for whole cell assays also poses many difficult challenges. Much of today's research is focused on calcium mobilization processes and visualization using fluorescent calcium indicators[31], highly-engineered yeast strains and cellular growth assays. These approaches, with the proper instrumentation, could serve as the basis for high throughput screening.

In *in vitro* systems, such specific reagents have been developed and utilized in the 1536 well format for screening combinatorial library collections; these include assays directed toward inhibitors of plasmepsin and carbonic anhydrase. It appears likely that the development of assay-specific reagents will coincide with instrumentation so that fully functioning, first generation ultra high-throughput screening assays will be in the laboratory environment within the next few years.

Acknowledgment

The authors gratefully acknowledge the excellent support provided by Drs. Ohlmeyer, Dillard, Dolle, Li, Chelsky, Appell and Carroll in the chemical and biological discovery program and by Dr. Affleck for his work on the HTS program. The authors also express their appreciation to Ms. Amy Pike and Ms. Debi Cohen for their careful preparation and editing of this manuscript.

REFERENCES

[1] H.M. Geysen, R.N. Meloen, S.J. Barleling, *Proc Nat'l Sci USA* 81, 3998 (1984)

[2] R.A. Houghton, *Proc Nat'l Sci USA* 82, 5131 (1985)

[3] A. Furka, F. Sebestyen, M. Asgedom, G. Dibo, *Int J Peptied Protein Res* 37, 487 (1991)

[4] K.S. Lam, S.E. Salmon, E.M. Hersh, V.J. Hruby, W.M. Kazmierki, R.J. Knapp, *Nature* 354, 82 (1991)

[5] S.H. DeWitt, J.S. Stankovic, M.C. Schroeder, D.M.R. Cody, M.R. Pavia, *Proc Nat'l Acad Sci USA* 90, 6909 (1993)

[6] B.A. Bunin, J.A. Ellman, *J. Am Chem Soc* 114, 10997 (1992)

[7] E.G. Horlbeck, J.J. Baldwin, *Synthesis of Combinatorial Libraries*, US Patent 5,663,046 (1997)

[8] C.L. Bummel, I.N.W. Lee, Y. Zhou, S.J. Benkovic, *Science* 264, 399 (1994)

[9] J.W. Metzger, C. Kempter, K.N. Wiesmuller, G. Jung, *Anal. Biochem* 219, 261 (1994)

[10] S. Brenner, R.A. Lerner, *Proc Nat'l Acad Sci USA* 89, 5381 (1992)

[11] M.N. Needels, D.G. Jones, E.N. Tate, G.L. Harokel, L.M. Koehersperger, W.J. Dower, R.W. Barrett, M.A. Gallop, *Proc Nat'l Acad Sci USA* 90, 10700 (1993)

[12] J.M. Kerr, S.C. Banville, R.D. Zuckermann, *J. Am Chem Soc* 115, 2529 (1993)

[13] V. Nikolauiev, A. Stierandova, V. Krchnak, B. Seligmann, R.E. Lam, S.E. Salmon, M. Lebl, *Pept Res* 6, 161 (1993)

[14] M.H.J. Ohlmeyer, R.N. Swanson, L.W. Dillard, J.C. Reader, G. Asouline, R. Kobayaski, M. Wigler, W.C. Still, *Proc Nat'l Acad Sci USA* 90, 10922 (1993)

[15] E. Roskamp, *IBC Conference, Oct 28-30* (1996) San Diego, CA USA

[16] X. Xiao, C. Zhao, H. Potash, M.P. Nova, *Angew Chem Int Ed* 36 (1997)

[17] K.C. Nicolaou, C. Zhao, X. Xiao, Z. Parandoosh, A. Senyei, M. Nova, *Angew Chem Int Ed* 34, 2289 (1995)

[18] E.J. Moran, S. Sarshar, J.F. Cargil, M.M. Shahbaz, A. Lio, A.M.M. Mjalli, R.W. Armstrong, *J. Am Chem Soc* 117, 10787 (1995)

[19] Z.J. Ni, D. Maclean, C.P. Homes, M.M. Murphy, B. Ruhland, J.W. Jacobs, E.M. Gordon, M.A. Gallop, *J. Med Chem* 39, 1601 (1996)

[20] H.M. Geysen, C.D. Wagner, W.M. Bodnar, C.J. Markworth, G.J. Parke, F.J. Schoen, D.S. Wanger, D.S. Rinder, *Chem Biol* 3, 679 (1996)

[21] E. Campian, F. Sebestyen, A. Furka, in *Innovation and Perspectives in Solid Phase Synthesis* (1994).Ed. E. R. Birmingham: Mayflower 469 (1994)

[22] B.J. Egner, S. Rana, H. Smith, N. Bouloc, J. Frey, W.S. Brocklesby, Bradley, *Chem Commun 1997*, 735 (1997)

[23] J.J. Baldwin, J.J. Burbaum, I. Henderson, M.H.J. Ohlmeyer, *J. Am Chem Soc* 117, 5588 (1995)

[24] J.J. Burbaum, M.H.J. Ohlmeyer, J.C. Reader, I. Henderson, L.W. Dillard, G. Li, T.L. Randle, N.H. Sigal, D. Chelsky, J.J. Baldwin, *Proc Nat'l Acad Sci* 92, 6027 (1995)

[25] K.C. Appell, T.D.Y. Chung, M.J.H. Ohlmeyer, N.H. Sigal, J.J. Baldwin, D. Chelsky, *J Biomol. Screening* 1, 27 (1996)

[26] C.D. Carroll, *Private Communcation*

[27] J.L. Sever, *J. Immunol* 88, 320 (1961)

[28] J.M. Ramsey, S.C. Jacobson, M.R. Knapp, *Nature Medicine*, 1, 1093 (1995)

[29] R. Rigler, *J. Biotechnology* 41, 177 (1995)

[30] S. Tomkins, P.A. Rota, J.C. Moore, P.E. Jensen, *J Immunol Methods* 103, 209 (1993)

[31] R.Y. Tsien, *Methods Cell Biol* 30, 127, (1989)

IDENTIFICATION OF PROTEASE INHIBITORS USING BIOCOMPATIBLE RESINS AND LIBRARY SYNTHESIS

Morten Meldal, Jörg Rademann, Morten Grøtli and Klaus Bock

Department of Chemistry, Carlsberg Laboratory
Gamle Carlsberg Vej 10. DK-2500 Valby, Copenhagen, Denmark

INTRODUCTION

There has been an explosion of interest in combinatorial chemistry, where the combinatorial approach allows the simultaneous preparation of millions of new compounds in a relatively small confined volume followed by decoding of the active structures in this small volume.[1] This aspect of combinatorial chemistry has led to a revolution in pharmaceutical industry and in material research, where it is suddenly possible to screen an incredible number of compounds in newly developed fast assays.[2,3] The combinatorial principle does not only involve the chemistry, but also the biological and physical conditions under which the compounds are tested. These have to keep pace with the development of the new chemical methodology for combinatorial synthesis in order to be useful for the testing of the large arrays of compounds synthesised and to be used quickly and effectively.[3] The analytical methods for structural analysis belong to another area, which has to develop in order to match the advances in synthesis.

Organic chemistry is traditionally carried out in solution and transfer of the reactions to solid phase is not without complication.[4] Many reactions will simply not occur when one of the reactants is a macroscopic resin molecule.[5] Therefore the solid phase resin is a determining factor for the success of combinatorial chemistry.[6] In the solid phase reactions chemically inert resins suited for harsh chemical reaction conditions such as carban- or carbonium ion reactions have been developed - e.g. the POEPOP and POEPS resins.[7] These resins are designed not only to be compatible with the chemical reactions but also with aqueous buffers used in enzyme assays. Thus, they swell in solvents ranging in polarity from toluene to aqueous salt solutions and may even be used in preparative solid-phase enzyme reactions.

The success in identifying potential protease inhibitors using a library approach is crucially dependent on the availability of such a solid support material, which is equally well suited for the organic chemical reactions required to assemble the library of interest as for the biological assay used to identify the active beads with potentially interesting compounds.[8,9]

THE SOLID PHASE

The character of the solid support has a major influence on the result of reactions[6] in solid phase organic chemistry (SPOC).[10,11] Since Merrifield introduced the concept using a crosslinked polystyrene solid support, other resins designed for peptide synthesis have been developed to overcome the initial limitations of the Merrifield resin. Among these are dimethyl acrylamide based resins[12,13] and the PEG (polyethylene glycol)[14] grafted polystyrene resin, TentaGel,[15] which have been used extensively for both SPOC and peptide synthesis.

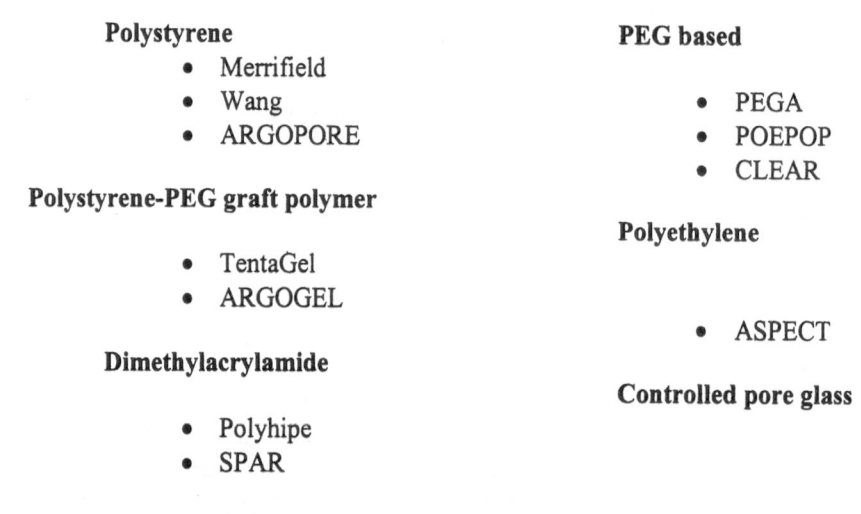

Polystyrene
- Merrifield
- Wang
- ARGOPORE

Polystyrene-PEG graft polymer
- TentaGel
- ARGOGEL

Dimethylacrylamide
- Polyhipe
- SPAR

PEG based
- PEGA
- POEPOP
- CLEAR

Polyethylene
- ASPECT

Controlled pore glass

Scheme 1. Types of commercial resins

In search for a support for peptide synthesis and subsequent bio-assay using resin bound peptide, the hydrophilic and flexible PEGA resins were developed as a superior polymer.[16] This hydrophilic PEG-polydimethyl acrylamide hybrid resin is compatible with a variety of solvents and has an open structure due to long PEG molecules acting as crosslinkers. PEGA was originally introduced as a versatile crosslinked support for peptide and glycopeptide synthesis. The crosslinked PEGA polymer matrix consists of *bis*-acryloyl amino terminated PEG chains polymerised to form a poly-acrylamide backbone. PEG is the major constituent of the polymer support and hence it is highly flexible and bio-compatible.[17] The polymer matrix was designed and synthesised in such a way that macromolecules such as enzymes can freely enter into the polymer network there by facilitating solid phase enzymatic reactions.[18] In a comparative study, it was shown that PEGA resin compares favourably to many of the existing polymer supports for peptide synthesis.[19] In fact, enzyme essays can be carried out with compounds attached to the support.[1]

However, a resin must be robust and compatible with a wide range of reaction conditions, and the content of amide bonds in the PEGA resin makes it unsuitable for this purpose. POEPS (polyethylene glycol-polystyrene based) and POEPOP (polyethylene glycol-polyoxypropylene based) resins which have a polar, open structure and yet are chemically inert have therefore been developed.[7]

FLUORESCENT-QUENCHED ASSAYS FOR SUBSTRATES AND INHIBITORS

The open structure of the biocompatible polyethylene glycol polyamide copolymer (PEGA resin) and more recent developments (POEPOP) of that type of resins gives clear advantages using such type of resins in the synthesis of inhibitor libraries for drug relevant proteases as will be discussed below. It has been demonstrated that conventional peptide synthesis resins generally do not permit the access of enzymes and other macromolecules into the interior of the polymeric support.[20]

Fluorescence based combinatorial enzyme analysis are performed by incorporation of the fluorescent 2-aminobenzoic acid at the anchoring point to the resin and a 3-nitrotyrosine at the far end of the molecule.[21] The 3-nitrotyrosine quenches the fluorescence effectively while the molecule is intact (FQ-substrate). If the molecule is a substrate and is cleaved by the enzyme under investigation, then the molecule begins to fluoresce. The resin beads with the active substrates are isolated and the structure of the substrate molecule determined by physical or chemical methods.

A complete subsite mapping of endoproteases by fluorescent quenched resin bound peptide libraries has been reported.[17,22] Furthermore, this method was modified for the identification of enzyme inhibitors of subtilisin Carlsberg,[8] and cruzipain[23] using D-amino acid containing peptide libraries.

LIBRARY GENERATOR

Production of libraries and testing of enzymes

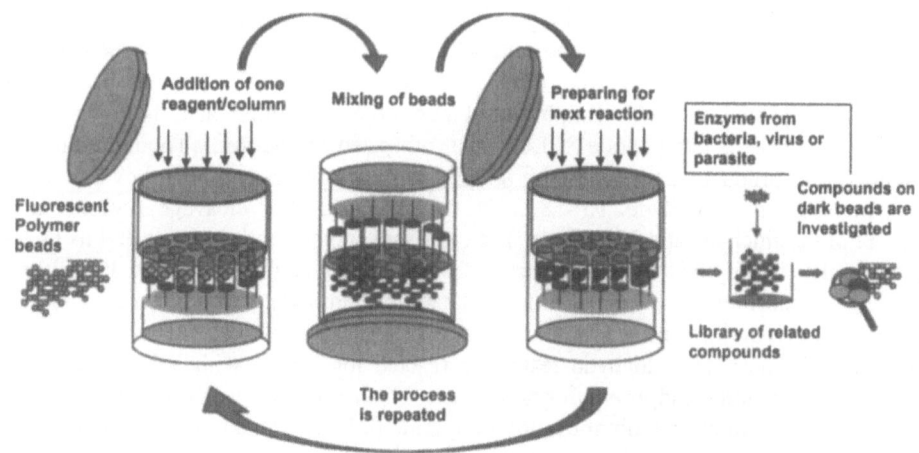

Combinatorial molecular libraries can be prepared in special equipment developed at the Carlsberg Laboratory, the so-called library generator.[24] The resin beads on which the chemical synthesis is carried out is split between the columns in the Teflon reactor of the library generator, and the first series of molecular building blocks are attached to the resin by a chemical reaction.

Different building blocks are added to each column and allowed to react. Excess solvent is added to the resin and the reactor, which is supplied with a lid, is turned upside down and the resin is thoroughly mixed by vigorous agitation. When the reactor is turned upright the resin splits uniformly between the columns. A functional group is chemically deprotected and the resin is now ready for chemical attachment of the second series of building blocks to the first ones on the resin. This process is repeated until the library synthesis is complete. The library is chemically deprotected and the quality of the synthesis is assessed by mass spectrometry or by NMR-spectroscopy on single beads. In the case of enzyme assays, substrate and inhibitor libraries contain a fluorescent probe at the attachment point to the resin and another group, which can quench the fluorescence at the far end of the molecule. Upon enzymatic cleavage of the substrate a substantial increase in fluorescence is observed and the molecules thereby function as indicators of the enzymatic activity [6,8]. Thus, the enzymes can be characterised and inhibitors can be developed. After the enzyme reaction has been quenched the beads are studied in a fluorescence microscope and active beads are isolated and subjected to structural analysis to afford potential substrates or inhibitors.

ORGANIC REACTIONS ON NEW RESINS

The above mentioned novel polymeric support, which exclusively consists of polyoxyethylene/ polyoxypropylenecopolymer (POEPOP),[7] shows a broad range of reaction conditions in combination with a solvent tolerance ranging from aqueous solutions to most of the commonly used organic solvents. This concept is extended to a polar resin suited to most kinds of organic reactions as well as Fmoc-based peptide methodology: the POEPOP-400 resin.

The new resin was subjected to a variety of different reaction protocols combining peptide and general organic chemistry in order to test reactions that might be useful in solid phase synthesis of peptide isosters. First POEPOP-400 was functionalised with the HMBA (4-hydroxymethyl-benzoate) linker. For this purpose the linker, which had been triethylsilyl-protected at the benzylic position, was coupled to the hydroxy resin with N-methylimidazole (MeIm) as base and with MSTN as condensing agent. Subsequently, the silyl-protecting group was removed with neutral TBAF (tetrabutylammonium fluoride)/acetic acid in THF. Employing Fmoc-protected amino acid building blocks, TBTU and NEM the tetrapeptide SLLG was synthesized on the resin. The terminal serine residue of this product was treated with 4 equivalents of sodium periodate in water for one hour to afford the aldehyde. RP-18-HPLC-MS analysis after cleavage revealed a 1:1 mixture of the aldehyde and its hydrate. The clear separation of the aldehyde and its hydrate indicates the stability of the two species under neutral and acidic conditions. Probably the observed equilibrium was established under the basic conditions during linker cleavage.

Reductive amination of compound the aldehyde was most efficiently achieved in a two-step procedure. The aldehyde resin was reacted for 1 hour with a 2:1:1 mixture of CH_2Cl_2, benzylamine and triethylorthoformate. The resin was washed with DMF and treated with a solution of sodium cyanoborohydride in DMF containing 1 % of acetic acid for 1 hour. The reaction was monitored with HPLC-MS analysis after cleavage and complete conversion of aldehyde to amine was achieved.

The basic addition of a nitroalkane was studied as an example for the reaction of CH-acidic compounds with the resin-bound aldehyde. Nitromethane was added to the aldehyde in a mixture with CH_2Cl_2 and piperidine at room temperature, the reaction consumed all starting material and after cleavage the alcohol was isolated as the main product in 60 % yield. The nucleophilic addition of silyl-protected carbanions was also investigated as an example for the reaction of stabilized carbanions. The Sakurai-reaction employing allyltrimethylsilan was performed with 10 equivalents of the nucleophile and under Lewis acid catalysis. The Lewis acids evaluated include trimethylsiloxytrifluormethanesulfonium (TMSOTf), $TiCl_4$, and $SnCl_4$. Smooth reaction under these conditions was only observed with tin(IV) as catalyst. The two diastereomers of the homoallylalcohol were obtained with a ratio of 2:1 and an overall yield of 69%. Both isomers could be easily separated by HPLC.

As a model reaction of resin-bound aldehyde with ylides the Wittig-Horner reaction was selected. The phosponate was deprotonated with n-butyllithium in toluene prior to the solid phase reaction. In a second step the ylide formed was added to the resin and allowed to react for two hours. The resin-bound ester was analysed with nanoprobe-MAS-NMR (^{1}H-NMR, 500 MHz). After cleavage the trans-olefinic acid was isolated as the only peptide product.

Finally, the reaction of aldehyde with metallated carbanions was investigated. The organocuprates are only weakly basic and were supposed to be compatible with the base-labile HMBA-linker. The organocuprate Me$_2$CuLi was formed at −78 °C from methyllithium and copper(I)iodide. After 30 min reaction of the aldehyde with the cuprate at room temperature, followed by cleavage off the resin a product was isolated having the mass of the expected alkyl-addition product. However, NMR-analysis after preparative cleavage revealed that instead of the expected addition an oxidation of the aldehyde to the N-terminal acid occurred. This observation was supported by the oxidation of the aldehyde employing neutrally buffered KMnO$_4$ The reaction with diethylzinc yielded the same oxidation product.

CONCLUSIONS

In summary, it has been demonstrated that the novel type of solid support is compatible with a broad range of organic reaction conditions from peptide chemistry to the more demanding reactions including the addition of ylides and metallo-organic compounds. Reactions were performed with good to excellent yields in aqueous solution as well as in organic solvents. The array of reactions can also be used for the preparation of peptide isosters. Furthermore, these resin types are fully compatible with biological assays in aqueous media with *eg.* enzymes in order to assess optimal substrates and potential inhibitors for proteases related to severe diseases.

ACKNOWLEDGMENTS

This work was carried out in the SPOCC Center and was supported by the Danish National Research Foundation and the Carlsberg Laboratory.

REFERENCES

1. M. Meldal,. Combinatorial solid phase assay for enzyme activity and inhibition, in: *Combinatorial Peptide Libraries* , C. Shmuel, ed., Humana Press, Totowa, New Jersey. (1998).

2. M.A. Gallop, R.W. Barrett, W.J. Dower, S.P.A. Fodor, and E.M. Gordon, Applications of combinatorial technologies to drug discovery. 1. Background and peptide combinatorial libraries, *J.Medical Chemistry*, 37:1233 (1994).

3. E.M. Gordon, R.W. Barrett, W.J. Dower, S.P.A. Fodor, and M.A. Gallop, Applications of combinatorial technologies to drug discovery. 2. Combinatorial organic synthesis, library screening strategies, and future directions, *J.Med.Chem.*, 37:1385 (1994).

4. E.K. Wilson, Combinatorial chemistry, *C&E News*, 24 (1997).

5. J. Eichler, J.R. Appel, S.E. Blondelle, C.T. Dooley, B. Dörner, J.M. Ostresh, E. Perez-Paya, C. Pinilla, and R.A. Houghten, Peptide, peptidomimetic, and organic synthetic combinatorial libraries, *Med.Res.Review*, 15:481 (1995).

6. M. Meldal. Properties of the solid support, in: *Methods in Enzymology 289, Solid-Phase Peptide Synthesis,* G. Fields, ed., Academic Press (1997).

7. M. Renil and M. Meldal, POEPOP and POEPS: Inert polyethylene glycol crosslinked polymeric supports for solid phase synthesis, *Tetrahedron Lett.*, 37:6185 (1996).

8. Meldal and I. Svendsen, Direct visualization of enzyme inhibitors using a portion mixing inhibitor library containing a quenched fluorogenic peptide substrate. 1: Inhibitors for subtilisin Carlsberg, *J.Chem.Soc., Perkin Trans.1*, 1591 (1995).

9. B. Seligmann, F. Abdul-Latif, F. Al-Obeidi, Z. Flegelová, O. Issakova, P. Kocis, V. Krchnak, K. Lam, M. Lebl, J. Ostrem, P. Safar, N. Sepetov, A. Stierandova, P. Strop, and P. Wildgoose, The construction and use of peptide and non-peptidic combinatorial libraries to discover enzyme inhibitors, *Eur.J.Medical Chem.*, 30:320S (1994).

10. P.H.H. Hermkens, H.C.J. Ottenheijm, and D.C. Rees, Solid-phase organic reactions II: a review of the literature nov 95-nov 96, *Tetrahedron*, 53:5643 (1997).

11. J.S. Fruchtel and G. Jung, Organic chemistry on solid supports, *Angew.Chem.Int.Ed.Engl.*, 35:17 (1996).

12. R. Arshady, E. Atherton, D.L.J. Clive, and R.C. Sheppard, Peptide synthesis. Part 1. Preparation and use of polar supports based on poly(dimethylacrylamide), *J.Chem.Soc.Perkin Trans.1*, 529 (1981).

13. E. Atherton, E. Brown, and R.C. Sheppard, A physically supported gel polymer for low pressure, continuous flow solid phase reactions. Application to solid phase peptide synthesis, *J.Chem.Soc., Chem.Commun.*, 1151 (1981).

14. S. Zalipsky, F. Albericio, and G. Barany. Preparation and use of an aminoethyl polyethylene glycol- crosslinked polystyrene graft resin support for solid-phase peptide synthesis, in: *Peptides 1985, Proc. Am. Pept. Symp.*, C.M. Deber, V.J. Hruby and K.D. Kopple, eds.,. Pierce Chemical Company, Rockford, Illinois (1986).

15. W. Rapp, L. Zhang, R. Häbish, and E. Bayer. Polystyrene-Polyoxyethylene graftcopolymers for high speed peptide synthesis, in: *Peptides 1988, Proc. Eur. Pept. Symp.*, G. Jung and E. Bayer, eds., Walter de Gruyter, Berlin (1989).

16. M. Meldal, PEGA: A flow stable polyethylene glycol dimethyl acrylamide copolymer for solid phase synthesis, *Tetrahedron Lett.*, 33:3077 (1992).

17. M. Meldal, F.I. Auzanneau, and K. Bock. PEGA, Characterization and application of a new type of resin for peptide and glycopeptide synthesis, in: *Innovation and Perspectives in Solid Pphase Synthesis*, R. Epton, ed., Mayflower Worldwide Limited, Kingswinford (1994)

18. M. Renil, M. Ferreras, J.M. Delaisse, N.T. Foged, and M. Meldal, PEGA supports for combinatorial peptide synthesis and solid-phase enzymatic library assays *J.Pept.Sci.*, 4:195 (1997).

19. M. Meldal. Multiple column peptide synthesis, developement and application, in: *Peptides 1992, Proc. Eur. Pept. Symp.*, C.H. Schneider and A.N. Eberle, eds., ESCOM, Leiden (1993).

20. J. Vagner, G. Barany, K.S. Lam, V. Krchnak, N.F. Sepetov, J.A. Ostrem, P. Strop, and M. Lebl, Enzyme-mediated spatial segregation on individual polymeric support beads: application to generation and screening of encoded combinatorial libraries, *Proc.Natl.Acad.Sci. USA*, 93:8194 (1996).

21. M. Meldal and K. Breddam, Anthranilamide and nitrotyrosine as a donor acceptor pair in internally quenched fluorescent substrates for endopeptidases - Multicolumn peptide synthesis of enzyme substrates for subtilisin carlsberg and pepsin, *Anal.Biochem.*, 195:141 (1991).

22. M. Meldal, I. Svendsen, K. Breddam, and F.I. Auzanneau, Portion-mixing peptide libraries of quenched fluorogenic substrates for complete subsite mapping of endoprotease specificity, *Proc.Natl.Acad.Sci.USA*, 91:3314 (1994).

23. M. Meldal, I. Svendsen, L. Juliano, M.A. Juliano, E. Del Nery, and J. Scharfstein, Inhibition of cruzipain visualized in a fluorescence quenched solid-phase inhibitor library. D-Amino acid inhibitors for cruzipain, cathepsin B and cathepsin L, *J.Pept.Sci.*, 4:83 (1998).

24. M. Meldal, Multiple column synthesis of quenched solid-phase bound fluorogenic substrates for characterization of endoprotease specificity *Methods: A Companoim to Methods in Enzymology*, 6:417 (1994).

COMBINATORIAL CHEMISTRY: THE ENLISTMENT AND APPLICATION OF SOLUBLE POLYMER SUPPORTS

Kim D. Janda

The Scripps Research Institute
Department of Chemistry
10550 N. Torrey Pines Road
La Jolla, California 92037

INTRODUCTION

Polymer-supported synthesis has become the method of choice for the preparation of peptides and oligonucleotides and largely responsible for the rapid expansion of combinatorial chemistry.[1-9] Although highly successful, the heterogeneous nature of solid-phase organic synthesis has been linked to problems including decreased reactivities, prolonged reaction times, and limited analytical techniques for reaction monitoring, all of which have been addressed by the use of soluble polymer supports.[8,10-27] The selection of a polymer support for liquid-phase synthesis (LPS) mainly depends upon its macromolecular solubility properties satisfying the requirements of a particular synthetic scheme. Considering the dearth of soluble supports currently available, the utility of LPS would be greatly enhanced by the availability of a larger and more diversified set of soluble supports. We detail report a new method of parallel polymer synthesis to produce materials with novel solubility characteristics tunable for LPS as well as potential applications in materials science.

DISCUSSION

The recent syntheses of combinatorial polymer libraries provided new materials for catalysis and potential biomedical applications.[28-30] However, we required an alternative methodology to provide a library of polymers that (1) are chemically stable to the variety of reaction conditions of LPS, (2) exhibit highly diverse solubility properties, and (3) allow identification and large-scale synthesis of novel members. Free radical polymerization of a large variety of vinyl monomers yielded robust materials.[31] Therefore, we developed bifunctional initiators **1** and **2** (Figure 1) capable of conducting two rounds of polymerization to produce a library of block copolymers in a combinatorial format.

Bifunctional initiators **1** and **2** contain both diazene and TEMPO (2,2,6,6-tetramethylpiperidinyl-1-oxy)[32] moieties that independently initiated free radical polymerization. Sequential polymerization at two different temperatures produced triblock copolymers tailored to the designers' needs (Figure 2). Furthermore, polymer molecular weights could be controlled by variation of monomer/initiator ratios. The formation of block copolymers was verified by size exclusion chromatography (SEC) of the polymers derived from styrene and **1** after each round of polymerization and following hydrolysis of the ester linkage between blocks.

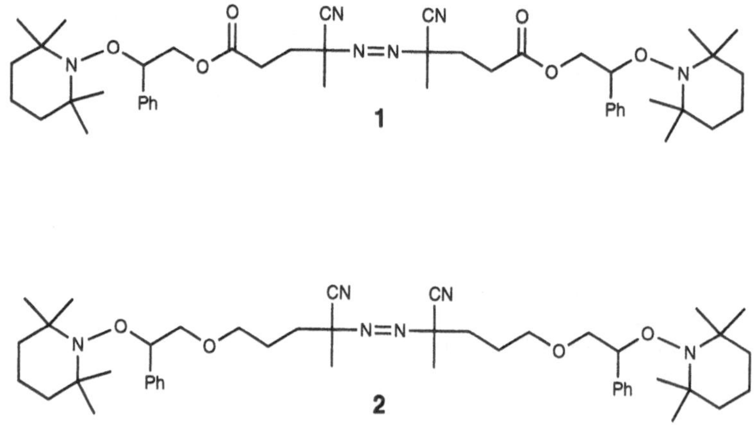
1
2

Figure 1. Bifunctional free radical initiators used in combinatorial polymer synthesis.

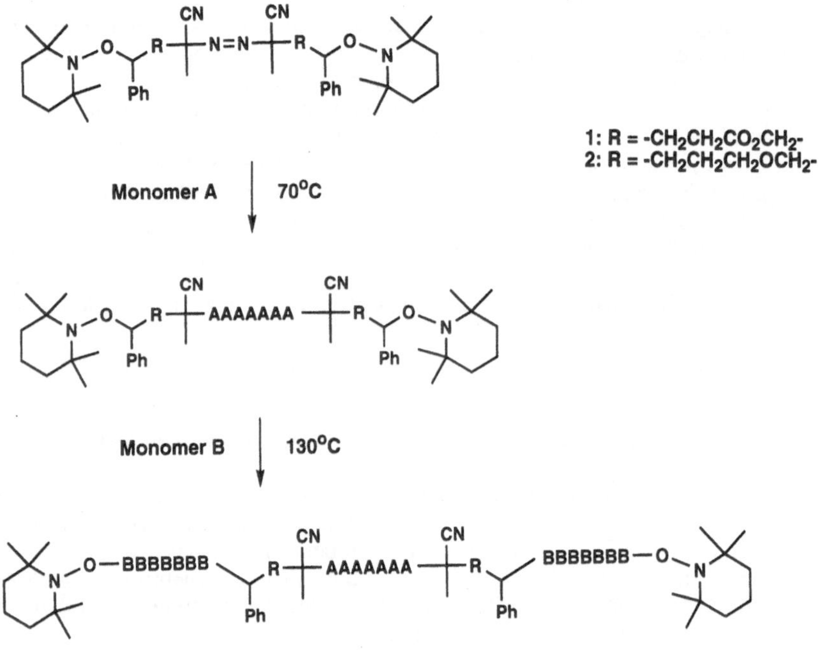
1: R = -CH2CH2CO2CH2-
2: R = -CH2CH2CH2OCH2-
Monomer A
70°C
Monomer B
130°C

Poly(B)-Poly(A)-Poly(B)

Figure 2. Methodology for the synthesis of triblock copolymers using bifunctional initiator **1** or **2**.

For LPS applications, four potential anchoring sites were uniformly distributed on the triblock copolymer. Selective functionalization of the polymer support with amino and/or hydroxyl groups was accomplished by LiAlH$_4$ reduction of the interior nitriles or by reductive cleavage (Zn/THF:AcOH:H$_2$O)[33] of the N-O bond of the TEMPO end groups (Figure 3). Loading capacity can be increased further through copolymerization of monomers containing appropriate linker functionalities during either round of polymerization.

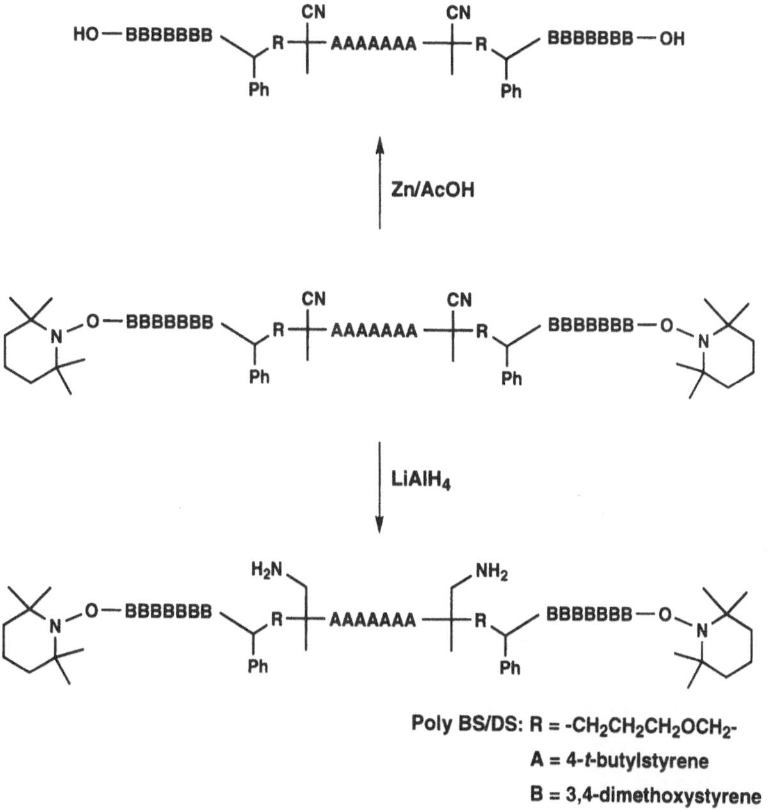

Poly BS/DS: R = -CH$_2$CH$_2$CH$_2$OCH$_2$-
A = 4-*t*-butylstyrene
B = 3,4-dimethoxystyrene

Figure 3. Selective functionalization of block copolymers to provide attachment sites for liquid phase synthesis applications.

Polymerizations using initiator **2**, containing an ether linkage, produced copolymers with the increased chemical stability desirable for some LPS applications. Applying a combinatorial format, we designed an apparatus in which copolymers were synthesized in parallel under conditions of uniform mixing and heating. Block copolymers in a 5 x 5 array were characterized by ^{1}H NMR, SEC, and solubility analyses. Typically, copolymers exhibited solubility properties intermediate of the two homopolymers corresponding to each block; however, some exceptions were observed.

All members of the synthesized polymer library are potential soluble supports for LPS and individually may function in different applications. For example, the copolymer containing outer blocks of poly(3,4-dimethoxystyrene) and an inner block of poly(4-*tert*-butylstyrene) (poly-**BS/DS**) dissolved in several organic solvents including both ether and acetone (unlike the corresponding homopolymers) and remained soluble in THF below room temperature. Thus, poly-**BS/DS** could be valuable for reaction conditions that precipitate the versatile support, poly (ethylene) glycol (PEG).

To assess the utility of this new soluble support for LPS, the kinetic behavior of poly-**BS/DS** was analyzed and compared with a low molecular weight compound. After functionalization of poly-**BS/DS** with amino groups (LiAlH$_4$, refluxing THF, 2 h), the rate of Schiff base formation of 4-dimethylamino-cinnamaldehyde[34-35] with either the polymeric **BS/DS** or 1-aminohexane was measured spectrophotometrically. Concentrations of stock solutions in CHCl$_3$ were measured using quantitative ninhydrin analyses[36] that reported a value of 0.14 mmol amine per gram polymer which compared favorably with the loading capacity of 0.2 mmol/g for MeO-PEG (molecular weight 5000). Remarkably, the calculated second order rate constants differed only slightly ($k_{polymer}$ = 0.49 L mol^{-1} hr^{-1} *vs.* $k_{aminohexane}$ = 0.69 L mol^{-1} h^{-1}) demonstrating that poly-**BS/DS** is an outstanding polymer support for LPS.

Attachment of a diphosphine ligand to poly-**BS/DS** in THF produced soluble polymeric ligand **3** which, in the presence of rhodium(I), mediated the catalytic reduction of 2-acetamidoacrylic acid at a rate an enantiomeric excess comparable to that of the unbound catalyst (Figure 4).[37-38] Precipitation of the polymer simplified the reaction workup and allowed recovery of the ligand for recycling. This result highlights the benefits of tunable solubility of polymeric supports in LPS, as PEG-supported phosphine would have been nearly insoluble under these reaction conditions.

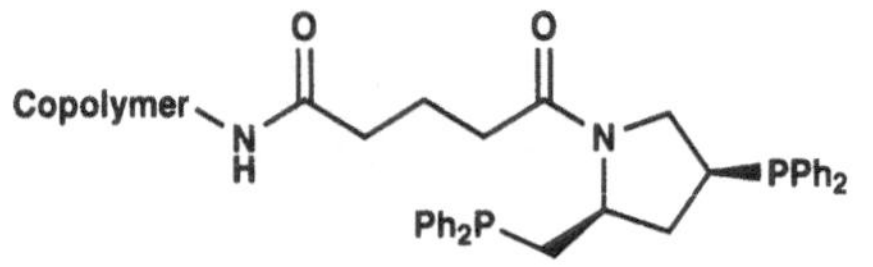

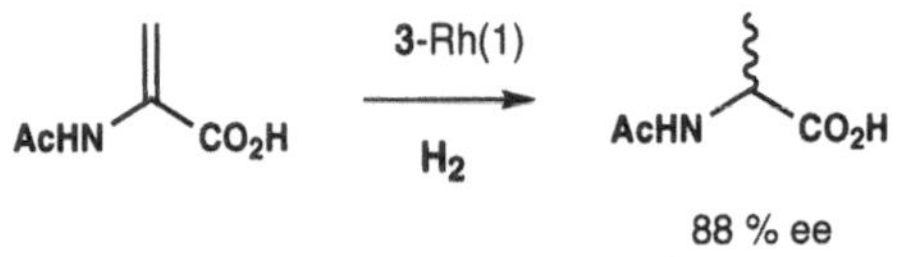

Figure 4. Polymer bound diphosphine ligand and hydrogenation of 2-acetamidoacrylic acid.

CONCLUSIONS

We used a combinatorial format to produce a spatially addressable library of triblock copolymers. These macromolecules contain differentiable anchoring sites, exhibit novel solubility characteristics, and expand the arsenal of polymer supports available for liquid-phase synthesis. Since polymer characteristics vary with monomer composition and block lengths, this methodology allows for the systematic modulation of macromolecular properties and provides materials that can be "fine tuned" for additional applications.

REFERENCES

1. I. Chaiken, K.D. Janda (Eds.) *Molecular Diversity and Combinatorial* American Chemical Society, Washington, D.C., (1996).
2. S.R. Wilson, A.W. Czarnik (Eds.) *Combinatorial Chemistry* (Eds.), John Wiley & Sons, New York, (1997).
3. J.S. Früchtel, G. Jung, Organic chemistry on solid supports, *Angew. Chem. Int. Ed. Engl.* 35:17 (1996).

4. F. Balkenhohl, C. von dem Bussche-Hünnefeld, A. Lansky, C. Zechel, Combinatorial synthesis of small organic molecules, *ibid.* 35:2288 (1996).
5. L.A. Thompson, J.A. Ellman, Synthesis and applications of small molecule libraries, *Chem. Rev.*, 96:555 (1996).
6. K.S. Lam, M. Lebl, V. Krchnák, The 'one-bead-one-compound' combinatorial method, *ibid.* 97:411 (1997).
7. A. Nefzi, J.M. Ostresh, R.A. Houghten, The current status of heterocyclic combinatorial libraries, *ibid.* 97:449 (1997).
8. M.C. Pirrung, Spatially addressable combinatorial libraries, *ibid.* 97:473 (1997).
9. D.J. Gravert and K.D. Janda, Organic synthesis on soluble polymer supports: liquid-phase methodologies, 97:489 (1997).
10. R.H. Andreatta and H. Rink, Zur problematik der peptidsynthese an tragern: beitrag eines neuen verfahrens mit loslichen tragem, *Helv. Chim. Acta.* 56:1205 (1973).
11. M. Mutter and E. Bayer in *The Peptides, Vol. 2* (Eds.: J. Mieinhofer, E. Gross), Academic, New York, pp. 285 (1979).
12. H. Han, M.M. Wolfe, S. Brenner and K.D. Janda, Liquid-phase combinatorial synthesis, *Proc. Natl. Acad. Sci. USA* 92:6419 (1995).
13. K.E. Geckeler, Soluble polymer supports for liquid-phase synthesis, *Adv. Polym. Sci.*, 121:31-79 (1995).
14. D.E. Bergbreiter and J.W. Caraway, Thermoresponsive polymer-bound substrates, *J. Am. Chem. Soc.* 118:6092 (1996).
15. P. Wentworth Jr., A.M. Vandersteen and K.D. Janda, Poly(ethylene glycol) (PEG) as a reagent support: the preparation and utility of a PEG-triarylphosphine conjugate in liquid-phase organic synthesis (LPOS), *Chem. Commun.* 759 (1997).
16. H. Han and K.D. Janda, Multipolymer-supported substrate and ligand approach to the Sharpless asymmetric dihydroxylation, *Angew. Chem. Int. Ed. Engl.* 36:1731 (1997).
17. E. Bayer, M. Mutter, Liquid-phase synthesis of peptides, *Nature (London)* 237:512 (1972).
18. G.M. Bonora, Polyethylene-glycol: a high-efficiency liquid-phase (HELP) for the large-scale synthesis of the oligonucleotides, *Appl. Biochem. Biotech.* 54:3 (1995).
19. J. Zhu, L.S. Hegedus, Incorporation of chromium aminocarbene comples-derived amino acids into soluble pol(ethylene glycol) (PEG)-supported peptides, *J. Org. Chem.* 60: 5831 (1995).
20. S.P. Douglas, D.M. Whitfield, J.J. Krepinsky, Polymer-supported solution synthesis of oligosaccharides using a novel versatile linker for the synthesis of D-mannopentaose, a structural unit of D-mannans of pathogenic yeasts, *J. Am. Chem. Soc.*, 117:2116 (1995).
21. H. Han, K.D. Janda, Azatides: solution and liquid-phase syntheses of a new peptidomimetic, *ibid.* 118:2539 (1996).
22. H. Han, K.D. Janda, Soluble polymer-bound ligand-accelerated catalysis: asymmetric dihydroxylation, *ibid.*, 118:7632 (1996).
23. W.K.C. Park, M. Auer, H. Jaksche, C.-H. Wong, Rapid combinatorial synthesis of aminoglycoside antibiotic mimetics: use of a polyethylene glycol-linked amine and a neamine-derived aldehyde in multicomponent condensation as a strategy for the discovery of new inhibitors of the HIV RNA Rev responsive element, *ibid.* 118:10150 (1996).
24. P. Wipf, S. Venkatraman, An improved protocol for azole synthesis with PEG-supported Burgess reagent, *Tetrahedron Lett.* 37:4659 (1996).
25. K.W. Jung, X. Zhao and K.D. Janda, A linker that allows efficient formation of aliphatic C-H bonds on polymeric support, *ibid.* 37:6491 (1996).
26. K.W. Jung, X. Zhao and K.D. Janda, Soluble polymer synthesis: an improved traceless linker methodology for aliphatic C-H bond formation, *ibid.* 38:977 (1997).
27. K.W. Jung, X. Zhao and K.D. Janda, Development of new linkers for the formation of aliphatic C-H bonds on polymeric supports, *Tetrahedron* 53:6645 (1997).
28. F.M. Menger, A.V. Eliseev, V.A. Migulin, Phosphatase catalysis developed via combinatorial organic chemistry, *J. Org. Chem.* 60:6666 (1995).
29. F.M. Menger, C. A. West, J. Ding, A combinatorially developed reducing agent, *Chem Commun.*, 633 (1997).

30. S. Brocchini, K. James, V. Tangpasuthadol, J. Kohn, A combinatorial approach for polymer design, *J. Am. Chem. Soc.* 119:4553 (1997).
31. G. Odian, *Principles of Polymerization*, 2nd ed., John Wiley & Sons, New York, pp 181 (1981).
32. C.J. Hawker, Molecular weight control by a living free-radical polymerization process, *J. Am. Chem. Soc.* 116:11185 (1994).
33. M.K. Georges, R.P.N. Veregin, P.M. Kazmaier, G.K. Hamer, Narrow molecular weight resins by a free-radical polymerization process, *Macromolecules* 26:2987 (1993).
34. D.L. Boger, J.A. McKie, An efficient synthesis of 1,2,9,9a-tetrahydrocyclopropa[c]-benz[e]indol-4-one (CBI): an enhanced and simplified analog of the CC-1065 and duocarmycin alkylation subunits, *J. Org. Chem.*, 60:1271 (1995).
35. D. Gargiulo, N. Ikemoto, J. Odingo, N. Bozhkova, T. Iwashita, N. Berova, K. Nakanishi, CD exciton chirality method: Schiff base and cyanine dye-type vhromophores for primary amino groups, *J. Am. Chem. Soc.* 116:3760 (1994).
36. V.K. Sarin, S.B.H. Kent, J.P. Tam, R.B. Merrifield, Quantitative monitoring of solid-phase peptide synthesis by the ninhydrin reaction, *Anal. Biochem.*, 117:147 (1981).
37. T. Masuda, J.K. Stille, Transition metal catalyzed asymmetric organic syntheses via polymer-attached optically active phosphine ligands. Synthesis of *R*-amino acids by hydrogenation with a polymer catalyst containing optically active alcohol sites, *J. Am. Chem. Soc.*, 100:268 (1978).
38. G.L. Baker, S.J. Fritschel, J.K. Stille, Transition-metal-catalyzed asymmetric organic synthesis via polymer-attached optically active phosphine ligands. 5. Preparation of amino acids in high optical yield via catalytic hydrogenation, *J. Org. Chem.* 46:2954 (1981).

TWO APPROACHES TO AUTOMATED SOLID PHASE SYNTHESIS OF SMALL ORGANIC COMPOUND ARRAYS

Michal Lebl[1,2], Viktor Krchňák[1], Georges Ibrahim[1], Jaylynn Pirez[1], Yidong Ni[1], Dave Podue[1], Petr Mudra[3], Vít Pokorný[3], and Karel Ženíšek[3]
[1]Trega Biosciences Inc.,
[2]Spyder Instruments Inc.,
9880 Campus Point Drive, San Diego, CA 92121, U.S.A.
[3]Institute of Organic Chemistry and Biochemistry,
Flemingovo 2, Praha 6, Czech Republic

INTRODUCTION

Combinatorial techniques (for reviews see e.g.[1]) require new methods for automation of synthetic processes. Solid phase synthesis is optimal for automation, since the complicating factor of unique behavior of different organic molecules is replaced by the predictable behavior of the solid support. Instruments available on the market today are relatively complicated and expensive. An instrument that would be rather simple, therefore inexpensive, and would allow each chemist to synthesize hundreds or thousands of compounds would be welcome by a number of medicinal chemists. Such an instrument would be used for the deconvolution of active compounds from biologically active mixtures, synthesis of arrays of compounds for general screening, or for compound optimization, so called "lead explosion".

RESULTS AND DISCUSSION

Even though solid phase synthesis brought about the potential of relatively simple processing of large arrays of synthetic vessels, some problems in the realization of a machine capable of parallel processing of multitude of samples remain. One of the basic problems is parallel separation of liquid and solid phases. Commercial solid phase synthesizers utilize filtration as the principle for separation of solid and liquid phase (for reviews see e.g.[2]). Filtration can lead to significant complications, especially in the case of multiple synthesizers, since the clogging of one vessel can result in overflowing of this particular vessel during the next solvent addition and distribution of the solid support from this vessel into neighboring ones. We have applied the principle of "surface suction" for removal of supernatant from the sedimented suspension of solid phase particles [3,4], in the design of our robotic synthesizer, in which up to 72 deep-well microtiterplates can be processed simultaneously [3]. The surface suction removal of liquid phase is based on the fact, that the flat-end needle lowered against the surface of the liquid while the suction is applied through the needle, does not disturb the bulk of the liquid and only the surface layer of the liquid is removed. In this way, the needle can be lowered very close to the sedimented resin without removing any solid support particles.

We have based the production of our libraries of single compounds on the combination of manual solid phase synthesis in "tea-bags"[5], combined with robotic processing of the resin distributed into individual wells of the microtiterplate. The synthesis of intermediates in the tea-bags is very flexible and a wide span of conditions can be used in the first several steps of the synthesis. The reaction conditions are limited only by the properties of the polypropylene mesh from which the tea-bags are constructed. Up to a thousand of tea-bags can be handled simultaneously and the resin from each tea-bag can be distributed into a microtiterplate's wells. Before distribution, a sample from each bag is cleaved and product is analysed by LC/MS. Only bags containing expected material in purity better than 85% (evaporative light scattering or UV detection) are used for the continuation of the synthesis. Individual wells of the microtiterplate then receive a different building block and/or reagent by simultaneous pipetting from a pre-prepared "master plate" in which an array of reactants was assembled. In this way, each intermediate is used for the synthesis of up to 96 individual compounds ("bag explosion" -- 1,000 bags can result in 96,000 individual compounds). The limitation of this approach is the necessity of "process friendliness" of the last step of synthesis (relatively stable reagents, temperature range from room temperature up to 80 degrees). The resin in the wells of microtiterplate is then incubated at the appropriate temperature and washed by the application of surface suction technique. After drying in vacuum, the plates are placed into the polypropylene chambers (Figure 1) and exposed to gaseous HF at room temperature for

Figure 1. Polypropylene vessels used for the cleavage of compounds from the resin in the atmosphere of gaseous HF. Left: Frontal view of open chamber; Right: Closed chambers loaded with microtiterplates. Capacity of one vessel is 18 deep-well microtiterplates.

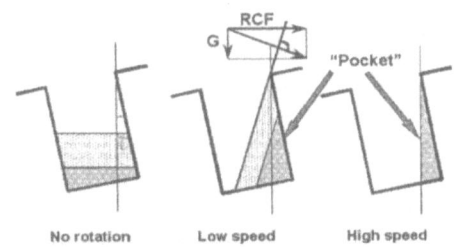

Figure 2. Formation of the pocket in the well of a tilted plate during centrifugation (direction: left to right). The solid support is collected in the pocket, while the liquid is expelled from the well. The liquid surface angle is perpendicular to the resulting force vector of the relative centrifugal force (RCF) and gravity (G).

2 hours. Gas HF is removed by nitrogen blowing and plates are transferred into dessicators for overnight evacuation. The plates are then transferred to the platform of Multiprobe 208 (Packard Canberra) and the product is extracted by repeated exposure to neat acetic acid. Acetic acid is a powerful extractant and allows simultaneous removal by lyophilization. Other solvents can be used for extraction, but the only alternative simultaneous way of removal is vacuum centrifugation (GeneVac). Every compound from the production is analysed by direct injection into mass spectrometer (one injection every 55 seconds) and 10% of the library is evaluated by HPLC with gradient elution.

However, this technique still does not allow processing an unlimited number of reaction vessels simultaneously - the number of processed vessels depends on the number of needles performing the suction. With 72 plates on the robotic surface, only 6,912 compounds can be synthesized in one batch.

We have found a simpler way for simultaneous processing of hundreds or thousands of reaction vessels. We call this new technique "tilted centrifugation". The principle of tilted centrifugation is shown in Figure 2. Resin suspended in the tilted flask placed at the perimeter of the centrifugal plate and spun, will not remain at the bottom of the flask. As the surface of liquid supernatant will move, the solid support layer will move as well. If the speed of rotation is increased, the centrifugal force created by rotation (which depends on the radius of rotation and the speed) combines with gravitation and the resulting force causes liquid surface to stabilize at the angle perpendicular to the resulting force vector. At the ratio of relative centrifugal force (RCF) to G of 3, the angle of the liquid surface will be about 61 degrees. If the speed is increased so that the ratio of these forces is more than 50, we will be getting close to the situation where RCF is infinity – therefore, the liquid (and resin layer) angle will be close to 90 degrees. The pocket created by the tilt should allow only solid phase to remain in the pocket and all of the liquid should be expelled. The pocket can be created in the vessel of basically any shape - flat bottom, U bottom, or V bottom vessel, as well as in the array of vessels, e.g. in the commonly used microtiterplates.

Situation of wells in microtiterplates placed on the perimeter of the centrifuge depends on the distance of the individual well from the axis of rotation. The volume of the "pocket" created by centrifugation in the wells closer to the axis is bigger than the volume of the "pocket" created in the wells more distant from the center of rotation. The volume of the pocket is not as important as the ratio of volumes of pockets in different wells of the microtiterplate. This ratio depends on the dimension of the centrifugal rotor, speed of the rotation, and the tilt of a plate. Wells placed on a rotor of very large diameter, or rotor spun very fast, will have an insignificant difference between forces exerted onto "inside" and "outside" wells. In the example given here, we were working with the tilt of 9 degrees, 350 rpm, and the diameter of centrifugal rotor of 48 cm. Under these conditions, the volume of the pocket in inner and outer wells differed by 8%, which we found to be an acceptable difference.

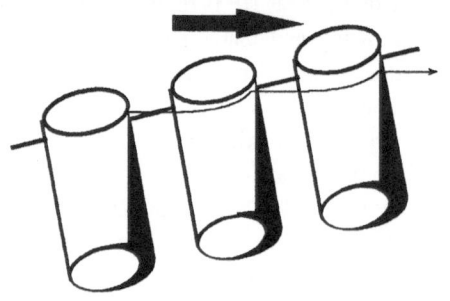

Figure 3. Trajectory of liquid removed by centrifugation from the well of the tilted microtiterplate. Liquid and/or resin expelled from the wells cannot contaminate neighboring wells, but is caught in the interwell space.

If the drilling of holes into inert material would create the array of wells, the liquid expelled from one well would inadvertently enter another well placed closer to the perimeter of the centrifuge. However, a 96 well shallow microtiterplate is actually composed of 96 small cylinders attached to a flat polypropylene sheet and connected by a thin "rib", creating thus an array of 96 round wells plus 117 interwell spaces. The liquid expelled by centrifugal force from the well comes into the interwell space, flies across this space and ends up on the outer wall of the adjacent well (see Figure 3). Then it flows along the well until it detaches and flies across another interwell space, eventually ending at the edge of the plate from where it flies onto the well of the centrifuge drum. We have tested the transfer of liquid and/or solid material from one well into another in several ways. We have loaded the wells with the amount of colorized solid support (resin) which exceeded the capacity of the pocket and observed the fate of the resin expelled from the well. Overflow of the resin ended in the interwell space and we have not observed any transfer of the resin beads into adjacent wells. In another experiment, we have analyzed products synthesized in all wells of the microtiterplate by HPLC and mass spectroscopy. We have not found any traces of contamination by liquid or solid transfer between wells in our model experiments. Figure 4 shows HPLC traces of substituted tetrahydroisoquinolinones synthesized in adjacent wells.

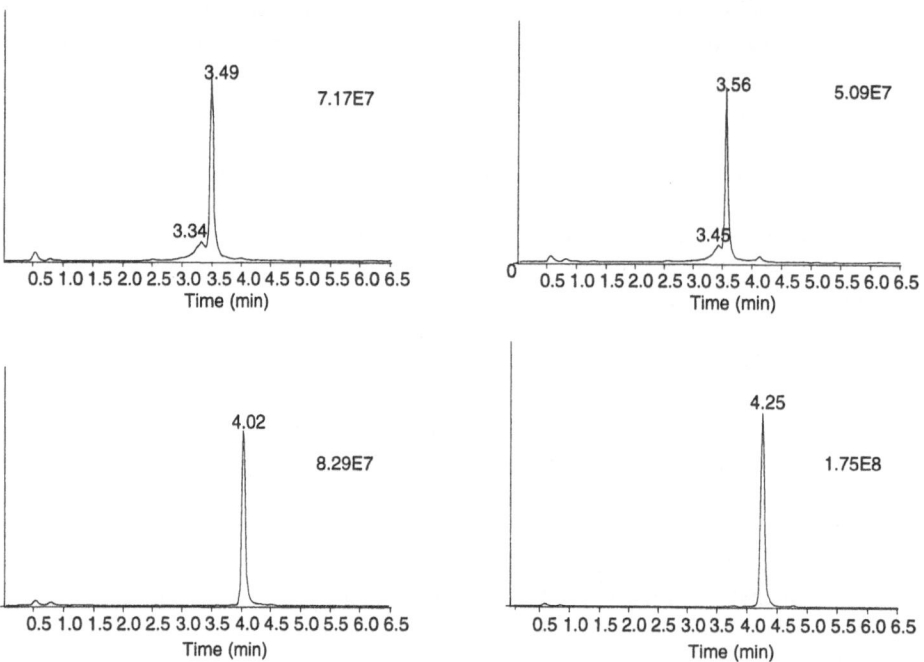

Figure 4. HPLC/MS (total ion current) traces of compounds synthesized in adjacent wells in the microtiterplate.

We have built the dedicated centrifuge with 8 positions for microtiter plates. This centrifuge is driven from the computer and all centrifugation parameters can be flexibly changed. A 96-channel distributor connected to 6 port selector valve performs the delivery of washing solvents and common reagents. The centrifuge was integrated with the Packard Multiprobe 104 liquid distribution system for the delivery of individual building blocks and reagents. Inclusion of the pipetting system allows us to perform the whole synthesis in a completely automatic regimen. Figure 5 shows the view of this instrument. This compact system can be easily enclosed in inert atmosphere.

The synthesis is performed in the following way. A microtiterplate with a slurry of solid support distributed into it is placed on the perimeter of a rotor with a permanent tilt of 9 degrees. The rotor is rotated at the speed required for complete removal of the liquid portion of the well content. After stopping the rotation, the microtiterplate is placed (rotor is turned) under the multichannel (96 channel) liquid delivery head. The solvent selector valve is turned into the appropriate position and the washing solvent is delivered by actuating the syringe pump. This operation is repeated until all plates are serviced. The rotor is spun at the speed at which the liquid phase is just reaching the edge of the well, wetting thus all solid support in the "pocket", and after reaching this speed, rotation is stopped. The cycle of slow rotation and stopping is repeated, thus, mixing the slurry of solid support in the liquid phase. After shaking for the appropriate time, the plates are spun at the high speed. The process of addition and removal of washing solvent is repeated as many times as many washes are required. The plates are then consecutively placed under the array of 96 openings in the centrifuge cover and appropriate building block solutions and coupling reagents are delivered by pipetting (Multiprobe 104) through the openings from the stock solutions placed on the centrifuge cover.

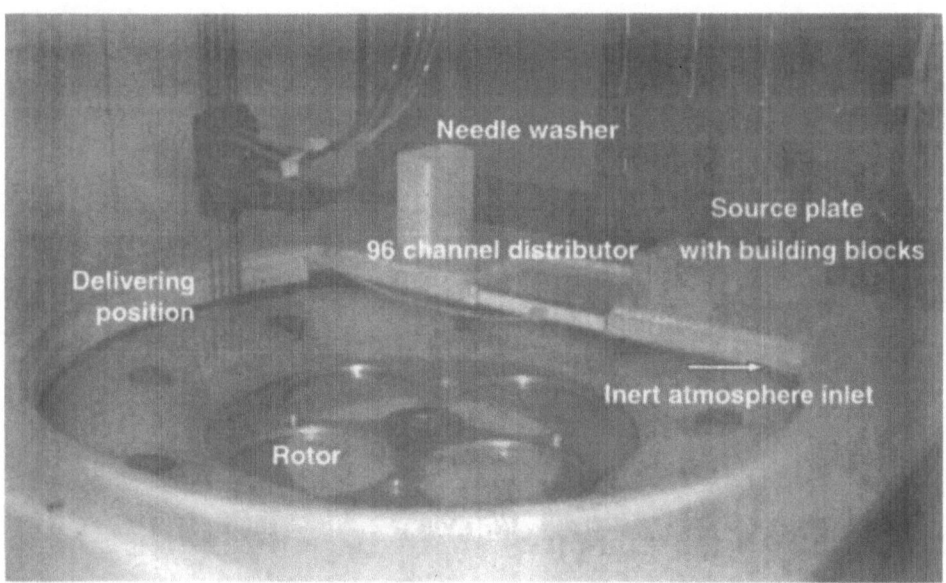

Figure 5. Detail of the centrifuge integrated with robotic liquid distributor Packard Canberra Multiprobe 104.

CONCLUSION

We have found surface suction and tilted centrifugation to be very effective and simple methods for liquid removal from a multiplicity of vessels. The surface suction principle is used for the preparation of large libraries in the single compound per well format. The polypropylene microtiterplates were found to be the ideal reaction vessels for tilted centrifugation based synthesis. The fact that tilted centrifugation is the only way for removal of liquids from unlimited number of reaction vessels simultaneously is suggesting its application in ultraminiaturized synthesizers.

REFERENCES

1. Leblova Z., Lebl, M. Compilation of papers in molecular diversity field. (1998). INTERNET World Wide Web address: http://www.5z.com.
2. Cargill, J.F. and Lebl, M. New methods in combinatorial chemistry: Robotics and parallel synthesis. *Curr.Opin.Chem.Biol.* 1:67 (1997).
3. Lebl, M. and Krchňák, V., Techniques for massively parallel synthesis of small organic molecules. In: Innovation and Perspectives in Solid Phase Synthesis & Combinatorial Libraries, edited by Epton, R. Birmingham: Mayflower Scientific Limited, 1998, in press.
4. Krchňák, V., Weichsel, A.S., Lebl, M. and Felder, S., Automated solid-phase organic synthesis in micro-plate wells. Synthesis of N-(alkoxy-acyl)amino alcohols. *Bioorg.Med.Chem.Lett.* 7:1013 (1997).
5. Houghten, R.A., General method for the rapid solid-phase synthesis of large numbers of peptides: Specificity of antigen-antibody interaction at the level of individual amino acids. *Proc.Natl.Acad.Sci.USA* 82:5131 (1985).

NEW TRENDS IN ELECTROPHILIC AMINATION: FROM SIMPLE ACHIRAL AND CHIRAL AMINES AND HYDRAZINES TO ADVANCED MATERIALS.

Pasquale Dembech,[1] Alfredo Ricci,[2] Giancarlo Seconi[1]

[1] Istituto dei Composti del Carbonio contenenti Eteroatomi e loro Applicazioni del CNR, Area di Ricerca di Bologna, Via P. Gobetti N°101, I-40129 Bologna, Italy.
[2] Dipartimento di Chimica Organica "A. Mangini", Università di Bologna, Via Risorgimento N°4, I-40136 Bologna, Italy.

INTRODUCTION

Among the modern methodologies for the construction of a new C-N bond, the electrophilic amination[1] is of high interest since it opens an unconventional entry to a wide range of amino derivatives a number of them not easily accessible through the conventional routes. According to this methodology, a reverse process, vs the delivery of an electron deficient nitrogen group "N^+" to a nucleophilic site, takes place with respect to the classical SN2-type reaction (Scheme 1).

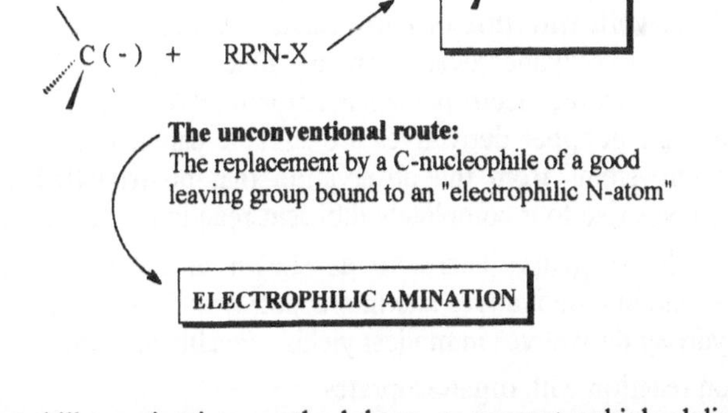

The conventional route:
The attack of a nucleophilic N-atom to an electrophilic C-atom bearing a leaving group' *via* ' SN2-type reaction

The unconventional route:
The replacement by a C-nucleophile of a good leaving group bound to an "electrophilic N-atom"

ELECTROPHILIC AMINATION

Scheme 1

In the electrophilic amination methodology a reagent which delivers the N-based functionality and acts as a synthetic equivalent of a "NH_2^+" synthon and an organometallic which provides a suitable source of a carbanion, are involved. Whereas until recently only organolithium and Grignard reagents have been used as carbanionic sources, since the original experiments by Sheverdina and Kocheshkov [2] the evolving utility of this reaction has been mainly related to the continuous invention of new reagents for the delivery of the amino group. The representative examples reported in Chart 1 outline that the presence of a good leaving group is a quite common feature of the reagents employed for this reaction even though also strained ring compounds such as oxaziridines can be employed for the electrophilic amination of both N- and C- nucleophiles. In several cases the advantages offered by these reagents in promoting the formation of new C-N bonds are balanced by drawbacks such as a non straightforward access, a low stability, or the cleavage the N-N bond when hydrazino derivatives are formed.

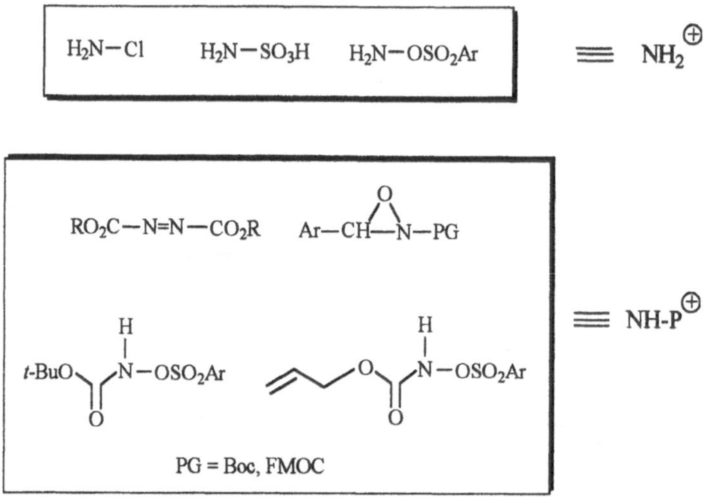

Chart 1

The formation of the C-N bond through electrophilic amination can also be approached by using simpler reagents for the delivery of the N functionality. Various O-substituted hydroxylamines have been tried for the conversion of organolithium derivatives into the corresponding primary amines.[1] The most consistently useful of these reagents, methoxyamine, suffers from disadvantages since a large excess of the organolithium compound is required. Moreover these reactions have only seldom been applied to the generation of synthetically appealing targets.

DISCUSSION

The combined use of simple starting materials, possibly commercially available, as the sources of the N-framework with that of fairly stable and easily accessible organometallics able to bear a wide range of ligands even contaning functional groups, has been taken into consideration for improving the electrophilic amination methodology.

The finding that organocopper derivatives are suitable candidates for being used in the electrophilic amination, stems from the observation that bis-trimethylsilyl hydroxylamine $Me_3SiNH-OSiMe_3$ gives rise to a completely different reaction outcome when reacted with different organometallic reagents. Therefore as shown in Scheme 2 while silylation is observed with organolithium derivatives and interaction with Grignard reagents does lead to the formation of hydroxy derivatives in modest yields, sizeable amounts of amino derivatives are generated[3] upon reaction with organocuprates.

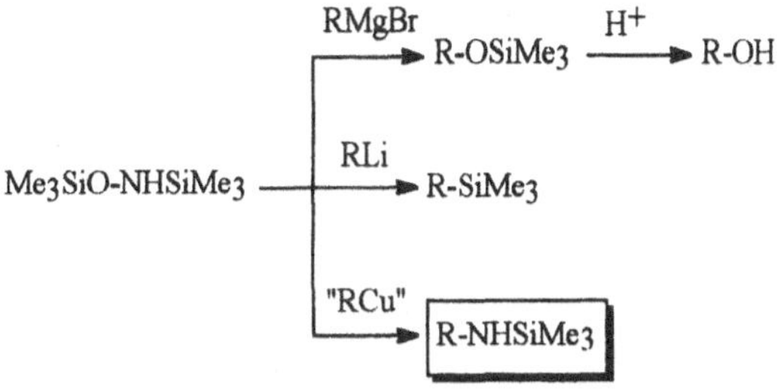

Scheme 2

Since bis-anionic homocyanocuprates $R_2Cu(CN)Li_2$ bear two anionic ligands capable of acting both as a base as well as the group to be delivered to a nitrogen moiety, and two metals (Li and Cu) having high affinity for nitrogen, they would meet particularly well the requirement for introducing an amino group to donor sites. Applied to a wide series of organocuprates Me_3SiNH-$OSiMe_3$ performs (Chart 2) as a suitable synthetic equivalent of the "NH_2^+" synthon. Worth noting this procedure can be applied to the synthesis of primary amino derivatives as well as to the preparation of the N-sila protected analogues by using hydrolytic or non-hydrolytic workup respectively. The latter procedure resulted mostly appropriate in those cases[4] where the free amine is unstable and for instance enabled the preparation and isolation of the previously unknown N-(trimethylsilyl)-2-aminothiophene.

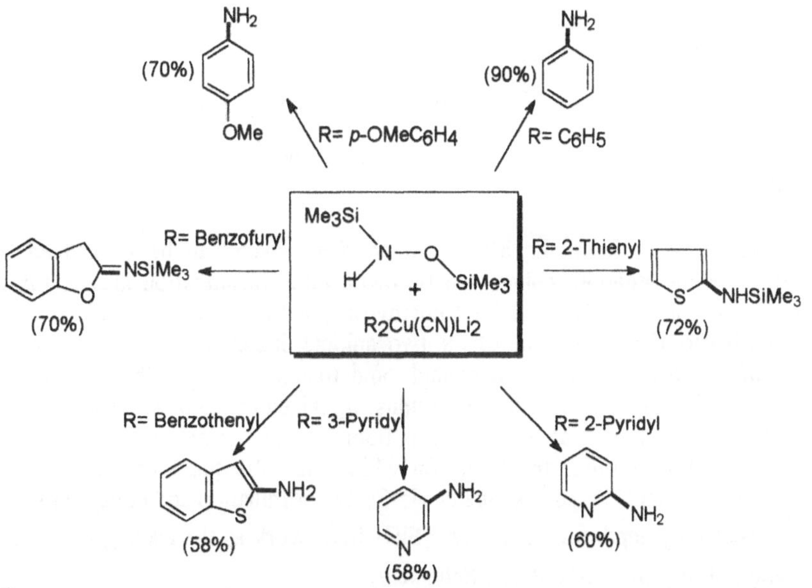

Chart 2

No detailed investigations have been performed on the mechanism of electrophilic amination, with the exception of the studies performed by Beak and coworkers.[5] They established that in the reaction between lithium alkoxyamides and an excess of alkyllithium the delivery of an amino group to a carbanionic center occurs through the displacement of a deprotonated alkoxyamine by an organolithium molecule. Even though the displacement process should formally involve the reaction of two anionic species, an interaction which should be repulsive, the organolithium are generally aggregated, and in the simplest case a dimer in which the entering carbon approaches from the back side of the nitrogen with polarization of the N-O bond, has been proposed (Scheme 3).

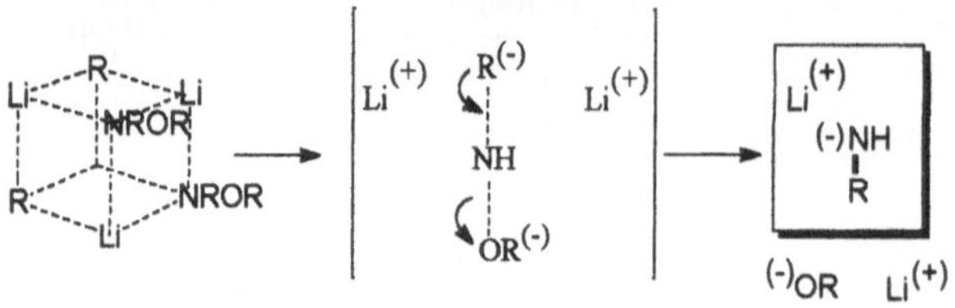

Scheme 3

On the other hand this increase of electrophilicity by metalation (!) has been generalized and explained by Boche[6] as due to a carbenoid, nitrenoid and oxenoid character of the metalated alcohols, hydroxylamines and hydroperoxides respectively, since calculations have revealed that the C-O, N-O, and O-O bonds in the lithiated derivatives are longer (Scheme 4) than the related bonds in the non lithiated counterparts.

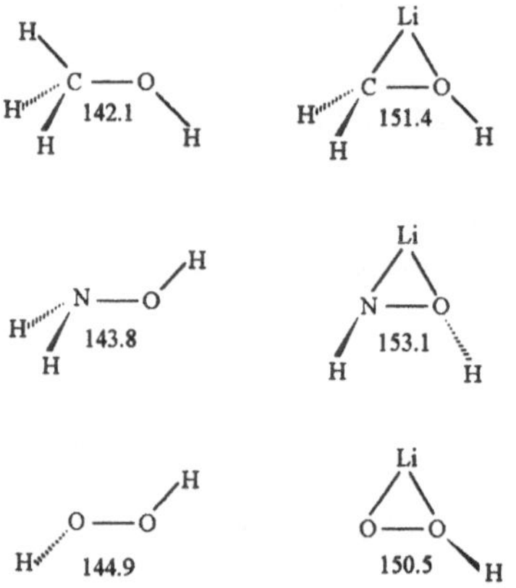

Scheme 4

Reaction pathways involving associated species, which can be envisioned in many processes involving organometallics, can be taken into consideration also in the reaction between organocyanocuprates and bis-trimethylsilyl hydroxylamine. Proton abstraction in the silylated hydroxylamine by one of the two anionic ligands in the cuprate gives rise to formation of a silylated lithium siloxyamide and to conversion of the bis-anionic into a mono-anionic cyanocuprate (Scheme 5), a highly organized mixed metal cluster consisting of a metal core to which each of the ligands is bound to two metals in a three-center two-electron bond. This identifies the frame on which the subsequent and final step occurs. Interception of the amide by such a cluster via Cu-N coordination and collapse of this newly formed aggregate by departure of the Me₃SiO- framework as the leaving group, leads[3] to the formation of the novel C-N bond (Scheme 5).

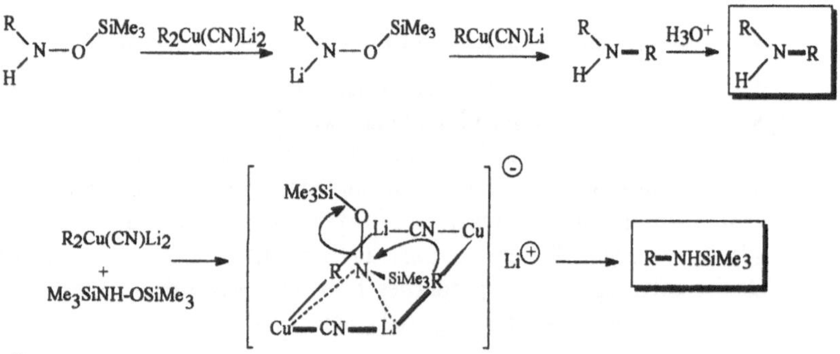

Scheme 5

100

Extension of the hydroxylamine-based protocol was exploited by using N-methyl-, N-isopropyl-, and N-*tert*-butyl-O-trimethylsilyl hydroxylamines.[7] These compounds resulted highly serviceable in the electrophilic amination towards organocuprates leading in satisfactory to good yields to a variety of N-alkyl-arylamines and -heteroarylamines. So far this appears to be the first straightforward and general method to synthetize secondary arylamines *via* N-arylation, which circumvents the restricted range of applicability of most of the previously reported multistep methodologies

The use of simple amines not bearing god leaving groups appears also viable following the observation that upon treatment of a bis-anionic organocyanocuprate with an equimolar amount of amine a clear solution is formed which is stable at room temperature for longtime; collapse into an heterogenoeus dark mixture occurs on the other hand upon saturation at low temperature with dioxygen, leading to the formation in good yields, of new amino derivatives.[8] The existence of an intermediate amidocuprate can be inferred based also on the literature data.[9] In these highly organized, fairly stable dimeric species which have a nearly planar structure with staggered organo and amido ligands occupying the corner positions, the starting amine becomes a "building block" of the whole molecular architecture. Since the N atom behaves as a non-transferrable cuprate ligand due to the strong Cu-N coordination, it conveys high stability to the cluster. The absence of leaving groups prevents a spontaneous decomposition, and an external reagent such as an oxidant is therefore needed to promote an intramolecular C/N coupling (Scheme 6).

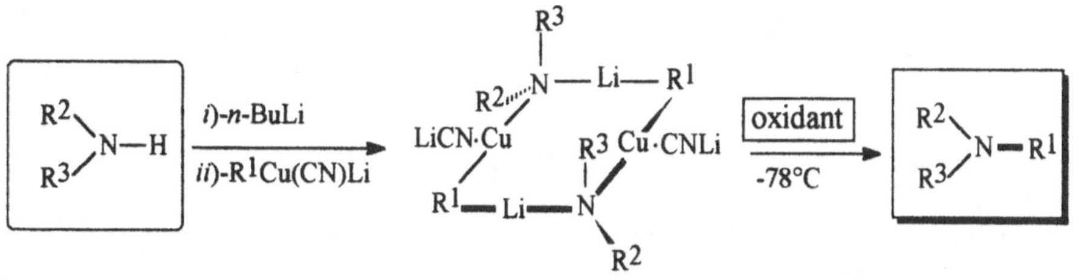

Scheme 6

The reaction in Scheme 6 competes with the well known inadvertent oxidation by dioxygen of organocuprates leading in the case of heterocuprates RR'CuLi to statistical ratios of the three possible products and therefore of low synthetic interest. Moreover in contrast with the frequently reported C-C bond forming reactions under oxidative coupling conditions which only recently have became synthetically appealing[10] leading to a new route to the biaryl nucleus, the cases of non-carbon ligand transfer with consequent formation of a new C-heteroatom bond, are rare. Products of unsymmetrical coupling in which a new C-N bond is formed, are obtained in good yields upon reaction of lithium amides with monoanionic cyanocuprates, after purging the amidocuprate solutions with molecular oxygen and employing equimolar amounts of the two reagents. A wide range of ligands can be employed in this reaction and representative examples are reported in Chart 3.

Besides N-vinylation and N-heteroarylation, N-arylation of amines is of high synthetic interest since arylamines are structural components in a variety of synthetic and natural, biologically active compounds. The oxidative coupling reaction of amidocuprates depicted in Scheme 6 and in Chart 4 and directed mostly towards the synthesis of tertiary amines, compares favorably with both known classical and more recent methodologies[11] in its simplicity, generality, mild reaction conditions and good yields.

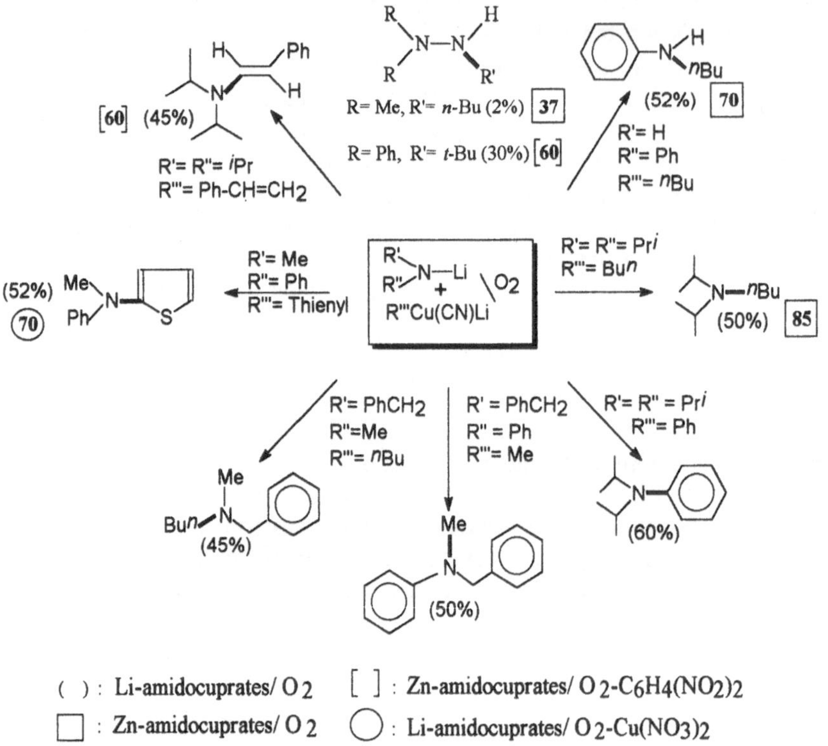

R= Me, R'= n-Bu (2%) **37**

R= Ph, R'= t-Bu (30%) **60**

[60] (45%)

R''= R''= iPr
R'''= Ph-CH=CH2

(52%) **70**

R'= H
R''= Ph
R'''= nBu

(52%) **70**

R'= Me
R''= Ph
R'''= Thienyl

R'= R''= Pr^i
R'''= Bu^n

(50%) **85**

R'= PhCH2
R''=Me
R'''= nBu

(45%)

R' = PhCH2
R'' = Ph
R''' = Me

(50%)

R'= R'' = Pr^i
R'''= Ph

(60%)

( ) : Li-amidocuprates/ O2   [ ] : Zn-amidocuprates/ O2-C6H4(NO2)2

□ : Zn-amidocuprates/ O2   ◯ : Li-amidocuprates/ O2-Cu(NO3)2

Chart 3

An insight into the mechanism of this reaction suggests the intervention of free radical intermediates and the observation of aminoxyl radicals allows to envision a mechanism in which aminyl radicals, not detectable under the reaction conditions, play the role of electrophiles (Scheme 7).

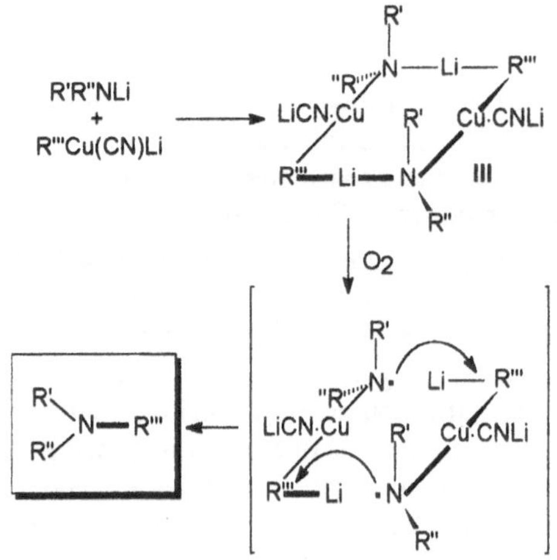

R'R''NLi
+
R'''Cu(CN)Li

III

O2

Scheme 7

Further improvement and optimization of the yields can be achevied,[12] as from the representative examples in Chart 4, through a subtle interplay between a higher stability of the cluster (Zn-amidocyanocuprate vs. Li-amidocyanocuprate) and higher efficiency of the oxidizing system exploited by adding substoichiometric amounts of co-oxidants to dioxygen.[13]

As far as the asymmetric version of this reaction a thorough study mainly concerning the diastereoselective electrophilic amination of β-hydroxy esters with, among others, *t*-butylazodicarboxylate (DBAD) and *t*-butyl-N-tosyloxycarbamate, has been carried in the last few years and recently reviewed.[14] Enantioselective electrophilic amination *via* hydroxylamines, previously unreported, has also been attempted following a procedure (Scheme 8) based on interaction between catecolboronate esters formed in the asymmetric hydroboration of vinylarenes and N-methyl-O-trimethylsilylhydroxylamine. The chiral amines were obtained[15] with high ee values (ca 90%) but in low yields due to the competing formation of alcohols indicating that the reagent can function[16] both as an O-nucleophile and as an N-nucleophile.

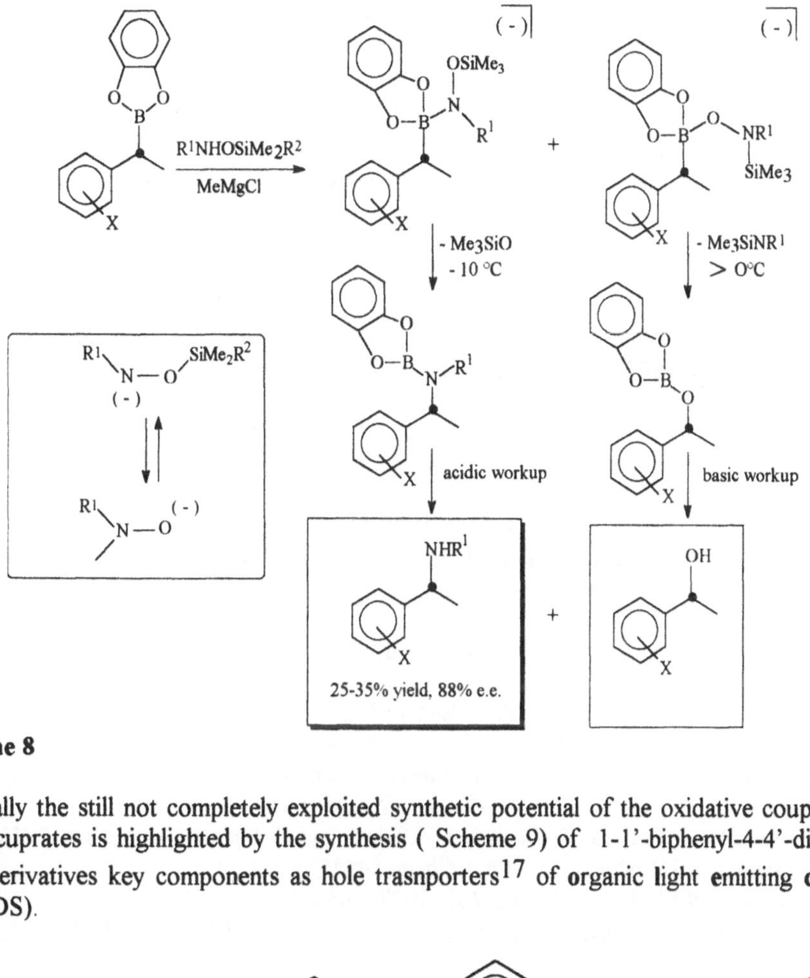

Scheme 8

Finally the still not completely exploited synthetic potential of the oxidative coupling of amidocuprates is highlighted by the synthesis (Scheme 9) of 1-1'-biphenyl-4-4'-diamine-type derivatives key components as hole trasnporters[17] of organic light emitting devices (OLEDS).

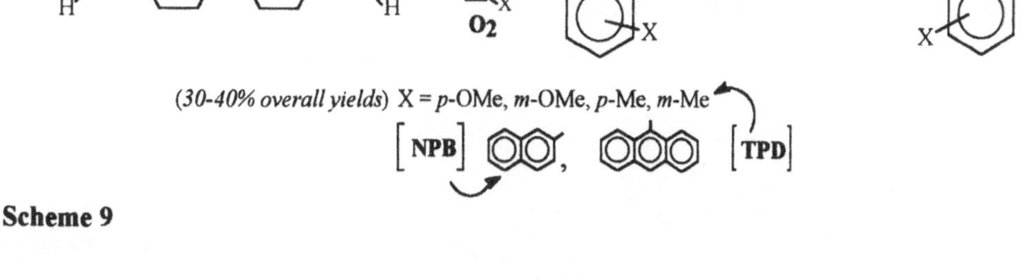

(*30-40% overall yields*) X = *p*-OMe, *m*-OMe, *p*-Me, *m*-Me

Scheme 9

This reaction[18] in which two new C-N bonds are formed (Scheme 9) in one step together with the mono N-arylated product which can recycled, offers great advantages in terms of yields and of operational simplicity with respect the Ullmann-based methodologies reported in the literature for these compounds.

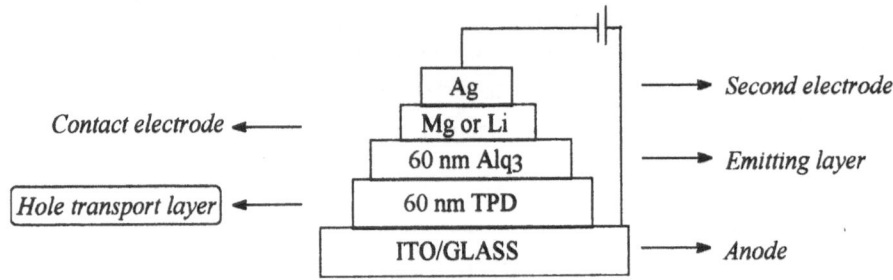

A waste array of polyamino derivatives used for building bright luminescent devices, has already been prepared according to the oxidative coupling protocol which is now focused on the synthesis of new "*starbust molecules*" of high interest as advanced materials.

REFERENCES

1)- E.F.V. Scriven, and K. Turnbull *Chem Rev.* 88: 298 (1988); M. Ay, and E. Erdik, *Chem.Rev.* 89: 1947 (1989); Y. Leblanc, and H. Mitchell *J. Org. Chem.* 59: 682 (1994).

2)- Z. Kocheskov, and N. Sheverdina, *J. Gen. Chem. USSR* 8: 1825 (1938).

3)- A. Casarini, P. Dembech, D. Lazzari, E. Marini, G. Reginato, A. Ricci, and G. Seconi, *J. Org. Chem.* 58:5620 (1993).

4)- D. Binder, G. Habison, and C.R. Noe, *Synthesis* 255 (1997).

5)- P. Beak, A. Basha, B. Kokko, and D.K. Loo, *J. Am. Chem. Soc.* 108: 6016 (1986).

6)- G.Boche, F, Bosold, and J.C.W. Lohrenz, *Angew. Chem. Int.Ed. Engl.* 33:1161 (1994).

7)- A. Ricci et al. in press.

8)- A. Alberti, F. Canè, P. Dembech, D. Lazzari, A. Ricci, and G. Seconi, *J. Org. Chem.* 61:1677 (1996)

9)- B.E. Rossiter, and M. Eguchi, *Tetrahedron Lett.* 31: 965 (1986).

10)- B.H. Lipshutz, K. Siegmann, and E. Garcia *J. Am. Chem. Soc.* 113: 8161 (1991).

11)- Rhee, H., and M.F. Semmelhack, *Tetrahedron Lett.* 34: 1395 (1993); T. Hattori, N. Hayashizaka, S. Myano, and J. Sakamoto, *Synthesis* 199 (1994).

12)- D. Brancaleoni, F. Canè, P. Dembech, A. Ricci, and G. Seconi, *Synthesis* 545 (1997).

13)- F. Asano, and J. Nishiguchi, *J. Org. Chem.* 54: 1531 (1989).

14)- C. Greck, and J.P. Genet, *Synlett* 741 (1997) and references therein.

15)- A.J. Blacker, J.M. Brown, D. Lazzari, F.I. Knight, and A. Ricci, *Tetrahedron* 53: 11411 (1997)

16)- P. Nowakowsky, and R. West, *J. Am. Chem. Soc.* 98: 5616 (1976); R. West, P. Boudjouk, and A. Matuszko, *J. Am. Chem. Soc.* 91: 5184 (1969).

17)- J. Shi, and C.W. Tang, *Appl. Phys.* Lett. 70: 1665 (1997).

18)- A. Ricci et al. unpublished results.

THE ATROPISOMER-SELECTIVE SYNTHESIS OF BIOLOGICALLY ACTIVE AND SYNTHETICALLY USEFUL CHIRAL BIARYLS

Gerhard Bringmann* and Stefan Tasler

Institut für Organische Chemie
der Universität Würzburg
Am Hubland
D-97074 Würzburg, Germany

INTRODUCTION

Axially chiral biaryls constitute a rapidly growing class of structurally, biosynthetically, and pharmacologically intriguing natural products, among them dimeric sesquiterpenes like mastigophorene A (**1**),[1] but also mixed, constitutionally unsymmetric biaryls like the naphthylisoquinoline alkaloids,[2] *e.g.* ancistrocladine (**2**)[3] and dioncophylline A (**3**),[4] some of them disposing of axial *and* centrochirality (Figure 1). As broad as their structural variety is the diversity of their sometimes most promising bioactivities, like nerve growth stimulating,[1] antimalarial,[5] or molluscicidal[6] properties. Furthermore, axially chiral biaryl reagents or ligands are of increasing value in stereoselective synthesis.[7]

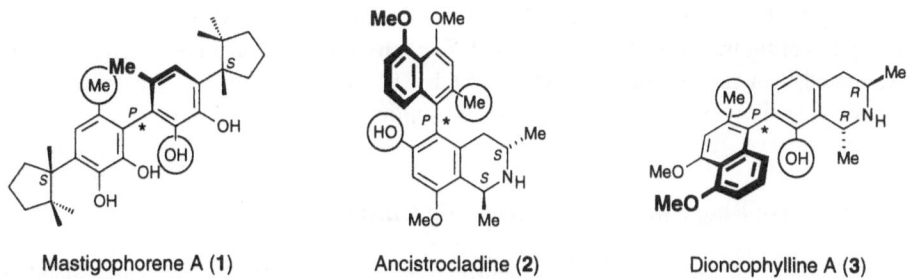

Mastigophorene A (1)   Ancistrocladine (2)   Dioncophylline A (3)

Figure 1.   Natural axially chiral biaryls: rewarding targets for stereoselective synthesis.

Despite the importance of axial chirality for biological activity, some natural biaryls are still published without considering the axis as a potential element of chirality.[8]

Moreover, in a contrast to the high standard of methods for the directed construction of stereogenic centers, good methods[9] for the regio- and stereoselective biaryl cross coupling are rare,[10,11] most of them suffering from low chemical or optical yields, especially for sterically strongly hindered biaryl products. And often only one particular atropisomer can be prepared and there is no possibility for a recycling of the undesired isomer. Here we present an efficient novel method for the regio- and stereoselective biaryl coupling that does fulfill these demands and is - in contrast to most of the other methods - applicable to the synthesis of a broad spectrum of natural and unnatural target molecules.

THE 'LACTONE METHODOLOGY'

The Basic Strategy

The concept is focused on a specific substitution pattern that is present in many natural biaryl systems, *viz.* with a C_1 unit and an oxygen function in opposite *ortho*-positions next to the axis (Figure 1).[12] According to our method the biaryl coupling is achieved intramolecularly, after prefixation of the aromatic moieties **4** and **5** *via* an ester bridge as in **6**, using exactly these C_1- and O-functionalities as the bridgeheads (Figure 2). This leads to favorable 6-membered ring lactones **7** in very high yields, even for sterically most demanding substituents next to the axis.[12]

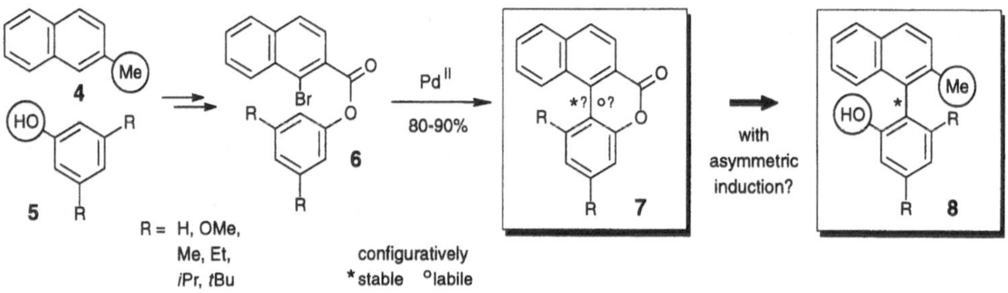

Figure 2.   The lactone coupling method: basic strategy.

With the biaryl axis thus established the structures (flat or helically distorted?) and dynamics (configuratively stable or unstable?) of the resulting lactones **7** remained to be investigated. But the key question was: would it be possible to cleave the bridge stereo-selectively, leading to only one atropisomer or - optionally - to the other one?

Atroposelective Ring Opening Reactions with Chiral Nucleophiles

Treatment of biaryl lactones **7** with chiral *O*-nucleophiles, *e.g.* with metallated mentholate (**9a**) or 9-phenylmentholate (**9b**),[11,13,14] indeed leads to *M*-**10** in up to 99:1 dr (Figure 3). The *P*-configured enantiomer of *M*-**10** can be attained by the use of the enantiomeric reagent, *ent*-**9a** or **b**. In each case the undesired atropo-diastereomer can be recycled by cyclization back to lactone **7** and renewed ring cleavage - chiral economy with respect to atropisomerism.[13]

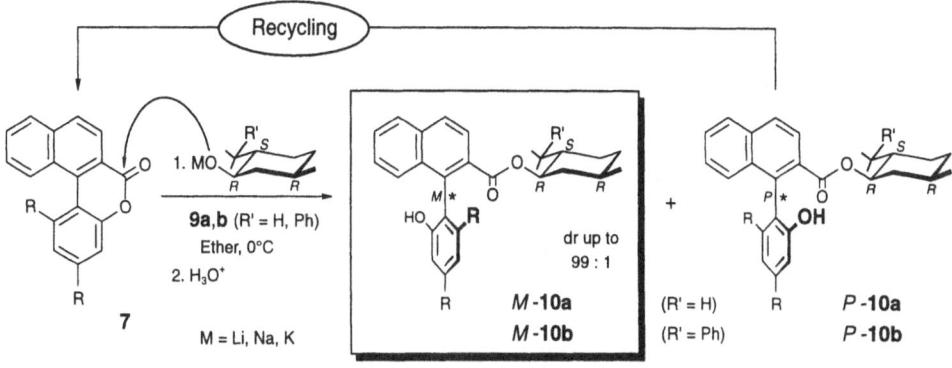

Figure 3. Atropo-diastereoselective ring opening with chiral *O*-nucleophiles.[*]

Likewise efficient are chiral *N*-nucleophiles like **11**,[11,13,15] leading to axially chiral amides **12**, again with high asymmetric inductions and with the option of preparing the other atropisomer by use of the enantiomeric reagent (Figure 4).

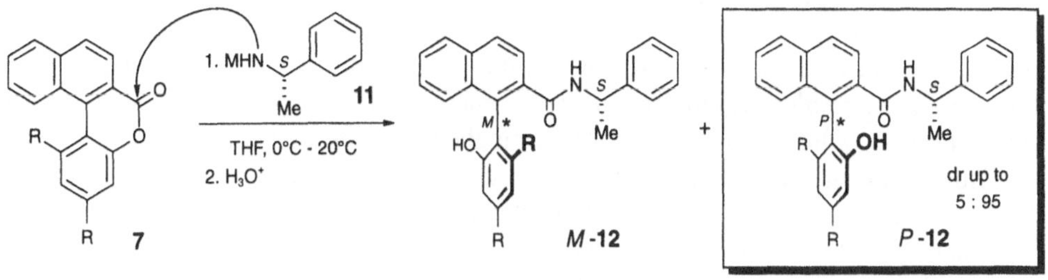

Figure 4. Atropo-diastereoselective ring opening with chiral *N*-nucleophiles.

Atropo-*enantioselective* ring opening reactions can be achieved, too, by using different chiral hydride transfer reagents, *e.g.* borane activated by oxazaborolidine (**14**) or '*S*-BINAL-H' (**15**),[16] giving excellent enantiomeric ratios (Figure 5). Again, by the choice of the enantiomeric reagent, either atropisomer can be prepared from the same precursor. In each of these cases, the product **13** is obtained enantiomerically pure through a single crystallization step.

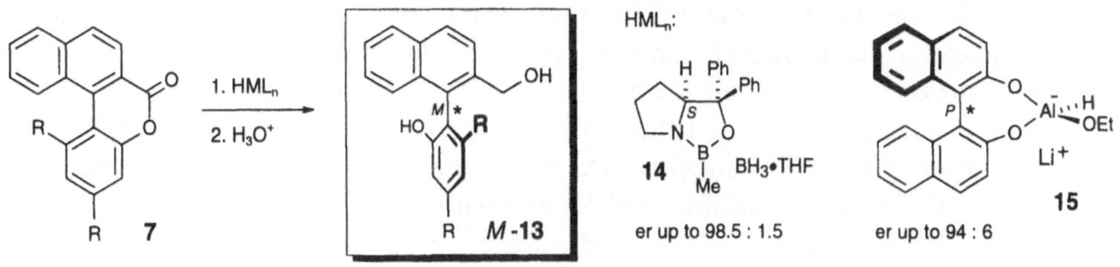

Figure 5. Atropo-enantioselective ring opening with chiral *H*-nucleophiles.

---

[*] Mind that the stereodescriptors *M* and *P* for stereogenic axes have arbitrarily been attributed for R = H, alkyl and are opposite for R = alkoxy!

For the assignment of the axial configuration of the ring cleavage products NMR, X-ray crystallography, and in particular the quantumchemical calculation of CD-spectra[17] as well as chemical methods were applied.

Mechanistic Course of the Ring Opening Reactions

Very attractive is the question about the origin of the high asymmetric inductions of these unprecedented and highly atroposelective cleavage reactions. Experimental and computational investigations[13,18] reveal the lactones **7** not to be flat, but helically distorted and therefore chiral, but configuratively unstable and thus a mixture of two rapidly interconverting helimeric enantiomers.

A nucleophile may attack the carbonyl function of **7** in four different ways (Figure 6): from either above or below and hence in an equatorial or an axial manner, for each of the two helimers *M*-**7** and *P*-**7**. An overall atroposelective ring cleavage reaction may result if, by the use of a *chiral* nucleophile, ideally only one of the two helimeric forms (*e.g. M*-**7**) is attacked and, as quantumchemical calculations suggest, only in an axial way, so that consequently one of the four possible stereoisomeric lactolate intermediates **16**, *e.g.* *R,M*-**16**, is formed predominantly. This should immediately burst open to give compounds *M*-**10**, **12** or **17** (dependent on the nucleophile used). By this way it should be possible to convert the entire racemic starting material into one single ring cleavage product in a *dynamic kinetic resolution*, provided that the isomerization barrier of **7** is low enough for a constant supply of the reactive helimer *M*-**7** from the remaining isomer *P*-**7**.

But even if the first step is fully stereoselective and only lactolate *R,M*-**16** is formed, the final stereochemical outcome is entirely open if this bridged biaryl easily helimerizes to its diastereomer *R,P*-**16** and then ring opens out of both of these *M*- and *P*-helimeric forms - a possible stereochemical leakage! The same stereochemical loss might result if the ring cleavage reaction is reversible because then the product (*M*-**10**, **12**, or **17**) might cyclize back to the lactolates *R,M*-**16** or *S,M*-**16** and thus, by helimerization to *R,P*-**16** and *S,P*-**16** respectively, would ultimately give rise to both atropisomeric forms of **10**, **12** and **17**.

There is no such risk for amides like **12** as resulting from *N*-nucleophiles,[19] nor should probably the corresponding esters undergo such a 'chemical atropisomerization', but it does take place for the analogous aldehydes **17**, as resulting from an attack of *H*-nucleophiles. Quantumchemical calculations[20] of the complete reaction course of the reduction reactions as well as experimental investigations[21] do show that there is a major stereochemical leakage at the level of the intermediate hydroxy aldehydes **17**, which are indeed configuratively unstable, since they isomerize *via* the different lactolate isomers **16**.

If, on the other hand, the experiments show that the overall reduction of the lactones **7** *via* the likewise configuratively unstable hydroxy aldehydes **17** does give excellent enantiomeric excesses, it can only mean
- that either the ring cleavage is sufficiently rapid (driven by larger residues R) and the intermediate aldehyde *M*-**17** is trapped immediately after its formation, thus by-passing the stereochemical leakage by reduction to *M*-**13**, which is configuratively stable;
- or that the system inevitably falls into this leakage and 'forgets' any stereochemical information possibly attained in the primary attack, in particular for sterically less hindered representatives; but then the high enantiomeric excesses actually observed must result from the *second* hydride attack, which would consequently be a dynamic kinetic resolution at the level of the racemic intermediate hydroxy aldehyde: ideally only one enantiomer of **17** would be reduced out of this population of rapidly interconverting isomeric aldehyde and lactol forms.

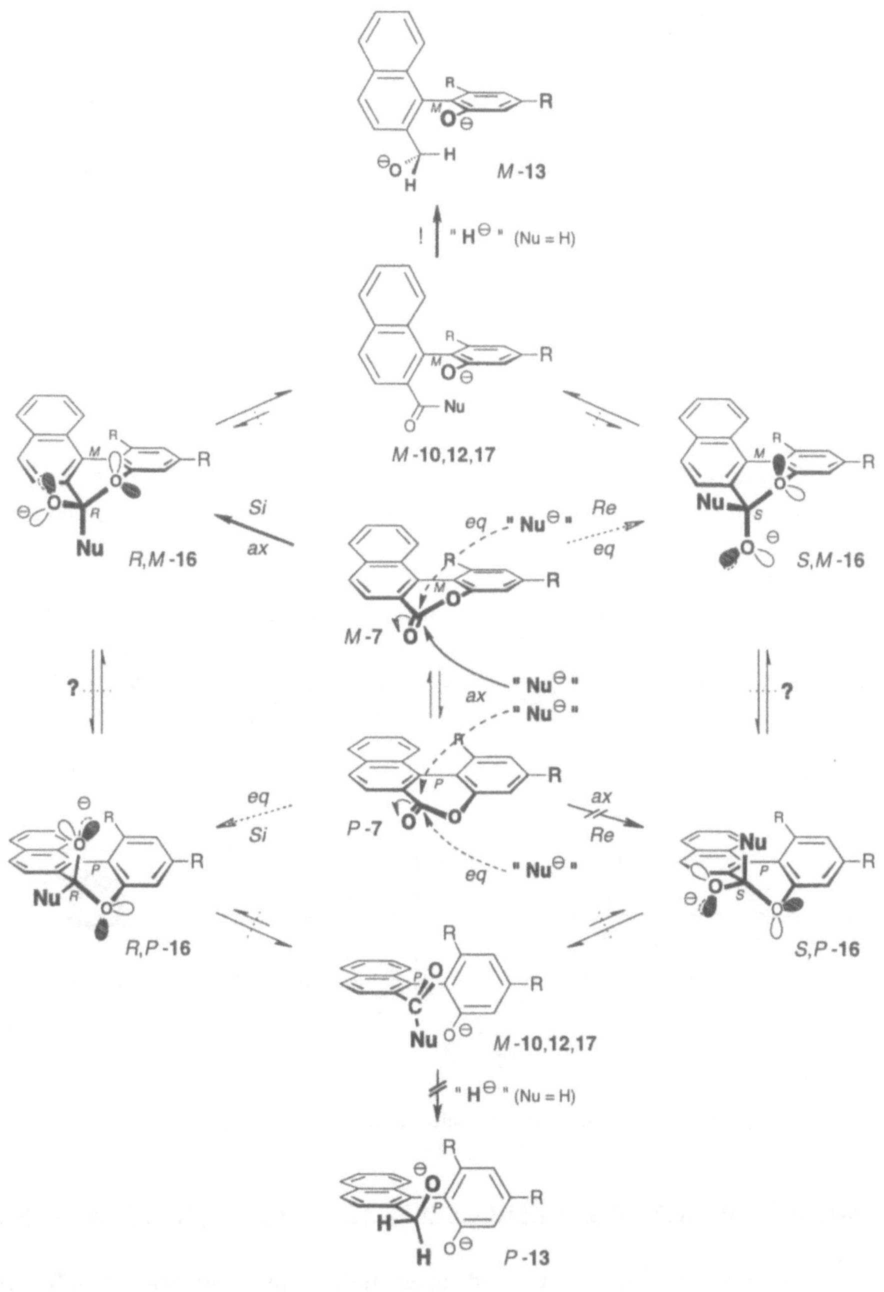

Figure 6. Mechanistic course of the atropisomer-selective ring cleavage of lactones **7** by chiral nucleophiles (descriptors of stereocenters for Nu = O-alkyl, H!).

OTHER SUBSTRATES FOR ATROPOSELECTIVE BIARYL SYNTHESES

This assumption, which is furthermore predicted by AM1 calculations,[20] suggests that the configuratively unstable hydroxy biaryl aldehydes **17** themselves should constitute promising novel substrates for atropo-enantioselective reductions. Indeed, at least for the stereochemically less hindered representatives, the atropo-enantioselective reduction yields the configuratively stable alcohols **13** by dynamic kinetic resolution (Figure 7) - another novel principle of stereoselective biaryl synthesis.[22-24]

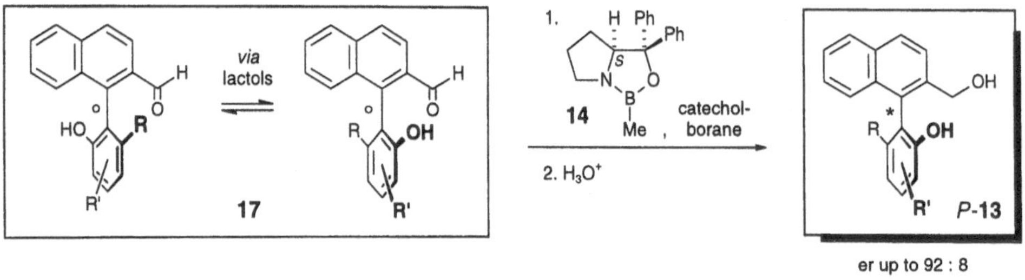

Figure 7.   Atropo-enantioselective reduction of biaryl hydroxy aldehydes **17**.

In contrast to the atropo-enantioselective ring cleavage of biaryl lactones **7** by dynamic kinetic resolution, a normal, non-dynamic kinetic resolution should be expected for the configuratively stable 7-membered lactones **18**. Indeed, these lactones can be reduced with very high relative rate constants, leading to enantiomerically pure C_2-symmetric diols of type **19** (Figure 8). The residual *P*-configurated lactone *P*-**18** of extremely high enantiomeric purity can be reduced to the corresponding diol *P*-**19** just by LiAlH$_4$ - or it can be racemized thermally and can then again be submitted to the kinetic deracemization - likewise a very efficient procedure.[25]

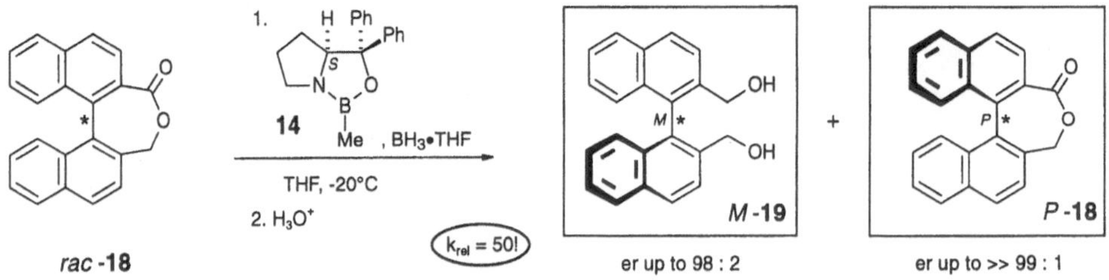

Figure 8.   Kinetic resolution of configuratively stable 7-membered lactones **18**.

PRINCIPAL OPTIONS FOR STEREOSELECTIVE CLEAVAGE REACTIONS

In principle, we have developed three fundamental strategies (**A-C**) for the stereoselective cleavage of biaryl lactones **7** (Figure 9):[15] Firstly, as shown above for the chiral *O*- and *N*-nucleophiles, it is possible to cleave the bridge with anionic (or metal-activated) nucleophiles (method **A**). A second option (method **B**) is the activation of the carbonyl group with a chiral Lewis acid, *e.g.* using a *S,S*-CHIRAPHOS-ruthenium fragment (X = S),[26] followed by the attack of a simple, achiral nucleophile. A third possibility (method **C**) is the activation of the phenolic leaving group by η^6-complexation, with the additional element of planar chirality allowing an internal asymmetric induction and thus again the use of achiral nucleophiles.

110

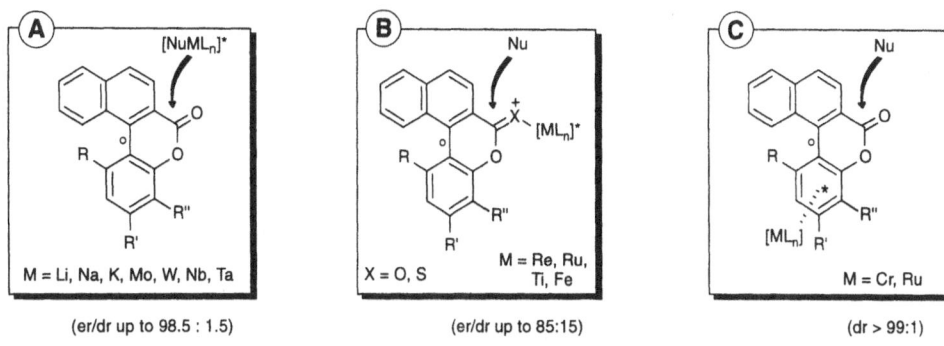

Figure 9. Three principal options for the metal-assisted atroposelective lactone cleavage.

One example of possibility **C** is the η^6-chromium complexed and therefore planar-chiral biaryl lactone **20**, which again consists of two helimeric forms rapidly interconverting at room temperature. It can be ring opened *e.g.* with NaBH$_4$ to give only one, now configuratively stable atropisomer **21** (Figure 10).[27] Extended quantumchemical calculations, here mainly by density functional (DF) methods, and crystallographic data gave detailed information on structure and dynamics of these interesting complexes.[28,29]

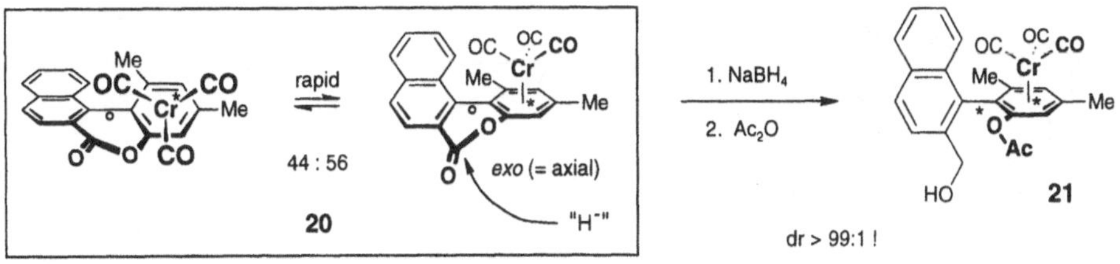

Figure 10. Activation and stereocontrol by η^6-Cr(CO)$_3$ complexation.

Unexpectedly different results were obtained with the cationic ruthenium complexes **22**: Firstly, the metal fragment is now on the distal naphthalene ring part, and secondly this compound displays a drastically differentiated reactivity towards different nucleophiles. In contrast to **20**, complex **22** does not give any lactone cleavage products **23** with hydride transfer reagents, whereas *O*-nucleophiles do lead to the desired ring cleavage reaction, again with an exclusive stereoselectivity (Figure 11).[30]

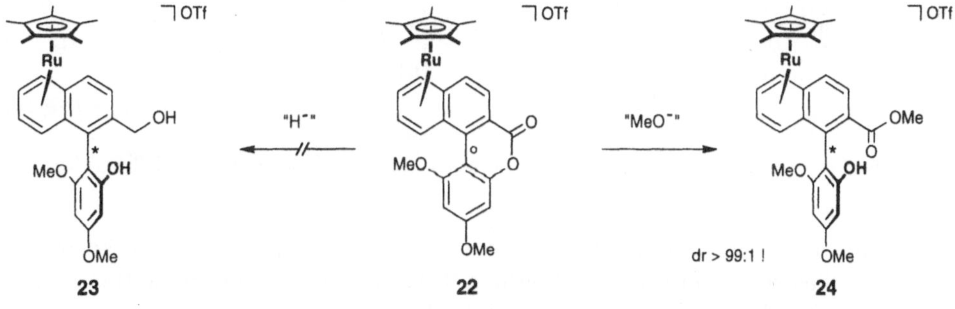

Figure 11. Highly atropisomer-selective cleavage of η^6-RuCp*-modified lactones.

Also in this case DF calculations gave valuable structural information, again confirmed by crystallographic data.[30] Using the Fukui function, which is defined within the density functional theory, an explanation of the divergent reactivity was elaborated: the calculations reveal the carbonyl function not to be one of the positions of maximum softness of the complex - and hence not a probable site for an attack by soft H-nucleophiles. On the other hand the carbonyl group, with the highest calculated positive charge, should readily be attacked by hard nucleophiles like methanolate, as also observed in the experiment.[30]

APPLICATION IN STEREOSELECTIVE BIARYL SYNTHESIS

Besides all these mechanistical and theoretical aspects, the deciding test for a new synthetic method is the applicability to the synthesis of concrete, functionalized natural products. As demonstrated for numerous cases, it is indeed a splendid procedure for the atroposelective preparation of a broad series of various natural biaryls, just one out of many examples being the atropo-divergent synthesis of dioncophylline A (3) and its likewise natural atropo-diastereomer, 7-epi-3 (Figure 12) - by ester-type prefixation of its molecular halves, 25 and 26, intramolecular coupling of 27 to the rapidly helimerizing lactone 28, and highly atropisomer-selective ring cleavage to give 3 or, optionally, 7-epi-3.[31]

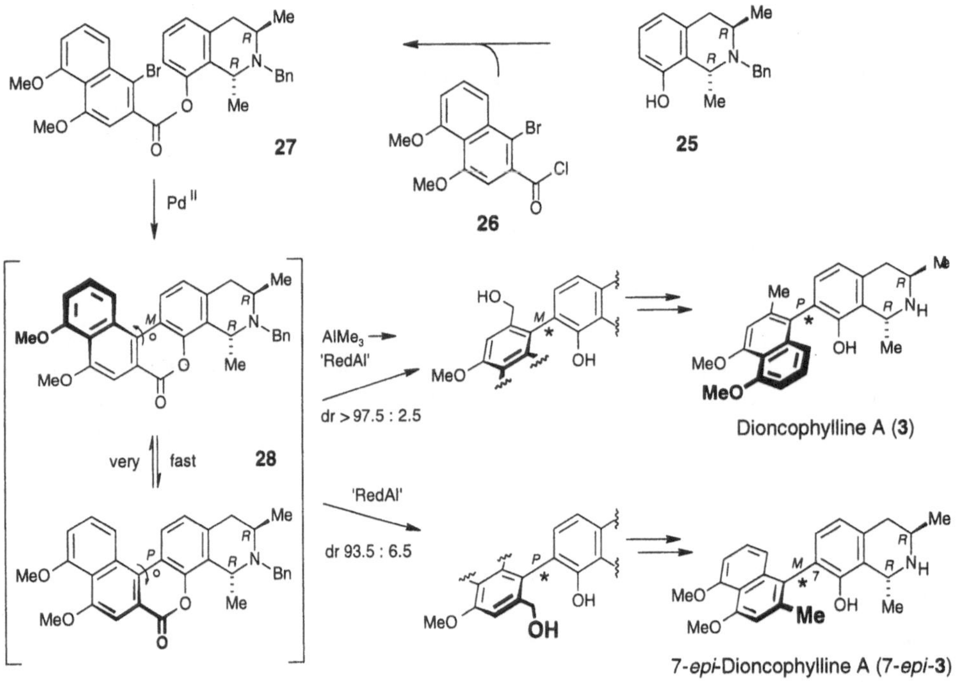

Figure 12. Application of the 'lactone concept' in natural product synthesis: dioncophylline A (3).

Figure 13 shows some of the numerous further natural products prepared by the lactone methodology - just a small selection. One of them is ancistrocladisine (29), a natural biaryl with two identical methoxy substituents next to the stereogenic axis - an extremely difficult problem for any other stereoselective coupling method, but easily solved by the lactone methodology.[32] Even biaryls without an oxygen function like dioncophylline

C (**30**)[33] or without a C_1 unit next to the biaryl axis like in korupensamine A (**31**),[34] which - at first sight - do not seem to fit into the lactone concept, have been synthesized by this efficient method. Besides naphthylisoquinolines also nitrogen-free biaryls have been synthesized, like bisorcinol **32**[35] and mastigophorene A (**1**, Figure 1).[36]

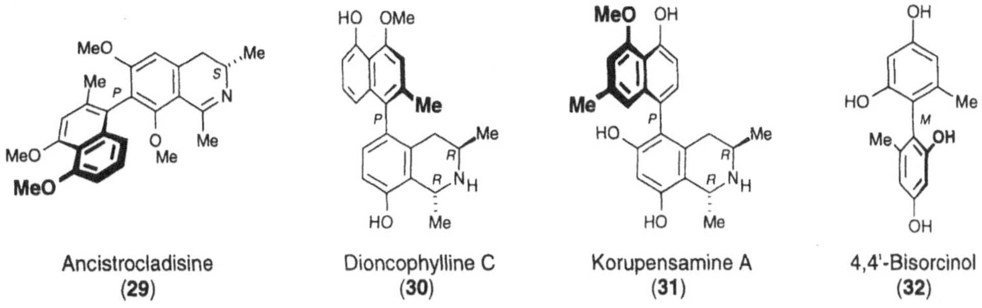

Figure 13. Natural biaryls prepared by the lactone methodology - a small selection.

A most rewarding advantage of our lactone methodology is that it gives rise to constitutionally symmetric and unsymmetric biaryls - and thus also to novel 'non-C_2 symmetric' reagents and ligands for asymmetric synthesis. As an example, the novel *O,N*-bidentate ligand **33**, which combines structural features of known catalysts with centro- or axial chirality, is easily prepared by atropo-enantioselective ring cleavage of **7** and subsequent conversion of the benzylic hydroxy group of **13** into the amino function (Figure 14). It catalyzes the diethylzinc addition to aldehydes in high er's (up to 99 : 1).[37]

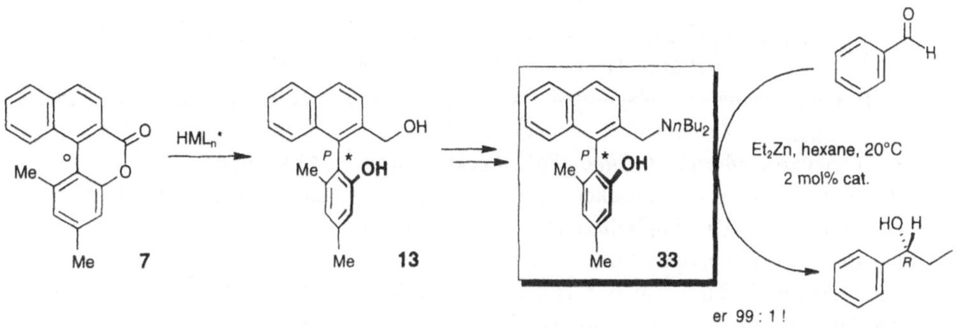

Figure 14. The bidentate ligand **33**: a selected example for novel 'non-C_2 symmetric' biaryl reagents and ligands available through the lactone methodology.

A broad variety of further new axially chiral, 'non-C_2 symmetric' biaryl reagents have meanwhile become available, with most different functionalities and catalytic activities.

CONCLUSION

In summary, our lactone methodology constitutes an efficient novel pathway to stereochemically homogeneous natural and unnatural biaryls: the required lactones are easily available by intramolecular coupling of the corresponding bromoesters and can be cleaved atropisomer-selectively with very high asymmetric inductions.

The method thus indeed fulfills the demands mentioned in the beginning, to a very high degree: It is a regioselective cross coupling that works perfectly even for very high steric hindrance. By subsequent modification of the *O*- and *C*-substituents, it gives rise to biaryls with a variety of substitution patterns. The procedure allows the optional preparation of both atropisomers from the same precursor, and a recycling of the undesired atropisomer is possible by re-cyclization to the lactone. Moreover, the method has proven its applicability to the synthesis of a broad series of concrete functionalized natural - and unnatural - target molecules - bioactive natural products and useful chiral reagents and ligands and certainly many further applications will succeed in the future.

ACKNOWLEDGMENTS

This work was supported by the Deutsche Forschungsgemeinschaft (SFB 347 'Selektive Reaktionen Metall-aktivierter Moleküle') and the Fonds der Chemischen Industrie. We are especially gratefull to all those who have contributed to this concept with skill and enthusiasm, their names can be seen from the literature citation. Furthermore we thank *R. Brückner* and *J. Hinrichs* for technical help in the preparation of the manuscript.

REFERENCES

1. Y. Fukuyama and Y. Asakawa, Novel neurotrophic isocuparane-type sesquiterpene dimers, mastigophorenes A, B, C and D, isolated from the liverwort *Mastigophora diclados*, *J. Chem. Soc., Perkin Trans. 1* 2737 (1991).

2. a) G. Bringmann and F. Pokorny, The naphthylisoquinoline alkaloids, in: *The Alkaloids*, G.A. Cordell, ed., Academic Press, New York 46:127 (1995);
    b) G. Bringmann, G. François, L. Aké Assi, and J. Schlauer, The alkaloids of *Triphyophyllum peltatum* (Dioncophyllaceae), *Chimia* 52:18 (1998).

3. T.R. Govindachari, K. Nagarajan, P.C. Parthasarathy, T.G. Rajagopalan, H.K. Desai, G. Kartha, S.L. Chen, and K. Nakanishi, Absolute stereochemistry of ancistrocladine and ancistrocladinine, *J. Chem. Soc., Perkin Trans. 1* 1413 (1974).

4. G. Bringmann, M. Rübenacker, J.R. Jansen, D. Scheutzow, and L. Aké Assi, On the structure of the Dioncophyllaceae alkaloids dioncophylline A ('triphyophylline') and '*O*-methyl-triphyphophylline', *Tetrahedron Lett.* 31:639 (1990).

5. G. François, G. Timperman, J. Holenz, L. Aké Assi, T. Geuder, L. Maes, J. Dubois, M. Hanocq, and G. Bringmann, Naphthylisoquinoline alkaloids exhibit strong growth-inhibiting activities against *Plasmodium falciparum* and *P. berghei in vitro* - structure-activity relationships of dincophylline C, *Ann. Trop. Med. Parasitol.* 90:115 (1996).

6. G. Bringmann, J. Holenz, L. Aké Assi, C. Zhao, and K. Hostettmann, Molluscicidal activity of naphthylisoquinoline alkaloids from *Triphyophyllum* and *Ancistrocladus* species, *Planta Med.* 62:556 (1996).

7. C. Rosini, L. Franzini, A. Raffaelli, and P. Salvadori, Synthesis and applications of binaphthylic C_2-symmetry derivatives as chiral auxiliaries in enantioselective reactions, *Synthesis* 503 (1992).

8. C. Ito, Y. Thoyama, M. Omura, I. Kajiura, and H. Furukawa, Alkaloidal constituents of *Murraya koenigii*. Isolation and structural elucidation of novel binary carbazolequinones and carbazole alkaloids, *Chem. Pharm. Bull.* 41:2096 (1993).

9. a) T.G. Gant and A.I. Meyers, The chemistry of 2-oxazolines (1985-present), *Tetrahedron* 50:2297 (1994); b) B.H. Lipshutz, F. Kayser, and Z.-P. Liu,

Asymmetrische Synthese von Biarylen durch intramolekulare oxidative Kupplung von Cyanocuprat-Zwischenstufen, *Angew. Chem.* 106:1962 (1994); *Angew. Chem. Int. Ed. Engl.* 33:1842 (1994); c) T. Watanabe and M. Uemura, Stereoselective synthesis of *O,O*-dimethylkorupensamine A *via* palladium(0)-mediated cross-coupling of a planar chiral (arene)Cr(CO)$_3$ complex with naphthylboronic acid, *J. Chem. Soc, Chem. Commun.* 871 (1998); d) K.S. Feldman and R.S. Smith, Ellagitannin chemistry. First total synthesis of 2,3- and 4,6-coupled ellagitannin pendunculagin, *J. Org. Chem.* 61:2606 (1996); and references cited therein.

10. G. Bringmann, R. Walter, and R. Weirich, The directed synthesis of biaryl compounds: modern concepts and strategies, *Angew. Chem.* 102:1006 (1990); *Angew. Chem. Int. Ed. Engl.* 29:977 (1990).

11. G. Bringmann, R. Walter, and R. Weirich, Synthesis of axially chiral compounds, part B.2.: Biaryls, in: *Methods of Organic Chemistry (Houben Weyl) 4. Ed.*, G. Helmchen, R.W. Hoffmann, J. Mulzer, and E. Schaumann, eds., Thieme, Stuttgart E21a:567 (1995).

12. G. Bringmann, T. Hartung, L. Göbel, O. Schupp, C.L.J. Ewers, B. Schöner, R. Zagst, K. Peters, H.G. von Schnering, and C. Burschka, Synthesis and structure of benzonaphthopyranones, useful bridged model precursors for stereoselective biaryl synthesis, *Liebigs Ann. Chem.* 225 (1992).

13. G. Bringmann and O. Schupp, Stereocontrolled 'twisting' of biaryl systems - a new pathway to axial chirality, *S. Afr. J. Chem.* 47:83 (1994).

14. G. Bringmann and M. Breuning, unpublished results.

15. G. Bringmann, M. Breuning, S. Busemann, J. Hinrichs, T. Pabst, R. Stowasser, S. Tasler, A. Wuzik, W.A. Schenk, J. Kümmel, D. Seebach, and G. Jaeschke, Metal-assisted synthesis and application of axially chiral biaryl systems, in: *Selective Reactions of Metal-Activated Molecules*, H. Werner and P. Schreier, eds.,Vieweg, Braunschweig 141 (1998).

16. G. Bringmann and T. Hartung, Atropo-enantioselective biaryl synthesis by stereocontrolled cleavage of configuratively labile lactone-bridged precursors using chiral *H*-nucleophiles, *Tetrahedron* 49:7891 (1993).

17. G. Bringmann and S. Busemann, Quantumchemical calculation of CD spectra: the absolute configuration of biologically active natural products, in: *Natural Product Analysis*, P. Schreier, M. Herderich, H.U. Humpf, and W. Schwab, eds., Vieweg, Wiesbaden 195 (1998).

18. G. Bringmann, H. Busse, U. Dauer, S. Güssregen, and M. Stahl, Structure and enantiomerization of helically twisted lactone-bridged biaryls: a theoretical study, *Tetrahedron* 51:3149 (1995).

19. G. Bringmann, S. Güssregen, D. Vitt, and R. Stowasser, The atropisomer-selective ring cleavage of helically distorted, configuratively unstable biaryl lactones with a chiral metallated *N*-nucleophile - the complete PM3 mechanistic course and its video presentation, *J. Mol. Model.* 4:165 (1998).

20. G. Bringmann and D. Vitt, Stereoselective ring-opening reaction of axially prostereogenic biaryl lactones with chiral oxazaborolidines: an AM1 study of the complete mechanistic course, *J. Org. Chem.* 60:7674 (1995).

21. G. Bringmann and T. Hartung, Synthesis and enantiomerization of a nonracemic 2-hydroxy-2'-biphenylcarbaldehyde, a probable intermediate in the atropo-enantioselective ring opening of biaryl lactones, *Liebigs Ann. Chem.* 313 (1994).

22. G. Bringmann, M. Breuning, H. Endress, D. Vitt, K. Peters, and E.-M. Peters, Biaryl hydroxy aldehydes as intermediates in the metal-assisted atropo-enantioselective reduction of biaryl lactones: structures and aldehyde-lactol equilibria, *Tetrahedron*, in press (1998).

23. G. Bringmann, D. Vitt, J. Kraus, and M. Breuning, The *ortho*-hydroxy-*ortho'*-formyl biaryl / lactol equilibrium: quantumchemical studies on structure and dynamics, *Tetrahedron*, in press (1998).

24. G. Bringmann and M. Breuning, The atropo-enantioselective reduction of configuratively unstable biaryl hydroxy aldehydes - a novel approach to axially chiral biaryls, *Synlett* 634 (1998).

25. G. Bringmann and J. Hinrichs, Efficient kinetic resolution of a racemic 7-membered biaryl lactone: enantioselective synthesis of 2,2'-dihydroxymethyl-1,1'-binaphthyl, *Tetrahedron: Asymmetry* 8:4121 (1997).

26. W.A. Schenk, J. Kümmel, I. Reuther, N. Burzlaff, A. Wuzik, O. Schupp, and G. Bringmann, Atropenantio-selective ring opening of biaryl thionolactones using [CpRu((*S,S*)-CHIRAPHOS)]⁺ as a chiral auxiliary, *Eur. J. Inorg. Chem.*, submitted.

27. G. Bringmann, L. Göbel, K. Peters, E.-M. Peters, and H.G. von Schnering, Synthesis, structure, dynamics, and first atropisomer-selective cleavage of the chromium tricarbonyl complex of a lactone-bridged biaryl, *Inorg. Chim. Acta* 222:255 (1994).

28. G. Bringmann, R. Stowasser, and D. Vitt, Local and non-local DF calculation of the structure of the helically twisted 1,3-dimethyl-{(η⁶-chromiumtricarbonyl)-benzo}-[b]naphtho[1,2-d]pyran-6-one: a comparison, *J. Organomet. Chem.* 520:261 (1996).

29. G. Bringmann, R. Stowasser, and L. Göbel, DF-studies on a 'stop and go' rotor: steric and electronic factors determining the regio- and stereochemical position of a η⁶-Cr(CO)₃ metal fragment on a helically distorted biaryl ligand, *J. Organomet. Chem.* 544:7 (1997).

30. G. Bringmann, R. Stowasser, A. Wuzik, D. Stalke, M. Pfeiffer, and W.A. Schenk, Application of the Fukui function: investigation of the reactivity of a [Cp*Ru] activated biaryllactone complex, *Organometallics*, in preparation.

31. G. Bringmann and J.R. Jansen, Stereocontrolled ring opening of *axially prostereogenic* biaryl lactones with hydrogen nucleophiles: directed synthesis of a dioncophylline A precursor and (optionally) its atropdiastereomer, *Synthesis* 825 (1991).

32. G. Bringmann and H. Reuscher, Atropdiastereoselective ring opening of bridged, 'axial-prostereogenic' biaryls: directed synthesis of (+)-ancistrocladisine, *Angew. Chem.* 101:1725 (1989); *Angew. Chem. Int. Ed. Engl.* 28:1672 (1989).

33. G. Bringmann, J. Holenz, R. Weirich, M. Rübenacker, C. Funke, M.R. Boyd, R.J. Gulakowski, and G. François, First synthesis of the antimalarial naphthylisoquinoline alkaloid dioncophylline C, and its unnatural anti-HIV dimer, jozimine C, *Tetrahedron* 54:497 (1998).

34. G. Bringmann and M. Ochse, unpublished results.

35. G. Bringmann, R. Walter, and C.L.J. Ewers, Diastereoselective ring opening of achiral bridged biaryls using chiral *O*- and *N*-nucleophiles: first atropo-enantioselective synthesis of (-)-4,4'-bis(orcinol), *Synlett* 581 (1991).

36. a) G. Bringmann, T. Pabst, S. Busemann, K. Peters, and E.-M. Peters, Atropo-enantioselective synthesis of a simplified analog of mastigophorenes A and B, *Tetrahedron* 54:1425 (1998); b) G. Bringmann and T. Pabst, unpublished results.

37. G. Bringmann and M. Breuning, Enantioselective addition of diethylzinc to aldehydes using novel axially chiral 2-aminomethyl-1-(2'-hydroxyphenyl)naphthalene catalysts; *Tetrahedron: Asymmetry* 9:667 (1998).

Radical reactions controlled by Lewis acids

Philippe Renaud,* Laura Andrau, Michèle Gerster and Emmanuel Lacôte

Université de Fribourg, Institut de Chimie Organique
Pérolles, CH-1700 Fribourg, Switzerland

Abstract: Complexation with Lewis acid allows to control the diastereoselectivity of radical reactions. Examples illustrating different strategies are presented. For instance, diastereoselectivity control based on size effects and chelation are shown. Lewis acid controlled radical cascade reactions are reported. Finally, the Lewis acid acceleration of a radical rearrangement (Surzur-Tanner rearrangement) is reported. In this last case, the Lewis acid influence the reaction rate by stabilizing a polarized transition state.

Introduction. Beginning in the 80's, an emergence of new free radical synthetic methods has been observed. Parallel to this development, the stereochemistry of radical reactions has been investigated and results exceeding the initial expectations have been obtained (ref. 1). Interestingly, the rules developed to rationalize the stereochemical outcome of radical reactions are very similar to the ones formerly developed for ionic and concerted reactions. However, due to the neutral nature of radicals, the effect of solvent and other possible complexing agents, which are crucial for ionic and concerted reactions, were not investigated until very recently (ref. 2). It has been well established that radical reactions can be rationalized by applying the frontier molecular orbital theory. Depending on the electronic character of the radical, SOMO-LUMO or SOMO-HOMO interactions are predominating. Therefore, reactivity as well as regio- and stereoselectivity should be influenced by the presence of Lewis acids. Recently, this hypothesis was confirmed and exceptional effects have been obtained by using complexing agents, mainly Lewis acids, on the stereochemical outcome of radical reactions. We wish to present here results concerning the use of Lewis acids in radical reactions with particular emphasis on the control of the stereochemistry and the reactivity (ref. 3).

Diastereoselectivity control: size effects. A first approach to control the stereochemistry of a radical reaction with Lewis acids consists of changing the relative size of one substituent at a stereogenic center by complexation. For example, during the study of the allylation reactions of cyclic sulfoxides, we discovered that protic solvents, which form hydrogen bonds with the oxygen atom of the S-O bond, enhance the stereoselectivity (ref. 4). Use of oxophylic Lewis acids give even better results (Scheme 1), exceptionally high levels of stereocontrol are reached with bulky aluminum derivatives such as methylaluminum di(2,6-di-*tert*-butyl-4-methyl)phenoxide (MAD) and methylaluminum di(4-bromo-2,6-di-*tert*-butyl)phenoxide (MABR) (ref. 5). For instance, the allylation of **1** without Lewis acid gives **2** in a modest *trans/cis* 82:18 ratio. However, the presence of 1.1 equivalents MAD or MABR enhances the diastereoselectivity to a *trans/cis* ratio >98:2 (ref. 6). Interestingly, the use of a less than stoichiometric amount of MABR (10 mol%) induced a clear enhancement of the diastereoselectivity (*trans/cis* 90:10).

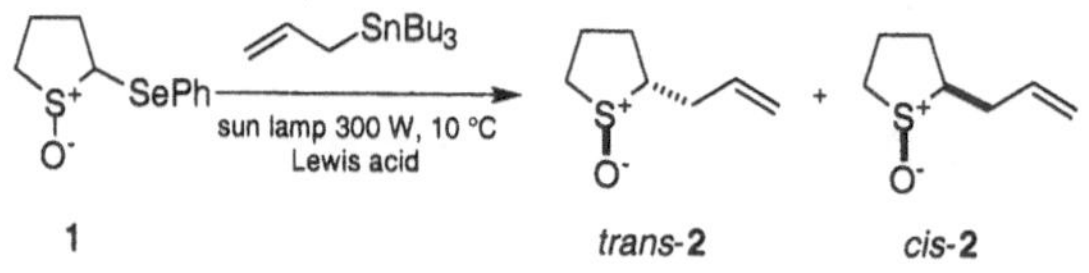

Lewis acid	yield	trans-2/cis-2
none	63%	82:18
MAD (1.1 equiv)	80%	>98:2
MABR (1.1 equiv)	57%	>98:2
MABR (0.1 equiv)	66%	90:10

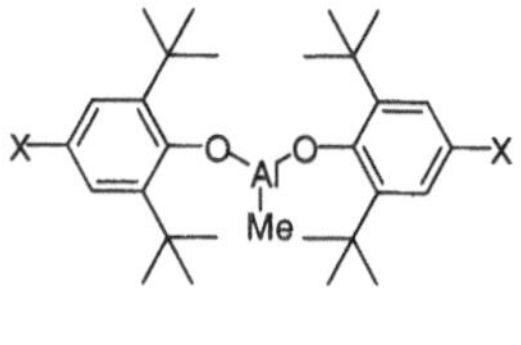

MAD (X = Me)
MABR (X = Br)

Scheme 1

The same approach was successfully applied to acyclic sulfinylated radicals (Scheme 2). In this case, the sense of the diastereoselectivity can be fully reversed by using bulky Lewis acids (ref. 7,8). A model based on minimization of allylic strain (**A** and **B**) was proposed to account for this stereochemical outcome.

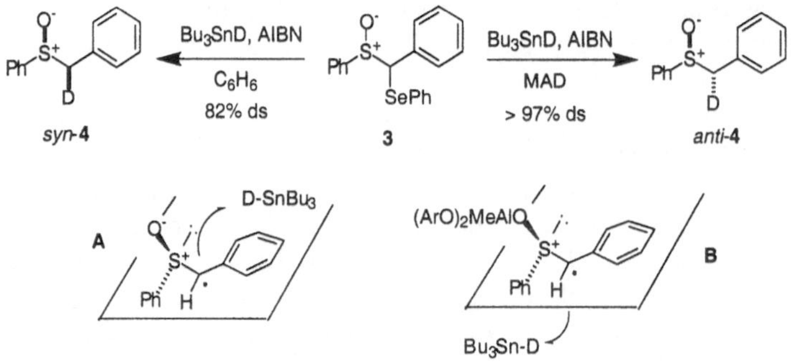

Scheme 2

This strategy is not limited to sulfoxides, similar results were obtained with cyclic and acyclic 2-hydroxyalkyl radicals (Scheme 3) (ref. 9,10). In this case, the use of methylaluminum diphenoxide leads to an intermediate aluminum alkoxide. The exceptional size of the aluminum based Lewis acids allows to reach extremely high ratio of diastereomers which cannot be obtained by using bulky protective groups. This point is illustrated in Scheme 3, the allylation of iodohydrin **5** furnished *trans*-**6** as a single diastereomer via an intermediate radical of type **C**.

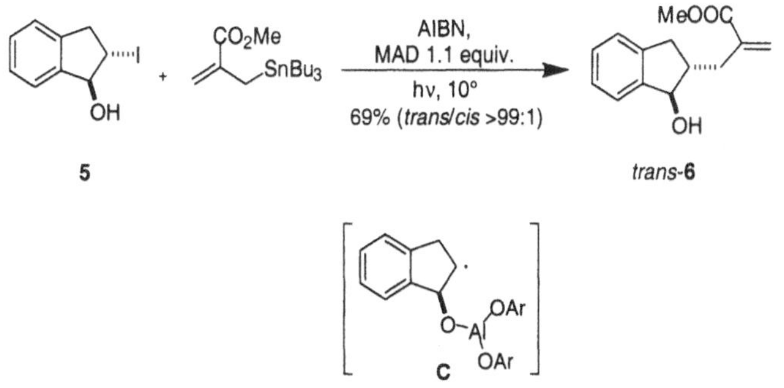

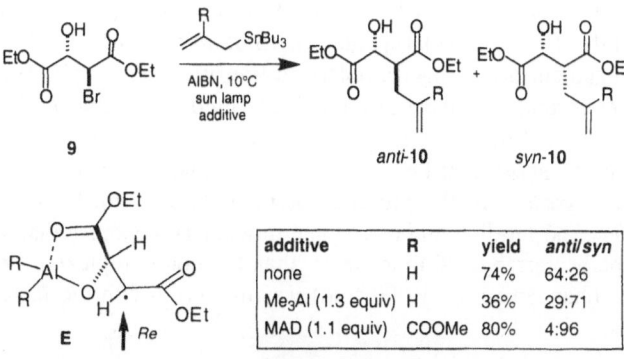

Scheme 3

Diastereoselectivity control: chelation. Another very effective way of controlling the conformation of a radical and as a consequence the diastereoselectivity of a radical reaction is based on chelation. The first example of this type with β-alkoxyester enolate radical was reported by Guindon (ref. 11). We have extended this approach to unprotected β-hydroxyesters using methylaluminum additives. Excellent stereocontrol is obtained for the allylation of simple β-hydroxyesters such as β-hydroxybutyrate **7** (Scheme 4) (ref. 12). These results are strongly related to the alkylation of the corresponding dianions developed by Frater and Seebach and gives preferentially the *anti* isomers via a cyclic intermediate of type **D** (ref. 13,14).

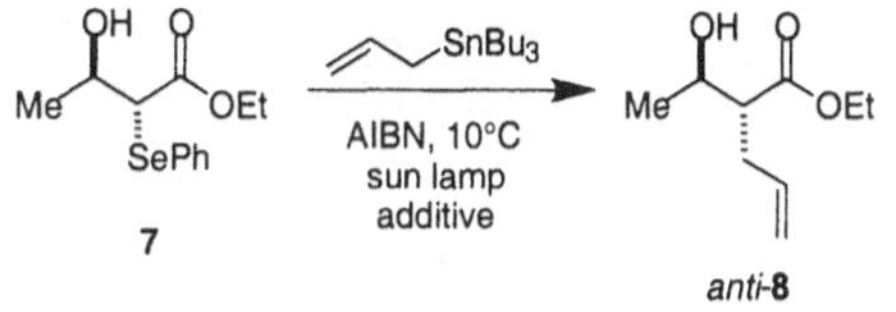

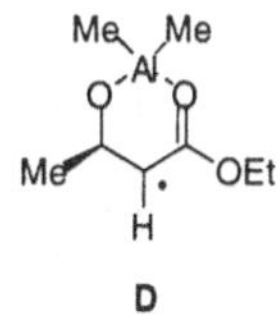

additive	yield	*anti/syn*
none	98%	63:27
Me₃Al	97%	95:5

Scheme 4

Interestingly, the allylation of the radical derived from malic acid gave in the presence of 1.1 equiv. MAD almost exclusively the *syn* isomer of **10** (Scheme 5). This has been explained by the preferential reaction of the cyclic 5-membered ring chelate **E** from the less hindered *Re* face. The conformation of **E** is controlled by minimization of allylic strain. It is interesting to note that the *syn* isomer cannot be obtained by the classical enolate alkylation (ref. 14).

additive	R	yield	*anti/syn*
none	H	74%	64:26
Me₃Al (1.3 equiv)	H	36%	29:71
MAD (1.1 equiv)	COOMe	80%	4:96

Scheme 5

119

A unique aspect of radical reactions is offered by the possibility of running cyclization reaction and other reaction cascades. The simple cyclization reaction of β-hydroxyesters depicted in Scheme 6 can be very efficiently controlled by Lewis acids. In the absence of Lewis acids, the cyclization of **11** gives **12** as a *cis/trans* 35:65 mixture. The addition of 1.1 equivalents of methlylaluminum diphenoxide provides the *cis* isomer with an excellent stereocontrol and a good yield. The same strategy can be applied to cascade reactions. For instance, treatment of the vinylcyclopropane **13** with tin hydride gives in the absence of Lewis acid the diquinane **14** as a complicated mixture of isomers. In the presence of methylaluminum derivatives, only two out of the 16 possible diastereomers are observed in satisfactory yields.

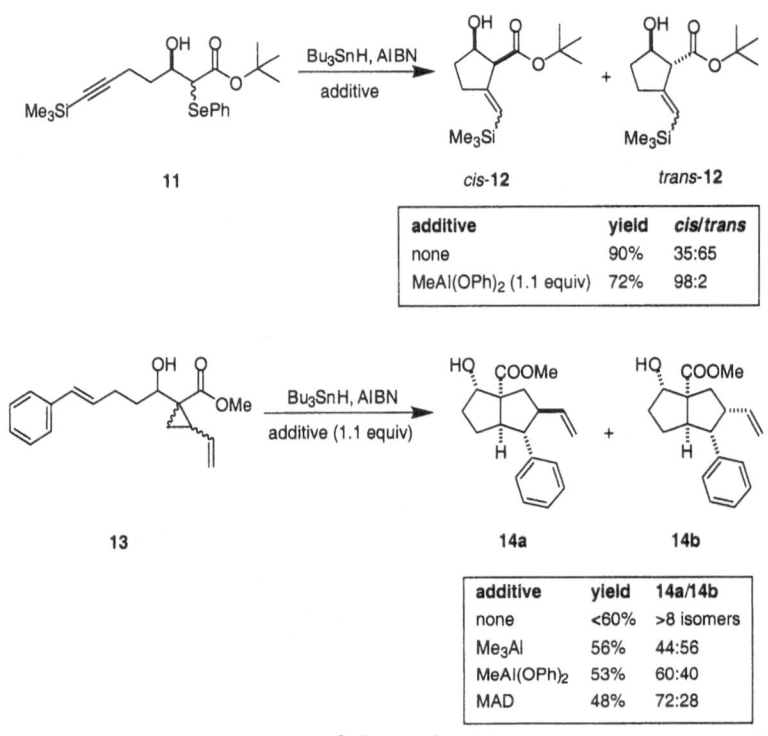

Scheme 6

additive	yield	cis/trans
none	90%	35:65
MeAl(OPh)$_2$ (1.1 equiv)	72%	98:2

additive	yield	14a/14b
none	<60%	>8 isomers
Me$_3$Al	56%	44:56
MeAl(OPh)$_2$	53%	60:40
MAD	48%	72:28

Reactivity control. As expected from the analysis of SOMO-HOMO and SOMO-LUMO interactions, Lewis acid can accelerate a radical addition by enhancing the electrophilic character of the radical or of the olefin. Several example of rate acceleration of a radical reaction belong to this category. A second strategy for a Lewis acid rate enhancement of a reaction is the stabilization of a polarized transition state by chelation with a Lewis acid. Very recently, we reported the first example of this approach (ref. 15). Indeed, the rate of the 1,2-acyloxy shift of β-(acyloxy)alkyl radicals (Surzur-Tanner rearrangement), a well known radical rearrangement occurring via polarized transition states (ref. 16), is enhanced in the presence of Lewis acid. A typical example is reported in Scheme 7. The salicylate **15** gives upon treatment with Bu$_3$SnH/AIBN the rearranged product **16** and the product of direct reduction **17**. In the absence of Lewis acid, **17** is the major product even under syringe pump addition of tin hydride. In the presence of methylaluminum bis(4-methylphenoxide) or scandium triflate/2,6-lutidine, the rearranged product **16** becomes major. This indicates that radicals of type **F** rearrange 20 times faster than the non-complexed species. An even more effective rate enhancement (> 1'000 fold) was observed with the lactate **18**. According to calculations (ref. 17), the Lewis acid stabilizes the transition state of this rearrangement. It also favors the three-membered ring transition state **G** over possible 5-membered ring transition states.

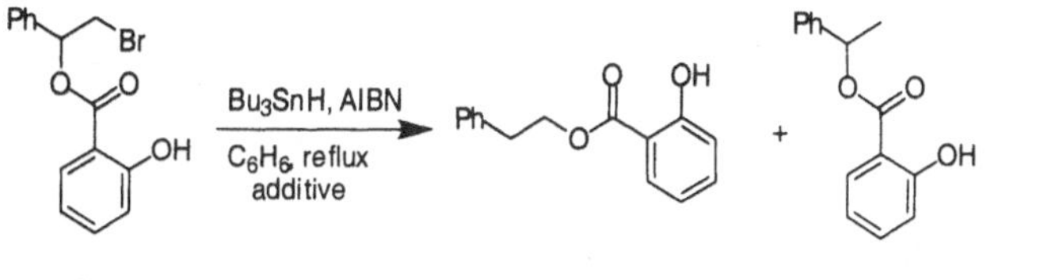

15

16 (rearrangement)   **17** (reduction)

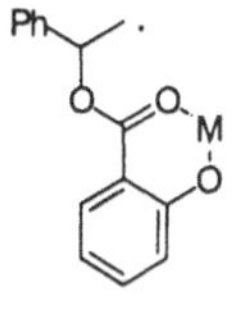

F (M = Al^{III} and Sc^{III})

additive	yield	16/17
none	78%	25:75
MeAl(Op-Tol)$_2$ (1.1 equiv)	73%	88:12
Sc(OTf)$_3$/2,6-lutiine (1.1 equiv)	76%	83:17

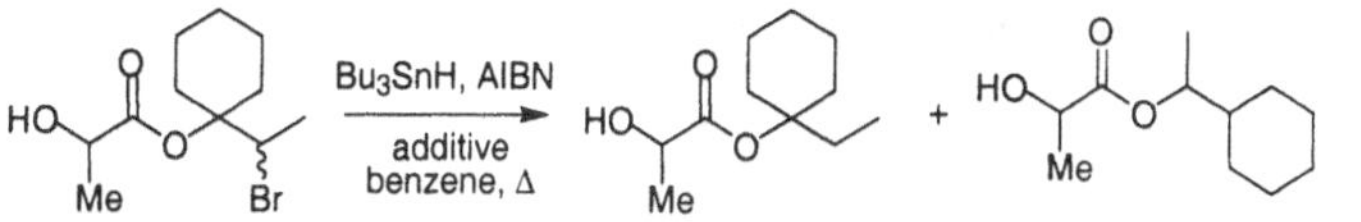

18

19 (reduction)   **20** (rearrangement)

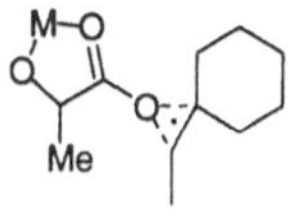

G

additive	yield	19/20
none	75%	<1:99
Sc(OTf)$_3$/2,6-lutiine (1.1 equiv)	54%	93:7

Scheme 7

Conclusions. The use of Lewis acids in radical reactions is opening new applications for this type of chemistry in synthesis. Excellent results with aluminum based Lewis acids have been obtained but as demonstrated by us and other, several other types of Lewis acids can also be used to control radical reactions. A particularly noteworthy point for future applications of radical reactions controlled by Lewis acids is their incorporation into reaction cascades. Indeed, formation of several bonds in a "one pot" procedure is typical of radical chemistry and highly attractive from a synthetic point of view. Finally, the possibility of rate acceleration of a radical reaction offers the possibility of catalysis and even enantioselective catalysis which is expected to become a major advance in this field.

Acknowledgment. The authors thank to the Swiss National Science Foundation who supported their work within the Project 20-45'755.95, 20-52'613.97 and the CHiral2 project 20-48'164.96. M. Gerster is very grateful to the Stipendienfonds der Chemischen Industrie Basel for financial support.

References

1)   Review: D. P. Curran, N. A. Porter, B. Giese. *Stereochemistry of Radical Reactions.* VCH, Weinheim (1995).

2)   Review: M. Gerster, P. Renaud, *Angew. Chemie Int. Ed.* **37**, in press (1998).

3)   Review: P. Renaud, *Chimia* **51**, 236 (1997).

4)   P. Renaud, M. Ribezzo, *J. Am. Chem. Soc.* **113**, 7803 (1991).

5)   S. Saito, H. Yamamoto, *Chem. Commun.* 1585 (1997).

6)   P. Renaud, N. Moufid, L. H. Kuo, D. P. Curran, *J. Org. Chem.* **59**, 3547 (1994).

7)   P. Renaud, T. Bourquard, M. Gerster, N. Moufid, *Angew. Chem. Int. Ed. Engl.* **33**, 1601 (1994).

8)   P. Renaud, T. Bourquard, P.-A. Carrupt, M. Gerster, *Helv. Chim. Acta* **81**, 1048 (1998).

9)   N. Moufid, P. Renaud, *Helv. Chim. Acta* **78**, 1001 (1995).

10)   N. Moufid, P. Renaud, C. Hassler, B. Giese, *Helv. Chim. Acta* **78**, 1006 (1995).

11)   Review: Y. Guindon, B. Guerin, J. Rancourt, C. Chabot, N. Mackintosh, W. W. Ogilvie, *Pure Appl. Chem.* **68**, 89 (1996).

12)   M. Gerster, L. Audergon, M. Moufid, P. Renaud, *Tetrahedron Lett.* **37**, 6335 (1996).

13)   Frater, G., *Helv. Chim. Acta* **62**, 2825 (1979).

14)   D. Seebach, D. Wasmuth, *Helv. Chim. Acta* **63**, 197 (1980).

15)   E. Lacôte, P. Renaud, *Angew. Chem. Int. Ed.* **37**, in press (1998).

16)   A. L. J. Beckwith, D. Crich, P. J. Duggan, Q. W. Yao, *Chem. Rev.* **97**, 3273 (1997).

17)   H. Zipse, *J. Am. Chem. Soc.* **119**, 1087 (1997).

DI-π-METHANE PHOTOREARRANGEMENT OF 2,3-DISUBSTITUTED BENZOBARRELENES AND BENZONORBORNADIENES: EFFECT OF SUBSTITUENTS OF OPPOSITE POLARITY ON THE REGIOSELECTIVITY AND SYNTHESIS OF BENZOPINANE SKELETON

Metin Balcı

Department of Chemistry
Middle East Technical University
06531 Ankara-Turkiye

INTRODUCTION

The di-π-methane rearrangement was shown to be a general photochemical process that occurs from excited states of molecules having two π-moieties bonded to a single atom-most generally carbon.[1] This mechanism is shown as outlined in Equation 1. In most cases the photochemical reaction proceeds from the excited singlet state in acyclic systems and from the excited triplet state in bicyclic systems.

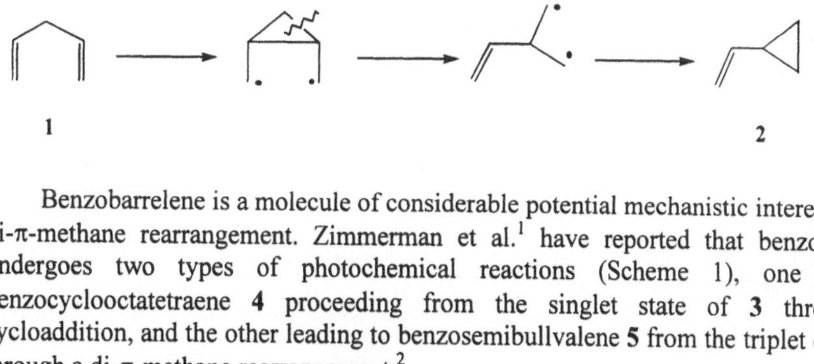

(Eq.1)

Benzobarrelene is a molecule of considerable potential mechanistic interest in view of di-π-methane rearrangement. Zimmerman et al.[1] have reported that benzobarrelene **3** undergoes two types of photochemical reactions (Scheme 1), one leading to benzocyclooctatetraene **4** proceeding from the singlet state of **3** through [2+2] cycloaddition, and the other leading to benzosemibullvalene **5** from the triplet excited state through a di-π-methane rearrangement.[2]

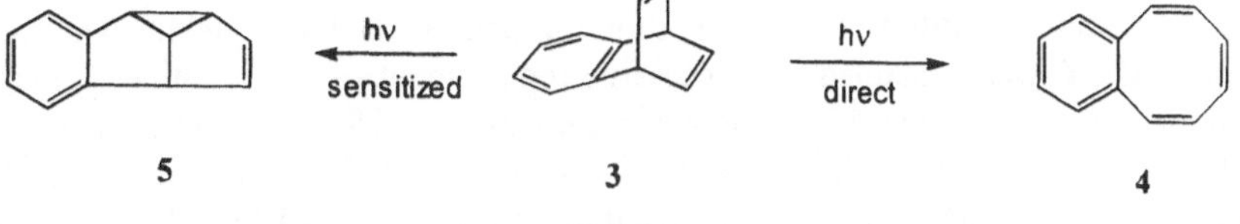

Scheme 1

Benzosemibullvalene **5**, in principle could be obtained by two-reaction pathways: benzo-vinyl and vinyl-vinyl bridging. Zimmerman showed clearly that vinyl-vinyl bridging is the main pathway. Introduction of a substituent on the vinyl group destroys the symmetry of the benzobarrelene skeleton, increasing the number of possible initial bonding modes to six di-π-methane rearrangements. The effect of one substituent on bridging and regioselectivity has been investigated by Bender,[3] Paquette[4] and Hemetsberger.[5] It has already been established that a cyano group controls the regiospecifity of the di-π-methane rearrangement. Labeling experiments of 5-cyanobenzobarrelene 5 reveal that the di-π-methane rearrangement proceeds via vinyl-cyanovinyl bridging between the less substituted carbon atoms because of the favorable stabilization of a biradical intermediate species by the cyano group (Scheme 2).

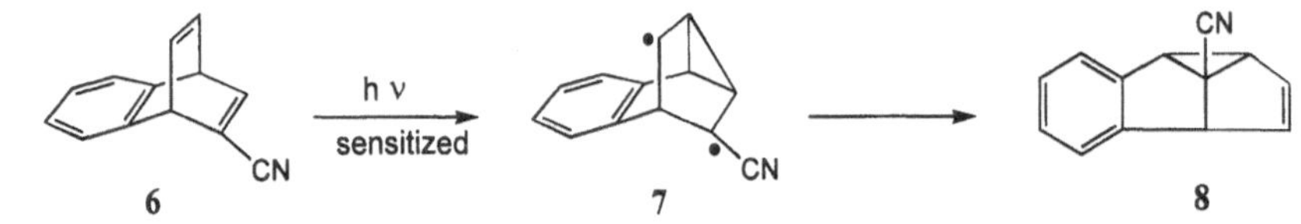

Scheme 2

In order to compare the effect of electron donating (methyl) and electron-accepting groups (cyano, aldehyde, ester) in the same molecule we have synthesized the unsymmetrically disubstituted benzobarrelene derivatives **13-16** and studied their photochemical behaviour (Scheme 3).

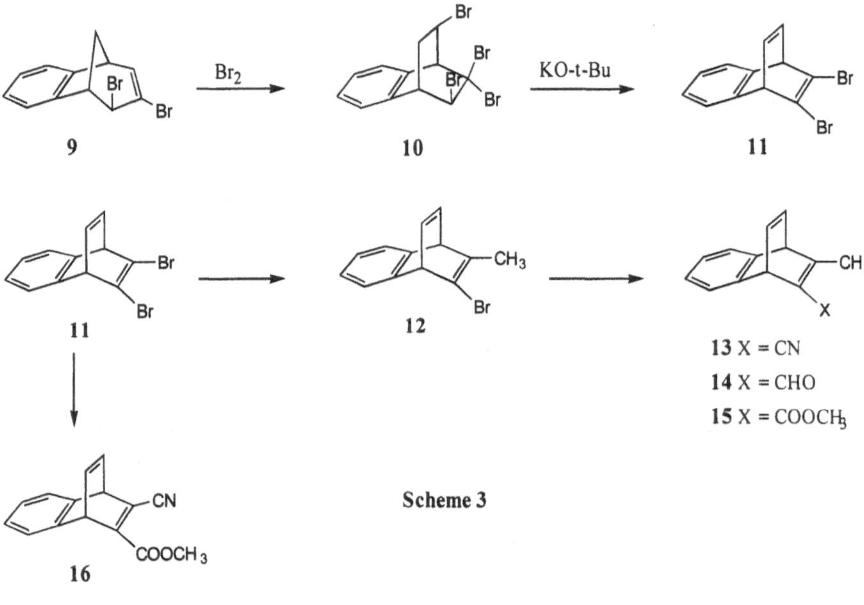

Scheme 3

The key compound, dibromobenzobarrelene **11** was synthesized by bromination of **9** at low temperatures followed by dehydrobromination in high yield.[6] The desired compounds **12-16** were obtained by successive replacement of the bromine atoms using conventional methods.[7] When **11** was treated with BuLi at − 78 °C, followed by quenching with CH₃I, **12** was formed with 84% yield. (Scheme 3). Methyl bromobarrelene **12** was allowed to react with cuprous cyanide to give the cyano compound **13** in 74% yield. On the other hand, metallation of **12** with BuLi followed by carboxylation afforded the corresponding acid, which upon esterification gave methoxycarbonyl methyl benzobarrelene **15** in high yield. Reduction of ester **15** with LiAlH₄ followed by CrO₃ oxidation in pyridine resulted in the formation of methyl aldehyde **14**. Cyano ester **16** was synthesized by successive replacement of bromine atoms in **11** by cyanide followed by carboxylation as described above.

Irradiation of **13** and **14** in both cases gave a mixture of two photoproducts **17/18** and **19/20** in a ratio of (7:3) and (9:1), respectively (Scheme 4). The products have been separated and characterized by ^{1}H, ^{13}C NMR and NOE measurements.

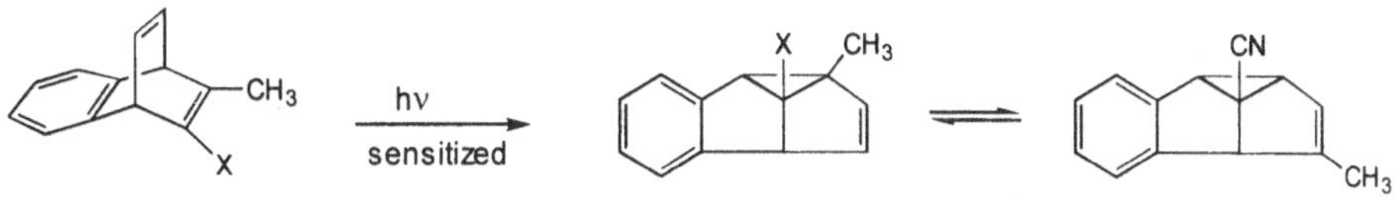

13 X = CN
14 X = CHO

17 X = CN
19 X = CHO

18 X = CN
20 X = CHO

Scheme 4

The photoproducts **17** and **19** are the results of a triplet state di-π-methane rearrangement of **13** and **14**. However, **18** and **20** cannot be formed by a di-π-methane rearrangement. They are secondary products. Irradiation of the major products **17** and **19** formed a reversible vinylcyclopropane-vinylcyclopropane interconversion to afford the isomers **18** and **20**. These observations indicate that only **17** and **19** are formed as primary products. Unsymmetrical substitution in the starting material **13** and **14** leads to two possible bridging modes: (i) vinyl-vinyl cyano (aldehyde) bridging, (ii) vinyl-vinyl methyl bridging. Examination of the products indicate that di-π-methane rearrangement is initiated by regiospecific vinyl-vinyl methyl bridging as observed by Bender[7] for cyanobenzobarrelene **6**. Probably, the strongly electron withdrawing cyano group in **21** and aldehyde group in **22** stabilize the diradical intermediates. Bordwell and Lynch[9] have calculated the relative radical satabilizing effects of the cyano and methyl groups and have shown that they are equal. In the present case, however, we observe only vinyl-vinyl bridging as in **21** and **22**.

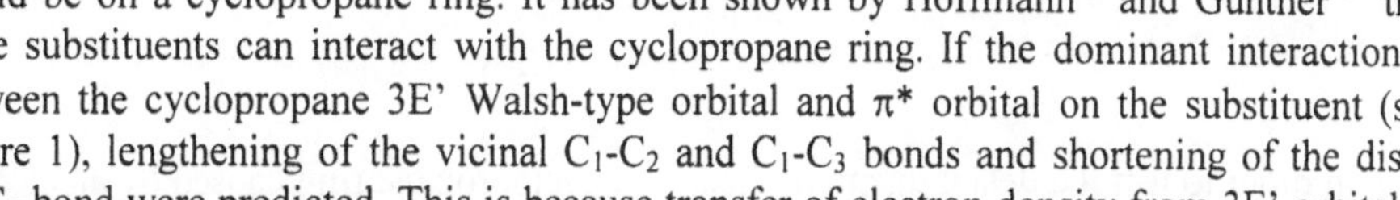

21   **22**   **23**   **24**

If vinyl-vinyl cyano (aldehyde) bridging were operative, the cyano (aldehyde) group would be on a cyclopropane ring. It has been shown by Hoffmann[10] and Günther[10] that these substituents can interact with the cyclopropane ring. If the dominant interaction is between the cyclopropane 3E' Walsh-type orbital and π* orbital on the substituent (see Figure 1), lengthening of the vicinal C_1-C_2 and C_1-C_3 bonds and shortening of the distal C_2-C_3 bond were predicted. This is because transfer of electron density from 3E' orbital to vacant π* orbital which decreases the antibonding electron density in the distal bond and decreases the bonding electron density in the vicinal bonds.

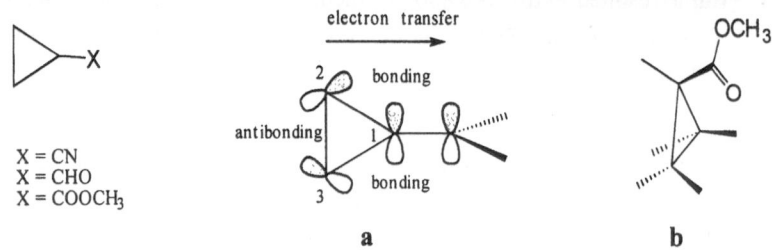

X = CN
X = CHO
X = COOCH$_3$

a   b

Figure 1. a) Interaction of cyclopropane 3E' Walsh-type orbital and π* orbital, b) Bisected conformation of a carbonyl group for maximum conjugation with the Walsh orbitals

Therefore, we propose that not only does the radical stabilizing effect of the substituents play an important role by determining the mode of the initial bridging, but also the destabilizing effect of the substituents with π–electron-withdrawing abilities on the vicinal bond of a cyclopropane ring is important. We assume that electron-withdrawing cyano and aldehyde groups destabilize the adjacent bonds in an initially formed cyclopropane.

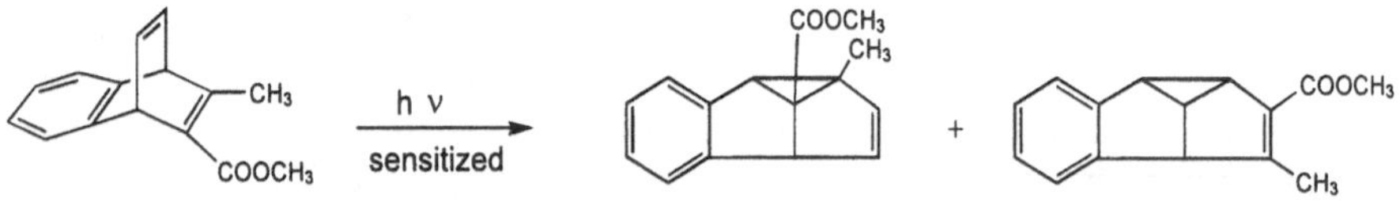

Scheme 25

Aceton sensitized irradiation of **15** gave contrary to our expectation two isomeric di-π-methane rearrangement products **25** and **26** derived from initial vinyl-vinyl methyl bridging **23** and vinyl-vinyl ester bridging **24**. It is well known that –COOH or –COOR groups can interact more strongly with Walsh orbital than the cyano group.[11] Therefore, only **25** was the expected product in this reaction. The formation of **26** can be explained on the basis of the conformational factors. For the maximum conjugation of the substituent with the Walsh orbital of cyclopropane, they must have a bisected conformation (Figure 1). Due to the free rotation of the C-C bond between the substituent and cyclopropane ring, the desired conformation can not be obtained. Bisected conformation of a carbonyl group can be forced by the synthesis of the following lacton **27**. Currently we are working on the synthesis of **27**.

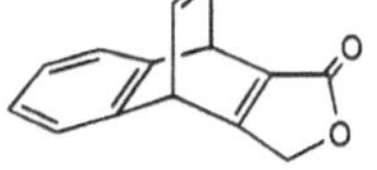

27

In order to test the destabilization effect of a cyclopropane ring caused by an electron withdrawing group we have synthesized **16** and submitted it to sensitized photochemical reaction and obtained only **28** as the isolable product (Scheme 6).

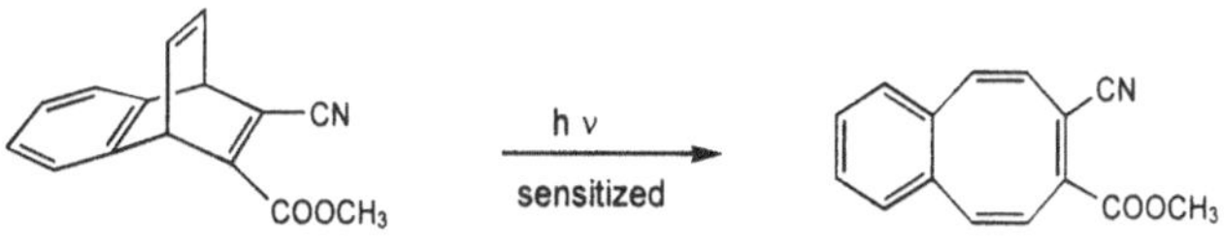

16   28

Scheme 6

Usually direct irradiation of benzobarrelene system leads to exclusive formation of benzocyclooctatetraene. This transformation occurs via an excited singlet state of benzobarrelene. Zimmerman et al. showed an initial benzo-vinyl cycloaddition prevails.[1b] However, introduction of substituents on the olefinic moiety may affect the efficiency of inter system crossing, the reactivity and modes of initial cycloadditions. In the system **16**, we assume that two electron-withdrawing substituents prevent the di-π-methane rearrangement and the system undergoes [2+2]-cycloaddition reaction. There are three possible modes of the initial [2+2]-cycloaddition modes: benzo-vinyl, benzo-vinyl(substituted), and vinyl-vinyl(substituted) [2+2]-cycloadditions. Benzocycloocta-tetraen derivative **28** can be formed either from the intermediate **29** vinyl-vinyl(substituted) cycloaddition or **30** benzo-vinyl(substituted) cycloaddition.

29   30

At this stage we can not distinguish between these possible cycloaddition modes. In both cases the unsubstituted vinyl group is involved in the formation of the intermediate **29** and **30**. It is possible that there is competition between [2+2]-cycloaddition reaction and the di-π-methane rearrangement so that we can not make any conclusion in terms of the reaction mechanism whether the electron-withdrawing groups in **16** prevent di-π-methane rearrangement or not. In order to find an answer to this question we have synthesized the corresponding benzonorbornadiene system **32** where only benzo-vinyl bridging can occur.

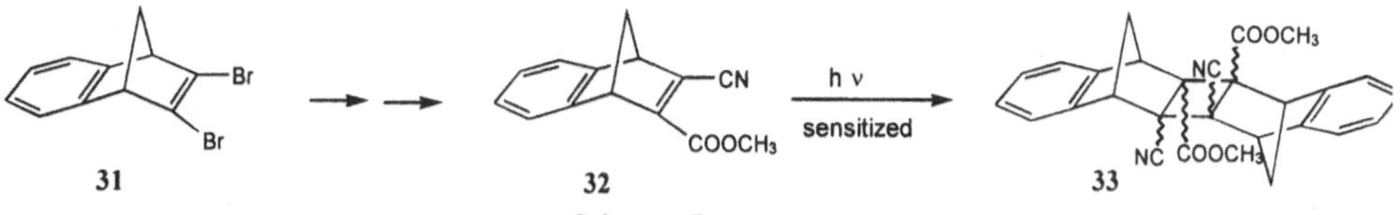

31   32   33

Scheme 7

The starting material **32** has been obtained by successive replacement of the bromine atoms in **31** as described for the synthesis of **16**.[12] The sensitized irradiation of **32** gave exclusively the dimer **33**. On the basis of these results we conclude that this system can not undergo a [2+2]-cycloaddition reaction. However, the fact that this system does not form any di-π-methane rearrangement product, indicates that the electron-withdrawing groups may play an important role and prevent the formation of the initially formed cyclopropane ring on which a substituent is attached. To support these arguments, further works are in progress.

APPLICATION OF THE DI-π-METHANE REARRANGEMENT TO SYNTHESIS OF BENZOPINANE SKELETON

Monoterpenes possessing the pinane skeleton **34** occur in the wood and leaf oil of many higher plants.[13] α-Pinene **35** and β-pinene **36** are the parents of most synthetic pinane derivatives. Several possible synthetic routes to this interesting natural product skeleton may be envisaged. These include rearrangement of a related bicyclic systems[14], ring contraction of a suitable diazo ketone[15] or photocyclization of a suitable diene or triene.[16] We have developed a new route leading to the bicyclo[3.1.1]heptane ring which could be used for pinane skeleton synthesis in the first instance. Especially we were interested for the synthesis of **37.**

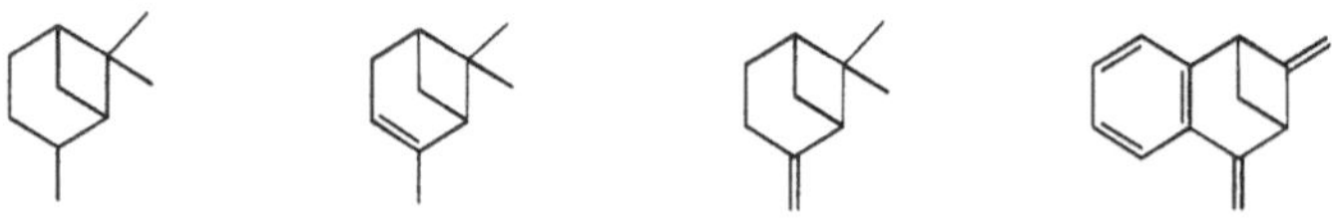

34 pinane   35 α-pinene   36 β-pinene   37

Our starting material, 2,3-dimethylene-1,2,3,4-tetrahydro-1,4-methano-naphthalene **38** was prepared as reported recently in the literature.[17] Treatment of **38** with 4-phenyl-*4H*-1,2,4-triazole-3,5-dione (PTAD) gave in quantitative yield, the urazole **39** (Scheme 8). The di-π-methane rearrangement of **39** conducted in acetone led to a single product **40** which was characterized by means of NMR spectra.[18]

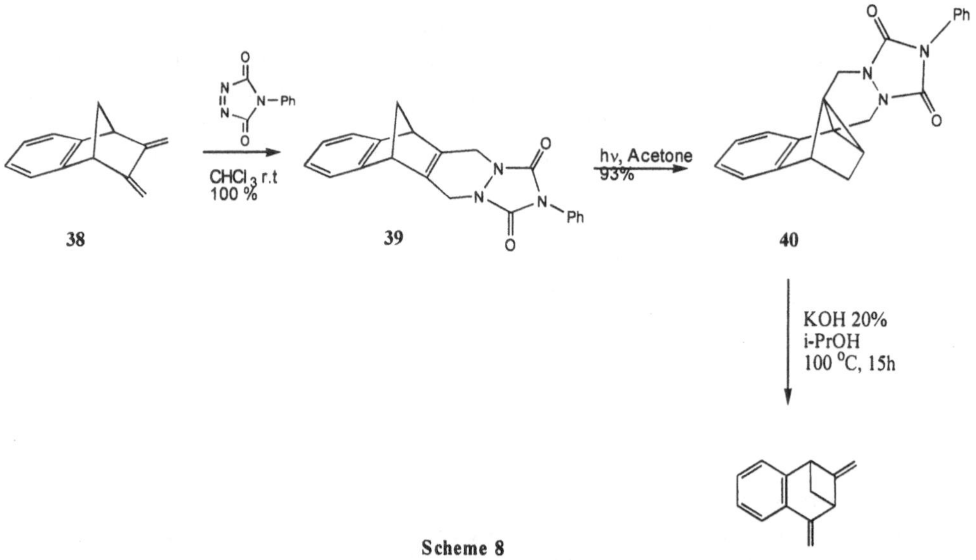

Scheme 8

The thermal and photochemical extrusion of N_2 from appropriate azo alkanes has served as a particularly convenient and effective method for the generation of authentic diradicals which have been postulated in photo-rearrangements.[19]

Hydrolysis of **40** and subsequent loss of N_2 were successfully realized by initial saponification with 20% KOH in i-PrOH and followed by oxidation with $CuCl_2$. After chromatography, we isolated the hydrocarbon **37** as the sole product in 26% yield (Scheme 8).

From a mechanistic viewpoint, the formation of this hydrocarbon **37** is the result of hydrolysis of the urazolring in **40** followed by dehydrogenation to give the labile diazo compound **42** which, in turn can undergo easily denitrogenation reaction under the given reaction conditions. The resulting diradical **43** easily forms the final product **37** by undergoing a *Grob*-type fragmentation reaction as shown below (Scheme 9).

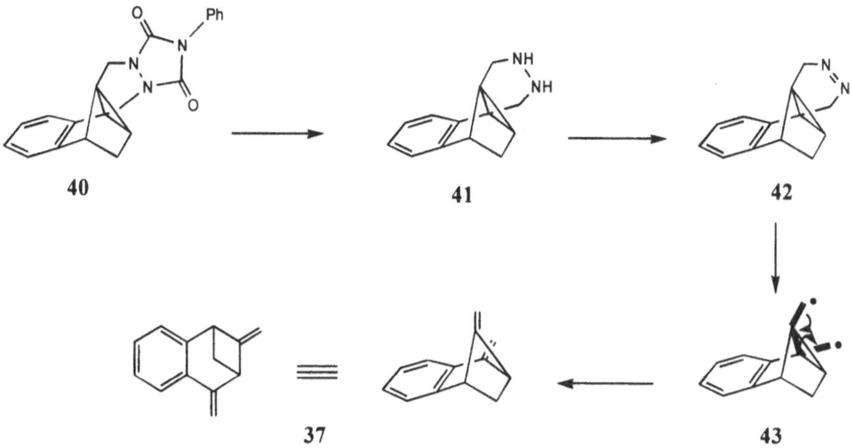

Scheme 9

The methodology described in this article opens up a new entry to the synthesis of hitherto unknown benzo-annelated pinane skeleton.

REFERENCES

1. H. E. Zimmerman, and D. Armesto, *Chem. Rev.* 96:3065 (1996); S. S. Hixon, P. S. Marino, and H. E. Zimmerman, *Chem. Rev.* 73:531 (1973); H. E. Zimmerman, R. S. Givens M. R. Pagni, *J. Am. Chem. Soc.* 90:6096 (1968); H. E. Zimmerman, The di-π-methane rearrangement in *CRC Handbook of Organic Photochemistry and Photobiology,* W. M. Horspool and P. S. Song, ed. CRC Press, New York (1995).
2. C. C. Liao, and P. H. Yang, Photorearrangement of benzobarrelenes and related analogues in *CRC Handbook of Organic Photochemistry and Photobiology,* W. M. Horspool and P. S. Song, ed. CRC Press, New York (1995).
3. C. O. Bender, D. W. Brooks, W. Cheng, D. Dolman, S. F. Oshea, and S. S. Shugarman, *Can. J. Chem.* 56:3027 (1978), C. O. Bender, E. H. King-Brown *J. Chem. Soc. Chem. Commun* .878 (1976); C. O. Bender, D. L. Bengston, D. Dolman, C. E. L. Herle, and S. F. Oshea, , *Can. J. Chem.* 60:1942, (1982); C. O. Bender, J. Wilson, *Helv. Chim. Acta 59:1469* (1976),
4. L. A. Paquette, A. Y. Ku, C. Santiago, M. D. Rozeboom, and K. N. Hook, *J. Am. Chem. Soc..* 101:5981 (1979).
5. H. Hemetsberger, and M. Nobbe, *Tetrahedron,* 44:67 (1988).
6. M. Balcı, O. Çakmak, and T. Hökelek *Tetrahedron,* 48:3163 (1992).

7.   R. Altundaş, and Metin Balcı, *Aust. J. Chem.* 50:787 (1997).
8.   C. O. Bender, and S. S. Shugarman, *J. Chem. Soc. Chem. Commun.* 934 (1974); *Can. J. Chem.* 56:3027 (1978), C. O. Bender, E. H. King-Brown *J. Chem. Soc. Chem. Commun.* 878 (1976).
9.   F. G. Bordwell, and T. Y. Lunch, *J. Am. Chem. Soc.* 111:7558 (1989).
10. R. Hoffmann, *Tetrahedron Lett.* 2907 (1970); H. Günther, *Tetrahedron Lett.* 2907 (1970); M. Balci, H. Fischer, H. Günther , *Angew Chem. Int. Ed. Engl.*19:301 (1980); M. Balci, *Turk. J. Chem.* 16:42 (1992).
11. W. J. Jorgenson, and L. Salem in *The Organic Chemist's Book of Orbitals* Academic Press, New York, (1973); R. Hoffmann, W. D. Stohrer, *J. Am. Chem. Soc.* 93:6941 (1971).
12. A. Daştan, B. Demirci, and M. Balci, unpublished results.
13. R. A. Raphael in *Chemistry of Carbon Compounds*, E. Rood, ed., Elsevier, Amsterdam, II A, p 314 (1953).
14. P. Yates, and R.J. Crawford, *J. Am. Chem. Soc.* 88:1562 (1966).
15. K.B. Wiberg, and B. A. Hess, *J. Org. Chem.* 31: 2270 (1966)
16. E. Wenkert, P. Bakuzis, R.J. Baumgarten, D. Doddrel, P. W. Jeffs, C.L. Leicht, R.A. Mueller, and A. Yoshikoshi, *J. Am. Chem. Soc.* **1970**, 92: 1617 (1970).
17. B. Atasoy, F. Bayramoglu, and T. Hökelek, *Tetrahedron;* 50:5753 (1994); N. Butler, R. A. Snow, *Can. J. Chem.* 53:256 (1975)
18. A. Altundaş, N. Akbulut, and M. Balcı 81:828 (1998).
19. H.E. Zimmerman, R.J. Boettcherr, N.E. Buehler, G.E. Keck, and M.G. Steimetz, *J. Am. Chem. Soc.* 98:7680 (1976); W. Adam, O. de Lucchi, and I. Erden, *J. Am. Chem. Soc.* 102:4806 (1980)

SIGHT-SEEING THE METALATED FLATLANDS. CARBANION-MEDIATED STRATEGIES FOR SYNTHETIC AROMATIC CHEMISTRY

Victor Snieckus[*]

Guelph-Waterloo Centre for Graduate Work in Chemistry and Biochemistry, University of Waterloo, Waterloo, ON, Canada N2L 3G1

* From July 1, 1998: Department of Chemistry, Queen's University, Kingston, ON, Canada K7L 3N6
fax: 613 545 2837; email: snieckus@chem.queensu.ca

The seminal mechanistic and synthetic investigations of Hoppe,[1] Hoffmann,[2] and Gawley[3] on the [1.2]-Wittig rearranagement of benzyl O-carbamates (Scheme 1) as well as the conceptual framework of the Complex Induced Proximity Effect (CIPE) championed by Beak and Meyers[4] and Klumpp[5] stimulated activity in our laboratories aimed to develop new methods for the regioselective construction of polysubstituted aromatics. Following the discovery of the anionic *ortho*-Fries rearrangement, the Directed remote Metalation (DreM) process was uncovered in the biaryl amide and O-carbamate series.[6] Although the homologus anionic Fries rearrangement was observed as early as 1985,[7] the generalization and application was questioned only recently.[8] A sequel query was the potential success of the carbamate equivalent of the Baker-Venkataraman reaction.

The homologous anionic Fries process (Scheme 2) was optimized either under LDA or LDA/TBDMSCl conditions, the latter of which allowed isolation of the OTBDMS product which resulted by migration from the initially generated *ortho*-tolyl silylated intermediate. Both phenyl acetamides, R = H and TBDMS, underwent smooth acid-catalyzed cyclization to the benzofuranones. The cases shown (Scheme 2) are illustrative of the scope, regiospecificity, and complimentarity of this reacion to the classical Friedel-Crafts process. The synthesis of isomeric naphthofuranones (Scheme 3) suggests potential extension to more complex systems.[8]

The powerful OCONEt2 Directed Metalation Group (DMG) was used to good advantage in the acylative variation of the Negishi cross coupling process (Scheme 4).[9] Thus metalation - zinc transmetalation - cross coupling led to *ortho*-acylated carbamates which, upon treatment with NaH (optimum base, LDA is less effective), afforded 4-hydroxybenzopyranones in good overall yields. Aside from constituting a regiospecific route to *ortho*-acyl phenol derivatives, this new version of the Baker-Venkataraman reaction furnishes substituted (e.g. OMe, Cl) benzopyranones which are not available by the standard Lewis-acid (LA) - mediated Fries rearrangement. The devised method was placed to the test in a synthesis of a coumarin natural product, recently isolated from *Colchicum decsaisnei* (Scheme 5). The synthesis of the unnamed coumarin, obtained in short order and high overall yield, also featured an alkylative Negishi coupling and a highly regioselective C-2/C-6 DoM process. There was only one problem and that was its structure! Comparison of physical and spectroscopic propeties of the synthetic material with those reported showed serious differences (Scheme 6). The identity of the natural product from *Colchicum decsaisnei* was established to be isoeugenetin methyl ether by comparison of spectral properties and synthesis via isoeuegenetin, isolated and synthesized almost 50 years ago.[10]

Anionic Chemistry of O-Carbamates. Regiospecific Construction of Aromatics / Heteroaromatics
[1, 2]- Wittig
Anionic ortho-Fries
DreM Anionic Fries
Hoppe (1982)
Hoffmann (1992)
Gawley (1993)
Sibi (1983)
Wang (1992)
Homologous Anionic Fries
Carbamate Baker-Venkataraman
Z = H, TMS, TBS
Miah, Kerr (1985)
Tsukazaki (1992)
Kalinin (1996)
Z = H, Alk, Ar
da Silva (1995)
Kalinin (1997)

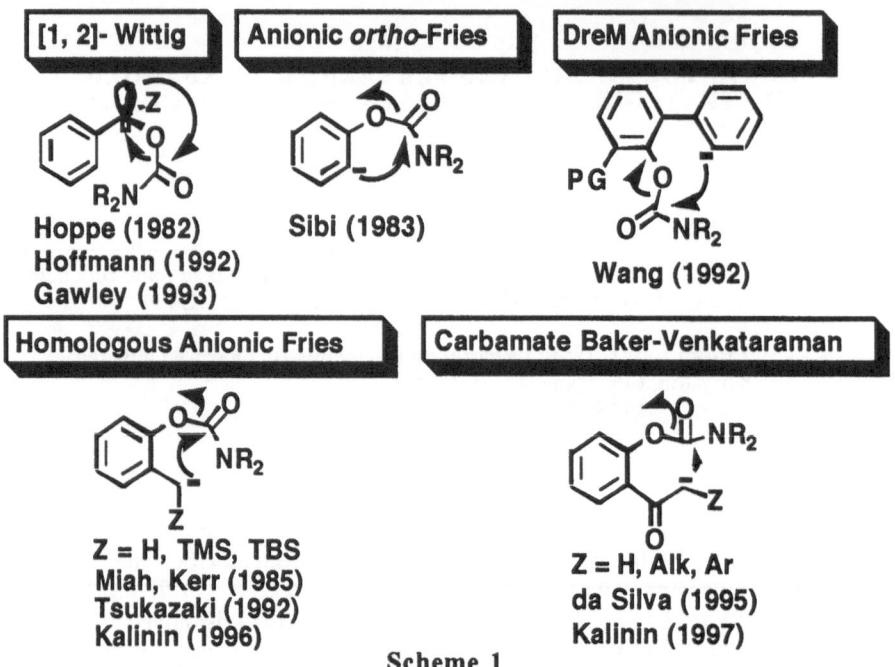

Scheme 1

Anionic Vinylogous ortho-Fries Rearrangement. Regiospecific Synthesis of Benzofuranones
LDA or
LDA / TBDMSCl
THF
-78oC or
0oC rt
HOAc
or
TsOH / PhH
or
TFA / Tol
R = H, TBDMS
C-5, C-6, C-7
(60 - 78%)
Z = OMe (69%)
Z = Cl (72%)
(95%)

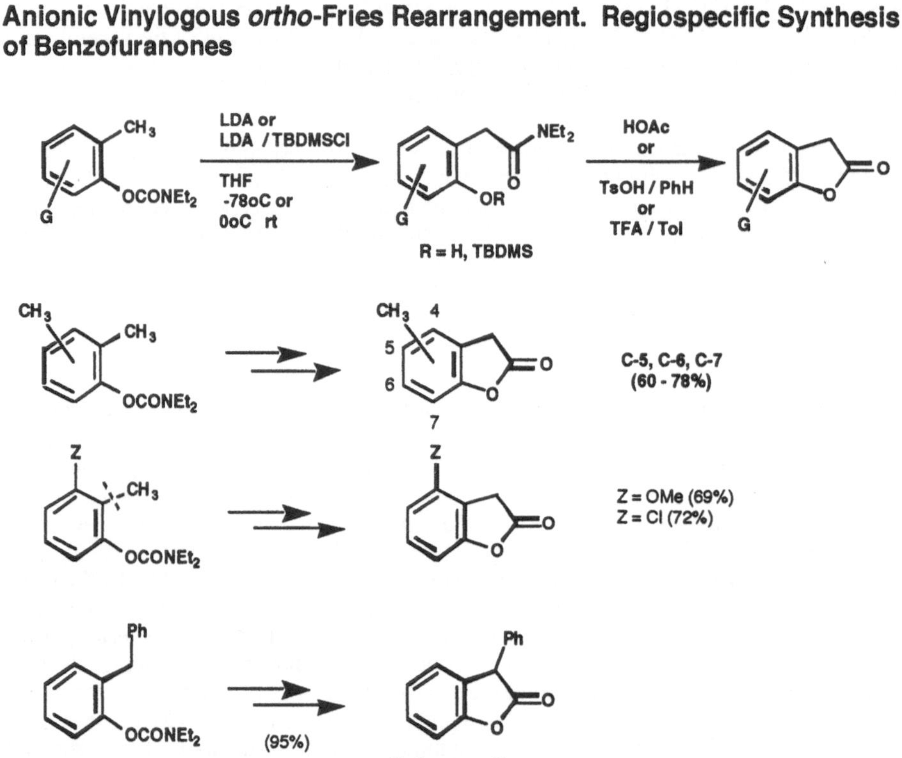

Scheme 2

Anionic O → C Carbamoyl Migration in 2-Methyl Aryl O-Carbamates. Synthesis of Naphthofuranones

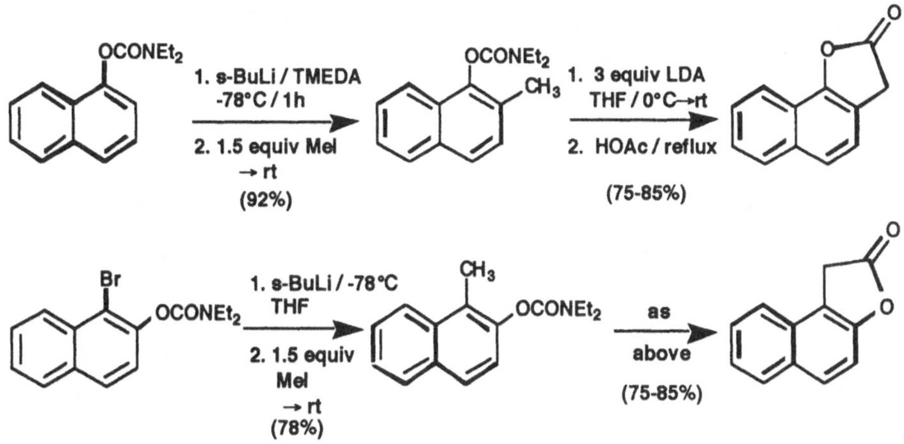

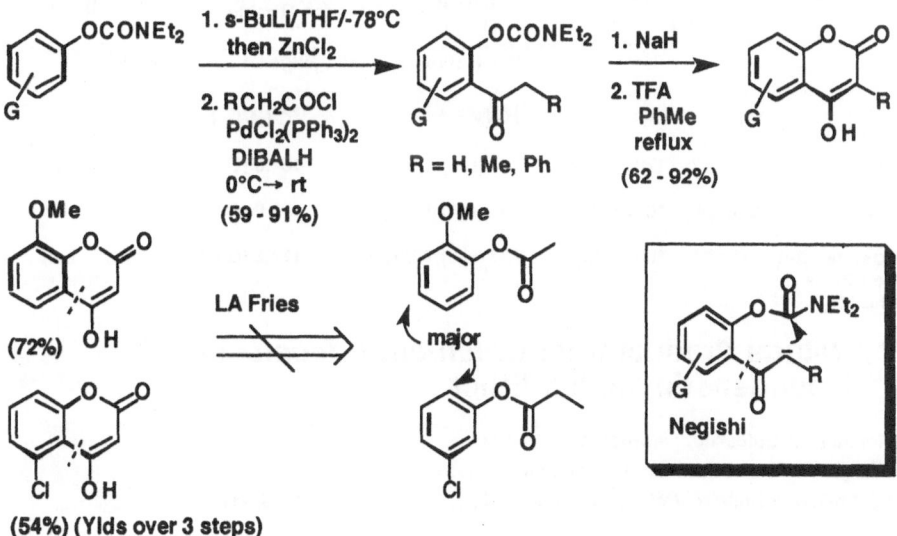

Scheme 3

Anionic *ortho* - Friedel-Crafts Equivalent. DoM-Xcoupl Connections. 4-OH Benzopyran-2-ones via Carbamoyl Version of the Baker-Venkataraman Reaction

Scheme 4

Coumarin Natural Product Total Synthesis *via* DoM - Negishi Xcoupl - Carbamoyl Baker - Venkataraman Sequence

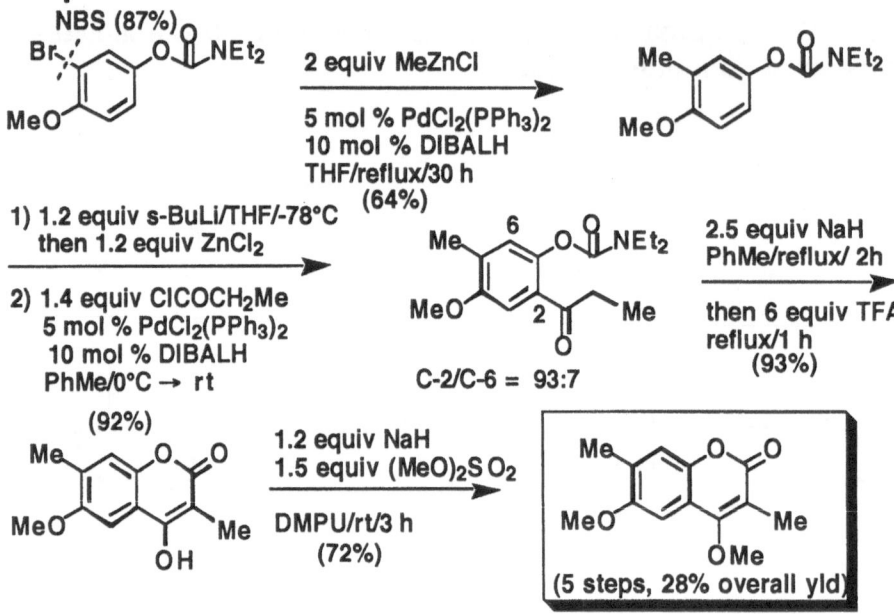

Scheme 5

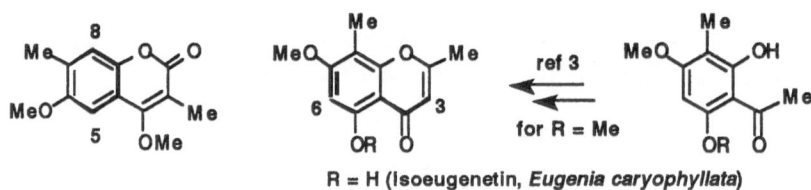

R = H (isoeugenetin, *Eugenia caryophyllata*)

	mp, °C	IR, cm^{-1}	^{1}H NMR, δ	^{13}C NMR, δ
Synthetic [1]	141-142	1704 (KBr)	7.13 (H-5), 7.00 (H-8)	164.6 (C=O), 16.6, 10.7 (2 Me)
Isolated [2]	amorph	1650 (CHCl₃)	6.45 (H-5), 5.96 (H-8)	178.0 (C=O), 19.7, 7.7 (2 Me)
Isoeugenetin Me ether[3] (R = Me)	173-174	1661 (KBr)	6.38 (H-6), 6.00 (H-3)	177.6 (C=O), 19.4, 7.3 (2 Me)

∴ Natural Product from *Colchicum decaisnei* IS Isoeugenetin Methyl Ether

[1] Kalinin and Snieckus, *Tetrahedron Lett* 1998

[2] Zarga and Atta-ur-Rahman, *Phytochemistry* 1991, *30*, 3081

[3] Schmid and Bolleter, *Helv. Chim. Acta* 1949, *32*, 1358; *ibib* 1950, *33*, 917

Scheme 6

Implementation of carbanion-mediated chemistry in total synthesis is illustrated (Scheme 7) by the vanquished antibiotic natural products phenanthroviridin aglycon, kinobscurinone related to the benzofluorene isolated from *Streptomyces murayamaensis*,[11] WS 95995 A,[12] and several naphthobenzopyranones related to the gilvocarcins.[13] The recently concluded[14] total synthesis of arnottin I, potentially related biosynthetically to arnottin II and chelerythrine (Scheme 8), retrosynthetically cascades to DreM of a naphthyl-aryl O-carbamate assisted by the OMe DMG effect; this, in turn, may be derived from the A/B- and D-ring cross coupling partners where the X and Y may be B, Zn, Mg and halogen, OTf combinations respectively or in inverted form. While a rational retrosynthetic step leads from the naphthalene A/B partner to a benzyne-furan cycloaddition, the issue of using a C-6,7-methylenedioxy precursor as required by arnottin I in strong base-mediated chemistry reared, as will be substantiated shortly, its ugly head. After considerable experimentation with many of the separate steps, the devised route for the C-6,7-OCH$_2$O A/B ring cross coupling partner was achieved in good overall yield (Scheme 9). To prepare for the potential treachery of the OCH$_2$O group, the C-6,7-di-O-i-Pr 2-iodonaphthalene was similarly obtained and both derivatives were subjected to Pd-catalyzed Negishi and Suzuki-Miyaura cross coupling reactions with diMeO-benzenes (Scheme 10, the Ba(OH)$_2$ mediated process is noteworthy).

As forecast above, treatment of the C-6,7-OCH$_2$O biaryl under several strong base conditions showed evidence for the formation of catechols, a result of the base lability of the methylenedioxy unit (Scheme 11). On the other hand, the corresponding C-6,7-di-i-OPr afforded the desired carbamoyl migration product in 68% yield under optimized two-portion LDA addition conditions. Acetic acid reflux then furnished the key naphthobenzopyranone (62% yield over two steps). Since the known selective lability of the O-i-Pr group to BCl$_3$ was apparently compromised by (the also known) amide and lactone carbonyl boron chelation leading to demethylated products, a protection (PG)-dePG end game was devised (Scheme 12). Thus the expediency of a deisopropylation-momylation sequence gave the naphthyl aryl O-carbamate (85% yield) which, when subjected to the crucial portionwise LDA reaction followed by TsOH treatment, produced the catechol albeit in 44% yield. A known CsF-mediated methylenation concluded the synthesis of arnottin I.

Antitumor Antibiotics. Phenanthroviridin Aglycon and Related Natural Products

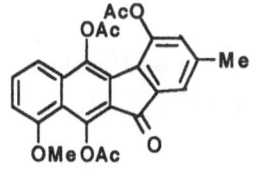

Phenanthroviridin Aglycon
Streptomyces viridiochromgenes

ex *S. murayamaensis*

WS 5995A
ex *S. auranticolor*

Gilvocarcin M Aglycon
ex *S. gilvotanareus*

Scheme 7

135

DoM - DreM - Xcoupl Connections. Retrosyn for Arnottin ex *Xanthoxylum arnottianum*

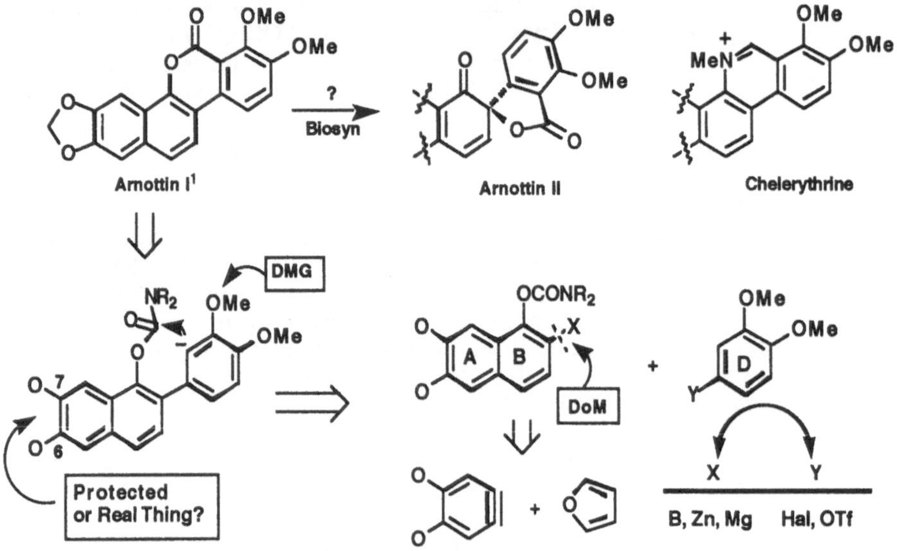
OMe
OMe
?
Biosyn
Arnottin I[1]
Arnottin II
MeN
Chelerythrine
DMG
NR2
OCONR2
X
A
B
D
DoM
7
6
Protected
or Real Thing?
X
Y
B, Zn, Mg
Hal, OTf

1. Isolation: Ishii, H. *et al* Yakugaku Zasshi 1977, *97*, 890.
   Structure: Ishii, H. *et al* J. Chem. Soc., Perkin Trans. 1 1993, 1019.

Scheme 8

DoM - DreM - Xcoupl Connections. The C6,7-OCH_2O A/B Ring Xcoupl Partner

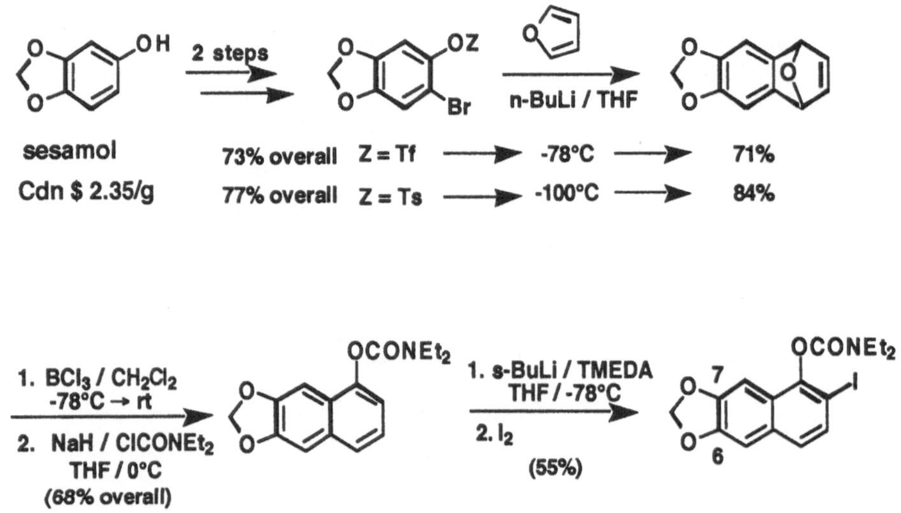
OH
2 steps
OZ
Br
n-BuLi / THF
sesamol
Cdn $ 2.35/g
73% overall Z = Tf → -78°C → 71%
77% overall Z = Ts → -100°C → 84%
1. BCl3 / CH2Cl2
-78°C → rt
2. NaH / ClCONEt2
THF / 0°C
(68% overall)
OCONEt2
1. s-BuLi / TMEDA
THF / -78°C
2. I2
(55%)
OCONEt2
7
6
I

Scheme 9

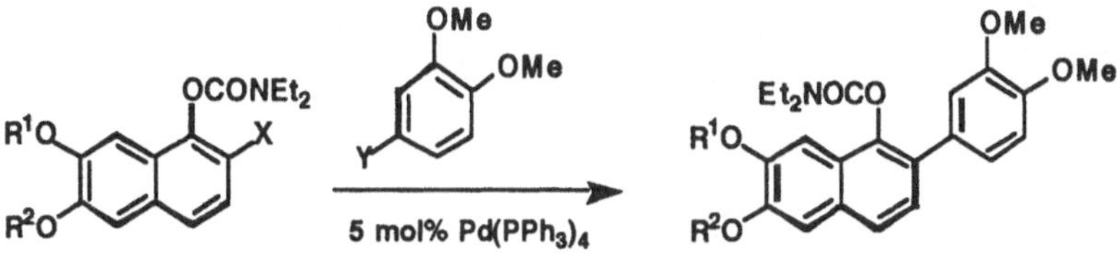

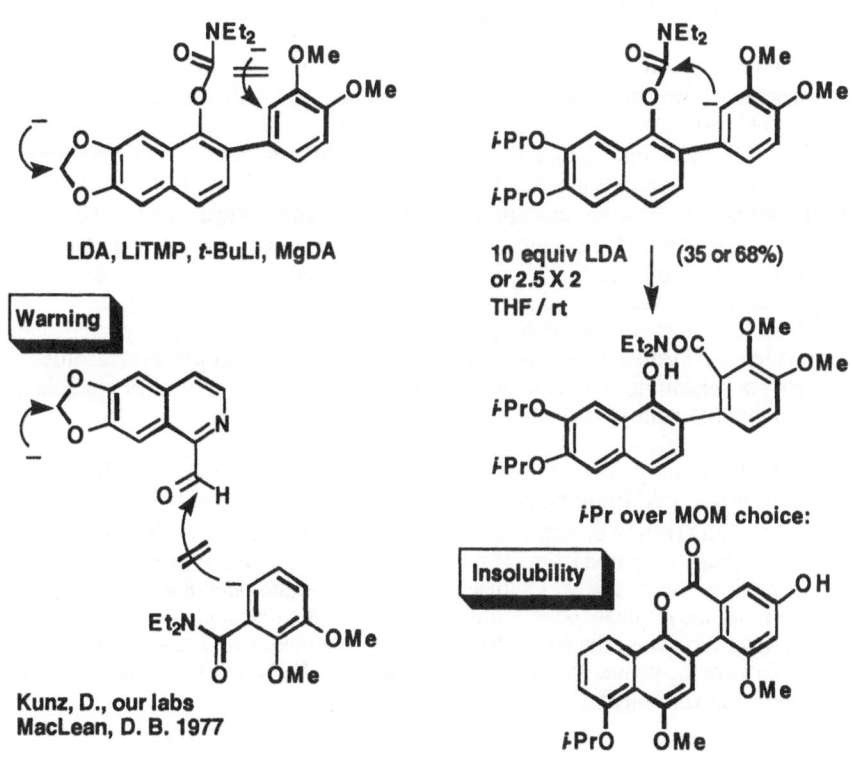

Scheme 10

R¹, R²	X	Y	Conditions	yld, %
OCH₂O	ZnCl	Br	THF / reflux	60
OCH₂O	I	ZnCl	THF / reflux	68
i-PrO	ZnCl	Br	THF / reflux	32
i-PrO	I	B(OH)₂	Ba(OH)₂·H₂O DME / reflux	> 95

LDA, LiTMP, *t*-BuLi, MgDA

Warning

Kunz, D., our labs
MacLean, D. B. 1977

10 equiv LDA
or 2.5 X 2
THF / rt

(35 or 68%)

i-Pr over MOM choice:

Insolubility

Scheme 11

137

Arnottin I. The PG - dePG End Game

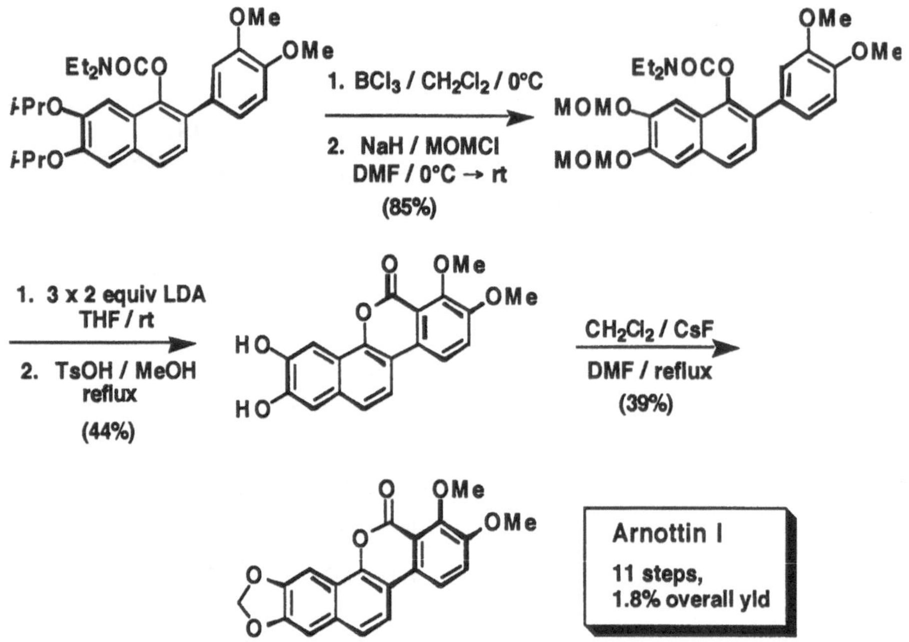

Scheme 12

The highly compact natural product plicadin isolated from *Psoralea plicata* (Scheme 13) posits a challenging target from the perspective of a combined carbamoyl Baker-Venkataraman (retron a,b), acylative Negishi cross coupling (retron c), and DoM (retron d) strategy.[15] In a further demonstration of the power of the OCONEt$_2$ DMG, the 1,3-di-O-carbamate (Scheme 14) was metalated with t-BuLi and quenched with senecioaldehyde to furnish, via O-to-O carbamoyl migration and S$_N$2' reactions, the 5-O-carbamoyl chromene derivative. This chromene, upon metalation - electrophile quench, gave excellent yields of C-6 substituted products. Aside from being the first demonstration of DoM chemistry on a chromene, this sequence, showing regiospecific metalation synergistically assisted by DMG[1] and DMG[2] and differential E$_2^+$ introduction may have general synthetic value. Following the previous protocol established for 4-hydroxychromone synthesis (Scheme 4), the chromene 5-O-carbamate was lithiated (Scheme 15), transmetalated with ZnCl$_2$, and subjected to Negishi cross coupling with a phenyl acetyl chloride to give the expected ketone (74% yield) which, when subjected to NaH followed by BCl$_3$ deisopropylation afforded the (isolable) triphenol ketoamide. PTSA treatment concluded the total synthesis of plicadin in 32% yield from the chromene O-carbamate.

Concurrently, a competitive retrosynthetic analysis was contemplated (Scheme 16) involving a carbamoyl DreM (retron a), combined Sonogarshira-Castro/Stephens (retron b), and simple Sonogashira (retron c) sequence.[15] The 6-acetylenic chromene O-carbamate (Scheme 17) was readily prepared from the corresponding iodo derivative (Scheme 13) by Sonogashira coupling. When this was subjected to the *ortho*-iodo phenol under Pd(0)-catalyzed conditions, the requisite benzofuran was obtained in modest yield. The O-carbamoyl DreM reaction was effected with LDA; direct treatment with HOAc at reflux efficiently (84%) gave the pentacycle which was treated with BCl$_3$ to afford plicadin (12% overall from the 6-iodochromene.

138

DoM - Transition Metal Catalysis Nexus

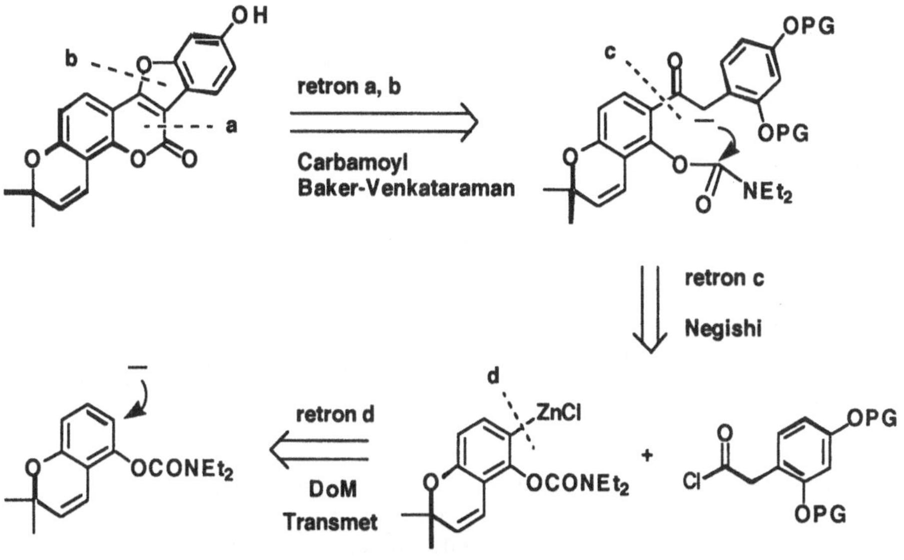

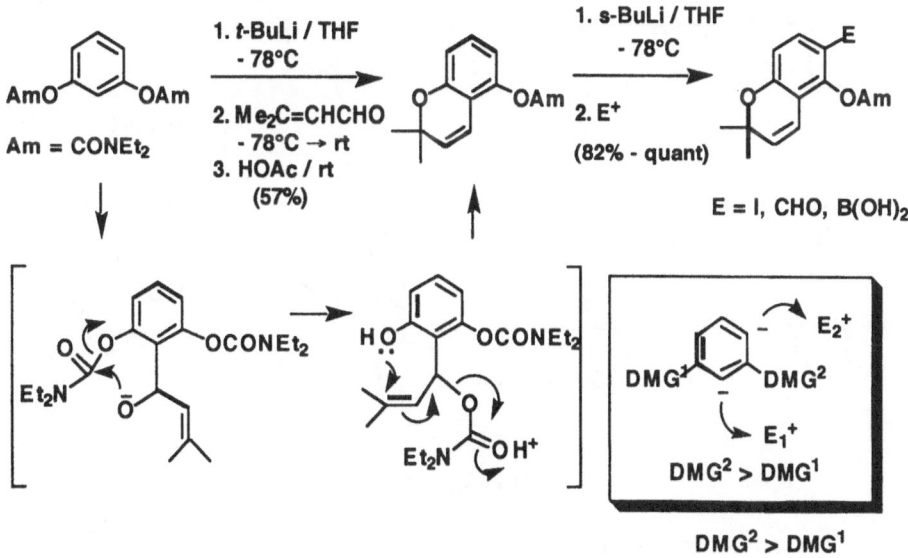

Scheme 13

DoM in Heterocycle Construction. The OCONEt₂ DMG for Regiospecific Synthesis of Polysubstituted Chromenes

Am = CONEt₂

1. *t*-BuLi / THF
   - 78°C

2. Me₂C=CHCHO
   - 78°C → rt

3. HOAc / rt
   (57%)

1. s-BuLi / THF
   - 78°C

2. E⁺
   (82% - quant)

E = I, CHO, B(OH)₂

DMG² > DMG¹

DMG² > DMG¹

Scheme 14

DoM / DreM - Transition Metal Catalysis in Multi-Oxygen Heterocycle Synthesis. Total Synthesis of Plicadin, ex *Psoralea plicata*

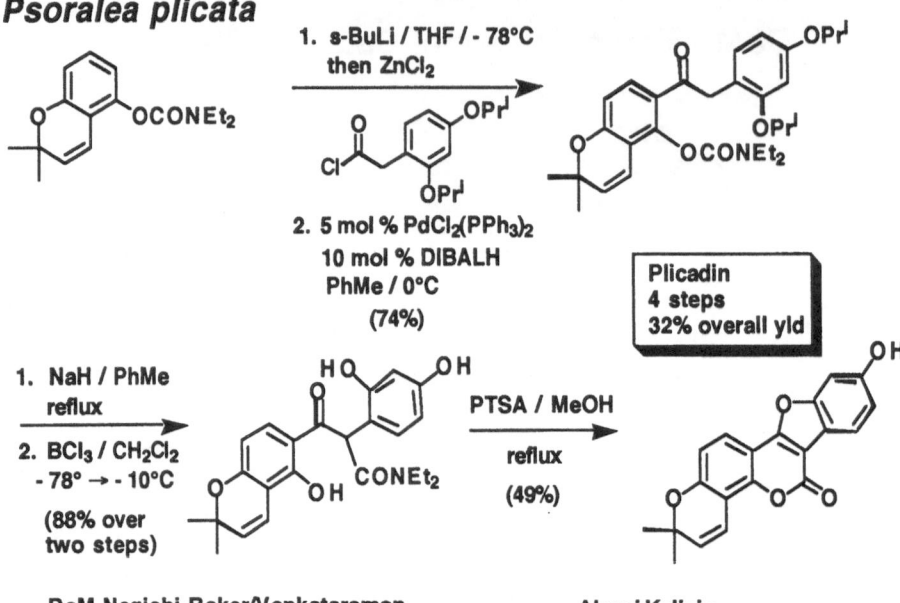

DoM-Negishi-Baker/Venkataraman

Alexei Kalinin

Scheme 15

DoM / DreM - Transition Metal Catalysis Nexus

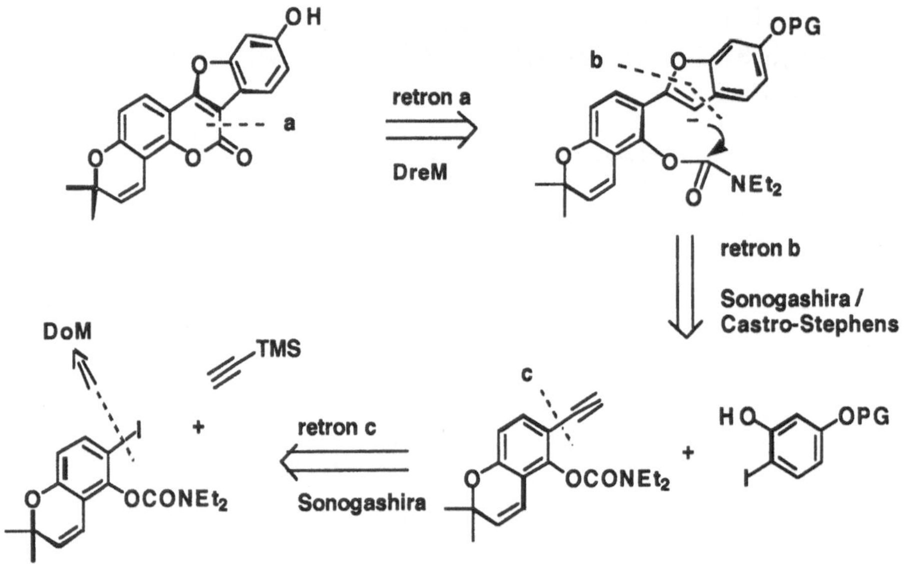

Scheme 16

The combination of the increasingly popular Grubbs metathesis protocol with the DoM strategy offers potentially powerful new methodology for annulation of medium and large carbo- and hetero-ring to aromatic moeties (Scheme 18). Retrosynthetically, DMG and latent DMG X substituents may be envisaged which are disected to diolefins with different m and n chains which, in turn, are derived from *ortho*-lithiated DMG benzene species. In-between DMG metalation may also be contemplated (box). In our first foray in this area, medium and large ring oxygen heterocycles annulated to an aromatic nucleus have been prepared by the DoM-metathesis synthetic link.[16] To illustrate this connection, the total synthesis of radularin A, a liverwort natural product, has been achieved (Scheme 19). The 1,2-diaryl ethane, readily obtained from 3,5-dimethoxybenzaldehyde, was lithiated, transmetalated to the corresponding Grignard reagent, and allylated. The MOM group was chemoselectively cleaved and the resulting phenol was methallylated to give the key Grubbs precursor. Considerable experimentation with Grubbs Ru and Schrock Mo catalysts led to the optimum conditions using the former reagent. Reductive decarbamoylation using Dibal afforded radularin A. As a further application of the DoM-metathesis link, the total synthesis of helianane, a marine product, begins with lithiation (Scheme 20), Cu transmetalation, and Michael addition with crotonaldehyde to give, after HCl hydrolysis, a phenyl propionaldehyde derivative. Wittig methylenation followed by reductive decarbamoylation afforded a phenol which, in an interesting reaction, was treated with acetone, chloroform, and base to install a gem-dimethyl acetic acid unit. Standard reductive/oxidative (LAH/PCC) manipulation provided an intermediate aldehyde which, upon Peterson olefination, gave the Grubbs reaction precursor. Application of Grubbs Ru catalyst followed by hydrogenation completed the synthesis of helianane.

DoM / DreM - Transition Metal Catalysis in Multi-Oxygen Heterocycle Synthesis. Total Synthesis of Plicadin ex *Psoralea plicata*

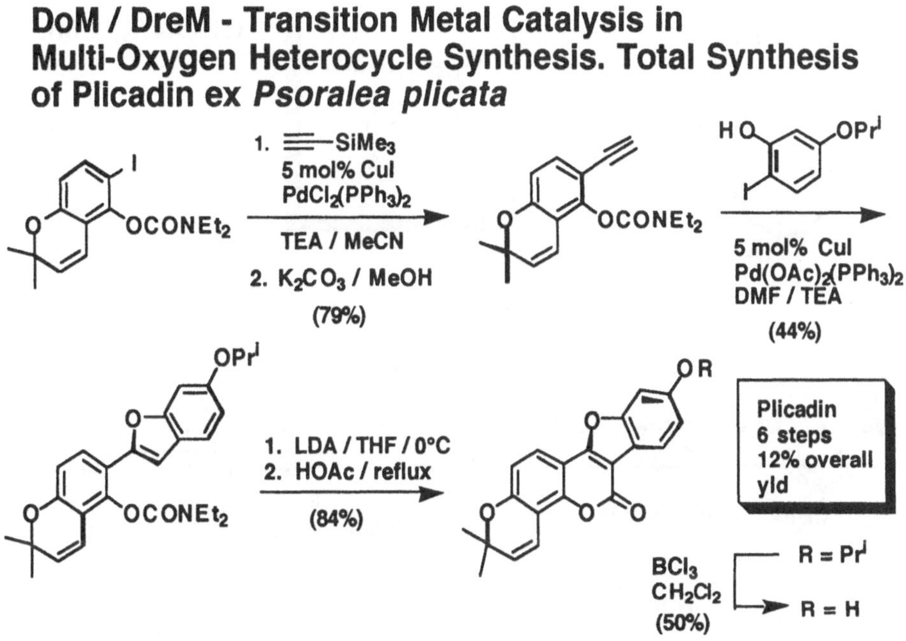

Scheme 17

Among other DoM-metathesis protocols under investigation in our laboratories is the sulfonamide variant (Scheme 21).[17] The requisite precursors with X = CH$_2$ and X = O, readily prepared by metalation technology, when subjected to Grubbs catalytic conditions, lead to previously unknown 8- and 9-membered ring cyclic sulfonamides in acceptable to high yields. The recent knowledge of regiospecific, condition dependent, *para*-methyl- or *ortho* -metalation (box) allows further contemplation for the synthesis of highly substituted systems of this unusual class of heterocycles.

Acknowledgement

I thank NSERC Canada and Monsanto for sustaining support of our work under the Industrial Research Chair and Research Grant programs. I am indebted to Dr. Prakash A. Patil for timely assistance with the preparation of this manuscript.

Synthetic Potential of DoM - Olefin Metathesis Links

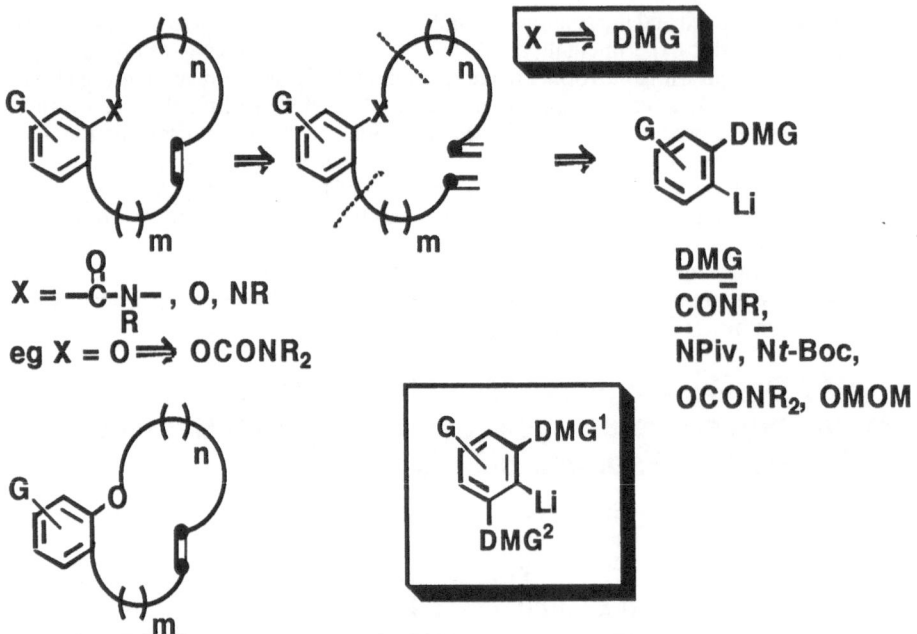

X = —$\overset{O}{\underset{R}{C}}$-N—, O, NR

eg X = O $\Rightarrow$ OCONR$_2$

$\underline{DMG}$

CONR,

$\overline{N}$Piv, $\overline{N}$t-Boc,

OCONR$_2$, OMOM

Scheme 18

A DoM-Olefin Metathesis Connection

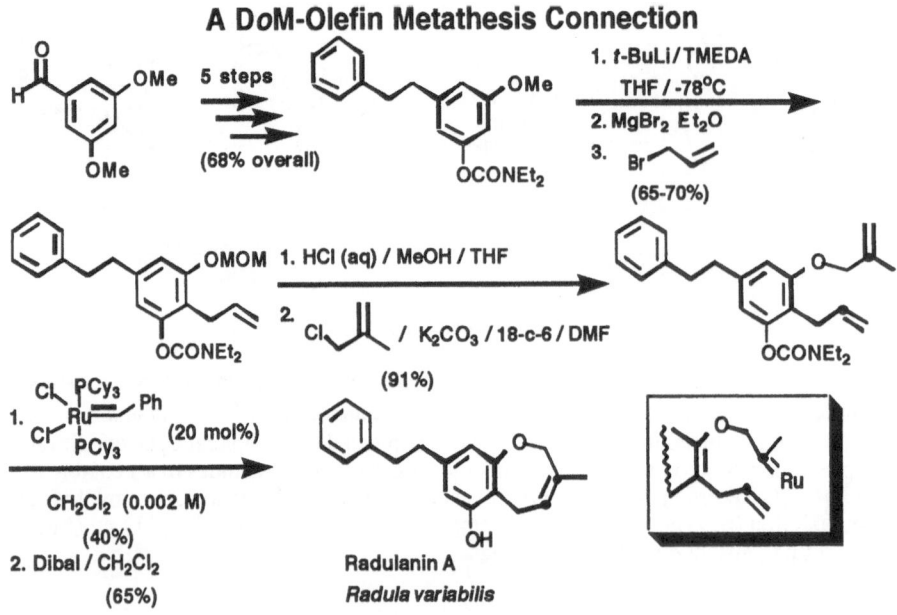

Radulanin A
Radula variabilis

Asakawa, Y. *et al. Phytochemistry* 1978, *17*, 2005.

Asakawa,Y. *et al. Phytochemistry* 1978, <u>17</u>, 2005.

Scheme 19

DoM-Metathesis Connection. Total Synthesis of (±)-Helianane

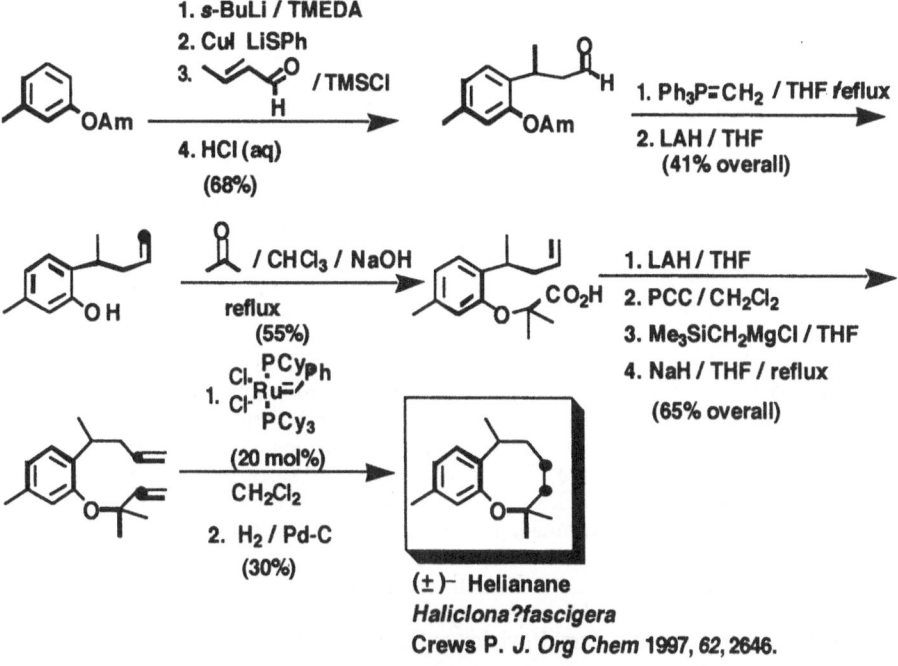

(±)- Helianane
Haliclona?fascigera
Crews P. *J. Org Chem* 1997, *62*, 2646.

Scheme 20

DoM - Metathesis Connection: Synthesis of Macrocyclic Sulfonamides

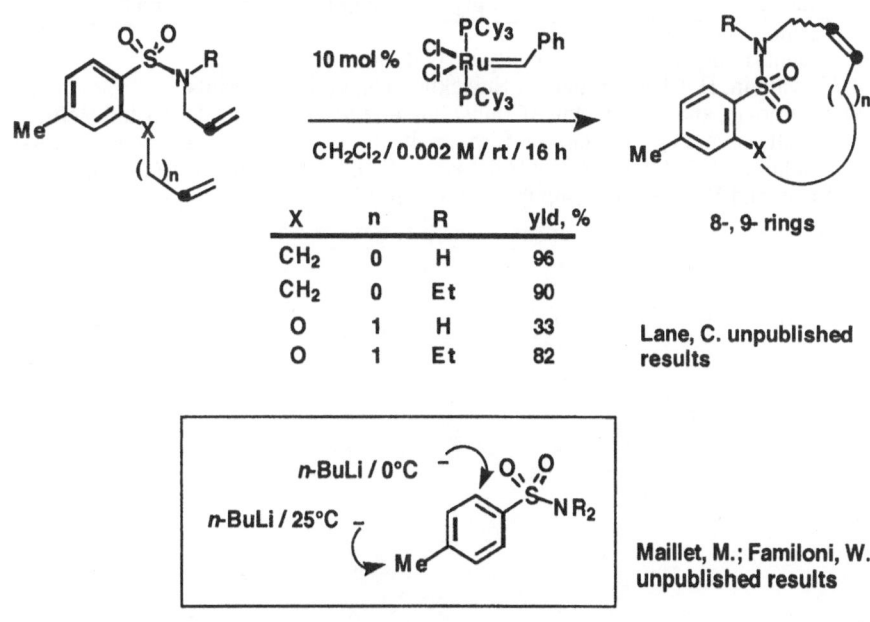

X	n	R	yld, %
CH₂	0	H	96
CH₂	0	Et	90
O	1	H	33
O	1	Et	82

8-, 9- rings

Lane, C. unpublished results

Maillet, M.; Familoni, W. unpublished results

Scheme 21

References

1. D. Hoppe, and A. Bronneke, Synthesis of 1-arylalkyl N,N-dialkylcarbamates by lithiation and substitution of benzylic-type carbamates, *Synthesis* 1045(1982).
2. R. Hoffmann, and R. Bruckner, A novel entry into Wittig rearrangements: A stereoselective [1,2]-Wittig rearrangement with inversion of configuration at the carbanionic center, *Chem. Ber.* 125:1957(1992).
3. P. Zhang, and R.E. Gawley, Directed metalation / Snieckus rearrangement of *O*-benzylic carbamates, *J. Org. Chem.* 58:3223(1993).
4. P. Beak, and A.I. Meyers, Stereo- and regio-control by complex induced proximity effects: Reactions of organolithium compounds, *Acc. Chem. Res.* 19:356(1986).
5. G.W. Klumpp, and M.J. Sinnige, Reaktionswarmen isomerer (lithioaryl)ether mit *s*-BuOH, *Tetrahedron Lett.* 27:2247(1986).
6. V. Snieckus, Directed *ortho* Metalation. Tertiary amide and *O*-carbamate directors in synthetic strategies for polysubstituted aromatics, *Chem. Rev.* 90:879(1990).
7. M.A.J. Miah, Ph.D. thesis, University of Waterloo (1985).
8. A.V. Kalinin, M.A.J. Miah, S. Chattopadhay, M. Tsukazaki, M. Wicki, T. Nguen, A.L. Coelho, M. Kerr, and V. Snieckus, Anionic homologus Fries rearrangement of *O*-(2-methylaryl)carbamates. A regiospecific route to benzo[b]furan-2(3H)-ones including an unnamed metabolite from *Helenium* species, *Synlett* 839(1997).
9. A. V. Kalinin, A. J. M. da Silva, C. C. Lopes, R. S. C. Lopes, and V. Snieckus, Directed *ortho* Metalation - Cross Coupling Links. Carbamoyl Rendition of the Baker-Venkataraman Rearrangement. Regiospecific Route to Substituted 4-Hydroxycoumarins, *Tetrahedron Lett.* in press (1998).
10. A.V. Kalinin, and V. Snieckus, 4,6-Dimethoxy-3,7-Dimethylcoumarin from *Colchicum decaisnei*. Total Synthesis by Carbamoyl Baker-Venkataraman Rearrangement and Structural Revision to Isoeugenetin Methyl Ether, *Tetrahedron Lett.* in press (1998).
11. S. Mohri, M. Stefinovic, and V. Snieckus, Combined directed *ortho*-, remote-metalation and cross-coupling strategies. Concise syntheses of the Kinemycin biosynthetic grid antibiotics Phenanthroviridin aglycon and Kinobscurinione, *J. Org. Chem.* 62:7072(1997).
12. B.-p. Zhao, Ph.D. thesis, University of Waterloo, 1993.
13. C.A. James, and V. Snieckus, Combined directed metalation - cross coupling strategies. Total synthesis of the aglycones of Gilvocarcin V, M and E, *Tetrahedron Lett.* 38:8149(1997).
14. C.A. James, Ph.D. thesis, University of Waterloo (1998).
15. A.V. Kalinin, B. Chauder, and V. Snieckus, Unpublished results.
16. M. Stefinovic, and V. Snieckus, Connecting directed *ortho* metalation and olefin metathesis strategies. Benzene-fused multiring-sized oxygen heterocycles. First syntheses of Radulanin A and Helianane, *J. Org. Chem.* 63:2808(1998).
17. C. Lane, and V. Snieckus, Unpublished results.

IODONIUM CHEMISTRY: MORE THAN A SIMPLE MIMICRY OF SOME TRANSITION METAL BASED ORGANIC TRANSFORMATIONS

José Barluenga and José M. González

Instituto Universitario de Química Organometálica
"Enrique Moles"-Unidad Asociada al C.S.I.C.
Universidad de Oviedo
Oviedo, 33071, Spain

INTRODUCTION

The formation of a complex between an alkyne and an organometallic reagent is at the outset of a wide set of metal-mediated transformations of this functional group.[1] Tipically, upon forming the complex the reactivity of the alkyne toward additional alkyne molecules is enhanced, enabling polimerization processes to take place.[2] The potential that this approach to carbon-carbon bond formation could offer for selective and efficiently preparing the homologous oligomers having lower molecular weight is a synthetic challenge. Thus, for instance, the cyclotrimerization reaction of alkynes has evolved into a mature synthetic methodology.[3] Although fundamental studies have been directed toward the formation of conjugated enynes by linear dimerization of alkynes catalyzed by transition metals;[4,5] the fact that many of these transformations show lack of regio- and/or stereocontrol have neverthelss precluded a further development of synthetic strategies based on this "atom economy"[6] process. Alternatively, sp^2-sp coupling reactions are commonly chosen for assembling this organic frame.[7] However, recent findings indicate that the cross-coupling of alkynes has a broader scope than that previously thought and suggest a renewed interest in the field.[8,9] Herein, are summarized some of our results on the preparation of enynes through novel linear coupling reactions of of heteroatom substituted alkynes promoted by bis(pyridine) iodonium (I) tetrafluoroborate (IPy_2BF_4). Furthermore, some other synthetic applications of the reagent will be outlined.

ABOUT THE IPy_2BF_4 REAGENT

The bis(pyridine) iodonium (I) tetrafluoroborate is a white solid, stable in air, reagent that can be safely stored for long periods of time without noticeable decomposition protecting it from exposure to the light. In early work,[10] it was shown that upon treatment with an acid the reagent - by protonation of the pyridine ligands- might act as an efficient source of iodonium ions in solution toward an alkene. Tetrafluoroboric acid was chosen for the standard activating step due to the fact that the low nucleophilicity of the tetra-fluoroborate counteranion should facilitate the incorporation of an externally added nucleophile avoiding a competitive nucleophilic attack. On this ground, a quite general procedure was established to yield products arising from the vicinal functionalization of an alkene by incorporation of iodine and a nucleophile (1). Several experimental protocols were developed, enabling a representative group of compounds to participate as nucleophiles.

$$R^1R^2C=CR^3R^4 + Nu + IPy_2BF_4 + 2HBF_4 \rightarrow R^1R^2C(Nu)-C(I)R^3R^4 + 2PyH^+BF_4^- \quad (1)$$

The products are obtained with regio- (according to Markonikov's rule for an addition of I[+]) and stereochemical control (resulting from an *anti* addition), that can be reassonably rationalize assuming a prior formation of an intermediate cyclic iodonium ion, from which ring-opening by nucleophilic attack leads to the final compounds. Interestingly, when none external nucleophile is added the associated tetrafluoroborate anion acts as a source of fluoride, resulting in a mild an efficient procedure to yield regio- and stereoselectively vicinal fluoroiodo compounds from simple olefins.[11] More recently, it has been shown that starting from terminal vinylsilanes the reaction offers an attractive synthetic manifold. Thus, in absence of other nucleophiles a clean an efficient stereospecific iododesilylation can be accomplished[12] but, in presence of alcohols, the reaction is directed to give the corresponding adducts, from which trisubstituted enolethers can be easily obtained by simple dihydroiodination with bases, opening a new and stereospecific access to trisubstituted enolethers containing silyl substituents.[13]

In a related manner, the IPy_2BF_4 reagent promotes addition reactions of iodine and several nucleophiles to alkynes giving the corresponding β-functionalized vinyl iodides, usually in good yields through regio- and stereoselctive processes.

A relevant synthetic feature of the reagent is their usefulness to promote stereoselctive carbocyclization reactions at low temperature; a reaction showing potential to furnish six-member rings arising by nucleophilic attack of the arene to the intermediate generated from either an starting alkene[14] or alkyne,[14,15] and the activated iodonium ion, in a novel Fridel-Crafts-like process (Figure 1). The related cyclization of α,ω-diolefins is also feasible forming carbocycles containing six-member rings.[14] The observed intermolecular arene-alkyne reaction, activating a C(aryl)-H bond and forming a carbocycle containing a versatile C(sp[2])-I bond, nicely complement the regiochemistry found for the Pd-catalyzed intermolecular arylation of a 4-aryl-substituted-1-alkyne,[16] a process requiring an *o*-C(aryl)-I bond in the arene component to trigger the reaction.

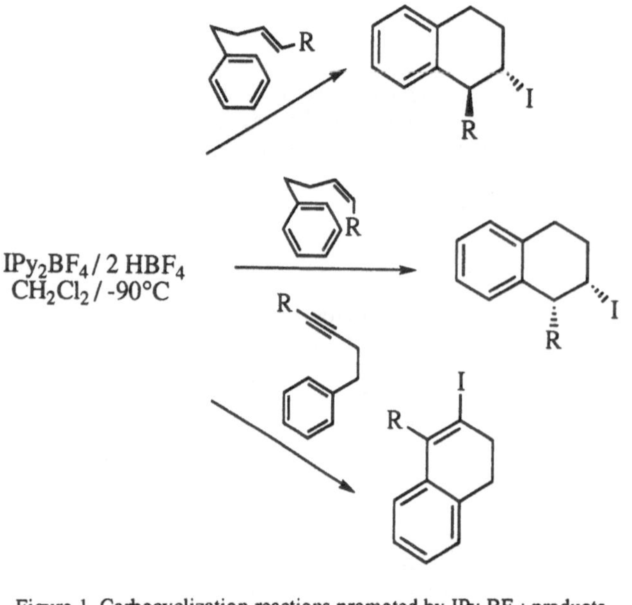

Figure 1. Carbocyclization reactions promoted by IPy_2BF_4: products formed upon arene capture of the intermediate iodonium ion.

The success of the intermolecular reaction to provide additional examples of carbon-carbon bond formation proved the usefulness of alkynes as alkylating agents in this Fridel-Crafts-like process,[17] a representative transformation of a major class of reaction that is subject of current research efforts.[18] Further early work on the reactivity of IPy_2BF_4 toward alkynes include an easy and general preparation of iodoalkynes.[19] In the following section, dimerization reactions of this class of functionalized alkynes and of related silylalkynes will be discussed.

COUPLING REACTIONS OF ALKYNES PROMOTED BY IPy₂BF₄

In this section are summarized some results concerning the application of IPy$_2$BF$_4$ to accomplish the dimerization of heteroatom substituted derivatives of terminal alkynes, in a selective manner. Examples, of both stoichiometric and catalytic process will be presented and attention will be paid to the synthetic potential that the newly formed functionality could have.

Dimerization of Iodoalkynes

Alkynyl iodides are useful and interesting compounds that keep on receiving synthetic attention.[20] Their utility in sp-sp coupling reaction is well documented;[21] however, the linear dimerization process of these substrata furnishing enynes remains as an elusive synthetic target.

We have reported that 1-iodo-1-alkynes can be smoothly coupled to give 1,1-diiodo-2,4-disubstituted-1,3-butenynes, dimers arising formally from a head-to-tail coupling mode, as exclusive oligomerization products upon reaction with IPy$_2$BF$_4$/2HBF$_4$ at low temperature.[22] The reaction requires that the alkyne bears appended an aryl or a vinyl substituent to take place. Those iodoalkynes featuring the former substitution pattern can be converted almost quantitatively into the corresponding dimers, while the yield ranges now only in the 30-40% for those derived from enynes, due to the occurrence of monomer side reactions -mainly competitive fluoroiodination-. On the basis of the experimental data, a reasonable mechanistic assumption is that the formation of the dimer would result in the liberation of an additional equivalent of "I$^+$". In keeping with this, the reaction should be catalytic in the iodonium source and, indeed, this possibility was tested. Significantly, a selective and efficient dimerization of iodoalkynes having an aromatic substituent attached to the acetylenic carbon takes place using only catalytic amounts of the IPy$_2$BF$_4$ reagent (Figure 2). To this regard only a 2% of IPy$_2$BF$_4$ was used to dimerize the p-MeOC$_6$H$_4$C≡CI at the 50 mmol scale. A comparison of the experimental conditions required for different 4-arylsubstituted derivatives to give efficient conversions into dimers suggest a reactivity trend compatible with an electrophilic reaction pathway. Crucial variables to control the reaction were concentration of the iodonium reagent, temperature and reaction time.

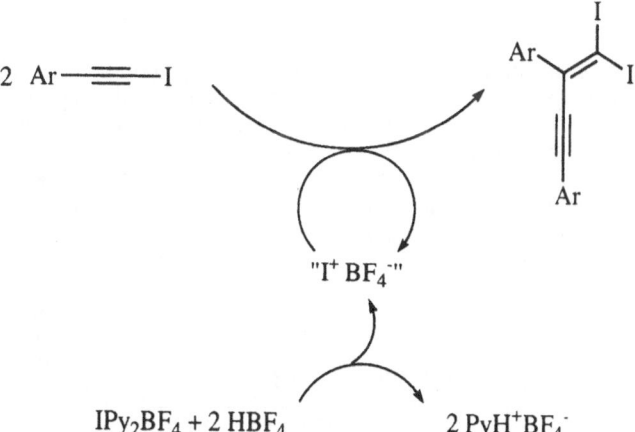

Figure 2. A schematic view of the head-to-tail dimerization
of iodoalkynes upon reaction with catalytic amounts of IPy$_2$BF$_4$.

This synthetic strategy provides a simple solution to accomplish the dimerization of this metal sensitive functionalized alkynes in the desired way. Besides control over the extension of the oligomerization reaction and regiochemistry, this non-destructive approach matches well requirements of a transformation following atom-economy criteria[6], and results in the elaboration of an interesting functionality. Among the main limitations of this

147

approach is the fact that iodoalkynes reacting in this process are so far restricted to those having as sp²-based unsaturation bounded to the alkyne. Nevertheless, this situation is also encountered in transition metal mediated dimerizations of alkynes, where a modification in the structure of the alkyne often results in dramatic changes in terms of reactivity.

Dimerization of Silylalkynes

An interesting challenge raised by the dimerization reaction of iodoalkynes previously described is to attempt to differentiate the two iodine atoms by chemical means to further develop its synthetic potential. Alternatively, we became interested in exploring the possibility of preparing enynes bearing two rich and well differentiate functionalities at C-1. In this frame, we chose to study the dimerization reaction of silylalkynes mediated by IPy_2BF_4 and, if it turned a feasible transformation, in analyzing the coupling mode, and the regio- and stereochemical outcome of the process. At low temperature, trimethylsilyl substituted terminal alkynes rapidly give rise to the formation of the corresponding iodoalkynes upon reacting with IPy_2BF_4 and variable amounts of HBF_4, this easy iododesilylation prevents the participation of these alkynes in dimerization reactions promoted by the iodonium reagent. However, the related (*tert*-butyldimethylsilyl)alkynes (TBDMS-alkynes) suffer a slower desilylatio, enabling a dimerization process to satisfactorily compete. After careful optimization of the experimental conditions, an efficient homocoupling of alkynylsilanes was accomplished,[23] by means of the reaction of aryl-substituted TBDMS-alkynes with IPy_2BF_4-HBF_4 (Figure 3).

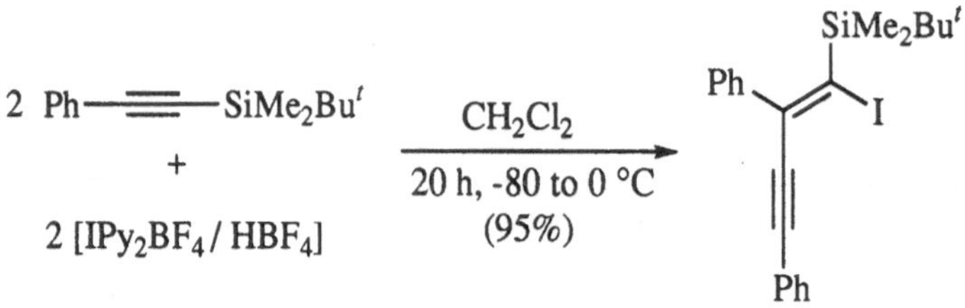

Figure 3. Representative example for the IPy2BF4 mediated head-to-tail homocoupling of alkynylsilanes at low temperature.

This new process furnished again dimers as single regio- and stereoisomers, without noticeable formation of further oligomerization products under the reported conditions. The fact that a new C(sp²)-I bond is created along the reaction adds synthetic interest to this homocoupling, and has been used to elaborate enediyne compounds.[23]

Furthermore, a different and remarkably selective reaction takes place starting from the same TBDMS-alkynes just by increasing the final reaction temperature and time,[23] giving (*E*)-1,4-diaryl-1,2-diiodo-1,3-butadiynes (Figure 4).

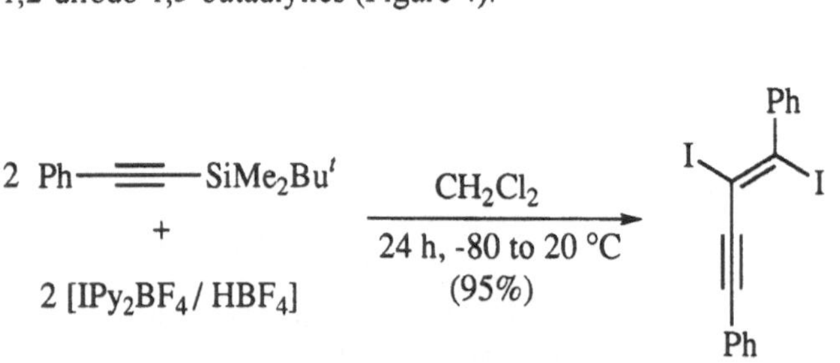

Figure 4. Selective dimerization of alkynylsilanes after room temperature reaction completion: synthesis of the products of a formal head-to-head coupling of 1-iodo-alkynes giving enynes.

These diiodo compounds (Figure 4) can also be clean and efficiently prepared from pure samples of 2,4-diaryl-1-iodo-1-(tert-butyldimethylsilyl)-1,3-enynes;[23] thus, showing those diiodinated dimers to be formed following a sequential transformation, for which the reaction depicted in Figure 3 takes place firstly. An iodolium ion has been characterized at low temperature (-30 °C) by means of several NMR experiments as an intermediate for the second transformation (Figure 5). Experiments with labels on the arene substituents show that the aryl group attached to the sp^2 carbon in the starting TBDMS-enyne rearranges along the reaction path, ending on a sp carbon in the final diiodo compound.

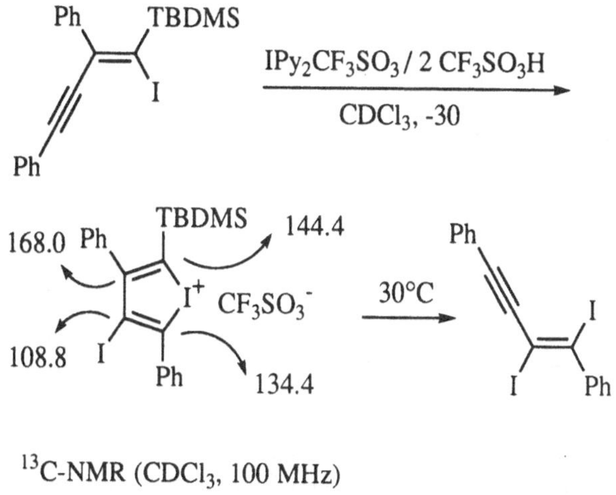

^{13}C-NMR (CDCl$_3$, 100 MHz)

Figure 5. Intermediate iodolium ion in the conversion of TBDMS-containing dimers into rearranged diiodocompounds.

Further work using different heteroatom-substituted alkynes is in progress, aimed to the activation of alkynes bearing aliphatic substituent to enter in these type of processes.

AROMATIC IODINATION WITH IPY$_2$BF$_4$

A different field of application for the reagent is in aromatic iodination processes. To this regard, an efficient and general method was established based on the unique capability of the IPy$_2$BF$_4$ reagent to iodinate a wide set of representative aromatic compounds at room temperature.[24] Thus, for instance, an efficient monoiodination of compounds -as opposite in electronic activation- as aniline or nitrobenzene can be equally accomplished using IPy$_2$BF$_4$. The products are formed in a regioselective manner, according to the general rules governing electrophilic aromatic substitution reactions. The iodinating efficiency of the reagent was tested in polyiodination of benzene.[25] Experimental conditions were traced to carry out these iodinations selectively, from the mono- up to hexaiododerivative.

This methodology is also useful to iodinate derivatives of the aminoacids tyrosine (Tyr) and phenylalanine (Phe). Using this approach, Tyr residues can be selectively iodine-labelled in small peptides containing Phe fragments.[26] Interestingly, this protocol for direct iodination is compatible with methionine-containing peptides, an aminoacid easily oxidized using alternative direct iodination reactions.

Further work exploring the feasibility of a solid phase adaptation of this chemistry and its implications is in progress.

REFERENCES

1. Davidson, J.L., Reactions of coordinated acetylenes, in: *Reactions of Coordinated Ligands*, P.S. Braterman, ed., Plenum Press, New York (1986).
2. D.L. Trumbo, C.S. Marvel, Polymerization of phenylacetylene using palladium (II) catalysts, *J. Polym. Sci., A* 25:1027 (1987).

3. K.P.C. Vollhardt, Cobalt-mediated [2+2+2]-cycloadditions: a maturing synthetic strategy, *Angew. Chem. Int. Ed. Engl.* 23:539 (1984); K.P.C. Vollhardt, Cobalt-mediated synthesis of polyheterocycles, *Lect. Heterocycl. Chem.* 9:59 (1987)

4. M.J. Winter, Alkyne oligomerization, in: *The Chemistry of Metal-Carbon Bond*, F.R. Hartley, S. Patai, eds., J. Wiley, New York (1985); D.B. Grotjahn, Transition metal alkyne complexes: transition metal-catalyzed cyclotrimerization, in: *Comprehensive Organometallic Chemistry II*, E.W. Abel, F.G.A. Stone, G. Wilkinson, eds., Pergamon Press, Oxford (1995)

5. For recent examples, see: M. Schäfer, N. Mahr, J. Wolf, H. Werner, Metal-initiated coupling of C_2 units to enynes and butatrienes: two different routes for the dimerization of 1-alkynes, *Angew. Chem. Int. Ed. Engl.* 32:1315 (1993), C. Bianchini, P. Frediani, D. Masi, M. Peruzzini, F. Zanobini, Regio- and stereoselective dimerization of phenylacetylene to (Z)-1,4-diphenylbut-3-en-1-yne by ruthenium(II) catalysis. Reaction mechanism involving intermolecular protonotation of σ-alkynyl by 1-alkyne, *Organometallics* 13:4616 (1994); T. Straub, A. Haskel, M.S. Eisen, Organoactinide-catalyzed oligomerization of terminal acetylenes, *J. Am. Chem. Soc.* 117:6364 (1995); C.S. Yi, N. Liu, Homogeneous catalytic dimerization of terminal alkynes by $C_5Me_5Ru(L)H_3$ (L = PPh_3, PCy_3, PMe_3), *Organometallics* 15:3968 (1996); M. Yoshida, R. F. Jordan, Catalytic dimerization of terminal alkynes by a hafnium carboranyl complex. A "self-correcting" catalyst, *Organometallics* 16:4508 (1997).

6. B.M. Trost, Atom-economy- a challenge for organic synthesis: homogeneous catalysis leads the way, *Angew. Chem. Int.Ed. Engl.*34:259 (1995)

7. R. Hara, Y. Liu, W.-H. Sun, T. Takahashi, Highly substitutrd enyne formation by coupling reaction of alkenylzirconium compounds with alkynyl halides, *Tetrahedron Lett.* 38:4103 (1997), and references therein.

8. B.M. Trost, M.T. Sorum, C. Chan, A.E. Harms, G. Rühter, Palladium-catalyzed additions of terminal alkynes to acceptor alkynes, *J.Am. Chem.Soc.* 119:698 (1997); B.M. Trost, M.C. McIntosh, An unusual selectivity in Pd catalyzed cross-coupling of terminal alkynes with "unactivated" alkynes, *Tetrahedron Lett.* 38:3207 (1997); B.M. Trost, M.J. Krische, Transition metal catalyzed cycloisomerizations, *Synlett* 1 (1998)

9. N. Chatani, N. Amishiro, S. Murai, A new catalytic reaction involving oxidative addition of iodo-trimethylsilane (Me_3SiI) to Pd(0). Synthesis of stereodefined enynes by the coupling of Me_3SiI, acetylenes and acetylenic tin reagents, *J. Am. Chem. Soc.* 113:7778 (1991); E. Shirakawa, H. Yoshida, T. Kurahashi, Y. Nakao, T. Hiyama, Carbostannylation of alkynes catalyzed by an iminophosphine-palladium complex, *J. Am. Chem. Soc.* 120:2975 (1998).

10. J. Barluenga, J.M. González, P.J. Campos, G. Asensio, IPy_2BF_4 a new reagent in organic synthesis: general method for the 1,2-iodofunctionalization of olefins, *Angew. Chem. Int. Ed. Engl.* 24:319 (1985).

11. J. Barluenga, P.J. Campos, J.M. González, J.L. Suárez, G. Asensio, Regio- and stereoselective iodofluorination of alkenes with bis(pyridine)iodonium (I) tetrafluoroborate, *J. Org. Chem.* 56:2234 (1991); A.P. Khrimian, A.B.DeMilo, R.M. Waters, N.J. Liquido, J.M. Nicholson, Monofluoro analogs of eugenol methyl ether as novel attractants for the oriental fruit fly, *J. Org. Chem.* 59.8034 (1994).

12. J. Barluenga, L.J. Alvarez-García, J.M. González, IPy_2BF_4 is also a useful reagent for stereospecific iodine-silicon exchange in open chain trimethylsilylalkenes, *Tetrahedron Lett.* 36:2153 (1995).

13. J. Barluenga, L.J. Alvarez-García, G.P. Romanelli, J.M. González, Stereospecific access to trisubstituted enol ethers from vinylsilanes, *Tetrahedron Lett.* 38:6763 (1997).

14. J. Barluenga, J.M. González, P.J. Campos, G. Asensio, Iodine-induced stereoselective carbocyclizations: a new method for the synthesis of cyclohexane and cyclohexene derivatives, *Angew. Chem. Int.Ed. Engl.* 27:1546 (1988).

15. M.B. Goldfinger, T.M. Swager, Fused-polycyclic aromatics via electrophile-induced cyclization reactions: applications to the synthesis of graphite ribbons, *J. Am. Chem. Soc.* 116:7895 (1994); M.B. Goldfinger, K.B. Crawford, T.M. Swager, directed electrophilic cyclizations: efficient methodology for the synthesis of fused polycyclic aromatics, *J. Am. Chem. Soc.* 119:4578 (1997).

16. Y. Zhang, E. Negishi, Palladium-catalyzed cascade carbometallation of alkynes and alkenes as an efficient route to cyclic and polycyclic structures, *J. Am. Chem. Soc.* 111:3454 (1989).

17. J. Barluenga, M.A. Rodríguez, J.M. González, P.J. Campos, Iodo-carbofunctionalization of alkynes with aromatic rings and IPy_2BF_4, *Tetrahedron Lett.* 29:4207 (1990).

18. M. Yamaguchi, A. Hayashi, M. Hirama, Ortho-vinylation and ortho-alkenylation of phenols, *J. Am. Chem. Soc.* 117:1151 (1995); M. Yamaguchi, Y. Kido, A. Hayashi, M. Hirama, Friedel-Crafts b-silylvinylations, *Angew. Chem. Int. Ed. Engl.* 36:1313 (1997).

19. J. Barluenga, J.M. González, M.A. Rodríguez , P.J. Campos, G. Asensio, An improved method for the synthesis of 1-iodoalkynes, *Synthesis.* 661 (1985).

20. H. Hopf, B. Witulski, Functionalized acetylenes in organic synthesis - the case of the 1-cyano- and the 1-halogenoacetylenes -, in: *Modern Acetylene Chemistry*, P.J. Stang, F. Diederich, e., VCH: Weinheim (1995).

21. C. Amatore, E. Blart, J.P. Genêt, A. Jutand, S. Lemaire-Audoire, M. Savignac, New-synthetic applications of water soluble acetate Pd/TPPTS catalyst generated *in situ*. Evidence for a true Pd(0) species intermediate, *J. Org. Chem.* 60:6829 (1995).

22. J. Barluenga, J.M. González, I. Llorente, P.J. Campos, 1-Iodoalkynes can also be dimerized: a new head-to-tail coupling, *Angew. Chem. Int. Ed. Engl.* 32:893 (1993).

23. J. Barluenga, I. Llorente, L.J. Alvarez-García, J.M. González, P.J. Campos, M. Rosario Díaz, S. García-Granda, Reactions of (*tert*-butyldimethylsilyl)alkynes with IPy$_2$BF$_4$: selective synthesis of novel head-to-head dimers, *J. Am. Chem. Soc.* 119:6933 (1997).

24. J. Barluenga, J.M. González, M.A. García-Martín, P.J. Campos, G. Asensio, Acid-mediated reaction of bis(pyridine)iodonium (I9 tetrafluoroborate with aromatic compounds. A selective and general iodination method, *J. Org. Chem.* 58:2058 (1993).

25. J. Barluenga, J.M. González, M.A. García-Martín, P.J. Campos, Polyiodination on benzene at room temperature. A regioselective synthesis of derivatives, *Tetrahedron Lett.* 34:3893 (1993).

26. J. Barluenga, M.A. García-Martín, J.M. González, P. Clapés, G. Valencia, Iodination of aromatic residues in peptides by reaction with IPy$_2$BF$_4$, *Chem. Commun.* 1505 (1996).

STEREOSELECTIVE SYNTHESES OF γ-LACTONES AND γ-ALKYLIDENE-BUTENOLIDES

Christian Harcken, Thilo Berkenbusch, Stefan Braukmüller, Andreas Umland, Konrad Siegel, Felix Görth, Frank von der Ohe, and Reinhard Brückner

Institut für Organische Chemie der Georg-August-Universität
Tammannstr. 2
D-37077 Göttingen, Germany

INTRODUCTION

Five-membered ring lactones abound in nature and have ever since challenged synthetic chemists.[1-3] They comprise butanolides (**1**), butenolides (**2**), α-alkylidenebutanolides (**3**), and γ-alkylidenebutenolides (**4**; Fig. 1). We have developed versatile routes to such compounds. They encompass only a few steps and ensure a high degree of stereocontrol at all stereogenic sp^3 centers and C=C bonds. In addition they allow to vary the substituents R, R´, and R´´ considerably. Therefore, we were able to synthesize butanolides and butenolides which differed structurally from one another distinctly.

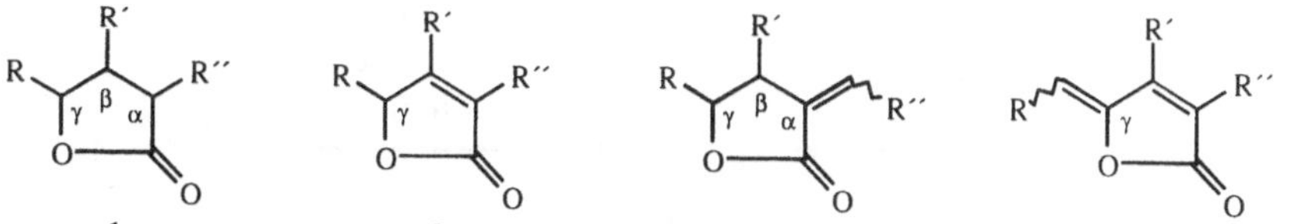

Figure 1. Types of γ-lactones synthesized in the present study.

Our retrosynthetic analysis traced back the butanolide targets **1-2** and the butenolide targets **3**, all of which are γ-chiral γ-lactones, to the hydroxybutanolides **6** and **7**, respectively (Fig. 2). The latters constitute lactonization products of the enantiopure diols originating from the Sharpless asymmetric dihydroxylation[4] ("AD") of the *trans*-configurated deconjugated esters **5** and **8**. Depending on whether these dihydroxylations are performed with AD mix α ™ or AD mix ™, lactones **6** and **7** are *S,S*- or *R,R*-configurated.[5]

The fourth kind of target lactones reached in the present study were the γ-alkylidenebutenolides *E*- and *Z*-**4**. They were traced back retrosynthetically to the γ-(α-

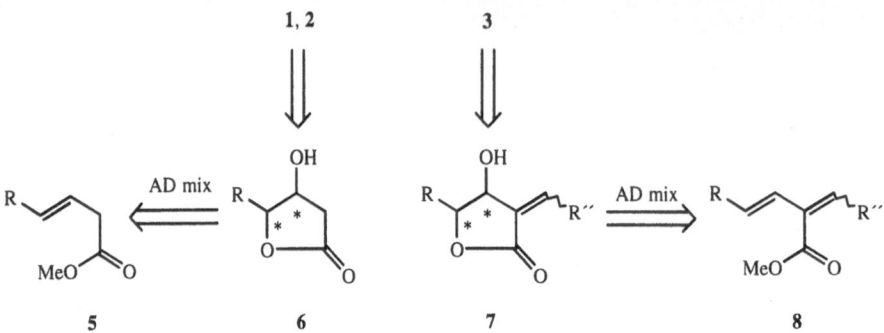

Figure 2. Our strategy for the preparation of γ-chiral butanolides/butenolides

hydroxyalkyl)butenolide isomers **9** and *epi*-**9**, respectively (Fig. 3). These isomers had to be subjected to *anti*-selective β-eliminations of water.[6] In doing so, the pairwise interconversion of **9** and *epi*-**9** and of *E*-**4** and *Z*-**4** had to be precluded and a competing *syn*-elimination, too.

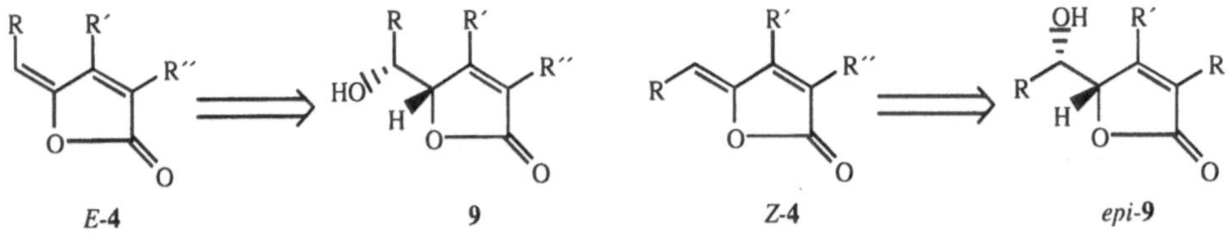

| *E*-4 | 9 | *Z*-4 | *epi*-9 |

Figure 3. Our strategy for the preparation of γ-alkylidenebutenolides.

ENANTIOSELECTIVE SYNTHESES OF γ-CHIRAL γ-LACTONES

Fig. 4-6 show three type-**1** butanolide syntheses performed via type-**2** butenolides.

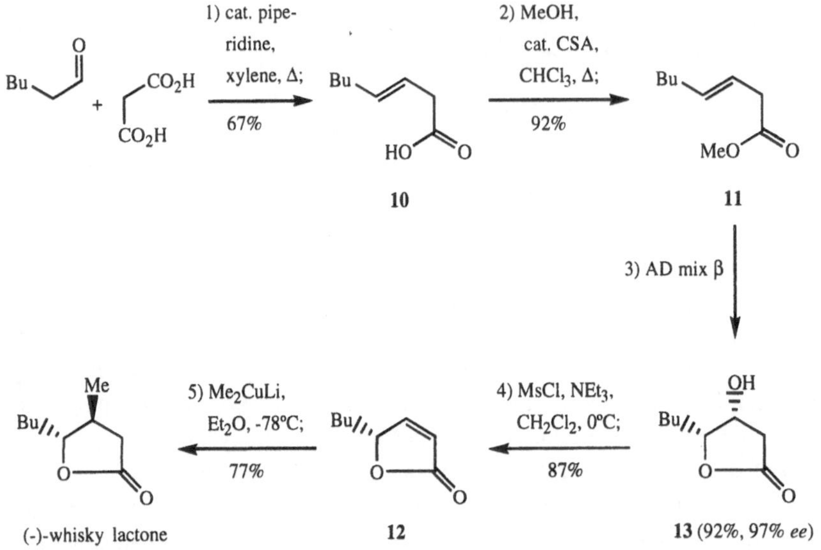

Figure 4. Five-step synthesis of whisky (or quercus) lactone (ref. 7).

154

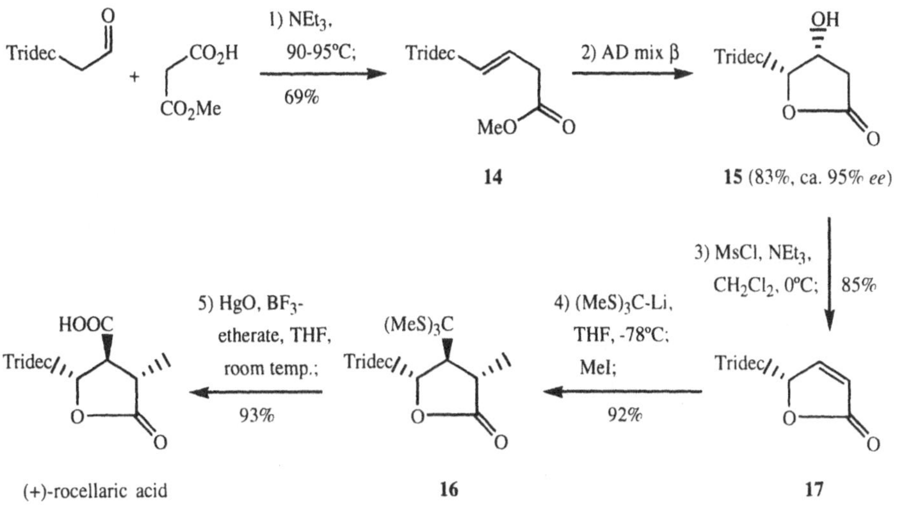

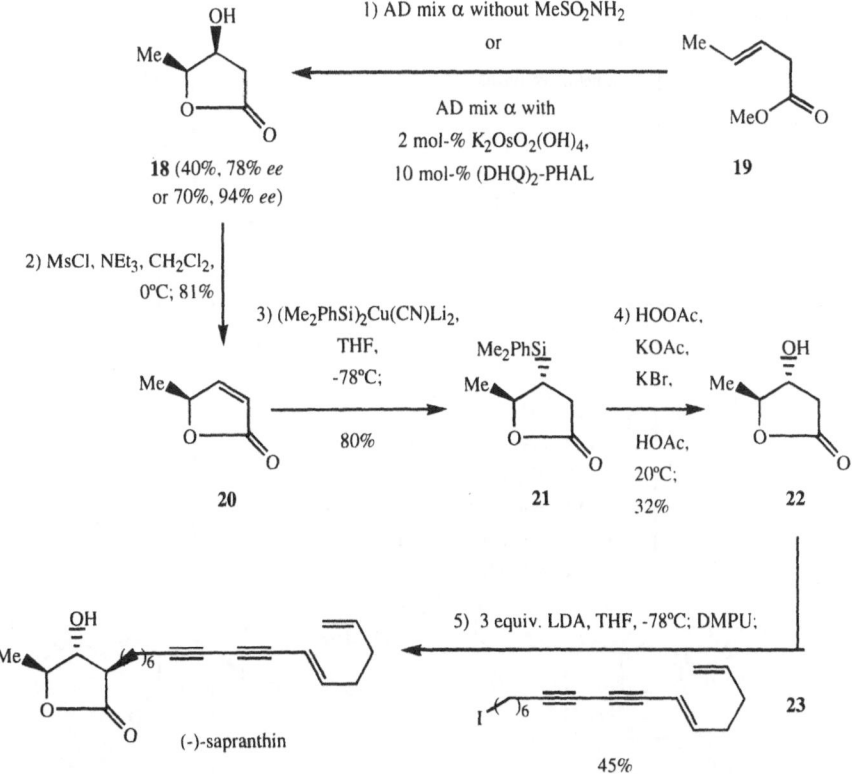

Figure 5. Five-step synthesis of rocellaric acid (ref. 8).

Fig. 4 exhibits the presently shortest synthesis of enantiopure whisky lactone. Acid **10** was prepared by a decarboxylative, deconjugating and essentially *trans*-selective Knoevenagel condensation between malonic acid and hexanal. The derived methylester **11** and AD mix β ™ furnished lactone **13** in 92% yield and with 97% *ee*. Dehydration and a *trans*-selective 1,4-addition Me_2CuLi to the resulting butenolide **12** provided the target compound.

Figure 6. Structure-proving synthesis of sapranthin (ref. 9; improved procedure for lactone **18**: ref. 10).

Fig. 5 represents a concise synthesis of dextrorotatory rocellaric acid. Here, substrate **14** of the dihydroxylation reaction emerged directly from a decarboxylative, deconjugating and now – gratifyingly – completely *trans*-selective Knoevenagel condensation between pentadecanal and malonic acid monomethylester. AD of substrate **14** with AD mix β ™ delivered lactone **15** with about 95% *ee*. Dehydration to the butenolide **17** occurred in the presence of mesyl chloride and triethylamine. Converting this butenolide into the final structure was based upon literature precedence. It was facilitated kinetically and yield-wise by introducing the COOH group via the 1,4-addition of Li–C(SMe)₃ and not via Li–C(SPh)₃.

Ester **19** and the usual amount of AD mix α ™ gave 40% lactone **18** with 78% *ee* (Fig. 6). Raising the osmium content to 2 mol-% and the amount of added chiral auxiliary to 10 mol-% increased the chemical yield to 70% and the *ee*-value to 94%. The OH group of lactone **18** was inverted through mesylation/β-elimination (→ butenolide **20**), 1,4-addition of a silyl cuprate (→ silyllactone **21**), and a Fleming oxidation (→ hydroxylactone **22**). The derived enolate dilithio-**22** was alkylated with iodide **23** *trans*-selectively. The ¹H- and ¹³C-NMR spectra of the resulting lactone and the sign and value of its optical rotation coincide with the data measured for natural sapranthin. These findings corrected the previously mis-assigned relative configuration of (-)-sapranthin and established its absolute configuration.[11]

STEREOSELECTIVE SYNTHESES OF γ-ALKYLIDENEBUTENOLIDES

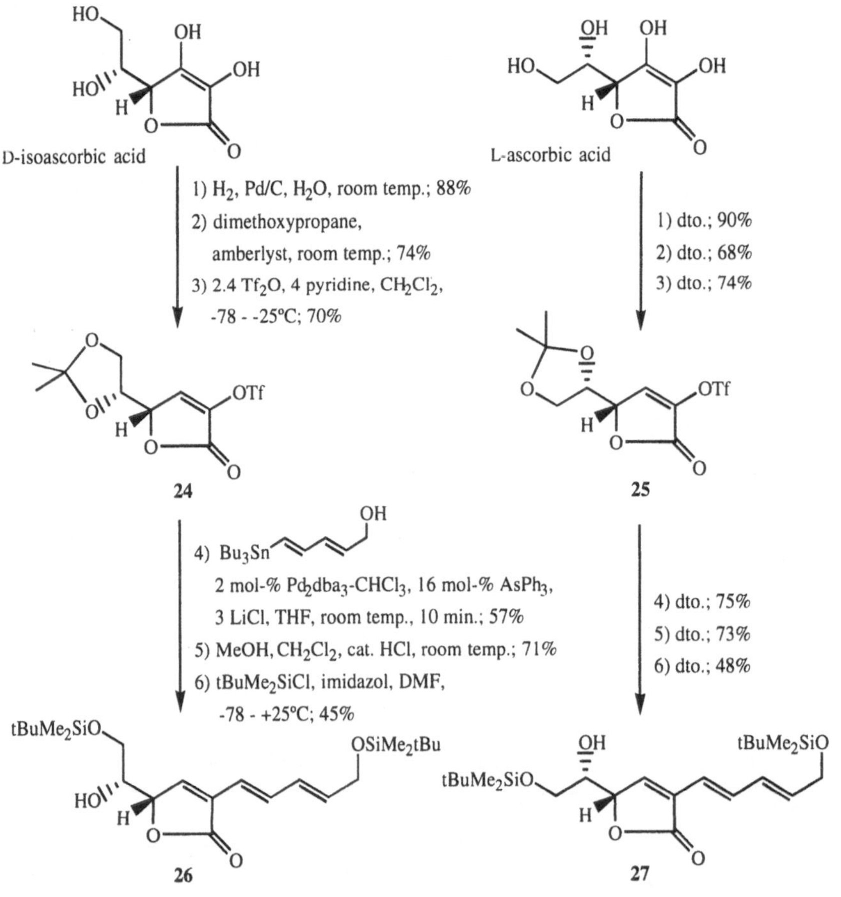

Figure 7. Preparation of stereopure model substrates for establishing the stereospecific γ-(α-hydroxyalkyl)-butenolide→γ-alkylidenebutenolide route (ref. 13).

Our strategy (Fig. 3) for attaining γ-alkylidenebutenolides *E*- and *Z*-**4** with configurationally homogenous $C_{exocyclic}=C_\gamma$ bonds required stereopure γ-(α-hydroxyalkyl)butenolide precursors **9** for the *E*-isomers and *ent*-**9** for the *Z*-isomers. Synthesizing these precursors with unambiguous relative configurations constitutes already a certain challenge. Therefore, we started our exploratory work from the commercially available γ-(α-hydroxyalkyl)butenolides isoascorbic acid and ascorbic acid possessing known configurations (Fig. 7).

These acids were transformed into the enol triflates **24** and **25**, respectively, as previously described.[12] Stille-couplings with *trans,trans*-Bu₃Sn–CH=CH-CH=CH-CH₂-OH followed. Subsequent cleavages of the acetonide rings and bis(*tert*-butyldimethylsilylations) rendered compounds **26** and **27**, respectively, as the elimination substrates proper. By treatment with triflic anhydride and pyridine (Fig. 8) they let us form the crucial $C_{exocyclic}=C_\gamma$ bonds *anti*-selectively (*ds* = 94:6). Desilylations provided „isotetrenolin" and „isolissoclinolide", i. e. γ-alkylidenebutenolides isomers of the antibiotics tetrenolin (active against Gram positive bacteria) and lissoclinolide (active against Gram negative bacteria).

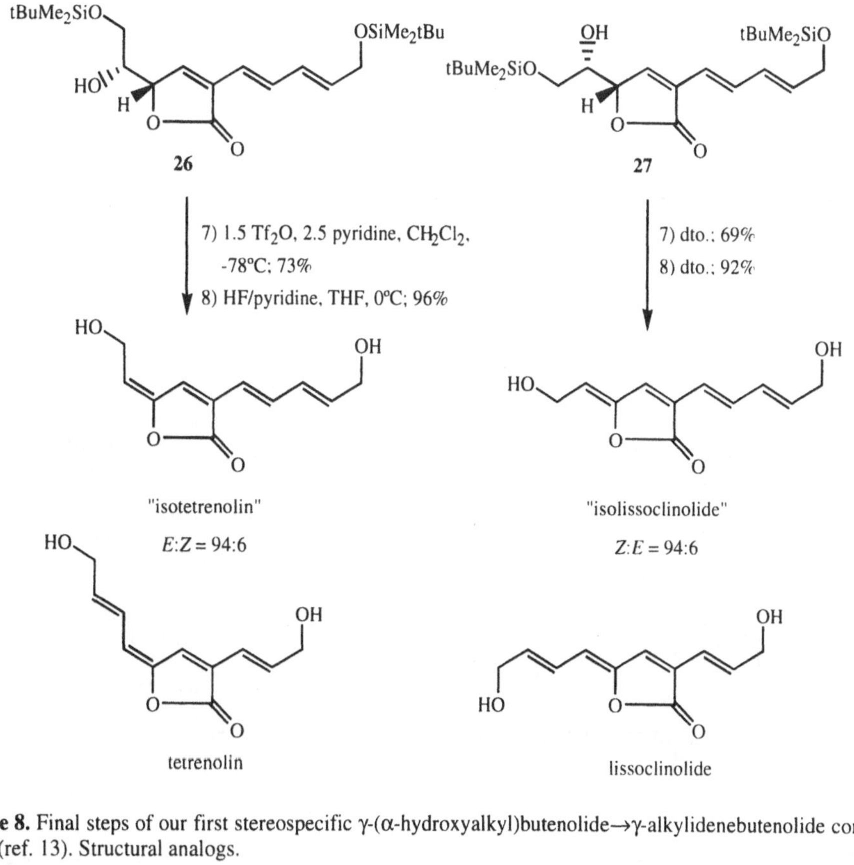

Figure 8. Final steps of our first stereospecific γ-(α-hydroxyalkyl)butenolide→γ-alkylidenebutenolide conversions (ref. 13). Structural analogs.

Fig. 9 reveals how a modification of the ascorbic acid → „isolissoclinolide" pathway (Fig. 7-8) allowed to synthesize lissoclinolide. In hydrogenated ascorbic acid, HBr engaged the primary OH group as well as the OH group in α position of the C=O bond in S_N1 brominations. The resulting diol was subjected to a twofold dehydration the second one of which formed the $C_{exocyclic}=C_\gamma$ bond with 99:1 *Z*-selectivity. The obtained allyl bromide and triphenylphosphine delivered phosphonium bromide **28**. A high-temperature Wittig reaction of the derived ylid, an ensuing Stille coupling, and a desilylation terminated our approach.

157

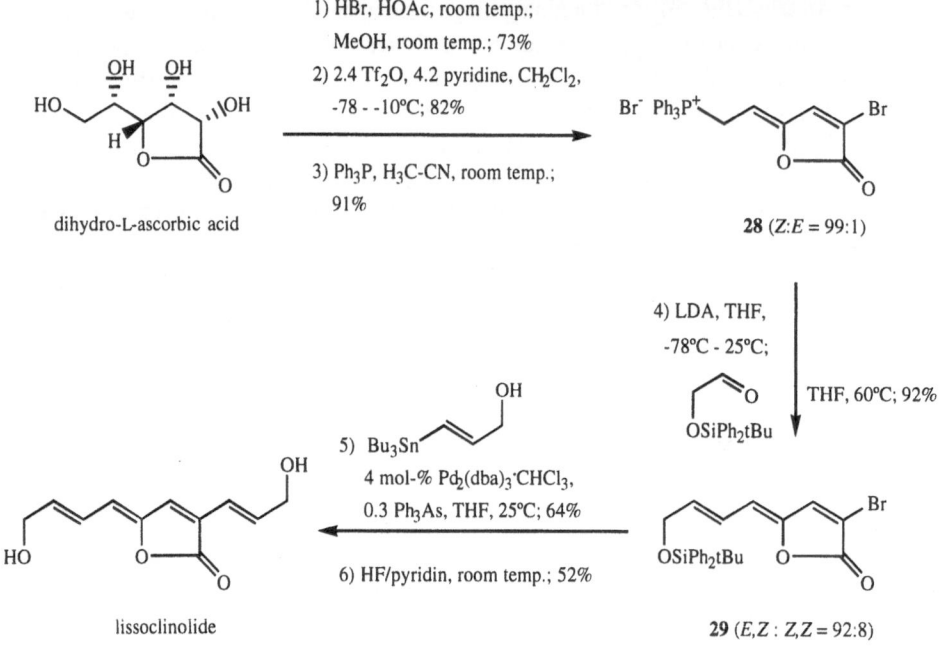

dihydro-L-ascorbic acid

1) HBr, HOAc, room temp.;
   MeOH, room temp.; 73%
2) 2.4 Tf$_2$O, 4.2 pyridine, CH$_2$Cl$_2$,
   -78 - -10°C; 82%
3) Ph$_3$P, H$_3$C-CN, room temp.;
   91%

28 (Z:E = 99:1)

4) LDA, THF,
   -78°C - 25°C;

THF, 60°C; 92%

OSiPh$_2$tBu

5) Bu$_3$Sn

4 mol-% Pd$_2$(dba)$_3$·CHCl$_3$,
0.3 Ph$_3$As, THF, 25°C; 64%

6) HF/pyridin, room temp.; 52%

29 (E,Z : Z,Z = 92:8)

lissoclinolide

Figure 9. First synthesis of lissoclinolide (ref. 14).

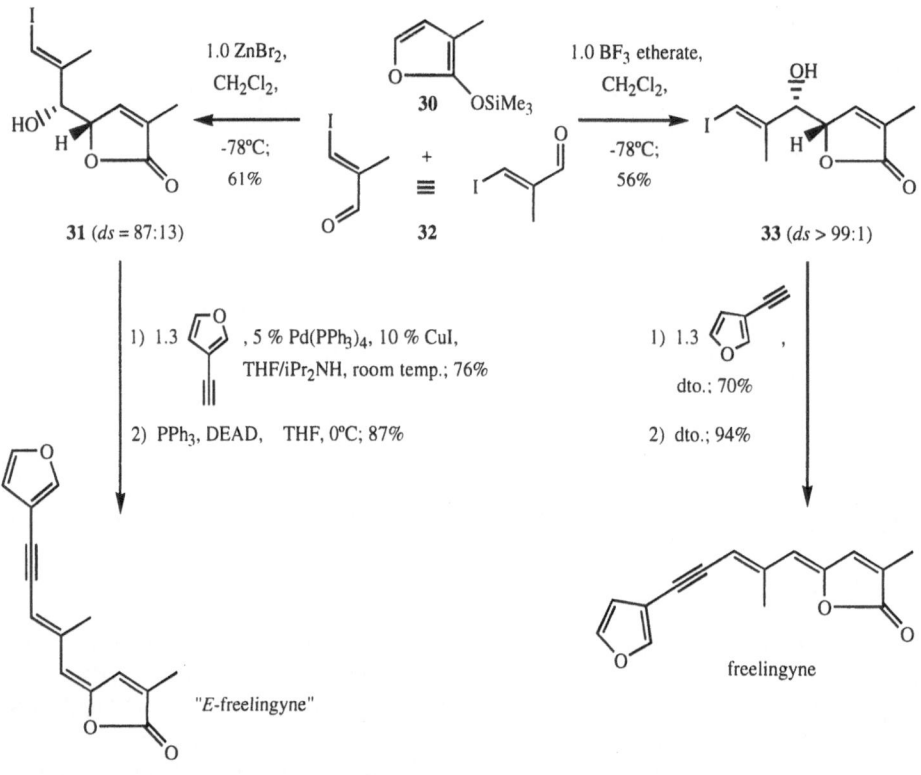

31 (ds = 87:13)

1.0 ZnBr$_2$,
CH$_2$Cl$_2$,

-78°C;
61%

30

+

≡

32

1.0 BF$_3$ etherate,
CH$_2$Cl$_2$,

-78°C;
56%

33 (ds > 99:1)

1) 1.3 ⌬ , 5 % Pd(PPh$_3$)$_4$, 10 % CuI,
   THF/iPr$_2$NH, room temp.; 76%

2) PPh$_3$, DEAD, THF, 0°C; 87%

1) 1.3 ⌬ ,

dto.; 70%

2) dto.; 94%

"*E*-freelingyne"

freelingyne

Figure 10. Vinylogous Mukaiyama aldol additions as a prelude to stereospecific γ-(α-hydroxyalkyl)buteno-lide→γ-alkylidenebutenolide conversions (ref. 15).

Of course, the γ-(α-hydroxylalkyl)butenolide precursors of γ-alkylidenebutenolides need not be derived from sugar lactones. For instance, we accessed the substitution pattern of the diterpene lactones shown in Fig. 10 by Mukaiyama aldol additions between the iodo-aldehyde **32** and the trimethylsiloxy-substituted furan **30**. In the presence of BF$_3$•OEt$_2$ we obtained the γ-(α-hydroxylalkyl)butenolide **33** as a single diastereomer and in the presence of ZnBr$_2$ mainly its epimer **31** (ds = 87:13). Couplings between butenolides **33** and **31** and 3-ethynylfuran followed by *anti*-eliminations of water mediated by DEAD/PPh$_3$ led to natural freelingyne (Z; ds = 92:8) and "E-freelingyne" (ds = 98:2), respectively.

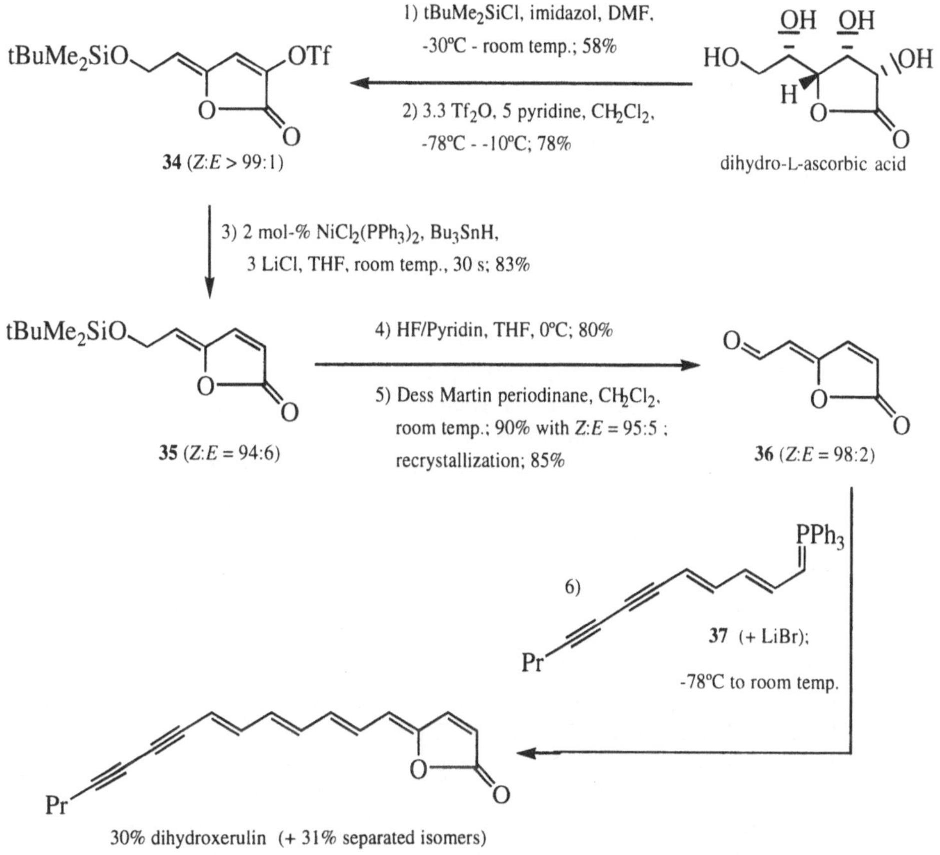

Figure 11. First total synthesis of dihydroxerulin, an inhibitor of cholesterol biosynthesis in man (ref. 16).

Dihydroxerulin (Fig. 11) is an inhibitor of cholesterol biosynthesis. In spite of being achiral and devoid of OH groups, it, too, was synthesized efficiently (11 steps, 6 steps in the longest linear sequence) from the optically active, multiply hydroxylated dihydroascorbic acid (Fig. 11). Step 2 of this synthesis combines the triflate forming elimination used in step 3 of Fig. 7 with the C$_{exocyclic}$=C$_\gamma$ bond forming elimination used in step 2 of Fig. 9. Here, we obtained the γ-alkylidenebutenolide triflate **34**. It was hydrogenolyzed under very mild conditions (→**35**). Desilylation and oxidation carried the material on to butenolide aldehyde **36**. A Wittig reaction with phosphorane **37** delivered dihydroxerulin in 30% yield. Its 800 MHz ^{1}H-NMR spectrum revealed that the C^8=C^9 bond of the natural product is *trans* substituted.

REFERENCES

1. Recent syntheses of non-racemic γ-chiral γ-lactones: W.-Y. Yu, C. Bensimon, and H. Alper, *Chem. Eur. J.* **1997**, *3*, 417-423; T. Chevtchouk, J. Ollivier, and J. Salaün, *Tetrahedron: Asymmetry* **1997**, *8*, 1011-1014; A.-M. Fernandez, J.-C. Plaquevent, and L. Duhamel, *J. Org. Chem.* **1997**, *62*, 4007-4014; S.-i. Fukuzawa, K. Seki, M. Tatsuzawa, and K. Mutoh, *J. Am. Chem. Soc.* **1997**, *119*, 1482-1483; M. P. Doyle, *Aldrichim. Acta* **1996**, *29*, 3-11.

2. Recent syntheses of non-racemic γ-chiral butenolides: S. M. Dankwardt, J. W. Dankwardt, and R. H. Schlessinger, *Tetrahedron Lett.* **1998**, *39*, 4971-4974; B. Figadère, J.-F. Peyrat, A. Cavé, *J. Org. Chem.* **1997**, *62*, 3428-3429; J. A. Marshall, M. A. Wolf, and E. M. Wallace, *ibid.* **1997**, *62*, 367-371; A. van Oeveren and B. L. Feringa, *ibid.* **1996**, *61*, 2920-2921.– Cf. also D. W. Knight, *Contemp. Org. Synth.* **1994**, *1*, 287-315.

3. Recent syntheses of γ-alkylidenebutenolides: R. Rossi, F. Bellina, and L. Mannina, *Tetrahedron Lett.* **1998**, *39*, 3017-3020 and *ibid.* 4933; F. Liu and E.-i. Negishi, *J. Org. Chem.* **1997**, *62*, 8591-8594; M. Kotora and E.-i. Negishi, *Synthesis* **1997**, 121-128; H. Mori, H. Kubo, H. Hara, and S. Katsumura, *Tetrahedron Lett.* **1997**, *38*, 5311-5312; J. Boukouvalas and F. Maltais, *Tetrahedron Lett.* **1995**, *36*, 7175-7176; M. A. Khan and H. Adams, *Synthesis* **1995**, 687-692.

4. Reviews: H. C. Kolb, M. S. VanNieuwenhze, and K. B. Sharpless, *Chem. Rev.* **1994**, *94*, 2483-2547. G. Poli and C. Scolastico, *Methoden Org. Chem. (Houben-Weyl) 4th ed. 1952-*, Vol. E21e (Eds.: G. Helmchen, R. W. Hoffmann, J. Mulzer, and E. Schaumann), Thieme, Stuttgart, **1995**, pp. 4547-4598.

5. Z.-M. Wang, X.-L. Zhang, K. B. Sharpless, S. C. Sinha, A. Sinha-Bagchi, and E. Keinan, *Tetrahedron Lett.* **1992**, *33*, 6407-6410 described three examples each for a **5→6** and for a **5→ent-6** conversion (R = Et, Hex, Ph; *ee* 92 - >99%) and Y. Miyazaki, H. Hotta, and F. Sato, *Tetrahedron Lett.* **1994**, *35*, 4389-4392 one example of a **5→3** conversion (R = SiMe$_3$, 86% *ee*). The principle *was* not recognized and/or pursued, though.

6. Related β-eliminations to γ-alkylidenebutenolides had been studied disregarding stereochemistry (F. Bohlmann and C. Zdero, *Chem. Ber.* **1966**, *99*, 1226-1228; M. Ito, Y. Hirata, Y. Shibata, and K. Tsukida, *J. Chem. Soc. Perkin Trans. I* **1990**, 197-199; D. Xu and K. B. Sharpless, *Tetrahedron Lett.* **1994**, *35*, 4685-4688; S. Y. Koo and J. Lerpiniere, *Tetrahedron Lett.* **1995**, *36*, 2101-2104) or finding no stereocontrol [J. Font, R. M. Ortuño, F. Sánchez-Fernando, C. Segura, and N. Terris, *Synth. Commun.* **1989**, *19*, 2977-2985; H. Itoh, *Noguchi Kenkyuscho Jiho* **1984**, 15-18 (cited from *Chem. Abst.* **1986**, *104*, 168723c); C. Di Nardo, L. O. Jeronic, R. M. Lederkremer, and O. Varela, *J. Org. Chem.* **1996**, *61*, 4007-4013]. Stereoselectivity combined with stereoconvergence was reported twice (J. Boukouvalas, F. Maltais, and N. Lachance, *Tetrahedron Lett.* **1994**, *35*, 7897-7900; J. Boukouvalas and F. Maltais, *Tetrahedron Lett.* **1995**, *36*, 7175-7176) and stereoselectivity combined with stereospecificity once (M. A. Khan and H. Adams, *Synthesis* **1995**, 687-692).

7. Harcken and R. Brückner, *Angew. Chem.* **1997**, *109*, 2866-2868; *Angew. Chem. Int. Ed. Engl.* **1997**, *36*, 2750-2752.

8. S. Braukmüller, *Diplomarbeit*, Universität Göttingen, **1998**.

9. C. Harcken, E. Rank (in part), and R. Brückner, *Chem. Eur. J.* **1998**, *3*, in press.

10. C. Harcken and R. Brückner, unpublished.

11. Similar syntheses: T. Berkenbusch and R. Brückner, *Tetrahedron* **1998**, *54*, in press.

12. I. Kalvinsh, K.-H. Metten, and R. Brückner, *Heterocycles* **1995**, *40*, 939-952.

13. F. C. Görth, A. Umland, and R. Brückner, *Eur. J. Org. Chem.* **1998**, 1055-1062.
14. F. C. Görth and R. Brückner, unpublished.
15. F. v. d. Ohe and R. Brückner, *Tetrahedron Lett.* **1998**, *39*, 1909-1910.
16. K. Siegel and R. Brückner, *Chem. Eur. J.* **1998**, *3*, 1116-1122.

LARGE SCALE CATALYTIC ASYMMETRIC HYDROGENATION, AN INDUSTRIAL PERSPECTIVE

Walter Brieden

LONZA LTD
Valais Works
CH-3930 Visp, Switzerland

INTRODUCTION

During the last two decades an increasing number of catalytic, enantioselective reactions have been developed.[1] Life science companies, which increasingly demand the preparation of enantiomerically pure products on large scale, have also fuelled the development of such methodologies with an emphasis on manufacturability.[2] Asymmetric hydrogenations of prochiral C=C, C=O and C=N double bonds, catalysed by chiral transition metal complexes are highly attractive and continue to be an efficient method for generating stereogenic centres in an enantioselective fashion.[1c] Critical factors which determine the outcome of an enantioselective hydrogenation on large scale are outlined in Figure 1. The chemistry of specific examples from LONZA's own experience with this technology is presented in this report.

- Efficient preparation of chiral complexes
- Catalyst productivity in terms of selectivity and activity
- Ease of ligand tuning
- Efficient substrate synthesis
- Qualitiy and stability of substrates

Figure 1. Critical factors for large scale catalytic, asymmetric hydrogenations.

BIOTIN

Biotin one of the water-soluble B vitamins, has found widespread applications in the market for health and nutrition. Fostered by its fundamental and commercial importance a number of elegant synthetic sequences have been developed during the past

Current Trends in Organic Synthesis
Edited by Scolastico and Nicotra, Kluwer Academic/Plenum Publishers, 1999

50 years.[3] The LONZA synthesis of biotin is outlined retrosynthetically (Figure 2) and is based on a strategy that takes advantage of an asymmetric hydrogenation of the tetra-substituted double bond in **3** to create two of the three contiguous stereogenic centres in biotin. For this key step it is conceivable that a diastereoselective hydrogenation could be achieved using the directing effect of a chiral auxiliary (R*) on a nitrogen atom. It was expected that the the bicyclic urea **3** should be accessible from tetronic acid (**4**), an in-house building block.[4] In the synthetic direction, the introduction of the biotin side is carried out by a selective reaction of a C4-Di-Grignard reagent to the chiral thiolactone **2**. To be able to achieve such a selective reaction the thiolactone **2** has to be slowly added to the Di-Grignard reagent in order to allow further functionalisation with CO_2.

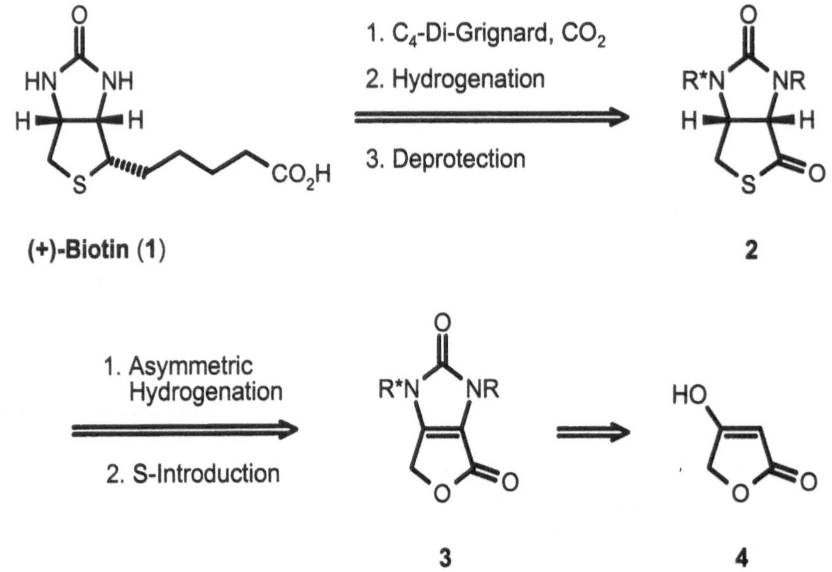

Figure 2. Retrosynthetis of (+)-biotin with asymmetric hydrogenation as key transformation.

Our synthesis commences with the introduction of the two nitrogen substituents onto the tetronic acid ring in order to set up the correct substitution for biotin (Figure 3). An electrophilic amination using a diazonium salt followed by a reaction with the chiral auxiliary (R)-α-methylbenzylamine provides **5**. The hydrogenation of intermediate **5** accomplishes the reduction of the diazo bond and the primary amine formed reacts spontaneously with a chloroformate to afford the bicyclic lactone **6**.

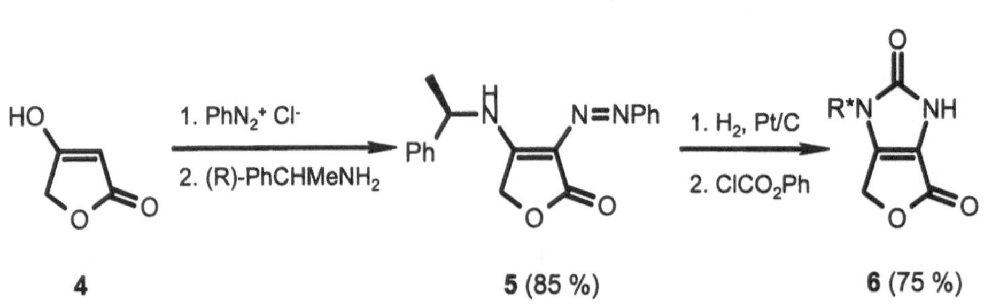

Figure 3. Synthesis of key intermediate **6** from tetronic acid (**4**).

Two options exist for the critical hydrogenation of the chiral bicyclic lactone **6**, namely, the use of heterogeneous or homogeneous catalysts.[5] Ten years ago we started with the development of a process using heterogeneous catalysts. Under optimised conditions we had a diastereoselectivity of 70 : 30 in the hydrogenation with a Rh-Al$_2$O$_3$ catalyst and obtained the desired diastereomer **7** in a yield of 58 %. Clearly, the key step needed to be improved. In this respect asymmetric hydrogenation with homogeneous catalysts could offer a distinct advantage. Selected results of the ligand screening with less than 50 complexes are listed in Figure 4.

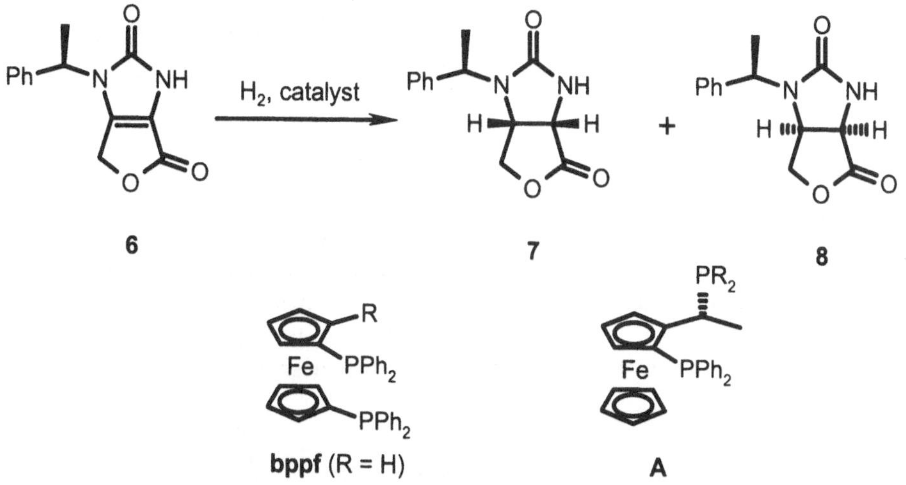

| R = H (bppf) | 44 : 56 | R = Ph | no reaction |
| R = (R)-CH(OAc)Me | 60 : 40 | **R = C$_6$H$_{11}$** | **94 : 6** |

Figure 4. Rh-catalyzed hydrogenation of the bicyclic imidazolidone **5**.

For homogeneous hydrogenation of **6**, neutral Rh-diphosphine complexes are the catalysts of choice.[5d] The reduction of the chiral lactone **6** with the achiral bppf ligand gave a selectivity of 44 : 56 in favour of the unwanted diastereomer whereas a substituted bppf derivative led to a selectivity of 60 : 40 (Figure 4). What about ligands with a different orientation of the phosphino groups on ferrocene?

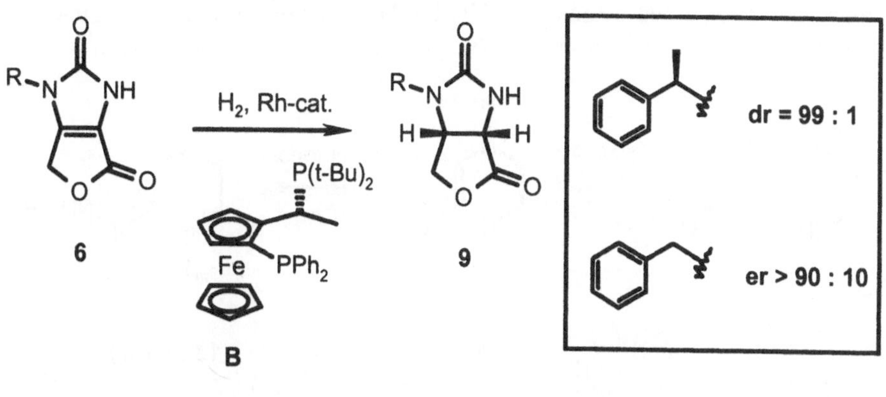

Figure 5. Enantioselective hydrogenation of **6**.

To our surprise, the bisdiphenylphosphino derivative **A** (R = Ph) did not act as a catalyst in the reduction of the tetrasubstituted double bond of **6**. The real breakthrough was achieved when the more basic ferrocene diphosphine **A** (R = C_6H_{11}), bearing a dialkyl-phosphino group, was employed. The selectivity increased to 94 : 6. An even better diastereomeric ratio of 99 : 1 could be achieved with the ferrocene diphosphine **B** (Figure 5). It is noteworthy that even a simple benzyl group on nitrogen still gives a selectivity of greater than 90 : 10. Here we have a truly enantioselective hydrogenation, because the substrate is prochiral and not homochiral as in reaction with the α-methylbenzyl substituent. The unique hydrogenation of a tetrasubstituted double bond using a chiral rhodium ferrocenyl diphosphine catalyst provides the biotin intermediate in a yield of > 95 %. The optimised conditions are suitable for an industrially viable process.[6]

PIPERAZINES

Optically active 2-substituted piperazines of type **10** (Figure 6) and its derivatives have found widespread application in the synthesis of a variety of pharmaceuticals.[6] Among a number of routes that can be envisaged to prepare such enantiomerically pure piperazines we focused on asymmetric hydrogenations of tetrahydropyrazines (**11**). Beside finding an highly productive catalytic system the substrate should be readily available.

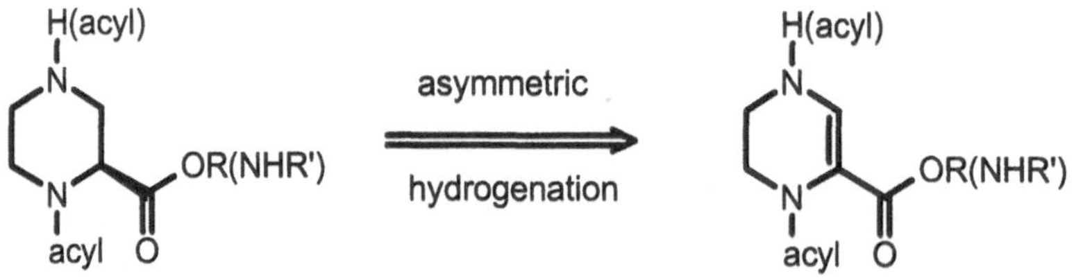

Figure 6. 2-Substituted piperazines via asymmetric hydrogenation of tetrahydropyrazines **11**.

Tetrahydropyrazine esters can be efficiently obtained by partial hydrogenation of the corresponding aromatic pyrazines.[7] While the partial hydrogenation of amides is difficult to achieve, because of overreduction, an alternative process was developed (Figure 7). The tetrahydropyrazine nitrile **12** is formed efficiently from reaction of ethylene diamine with glyoxal and sodium cyanide. Further functionalisation in a Ritter reaction allows the preparation of the trifluoroacetate protected amide **13** in 80 % yield.[6]

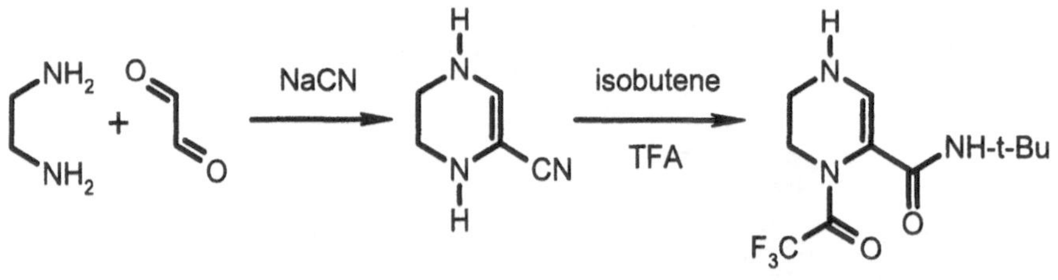

Figure 7. Synthesis of 2-substituted tetrahydropyrazine by a combination of Strecker and Ritter reactions.

The tetrahydropyrazines **11**, which are structurally related to the imidazolidine **6** in terms of configurational arrangements of substituents on the C=C double bond, are hydrogenated under similar conditions. As demonstrated in Figure 8, bisacetamides of tetrahydropyrazine esters and amides can be hydrogenated with a selectivity of > 95 % ee when ferrocene **B** is employed. In all cases cationic rhodium complexes exhibit higher activity in the enantioselective reduction than the neutral ones.

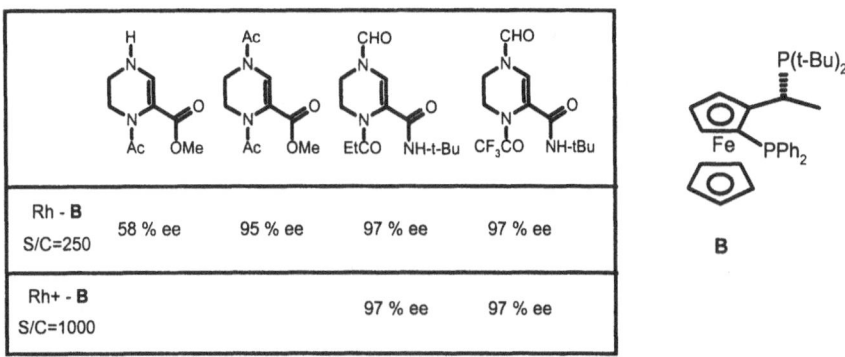

Figure 8. Enantioselectivities in the hydrogenation of tetrahydropyrazine esters and amides.

From an economical point of view the most attractive approach would be a direct asymmetric hydrogenation of the aromatic pyrazines. Indeed, under standard conditions a variety of pyrazine esters and amides can be hydrogenated with impressive enantio-selectivities (up to 77 % ee) for such a novel transformation (Figure 9).[8]

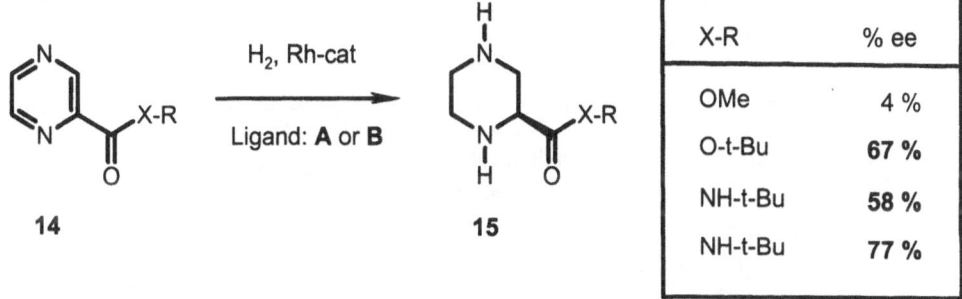

Figure 9. Direct asymmetric hydrogenation of pyrazine esters and amides.

3-QUINUCLIDINOL

3-Quinuclidinol is found in a number of biologically active compounds and an enzyme catalysed kinetic resolution[9] is used to supply the enantiomerically pure alcohol **17**. The inherent disadvantage of this resolution is the loss of the undesired enantiomer, as its recycling is too difficult to achieve. After an intensive ligand screening it was discovered that the hydrogenation of 3-quinuclidinone hydrochloride using the best conditions found (Rh-catalyst, ligand **B**, S/C = 10 000) gave **17** in only 24 % ee.[10] Since we were not successful in improving the enantioselectivity by a ligand modification, another approach, the modification of the substrate, was undertaken.

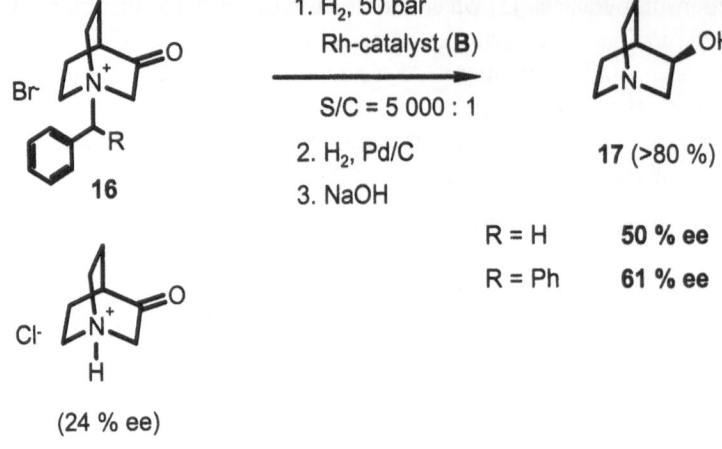

Figure 10. Enantioselective hydrogenation of prochiral 3-quinuclidinones 16.

It was shown that the benzyl ammonium salts 16 can be hydrogenated with higher selectivities of 50 and 61 % ee. After the homogeneous hydrogenation the benzyl groups are removed by hydrogenolytic cleavage with Pd/C as a catalyst.

Dextromethorphan

Dextromethorphan is an important bronchodilating agent which is on the market as an antitussive drug. Figure 11 shows a part of the classical synthetic pathway with a selective C=N reduction by NaBH$_4$ of the critical hexahydroisoquinoline 18, leading to the formation of racemic octahydroisoquinoline (rac-19).

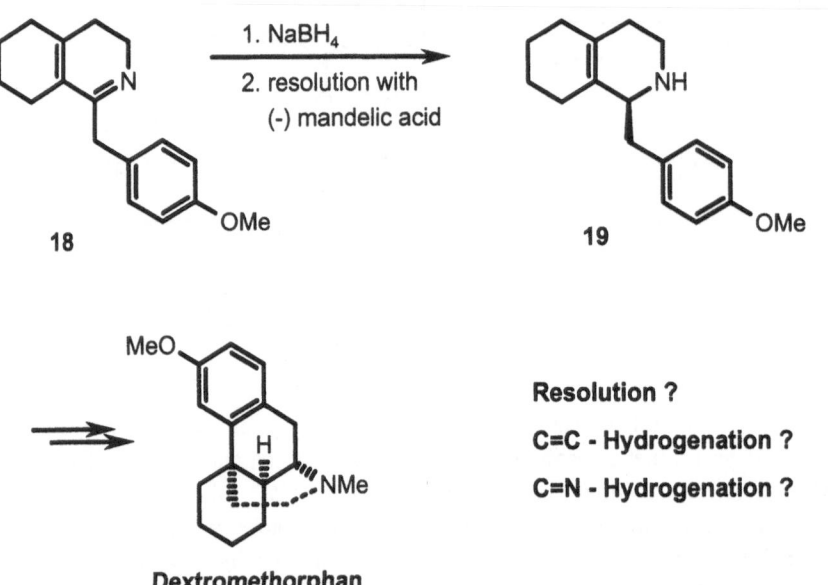

Figure 11. Synthetic pathway for the production of dextromethorphan.

Finally, after resolution of the racemic mixture, enantiomerically pure **19** is converted to the morphinan skeleton by a stereospecific cyclisation. The classical resolution is associated with time- and volume-consuming recycling operations for solvents, the undesired enantiomer, and the resolving agent. A more elegant way to the preparation of **19** would be via asymmetric hydrogenation. Two different approaches can be envisaged: either the chemo- and enantioselective reduction of the C=N double bond of **18**, or an enantioselective hydrogenation of the corresponding N-acyl enamines. The latter approach has been described in the literature, but apparently only the Z-configurated enamides are hydrogenated, whereas the synthesis of the substrates always leads to the formation of Z- and E-enamides of **18**.[11] We decided to tackle the problem in the most direct way and developed an enantioselective imine hydrogenation of **18**.[12] In the initial screening moderate ee's were obtained, but fine tuning of ferrocene diphosphines with respect to the electronic and steric requirements led to the optimised conditions (Figure 12). The hexahydroisoquinoline in form of the stable monophosphate is hydrogenated in a biphasic reaction mixture to provide **19** in 89 % ee. In course of subsequent steps towards dextromethorphan further enrichment of the enantiomeric purity takes place.

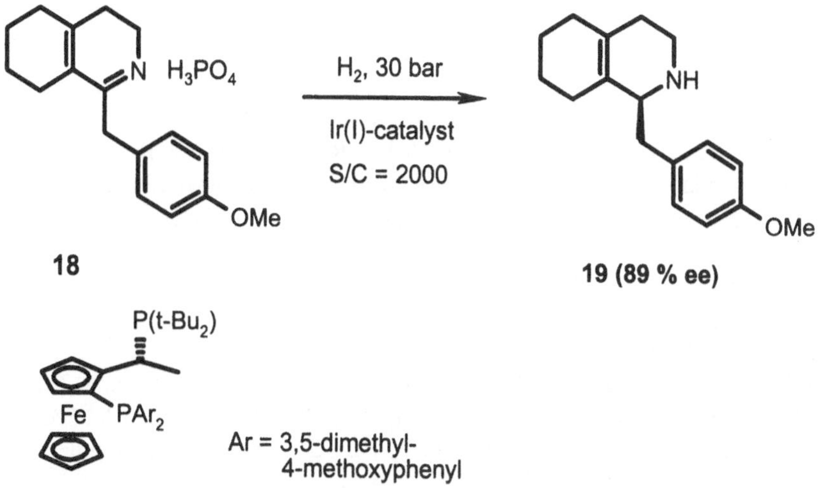

Figure 12. Optimised asymmetric imine hydrogenation of **18**.

Conclusion

Selected examples of chiral heterocycles have been used to illustrate the advantage of one particular technology, namely the asymmetric hydrogenation of double bonds. Reasons why the asymmetric hydrogenation technology is still under-utilised in industry have been outlined in Figure 13.

- Model reactions and not industrially relevant substrates are studied
- Tight time frames for process development and implementation
- Large investment is frequently required
- Low tolerance for structural diversity in substrates
- Lack of efficent substrate preparation

Figure 13. Reasons which might explain why asymmetric hydrogenation technology is still under-utilised in industry.

REFERENCES

1a. I. Ojima. *Catalytic Asymmetric Synthesis*, VCH Publishers Inc., New York (1993).

1b. D.J. Ager, M.B. East. *Asymmetric Synthetic Methodology*, CRC Press, Inc., Boca Raton (1996).

1c. R. Noyori. *Asymmetric Catalysis in Organic Chemistry*, New York (1994).

2. I.W. Davis, P.J. Reider, Practical asymmetric synthesis, *Chemistry and Industry* 3 June:412 (1996).

3. P.J. De Clercq, Biotin: a timeless challenge for total synthesis, *Chem. Rev.* 97:1755 (1997).

4. T. Meul, R. Miller, L. Tenud, Ökonomische Herstellung von Feinchemikalien: Tetronsäure, *Chimia* 41:73 (1987).

5a. J. Mc Garrity, L. Tenud, EP Appl. EP 273 270; CA 110: 75 168j (1988).

5b. M. Eyer, R. Fuchs, , J. Mc Garrity, EP Appl. EP 602 653; CA 121: 230 774b (1994).

5c. M. Eyer, R.E. Merrill, PCT Int. Appl. WO 94 24 137; CA 122: 56 043w (1994).

5d. J. Mc Garrity, F. Spindler, R. Fuchs, M. Eyer, EP Appl. EP 624 587; CA 122: 81 369q (1994).

6. W. Brieden, Taking the right route to the manufacture of enantiomerically pure fine chemicals, *Proceedings of the Chiral USA '97 Symposium*, p. 45.

7a. E. Felder, S. Maffei, S. Pietra, D. Pitré, Über die katalytische Hydrierung von Pyrazincarbonsäuren, *Helv. Chim. Acta* 43:888 (1960).

7b. K. Rossen, S.A. Weissman, J. Sager, D. Askin, R.P. Volante, P.J. Reider, Asymmetric hydrogenation of tetrahydropyrazines: synthesis of (S)-piperazine-2-tert-butylcarboxamide, an intermediate in the preparation of the HIV protease inhibitor Indinavir, *Tetrahedron Lett.* 36:6419 (1995).

8. R. Fuchs, EP Appl. EP 803 502; CA 128: 13 286 (1997).

9. D.C. Muchmore, US Patent 5 215 918 (1993).

10. W. Brieden, EP Appl. EP 785 198; CA 127: 176 346 (1997).

11a. M. Kitamura, Y. Hsiao, M. Ohta, M. Tsukamoto, T. Ohta, H. Takaya, R. Noyori, General asymmetric synthesis of isoquinoline alkaloids. Enantioselective hydrogenation of enamides catalyzed by BINAP-Ruthenium(II) complexes, *J. Org. Chem.* 59:297 (1994).

11b. B. Heiser, E.A. Broger, Y. Crameri, New efficient methods for the synthesis and in-situ preparation of Ruthenium(II) complexes of atropisomeric diphosphines and their application in asymmetric catalytic hydrogenations, *Tetrahedron: Asymmetry* 2:51 (1991).

12a. O. Werbitzky, PCT Int. Appl. WO 97 03 052 (1997).

12b. O. Werbitzky, Technical synthesis of a morphine alkaloid through an enantioselective hydrogenation of a cyclic imine, *Proceedings of the Chiral Europe '97 Symposium*, p. 37.

CAMPHOR-DERIVED 2-STANNYL-*N*-BOC-1,3-OXAZOLIDINE: A NEW VERSATILE CHIRAL FORMYLANION EQUIVALENT

Lino Colombo, Marcello Di Giacomo

Dipartimento di Chimica Farmaceutica, Università di Pavia
Via Taramelli, 12 -I-27100 Pavia (Italy)

INTRODUCTION

A method that employs stoichiometric quantities of a chiral auxiliary for the preparation of highly enantiomerically enriched compounds might in principle appear to be non competitive with an asymmetric catalytic procedure. This would seem particularly true for the synthesis of vicinal diols and α-aminoacids in light of the impressive achievements in the asymmetric dihydroxylation of olefins,[1] the resolution of racemic 1,2-diols[2] and the hydrogenation of α,β-didehydro α-aminoacids.[3] However, an auxiliary based protocol could offer some advantage over existing procedures if the level of stereoinduction is uniformly high, regardless of structural variation both of the substrate and the reagents. Another important feature that can raise the value of an auxiliary based method is the possible exploitation of the same auxiliary for different synthetic applications.

We recently introduced the 2-tributylstannyloxazolidine **1**[4] as an efficient chiral template for asymmetric nucleophilic formylation reactions.[5] From this common intermediate either 2-acyl- or 2-formyloxazolidines **2a** could be derived in a highly diastereoselective fashion. Addition of Grignard reagents or metal hydride reduction under chelation control afforded the corresponding alcohols **3a** with very high asymmetric induction in all cases tested. Thus the criterion of uniformly high stereoselection is met by our method. The absence of racemization and the mild conditions used in the removal of the auxiliary render the methodology practical, whether it is for the preparation of α-monosubstituted, or for the preparation of α,α-disubstituted α-hydroxy aldehydes. Furthermore a conceptually similar sequence allowed for the extention of the same methodology to the asymmetric synthesis of α-amino aldehyde acetals. Only a limited number of examples related to this modification have been tested thus far, but the preliminary results seem to attest to a wide applicability.

1   2a X=O; 2b X=NNMe$_2$ 3a,b

RESULTS AND DISCUSSION

The design of the auxiliary **1** has evolved from initial efforts to devise a novel chiral formylanion equivalent. Our original idea, following previous work by Quintard[6] and Shiner,[7] was to use C_2-symmetric 2-stannyl-1,3-dioxolanes[8] as stable precursors of the corresponding 2-lithio derivatives, the actual formylanion equivalents, whose reactions with aldehydes would have had to afford stereoselectively α-hydroxy-1,3-dioxolanes. Uniformly poor stereoinduction and resistance to any attempt to remove the chiral auxiliary precluded the usefulness of these reagents. Thus our attention turned to the more readily hydrolyzable oxazolidines. Moreover, even if the addition reactions of 2-lithiooxazolidines to aldehydes were to have proved non-selective, oxidation of the resulting alcohols would have afforded the corresponding acyl derivatives whose reactions with nucleophiles could have stood a good chance of being stereoselective. Indeed Scolastico[9] and Hoppe[10] have independently reported stereoselective nucleophilic additions to 2-acyloxazolidines derived from N-tosyl-norephedrine.

On the basis of these considerations the hitherto unknown 2-stannyloxazolidine **1** emerged as the most promising candidate. The N-t-Boc group, instead of the N-tosyl protecting function, should have secured milder conditions in the auxiliary removal step. The choice of the camphane skeleton was suggested by our previous experience with the transacetalyzation reaction of 1,2-camphandiol with $(EtO)_2CHSnBu_3$ (**4**), which we have already shown to have given diastereomerically pure 2-stannylacetals. In actuality, reaction of the N-t-Boc aminoalcohol **5**[11] with **4** afforded the desired 2-stannyloxazolidine **1** as a single isomer in a yield of 75%. The configurational assignment of the newly created stereocenter as S was assured by n.O.e. experiments. Treatment with n-BuLi at -78°C, followed by reaction with a range of representative aldehydes, gave secondary alcohols **6** as mixtures of diastereoisomers epimeric at the new stereocenter. Oxidation of these alcohols with the Dess-Martin periodinane[12] afforded acyloxazolidines **7** in yields as high as 90%.

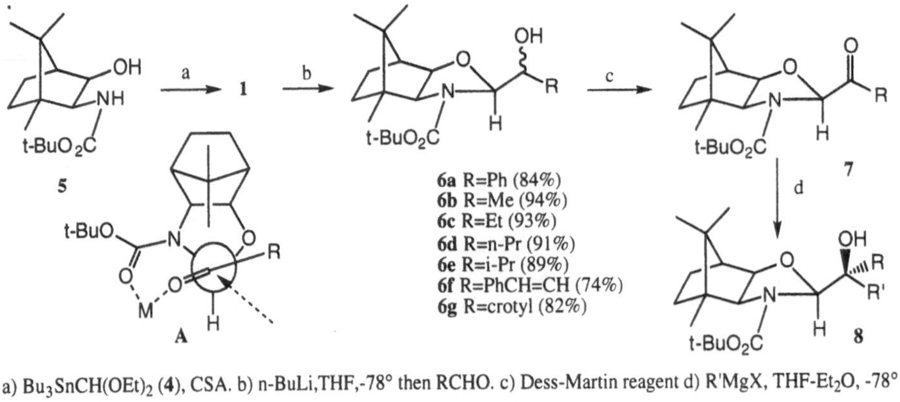

6a R=Ph (84%)
6b R=Me (94%)
6c R=Et (93%)
6d R=n-Pr (91%)
6e R=i-Pr (89%)
6f R=PhCH=CH (74%)
6g R=crotyl (82%)

a) $Bu_3SnCH(OEt)_2$ (**4**), CSA. b) n-BuLi,THF,-78° then RCHO. c) Dess-Martin reagent d) R'MgX, THF-Et$_2$O, -78°

Scheme 1. Preparation of alcohols **8**.

The data reported in scheme 1 clearly demonstrate the general applicability of this procedure, regardless of the nature of the aldehyde employed. The acyl derivatives **7**, reacting with a great variety of Grignard reagents at -78°C, gave the tertiary alcohols **8** as single stereoisomers, in all cases tested, in yields which were good to excellent. Interchanging of R and R' groups as components of the aldehyde and Grignard reagent, gave products of opposite configuration at the carbinol center, with similar efficiency in terms of yield and stereoselection. Ultimately this leads to the obtainment of both enantiomers of the final α-hydroxyaldehydes through the assistance of only one antipodal auxiliary.

Removal of the chiral auxiliary was efficiently effected by treatment with excess t-BuOK in hot THF followed by addition of water and final acid treatment. The α-hydroxyaldehydes

9 produced were isolated and immediately subjected to $NaBH_4$ reduction which uneventfully gave diols **10**. Comparison of the optical rotations with reported values allowed for assignment of absolute configuration. Enantiomeric excess was higher than 96% in each case. The sense and the uniformly high degree of asymmetric induction are consistent with the intervention of a transition state, conformationally constrained by internal chelation of both carbonyl oxygens with magnesium (**A** in scheme 1).

Yield from **8** : 83-98%
e.e. ≥ 96%

The preceding findings prompted an analysis of the level of stereoinduction that would be attainable upon addition of hydride reagents. Efficient and selective reduction of the carbonyl function would pave the way for the extension of the methodology to the preparation of α-hydroxyaldehydes with a secondary, instead of a tertiary, α-stereocenter. The high efficiency of chelation effects as primary stereocontrol factors on hydride reagent additions was confirmed by the high level of stereoselection reached through the use of a non-coordinating solvent, such as toluene, and a reducing agent capable of chelation, such as lithium Selectride® (Table 1). Further confirmation of the notion that a coordinating metal ion was necessary for high levels of stereoselection to be achieved came from the notable increase in stereoselection effected by precomplexation of 2-acyloxazolidines with $TiCl_4$, before addition of the non-chelating nucleophile DIBAL.

Table 1. Reduction of 2-acyl oxazolidines **7**.

R	Exp. conditions	11:12 Ratio	Yield (%)
CH_3	$NaBH_4$, $MeOH/Et_2O$, 0°C	75:25	95
$n-C_4H_9$	$NaBH_4$, $MeOH/Et_2O$, 0°C	76:24	98
	Li-Selectride®, toluene, -78°C	>98:2	90
	DIBAL, CH_2Cl_2, -78°C	90:10	95
	$TiCl_4$, DIBAL, CH_2Cl_2, -78°C	>98:2	94
$c-C_6H_{11}$	$NaBH_4$, $MeOH/Et_2O$, 0°C	50:50	98
	Li-Selectride®, toluene, -78°C	>98:2	87
	$LiBH_4$, Et_2O, -15°C	80:20	90
$i-C_3H_7$	$NaBH_4$, $MeOH/Et_2O$, 0°C	64:36	87
C_6H_5	$NaBH_4$, $MeOH/Et_2O$, 0°C	93:7	98
	$LiBH_4$, Et_2O, -15°C	>98:2	91
$t-C_4H_9$	$NaBH_4$, $MeOH/Et_2O$, 0°C	>98:2	84

The stereoinduction was shown to be still in the same direction as in the Grignard additions. Distinction between the two diastereoisomers was first accomplished by the observation that all major stereoisomers had a similar coupling constant of the C2 oxazolidine proton (1.2-1.8 Hz), differing by more than 4 Hz from that shared by the other isomer (6.2-6.7 Hz). Molecular mechanics calculations using both MM2* and AMBER* force fields[13] gave calculated constants in excellent agreement with the experimental value, allowing for the configurational assignments to be determined only by inspection of the NMR spectra. This protocol was utilized rather extensively in the subsequent segments of the investigation. Final confirmation of these data derived from comparison of the sign of optical rotation of final diols (*vide infra*) with literature values.

Whilst all the results seemed to be self-consistent when a chelating reducing agent was involved, a seemingly odd situation emerged from experiments using NaBH$_4$. A chelated transition state, though less tight than in the Li$^+$ counterion case, can be excluded on the basis of the following considerations: 1) although the major isomer is the same as that produced by Li-Selectride®, the level of stereoinduction greatly varies as a function of the acyl substituent size; 2) the decrease of stereoselection changing from LiBH$_4$ to NaBH$_4$ strongly suggests the intervention of a different transition state; 3) an alternative chelation model involving the oxazolidine oxygen can be excluded as it would produce an inverted diastereomeric ratio.

A simplistic rationale of these results would invoke a Felkin-Ahn[14] transition state geometry that positions the oxazolidine oxygen atom orthogonal to the ketone carbonyl group. This would stand in contrast to previous studies on *N*-tosyl-2-acyloxazolidines,[10] *N*-tosyl-2-acyloxazine[15] and *N*-Boc-2-acyl oxazolidines,[16] where a Felkin model with the nitrogen atom perpendicular to the carbonyl group was hypothesized.

It interested us to gain deeper insight into this issue. What is not readily explained simply by the stereoelectronic effects underlying the Felkin model, is that the level of stereoinduction drops upon the increase of the acyl substituent size from primary to secondary, and then dramatically increases in the case of bulkier phenyl or ter-butyl groups. The above results suggest an influence on the transition state stability of a steric interaction between the acyl substituent and the carbamate ter-butoxycarbonyl appendage. If this assumption were to be demonstrated, an explanation of the high selectivity in NaBH$_4$ reduction of sterically demanding acyl substituents would be at hand.

In the competition between the two possible Felkin transition states **13** and **14**, the latter, leading to the minor diastereoisomer, should be disfavored to a greater extent as the size of the R group is increased, because the two interacting groups are closer to each other than in the transition state **13**. Reduction in the size of the carbamate group should lead to a smaller steric interaction, thus lowering the energy difference between the two competing transition states and consequently the level of stereoinduction.

In order to test this hypothesis, we prepared the *N*-methoxycarbonyl derivative **15**, whose reduction with NaBH$_4$ was shown to display a selectivity still in the same direction but greatly diminished (57:43 vs. 93:7).

The question then arose as to what kind of effects were responsible for the decrease in selectivity, changing the degree of substitution of the carbonyl α'-substituent from tertiary to either secondary or primary. Our working hypothesis to rationalize this second critical issue

174

was based on the assumption that a major role for the stereoinduction was played by the attack direction of the nucleophile. Taking into account that the trajectory of the incoming nucleophile can deviate from the plane normal to the carbonyl group (Heathcock-Flippin-Lodge angle)[17] as a function of the relative bulkiness of the subsituents, we can assume that the trajectory of the approaching hydride lies closer to the α'-carbon than to the oxazolidine stereocenter. If this were the case, the stereoselectivity should be strongly influenced by the size and the conformational arrangement of the groups bound to the α' non-stereogenic carbon.

(Heathcock, Lodge, Flippin angle)

An indirect way of pointing out such an effect would be that of placing a stereogenic center in the α' position adjacent to the carbonyl and of testing the extent of asymmetric induction exerted by this centre with respect to the oxazolidine C2. To this end we prepared both ketones **18** and **19** from the two enantiomers of 2-phenylpropionaldehyde, following the same procedure outlined above. Separate reduction of the two ketones **18** and **19** gave, in both cases, a mixture of two diastereoisomers in the same 4:1 ratio. The absence of matching/ mismatching effects[18] exerted by the two stereocenters flanking the carbonyl group is a preliminary indication that only one of them is effective in the stereoinduction process. Moreover, the fact that the predominant isomers have opposite configuration at the carbinol center demonstrates that the chiral moiety derived from 2-phenylpropionaldehyde controls the stereochemical outcome of the reaction.

As a further application of the chiral auxiliary **1**, we thought that the same hydroxyalkyl oxazolidines **12**, previously obtained as minor components by reduction, could be stereoselectively produced by organometal additions to the hitherto unknown formyloxazolidine **21** under conditions favouring intramolecular chelation. The aldehyde **21** could be readily prepared from the 2-stannyloxazolidine **1** in a yield as high as 83% yield, by treatment with n-BuLi followed by rapid addition of excess DMF. Initial experiments involving addition of either Grignard or organolithium reagents in THF solution resulted in the formation of two epimeric alcohols with marginal selectivity. When using less coordinating solvents, only a modest improvement of the diastereomeric ratio was achieved. Finally, precomplexation of the formyloxazolidine **21** with a Lewis acid as strong as MgBr$_2$ or TiCl$_4$ in dichloromethane solution, proved effective to furnish Grignard addition products in a highly diastereoselective fashion (**Table 2**).

Table 2. Addtion of organometal reagents to aldehyde **21**.

RM	Exp. conditions	Yield (%)	12:11 ratio
n-C_4H_9MgCl	THF,-78°C	87	60:40
	Et_2O,-78°C	85	65:35
	CH_2Cl_2,-78°C	90	70:30
	$MgBr_2$, CH_2Cl_2,-78°C	63	>98:2
	$TiCl_4$,CH_2Cl_2,-78°C	84	>98:2
n-C_4H_9Li	THF,-78°C	82	50:50
c-$C_6H_{11}MgBr$	$TiCl_4$,CH_2Cl_2,-78°C	78	>98:2
	$MgBr_2$, CH_2Cl_2,-78°C	54	>98:2
C_6H_5MgBr	$TiCl_4$,CH_2Cl_2,-78°C	80	>98:2
$AllylSn(Bu)_3$	$TiCl_4$,CH_2Cl_2,-78°C	90	>98:2
t-C_4H_9MgBr	$TiCl_4$,CH_2Cl_2,-78°C	75	>98:2

Benzoylation of alcohols **12** afforded the corresponding benzoyl derivatives which underwent hydrolysis by treatment with 4M HCl in EtOAc. The α-benzoyloxyaldehydes **22** thus obtained were isolated and, without purification, reduced to the corresponding monoprotected diols **23** with $NaBH_4$. Comparison of the optical rotation values of deprotected diols, obtained by $LiAlH_4$ reduction, with compounds of known configuration, allowed a secure configurational assignment. NMR analysis of the Mosher esters[19] of **23** showed an enantiomeric excess higher than 96% thus showing that the auxiliary removal proceeded with no racemization of the α-stereocenter in the cases examined thus far.

The highly yielding and stereoselective preparation of the formyloxazolidine **21** paved the way for a further synthetic application. It seemed plausible that organometal addition reactions could entail a similar degree of stereoselection by substituting a CN double bond for the carbonyl group. If successful, this modification would constitute a new method for the asymmetric synthesis of α-amino aldehydes.[20]

Our first attempts to put this plan into effect involved the use of imine derivatives but low yields and poor stereoselection thwarted such efforts. We then turned our attention to the N,N-dimethylhydrazone derivative **24** because of ease of preparation and ample literature precedents as to the use of this functional group in stereoselective reactions with nucleophiles.[21]

Reaction of the aldehyde **21** with N,N-dimethylhydrazine afforded in quantitative yield the hydrazone **24** as a crystalline solid amenable to X-ray analysis. Inspection of the solid state structure (**Fig. 1,** left) clearly showed that one face of the CN double bond was shielded by the carbamate t-butyl group. Although one must be aware of the inadequacy in drawing conclusions from a ground state structure in the solid state to a reactive conformation *in solution*, a high degree of stereoselectivity would be expected if the solid state conformation were operative. Following these considerations, the configuration of the stereocenter created by addition of a nucleophile, should be S as resulting from the reagent attack on the *Si* face. (**Fig.1,** right) The same stereochemical outcome would result if a bis-chelated transition state were involved.

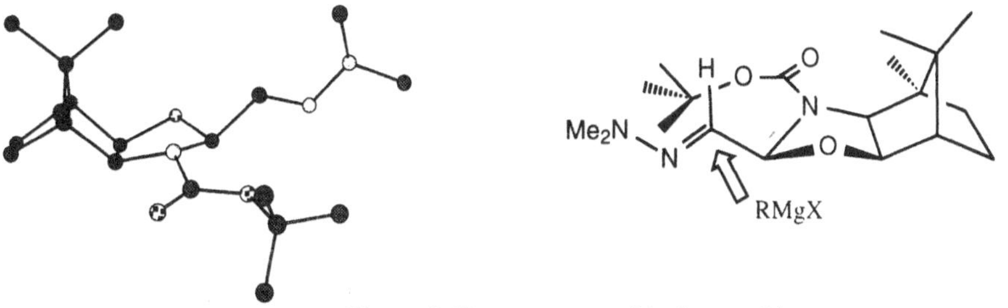

Figure 1. X-ray structure of hydrazone 24.

Preliminary experiments with i-PrMgCl, aimed at finding the best reaction conditions, showed the use of a non-coordinating solvent, such as toluene, and an excess Grignard reagent at -15°C, to be essential for the completion of the reaction. However, under these conditions, the desired hydrazine 25 was accompanied by approximately the same amount of the imidazolidinone 26. This compound was most plausibly derived from intramolecular nucleophilic attack by the metalated hydrazine nitrogen on the urethane carbonyl group. Precomplexation of the hydrazone with etherate BF₃ enabled the reaction to be complete at -78°C, by accelerating the Grignard addition to a much greater extent than the cyclization reaction that brings about the formation of the imidazolidinone.

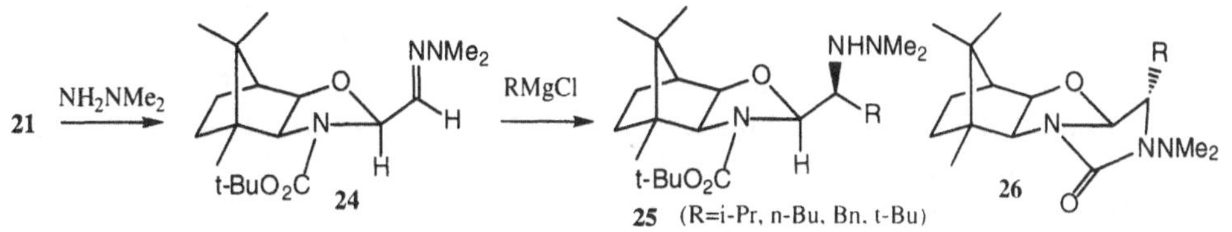

The Grignard reagents examined thus far gave products in uniformly high diastereomeric ratio (≥97:3), and non-optimized yields ranging from 73% to 91%. Hydrogenolysis of the N-N bond was best effected by treatment with W2 Raney Nickel in refluxing methanol, affording the corresponding amino derivatives 27 in quantitative yields. We then selected the benzyl derivative 27 (R=Bn) as a test compound for experimenting auxiliary removal, and for assigning absolute configuration of the amino center by correlation with known compounds. The amino function was protected as a carbobenzyloxy derivative, and the oxazolidine auxiliary was unmasked by reaction with methanolic HCl, giving the dimethylacetal 28 which was shown to have S configuration, as predicted, by comparison of its sign of optical rotation with the literature value. In the ^{1}H-NMR spectrum of the Mosher amide obtained from the N-deprotected product, no contamination with a second stereoisomer could be discerned indicating an enantiomeric excess higher than 96%.

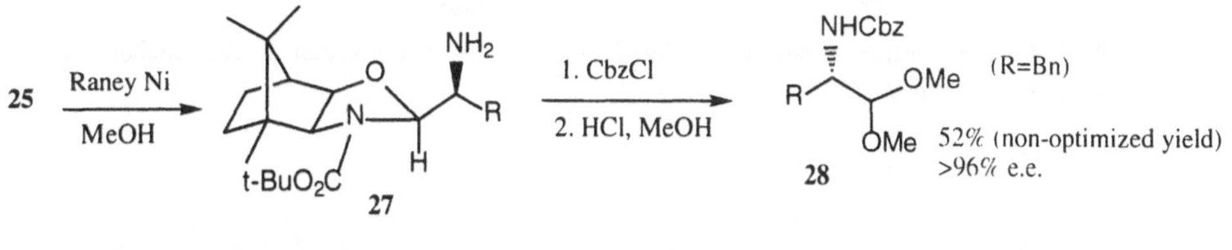

Racemization-free methods for the conversion of N-protected α-amino acetals into the corresponding aldehydes have been already reported, but only for limited cases.[22] Future work will then focus on searching for the amine protecting group most suitable for a racemization-free conversion of acetals 28 into the corresponding aldehydes. The extension of the methodology to hydrazones, derived from 2-acyl oxazolidines, will be also investigated.

ACKNOWLEDGMENTS

We thank MURST (Rome) and Università di Pavia for granting. Dr. M. Zema (C.G.I., University of Pavia) is acknowledged for the X-ray structure determination.

REFERENCES

1. Johnson, R. A.; Sharpless, K. B. in :*Catalytic Asymmetric Synthesis*; Ojima, I., ed. VCH Publishers: Weinheim, 1993; p 227.

2. Tokunaga, M.; Larrow, J. F.; Kakiucki, F.; Jacobsen, E. N.; *Science* **1997**, *277*, 936.

3. (a) Williams, R. M. *Synthesis of Optically Active α-Amino Acids*; Pergamon Press: Oxford, 1989; p 230. (b)Takaya, H.; Ohta, T.; Noyori, R. in: *Catalytic Asymmetric Synthesis*; Ojima, I., ed. VCH Publishers: Weinheim, 1993; p 1.

4. Colombo, L.; Di Giacomo, M.; Brusotti, G.; Milano, E. *Tetrahedron Letters* **1995**, *36*, 2863.

5. Dondoni, A.; Colombo, L. In *Advances in the Use of Synthons in Organic Chemistry*, Dondoni, A., ed.; JAI Press, London, 1993, p 1.

6. (a) Quintard, J.-P.; Elissondo, B.; Mouko-Mpegna, D. *J. Organomet. Chem.* **1983**, *251*, 175. (b) Quintard, J.-P.; Elissondo, B.; Jousseaume, B. *Synthesis* **1984**, 495.

7. Shiner, C. S.; Tsunoda, T.; Goodman, B. A.; Ingham, S.; Lee, S. *J. Am. Chem. Soc.* **1989**, *111*, 1381.

8. Colombo, L.; Di Giacomo, M.; Brusotti, G.; Delogu, G. *Tetrahedron Letters* **1994**, *35*, 2063.

9. (a) Manzoni, L.; Pilati, T.; Poli, G.; Scolastico, C. *J. Chem. Soc. Chem. Comm.* **1992**, 1027. (b) Poli, G.; Maccagni, E.; Manzoni, L.; Pilati, T.; Scolastico, C. *Tetrahedron*, **1997**, *53*, 1759.

10. (a) Frieboes, K. C.; Harder, T.; Aulbert, D.; Strahringer, C.; Bolte, M.; Hoppe, D. *Synlett* **1993**, 921. (b) Harder, T.; Lohl, T.; Bolte, M.; Wagner, K.; Hoppe, D. *Tetrahedron Letters* **1994**, *35*, 7365.

11. Kouklovsky, C.; Pouilhé, A.; Langlois, Y. *J. Am. Chem. Soc.* **1990**, *112*, 6672.

12. (a) Dess, D. B.; Martin, J. C. *J. Org. Chem.* **1983**, *48*, 4155. (b) Dess, D. B.; Martin, J. C. *J. Am. Chem. Soc.* **1991**, *113*, 7277. (c) Ireland, R. E.; Longbin, L. *J. Org. Chem.* **1993**, *58*, 2899.

13. MacroModel V3.5X: Mohamadi, F.; Richards, N. G. J.; Guida, W. C.; Liskamp, R.; Caufield, C.; Chang, G.; Hendrickson, T.; Still, W. C. *J. Comput. Chem.* **1990**, *11*, 440.

14. Chérest, M.; Felkin, H. *Tetrahedron Letters* **1968**, 2205. (b) Anh, N. T.; Eisenstein, O. *Nouv. J. Chimie* **1977**, *1*, 61.

15. Ko, K.-Y.; Park, J.-Y. *Tetrahedron Letters* **1997**, *38*, 407.

16. Agami, C.; Couty, F.; Mathieu, H. *Tetrahedron Letters* **1998**, *39*, 3505.

17. (a) Heathcock, C. H.; Flippin, L. A. *J. Am. Chem. Soc.* **1983**, *105*, 1667. (b) Lodge, E. P.;Heathcock, C. H. *J. Am. Chem. Soc.* **1987**, *109*, 2819.

18. Masamune, S.; Choy, W.; Petersen, J. S.; Sita, L. R. *Angew. Chem. Int. Ed. Engl.* **1985**, *24*, 1.

19. Dale, J. A.; Mosher, H. S. *J. Am. Chem. Soc.* **1973**, *95*, 512.

20. (a) Jurczak, J.; Golebiowski, A. *Chem. Rev.* **1989**, *89*, 149. (b) Reetz, M. *Angew. Chem. Int. Ed. Engl.* **1991**, *30*, 1531.

21. Enders, D.; Reinhold, U. *Tetrahedron: Asymmetry* **1997**, *8*, 1895.

22. (a) Lubell, W.; Rapoport, H. *J. Org. Chem.* **1998**, *54*, 3824. (b) Muralidharan, K. R.; Mokhallalati M. K.; Pridgen L. N. *Tetrahedron Letters* **1994**, *35*, 7489.

EFFECT OF COUNTER-CATIONS ON AGGREGATION AND CHIRAL DISCRIMINATION : APPLICATION TO HIGHLY DIASTEREO- AND ENANTIO-SELECTIVE SYNTHESES.

A. Solladié-Cavallo

Department of Fine Organic Chemistry, ECPM/University L. Pasteur, 1 rue B. Pascal, 67008-Strasbourg, France. Tel. +33 3 88 41 68 32 ; Fax +33 3 88 61 65 31 ; E-Mail ascava@chimie.u-strasbg.fr.

INTRODUCTION

Various methods involving alkylations, reductions, aldolization and formation of epoxides or cyclopropanes have been optimized using effects of counter-cations and aggregation. The chiral auxiliaries are used in stoichiometric amounts but they are recovered and re-used.

I - 8-PHENYLMENTHOL

Reduction

Although 8-phenylmenthol is usually a powerful chiral auxiliary, reduction of **1** had been performed with only 50% d.e. using racemic L-selectride.[1] Changing from Li to K (to fit better with the bite-angle) and from chiral but racemic [Et(Me)-CH]$_3$B to non chiral (Bu)$_3$B allowed to increase the diastereoselectivity up to ≥ 96% d.e.[2] (determined by NMR). The use of reagent **2**[3] then provided a 4-step preparation of enantiomerically pure β-bloquers having the desired *R* configuration.

The chiral auxiliary is recovered and re-used.

2 : LiAl(RNH)$_4$ with R = t-Bu ; i-Pr ; Et ; Me ('solid' and storable methyl amine)

Alkylation

Alkylation of chiral alkyl phenylacetate provides usually low d.e. because of competition between kinetic and thermodynamic control (E and Z enolates). The use of a phosphazene base leading to formation of the Z enolate only (thermodynamic control) provided 86 to 96% d.e. upon alkylation of **3** with various alkylating agents[4] and thus a 4-step-route to antiinflammatory compounds.

The chiral auxiliary is recovered and re-used.

II - HYDROXYPINANONE

Aldolization

When obtained *via* Li/Ti exchange[5] , the "titanium enolate" derived from iminoglycinate **5** leads to complex mixtures upon reaction with aldehydes while 'pure titanium enolate' directly generated from **4** and ClTi(OR)$_3$/NEt$_3$ leads to clean and highly diastereoselective aldol reactions[6].

The best conditions until now are 1.5 equiv. of monochlorotitanium-tri-alcoholate and 2 equiv. of triethylamine. These conditions provide a clean reaction and only one diastereomer (**I** = 95-98%, from 200 MHz ^{1}H-NMR of the crude products of the reactions) among the four possible.

After hydrolysis the (*2R3R*) configuration of this major isomer **I** was assigned on the basis of :

> -the sign of the $[\alpha]_D$ + X-ray for known ***allo*-threonine** (R = Me)[7]
> -the sign of $[\alpha]_D$ + ^{1}H-NMR pattern for known **chloramphenicol base**[7] (obtained after further reduction of the ester group) using authentical samples from Roussel-Uclaf (R = pNO$_2$C$_6$H$_4$)

-the sign of $[\alpha]_D$ + ^{1}H-NMR pattern for known **D-*erythro*-sphingosine**[6] (obtained after reduction of the ester group), (R = CH=CHC$_{13}$H$_{27}$).

This high *erythro(anti)*-selectivity can be rationalize on the basis of a dimer of type **D**.

Applications have been done to the preparation of other hydroxy esters involved in important antibiotics (Vancomycin, Aridicin A...).[8]

Switching from dynamically stable titanium enolates to "naked" enolates by using a phosphazene base allowed to invert the diastereoselectivity toward the *syn(threo)*-isomer (**II** or **III** = 85%).

The chiral auxiliary is recovered and re-used.

III - OXATHIANE

Formation of epoxides

It has been found that the chiral sulfide **6** is an efficient chiral auxiliary able to provide epoxides and cyclopropanes in high yield and high enantiomeric purities.

While phase-transfer conditions (Li, Na or KOH 50%/CH$_2$Cl$_2$/BuEt$_3$NCl) lead to mono-aryl epoxides and *trans*-diaryl epoxides with no more than 70% e.e[9] the use of aprotic conditions (NaH/THF or NaH/CH_2Cl_2) allowed to reach 96-99.9% e.e[10] (chiral chromat.).

Then the used of a phosphazene base (EtP_2) shortened the reaction time from 1 or 2 days to 15 to 30 mn and the e.e % determined by ^{1}H-NMR were > 96%.

The chiral auxiliary is recovered and re-used.

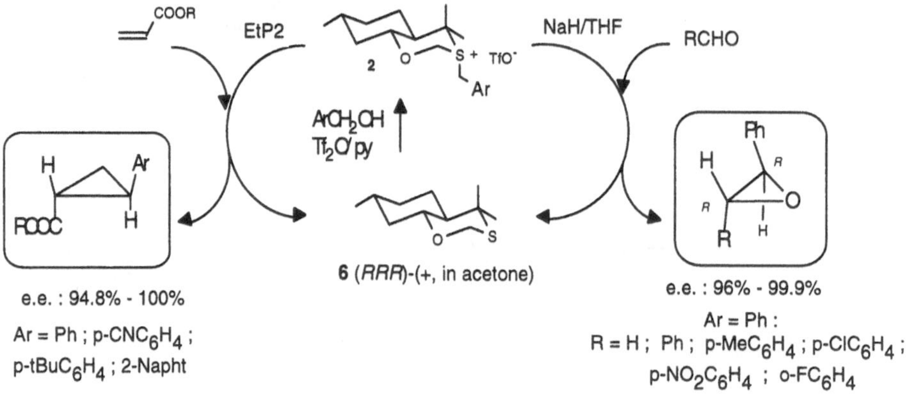
COOR
EtP2
NaH/THF
RCHO
2
S + TfO⁻
Ar
ArCH2OH
Tf2O/py
H
Ar
ROOC
H
Ph
H
R
R
O
R
H
6 (RRR)-(+, in acetone)
e.e. : 94.8% - 100%
Ar = Ph ; p-CNC6H4 ;
p-tBuC6H4 ; 2-Napht
e.e. : 96% - 99.9%
Ar = Ph :
R = H ; Ph ; p-MeC6H4 ; p-ClC6H4 ;
p-NO2C6H4 ; o-FC6H4

Formation of cyclopropanes

Disubstituted cyclopropanes having the same e.e % (94.8-100% , chiral chromat.) have also been obtained using EtP_2 in CH_2Cl_2.[11]

The chiral auxiliary is recovered and re-used.

REFERENCES

1. P. Whitesell et al. *JCS Chem. Commun.* **1983**, 802.
2. ASC, M. Bencheqroun *Tetrahedron : Asymm.* **1991**, *2*, 1165.
3. ASC, M. Bencheqroun *J. Org. Chem.* **1992**, *57*, 5831.
4. ASC, A.G. Csaky et al. *J. Org. Chem.* **1994**, *59*, 5343.
5. P. Viallefont et al. *Tetrahedron* **1988**, *44*, 5319.
6. ASC, J.L. Koessler *J. Org. Chem.* **1994**, *59*, 3240.
7. ASC, J.L. Koessler et al. *Gazzetta Chim. It.* **1996**, *126*, 173.
8. ASC, T. Nsenda *Tetrahedron Lett.* **1998**, *39*, 2191.
9. ASC, A. Adib *Tetrahedron* **1992**, *48*, 2453.
10. ASC, A. Diep-Vohuule et al. *Tetrahedron : Asymm.* **1996**, *7*, 1783.
11. ASC, T. Isarno et al. *Angew. Chem. Int. Ed. Engl.* **1998**, *37* (june).

CATALYTIC ENANTIOSELECTIVE REACTIONS OF ALDIMINES USING CHIRAL LEWIS ACIDS

Shū Kobayashi

Graduate School of Pharmaceutical Sciences, The University of Tokyo, Hongo, Bunkyo-ku, Tokyo, 113-0033

MANNICH-TYPE REACTIONS

Asymmetric Mannich-type reactions provide useful routes for the synthesis of optically active β-amino ketones or esters, which are versatile chiral building blocks in the preparation of many nitrogen-containing biologically important compounds. While several diastereoselective Mannich-type reactions have already been reported, very little is known about the enantioselective versions. In addition, asymmetric Mannich-type reactions using small amounts of chiral sources have not been reported, when we started these research efforts.

Our approach is based on chiral Lewis acid-catalyzed reactions. Asymmetric reactions using chiral Lewis acids are of great current interest as one of the most efficient methods for the preparation of chiral compounds.[1] While rather rapid progress has been made on the enantioselective reactions of carbonyl compounds using chiral Lewis acids (aldol reactions, allylation reactions, Diels-Alder reactions, etc.), very few examples have been reported for their aza analogues. We thought that this was due to two main difficulties. First, many Lewis acids are deactivated or sometimes decomposed by the nitrogen atoms of starting materials or products, and even when the desired reactions proceed, more than stoichiometric amounts of the Lewis acids are needed because the acids are trapped by the nitrogen atoms. Secondly, aldimine-chiral Lewis acid complexes are rather flexible and often have several stable conformers (including E/Z isomers of aldimines), while aldehyde-chiral Lewis acid complexes are believed to be rigid. Therefore, in the additions to aldimines activated by chiral Lewis acids, plural transition states would exist to decrease selectivities. In order to solve these problems, we first screened various metal salts in the achiral reactions of aldimines with silylated nucleophiles. After careful investigation of the catalytic ability of the salts, we found unique characteristics in zirconium (IV) (Zr (IV)) and decided to design a chiral Lewis acid based on Zr (IV) as a center metal.[2,3] On the other hand, as for the problem of the conformation of the aldimine-Lewis acid complex, we planned to utilize a bidentate chelation (*vide infra*).[4]

A chiral zirconium catalyst was prepared *in situ* according to Scheme 1. In the presence of 20 mol% of catalyst **1**, aldimine **3** prepared from 1-naphthaldehyde and 2-aminophenol was treated with the ketene silyl acetal derived from methyl isobutylate (**4**) in dichloromethane at -15 °C. The reaction proceeded smoothly to afford the corresponding adduct in a quantitative yield, and the enantiomeric excess of the product was 34%. The ee was improved to 70% when 1-methylimidazole (NMI) was used as an additive. Moreover, the ee was further improved when catalyst **2** using (*R*)-6,6'-dibromo-1,1'-bi-2-naphthol ((*R*)-Br-BINOL) was employed, and the desired adduct was obtained in a 95% ee when the reaction was carried out at -45 °C. It should be noted that the same high level of ee was obtained when 2 mol% of the catalyst was employed. We then tested other aldimines and silyl enolates and the results are summarized in Table 1.[5] Not only aldimines derived from aromatic aldehydes but also aldimines from heterocyclic and aliphatic aldehydes worked well in this reaction, and good to high yields and enantiomeric excesses were obtained. Similar high levels of ees were also obtained when the silyl enol ether derived from S-ethyl thioacetate (**5**) was used. The N-substituent of the product was easily removed according to Scheme 2. Thus, methylation of the phenolic OH of **6** using methyl iodide and potassium bicarbonate, and deprotection using cerium ammonium nitrate (CAN)[6] gave β-amino ester **7**. The absolute configuration assignment was made by comparison of the optical rotation of **7** with that in the literature.[7]

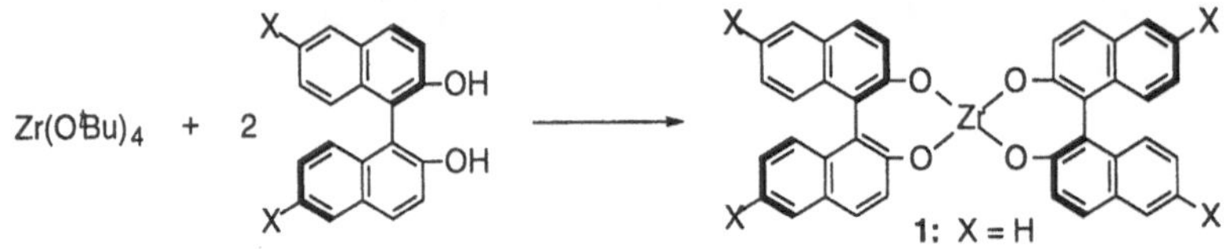

Scheme 1. Preparation of the Catalyst

Table 1. Catalytic Enantioselective Mannich-type Reactios

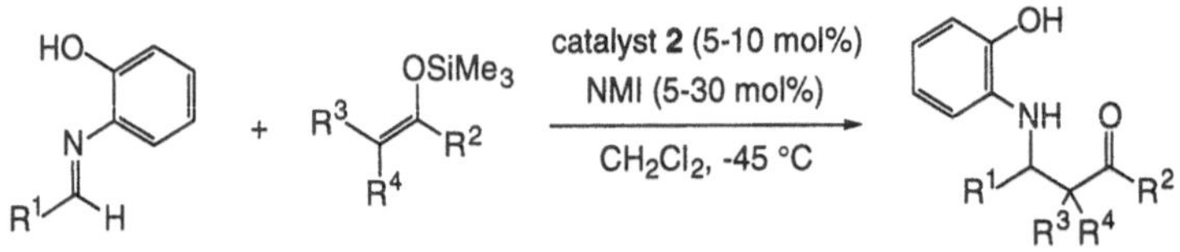

R^1	silyl enolate	yield/%[a]	ee/%[b]
Ph	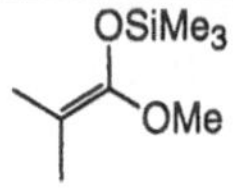 OSiMe$_3$, OMe (**4**)	70	87
p-Cl-Ph	**4**	86	83
1-naphthal (**3**)	**4**	quant.	92
Ph	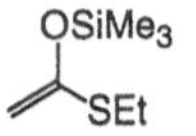 OSiMe$_3$, SEt (**5**)	78	88
p-Cl-Ph	**5**	88	86
1-naphthal (**3**)	**5**	quant.	>98
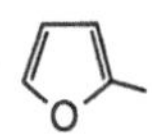 (furyl)	**5**	89	89
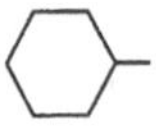 (cyclohexyl) c	**5**	45	80

[a] Isolated yields after acidic work-up. [b] Determined by HPLC analyses. [c] The aldimine prepared from cyclohexanecarbox-aldehyde and 2-amino-3-methylphenol was used.

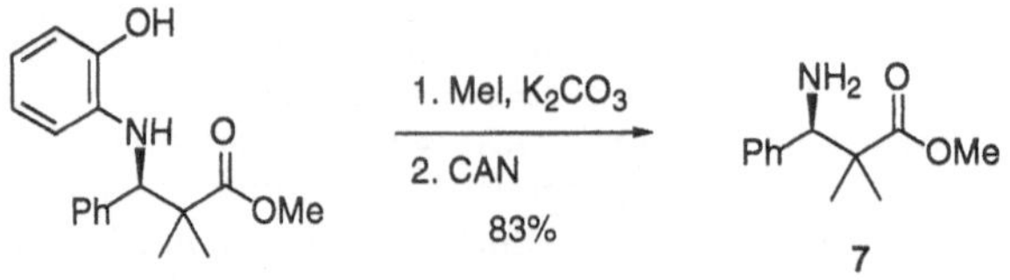

Scheme 2. Conversion to β-Amino Ester

184

SYNTHESIS OF *SYN*- AND *ANTI*-AMINO ALCOHOLS

We then performed catalytic diastereo- and enantioselective Mannich-type reactions of α-alkoxy enolates with aldimines for the synthesis of chiral β-amino alcohols (Scheme 3).[8] First the reaction of aldimine **8** with α-TBSO-ketene silyl acetal **9a** was tested using 10 mol% of zirconium catalyst **2**. The reaction proceeded smoothly to afford the corresponding α-alkoxy-β-amino ester in a 76% yield with moderate *syn*-selectivity and the enantiomeric excess of the *syn*-adduct was proven to be less than 10%. We then screened various reaction conditions. It was found that when 1,2-dimethylimidazole (DMI) was used instead of NMI, the selectivity increased dramatically. Moreover, the diastereo- and enantioselectivities were improved when the reaction was carried out at -78 °C. The *O*-substituents of ketene silyl acetals and solvents also influenced the yield and selectivity, and finally the best result (quant, *syn/anti* = 96/4, *syn* = 95% ee) was obtained when the reaction was carried out in toluene using ketene silyl acetal **9b(E)**. It was also interesting from a mechanistic point of view that geometrically isomeric ketene silyl acetal **9b(Z)** also gave excellent diastereo- and enantioselectivities (*syn/anti* = >99/1, *syn* = 96% ee). Other substrates were tried and the results are shown in Table 2. In all cases, the desired adducts including *syn*-β-amino alcohol units were obtained in high diastereo- and enantioselectivities.

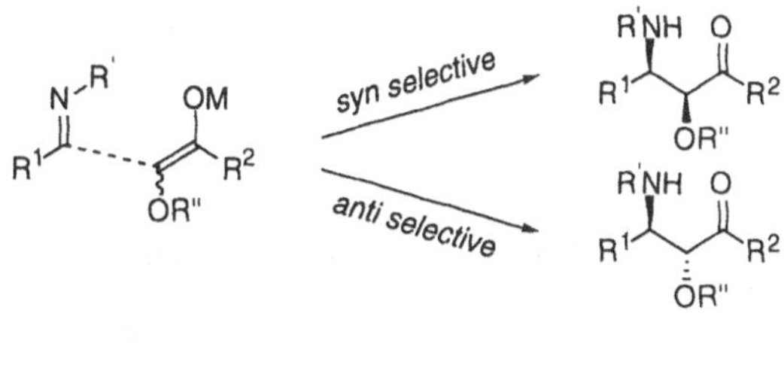

Scheme 3. Chiral β-Amino Alcohol Synthesis

On the other hand, it was found that *anti*-β-amino alcohol derivatives were obtained by the reaction of aldimine **8** with α-BnO-ketene silyl acetal **9c** under the same reaction conditions.[9] Namely, in the presence of 10 mol% of the above catalyst, aldimine **8** reacted with **9c** smoothly to give the corresponding adduct quantitatively with *anti* preference, and the enantiomeric excess of the *anti*-adduct was 95%. It was exciting that both *syn*- and *anti*-amino alcohol units were prepared by simply choosing the protective groups of the α-alkoxy parts of the silyl enolates. Several aldimines were then tested and the results are summarized in Table 3. In most cases, the desired *anti*-adducts were obtained in high yields with high diastereo-

Table 2. Synthesis of *Syn*-Amino Alcohol Units

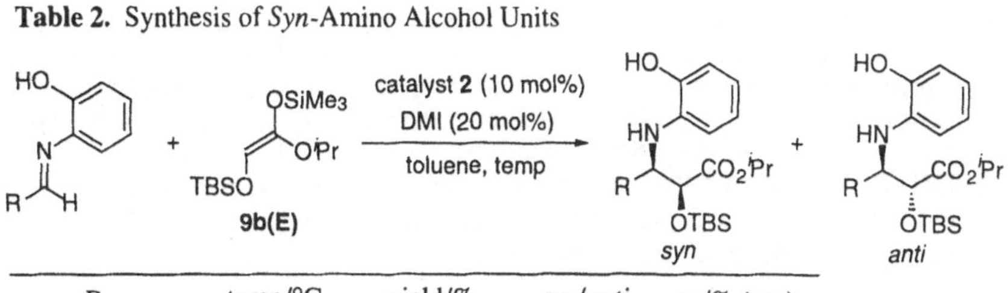

R	temp/°C	yield/%	syn/anti	ee/% (syn)
Ph (**8**)	-78	quant	96/ 4	95
1-naphthyl	-78	65	>99/ 1	91
2-furyl	-45	68	82/18	92
p-ClPh[a)]	-78	73	92/ 8	98

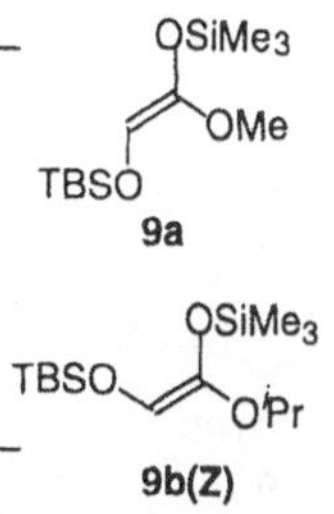

a) Dichloromethane was used as solvent, and 30 mol% of DMI was added.

185

and enantioselectivities. While higher diastereoselectivities were obtained using a ketene silyl acetal derived from a p-methoxyphenyl (PMP) ester, higher enantiomeric excesses were observed in the reactions using a ketene silyl acetal derived from isopropyl or cyclohexyl ester. In the reaction of the aldimine derived from cyclohexanecarboxaldehyde, use of 2-amino-3-methylphenol instead of 2-aminophenol was effective in affording the corresponding *anti*-adduct in high selectivities.

Table 3. Synthesis of *Anti*-Amino Alcohol Units

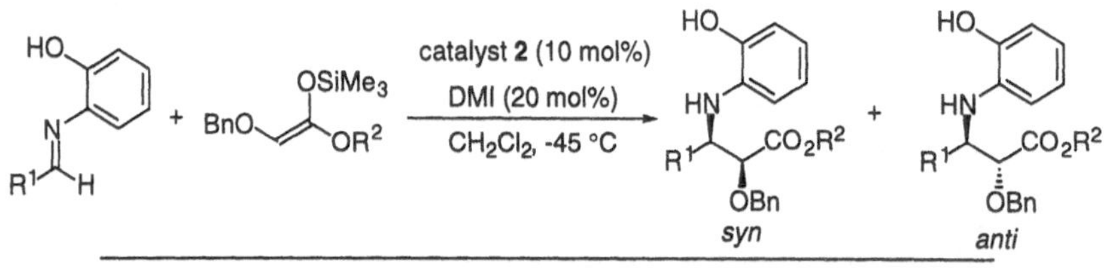

R^1	R^2	yield/%	syn/anti	ee/% (anti)
Ph (**8**)[a]	^{i}Pr (**9c**)[e]	quant	32/68	95
Ph (**8**)[b]	PMP[f,g]	91	6/94	80
1-naphthyl[c]	c-C$_6$H$_{11}$[e]	80	8/92	96
2-furyl[b]	PMP	68	13/87	80
p-ClPh[a,c]	^{i}Pr (**9c**)	quant	43/57	91
p-ClPh[b]	PMP	72	8/92	76
c-C$_6$H$_{11}$[d]	c-C$_6$H$_{11}$	41	18/82	92

[a]DMI (30 mol%) was used. [b]NMI (20 mol%) was used. [c]The reaction was carried out at -78 °C. [d]The imine was prepared from cyclohexanecarboxaldehyde with 2-amino-3-methylphenol *in situ* in the presence of MS4A. [e]E/Z = <1/>99. [f]E/Z = 4/96. [g]PMP = p-methoxyphenyl.

In order to demonstrate the utility of these reactions, we undertook the synthesis of (2R,3S)-3-phenylisoserine•hydrochloride (**10**), which is a precursor of the C-13 side-chain of taxol, known to be essential for its biological activity.[10] The key catalytic asymmetric Mannich-type reaction of aldimine **8** with ketene silyl acetal **9b(E)** using the chiral zirconium catalyst prepared using (S)-6,6'-dibromo-1,1'-bi-2-naphthol proceeded smoothly in toluene at -78 °C to afford the corresponding *syn*-adduct quantitatively in excellent diastereo- and enantioselectivities (*syn/anti* = 95/5, *syn* = 94% ee). Methylation (MeI, K$_2$CO$_3$) of the phenolic OH of the adduct (**10**) and deprotection using CAN gave β-amino ester **11**. Hydrolysis of the ester and deprotection of the *t*-butyldimethylsilyl (TBS) group were performed using 10% HCl to afford **12** quantitatively.

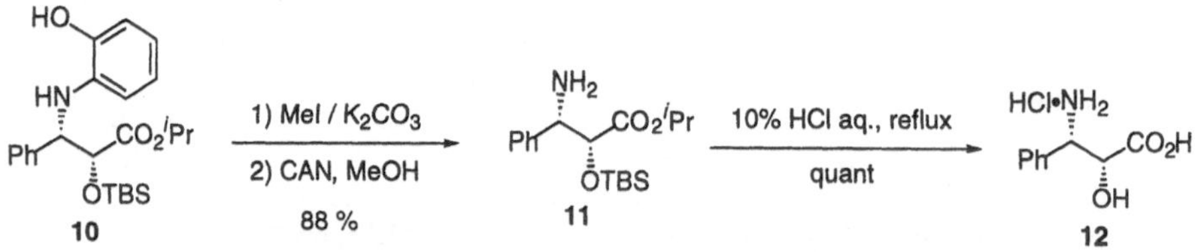

Scheme 4. Synthesis of (2R,3S)-3-Phenylisoserine•hydrochloride (**12**)

186

AZA DIELS-ALDER REACTIONS

Asymmetric aza Diels-Alder reactions provide a useful route to optically active heterocyclic compounds such as piperidines, tetrahydroquinolines, etc.[11] Although successful examples of diastereoselective approaches have been reported, there have been few reports on enantioselective reactions. Yamamoto et al. reported elegant enantioselective aza Diels-Alder reactions of aldimines with Danishefsky's diene using chiral boron compounds, however, stoichiometric amounts of chiral sources were needed.[12] Quite recently, we have reported the first example of the catalytic enantioselective aza Diels-Alder reactions of azadienes using a chiral lanthanide catalyst.[4] While high diastereo- and enantioselectivities were attained in the reaction of N-aryl aldimines (azadienes) with dienophiles, the products obtained were rather limited to 8-hydroxytetrahydroquinoline derivatives.

Chiral zirconium compound 2 was prepared from $Zr(O^tBu)_4$, 2 eq. of (R)-Br-BINOL, and 2-3 eq. of a ligand, and the model reaction of the aldimine derived from 1-naphthaldehyde and 2-aminophenol (3) with Danishefsky's diene (13)[13] was investigated. It was found that ligands and solvents influenced the yields and enantioselectivities strongly. For ligands, NMI gave the best result. When the chiral catalyst (10 mol%) was prepared in dichloromethane, the desired aza Diels-Alder reaction of aldimine 3 with diene 13 proceeded smoothly, but the enantiomeric excess of the adduct was only 40%. On the other hand, the enantioselectivity was improved to 61% ee, when the catalyst was prepared in benzene by stirring for 1h at rt, the mixture was evaporated to remove benzene and ^tBuOH under reduced pressure, and the reaction was then carried out in dichloromethane. While use of a bulky diene (4-t-butoxy-2-trimethylsiloxy-1,3-butadiene)[14] decreased the selectivity in this case, a higher enantiomeric excess was obtained when the catalyst was prepared in toluene. The best result was finally obtained when the preparation of the catalyst and the successive reaction was carried out in toluene (without removing the solvent).

We then examined the effect of the metals used (Table 4). Our preliminary screening proved that zirconium compounds gave excellent catalytic abilities for the activation of aldimines. Other group 4 metals, titanium and hafnium, were screened, and it was found that a chiral hafnium catalyst gave high yields and enantioselectivities in the model reaction of aldimine 3 with Danishefsky's diene (13), while lower yields and enantiomeric excesses were obtained using a chiral titanium catalyst.

Several examples of the catalytic aza Diels-Alder reactions using the chiral zirconium catalyst were examined. In most cases, high chemical yields and good to high levels of enantioselectivities were obtained in the presence of 5-20 mol% of the chiral catalyst.[15] 4-Methoxyl-3-methyl-2-trimethylsiloxy-1,3-butadiene (14)[16] also worked well under standard conditions and the desired 2,3-dihydro-4-pyridone derivatives were obtained in high yields with high enantioselectivities. As for the R^1 group in Scheme 5, ortho-substituted aromatics gave higher selectivities. For example, while the aldimine derived from benzaldehyde reacted

Table 4. Effect of Metals in Aza Diels-Alder Reactions

catalyst = $M(O^tBu)_4$ + 2 (R)-Br-BINOL + 3 NMI

M	x/mol%	yield/%	ee/%
Zr	10	86	82
Zr	20	96	88
Hf	10	89	73
Hf	20	96	84
Ti	10	68	39
Ti	20	70	62

with **14** to afford the corresponding adduct in a 65% ee, a 74% ee of the pyridone derivative was obtained in the reaction of the aldimine derived from *o*-tolualdehyde with **14** under the same reaction conditions. The aldimine derived from 2-thiophenecarboxyaldehyde reacted with **13** smoothly to give the corresponding pyridone derivative in a high yield with a good enantiomeric excess. In the reaction of the aldimine derived from cyclohexanecarboxaldehyde with **14**, a low enantiomeric excess of the adduct was observed under standard reaction conditions. It was exciting to find that the enantiomeric excess of the corresponding pyridone derivative was improved to 86% ee, when the aldimine derived from cyclohexanecarboxaldehyde and 2-amino-3-methylphenol was used.

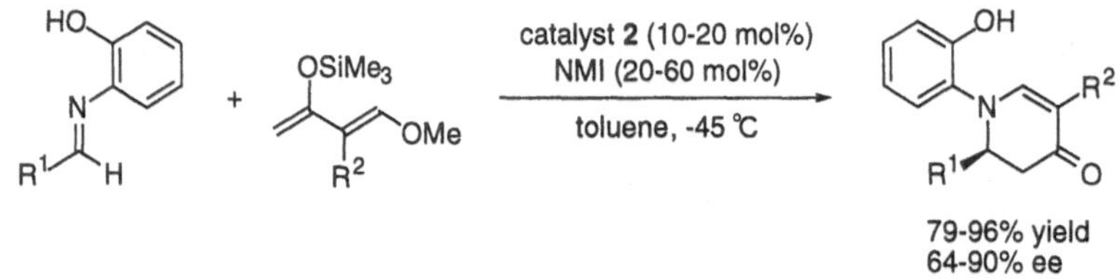

79-96% yield
64-90% ee

Scheme 5. Catalytic Enantioselective Aza Diels-Alder Reactions

CATALYTIC ENANTIOSELECTIVE SYNTHESIS OF α-AMINO NITRILES USING A NOVEL ZIRCONIUM CATALYST

α-Amino nitriles are useful intermediates for the synthesis of amino acids and nitrogen-containing heterocycles such as thiadiazoles, imidazoles, etc.[17] The Strecker-type reactions of aldimines with cyanide sources provide one of the most efficient methods for the preparation of α-amino nitriles, and several diastereoselective approaches for the synthesis of optically active α-amino nitriles have been reported.[18] In 1996, Lipton *et al.* reported the first catalytic enantioselective Strecker-type reactions using a dipeptide ligand as a catalyst.[19] Although efficient catalytic reactions provide α-amino nitriles derived from benzaldehyde derivatives in high enantioselectivities, low selectivities were observed in the reactions of aldimines derived from aliphatic and heterocyclic aldehydes.

We intended to use a zirconium catalyst in asymmetric Strecker-type reactions. In the presence of 10 mol% of a zirconium catalyst, which was prepared from $Zr(O^tBu)_4$, 2 equiv of (*R*)-6-Br-BINOL, and 3 equiv of NMI, aldimine **8** was treated with Bu_3SnCN^{20} in dichloromethane at -45 °C. The reaction proceeded smoothly to afford the corresponding α-amino nitrile in a 70% yield with 55% ee. After several reaction conditions were examined, the best results (92% yield, 91% ee) were obtained when the reaction was carried out in benzene-toluene (1:1) using a chiral zirconium catalyst, prepared from 1 equiv of $Zr(O^tBu)_4$, 1 equiv of (*R*)-6-Br-BINOL and (*R*)-3,3'-dibromo-1,1'-bi-2-naphthol ((*R*)-3-Br-BINOL),[21] and 3 equiv of NMI. Use of other solvents slightly decreased the selectivity. The free hydroxyl group of the aldimine was important in obtaining both high yield and high selectivity. Actually, when the aldimine prepared from aniline or 2-methoxyaniline was used under the same reaction conditions, the corresponding α-amino nitrile derivatives were obtained in much lower yields and lower ee's (aniline, 29% yield, 1% ee; 2-methoxyaniline, 45% yield, 5% ee).

It was very interesting to find that use of a mixture of (*R*)-6-Br-BINOL and (*R*)-3-Br-BINOL gave the best results. We then carefully examined the structure of the zirconium catalyst, and it was indicated from NMR studies that a zirconium binuclear complex (**15**) was formed under the conditions used (Scheme 6). The binuclear complex consists of two equivalents of zirconium, (*R*)-6-Br-BINOL, and NMI, and one equivalent of (*R*)-3-Br-BINOL. The structure was indicated to be very stable and very likely because the complex was formed even when different molar ratios of $Zr(O^tBu)_4$, (*R*)-6-Br-BINOL, (*R*)-3-Br-BINOL, and NMI were combined. Actually, formation of **15** was confirmed by ^1H and ^{13}C NMR spectra when 1 equiv of $Zr(O^tBu)_4$ and (*R*)-6-Br-BINOL, 0.5-1 equiv of (*R*)-3-Br-BINOL, and 2-3 equiv of NMI were combined.

We then examined several examples of the Strecker-type reactions, and the results are summarized in Table 5. Aldimines derived from various aromatic aldehydes as well as aliphatic and heterocyclic aldehydes reacted with Bu_3SnCN smoothly to afford the

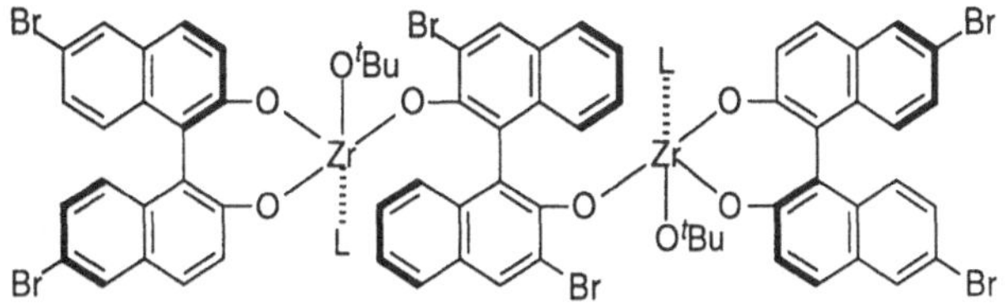

Scheme 6. A Novel Chiral Zirconium Catalyst (**15**) (L = NMI)

corresponding α-amino nitrile derivatives in high yields with high enantiomeric excesses. Since both enantiomers of the chiral sources (6-Br-BINOL and 3-Br-BINOL) are readily available, both enatiomers of α-amino nitrile derivatives can be easily prepared according to this protocol. In addition, it is noteworthy that Bu₃SnCN has been successfully used as a safe cyanide source. Bu₃SnCN is stable in water and no HCN produces. This is in contrast to trimethylsilyl cyanide (TMSCN) that easily hydrolyzes to form HCN even in the presence of a small amount of water. After the reaction was completed, all tin sources were quantitatively recovered as bis(tributyltin) oxide, which was already reported to be converted to tributyltin chloride[22] and then to Bu₃SnCN.[20,23]

Table 5. Catalytic Enantioselective Strecker-Type Reactions

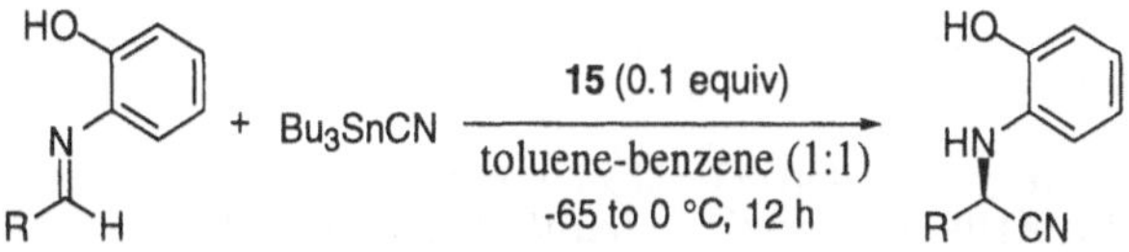

R	yield/%	ee/%	R	yield/%	ee/%
1-Nap	98	91	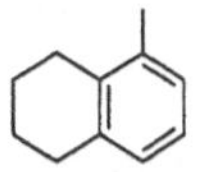 (tetrahydronaphthyl)	85	87
Ph	92	91	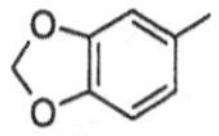 (methylenedioxyphenyl)	89	80
p-ClPh	90	88	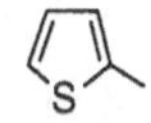 (methylthienyl)	89	92
p-MeOPh	97	76	Ph(CH₂)₂	55	83[c]
o-MePh	96	89[a]	i-Bu	79	83[c]
o-MePh	93	89 (S)[b]	C₈H₁₇	72	74[c]

[a]When 0.05 equiv of **15** was used, 94% yield and 87% ee were obtained. [b]0.1 equiv of *ent*-**15** was used. [c]The imine was prepared from the corresponding aldehyde and 2-amino-3-methylphenol *in situ* in the presence of MS 4A.

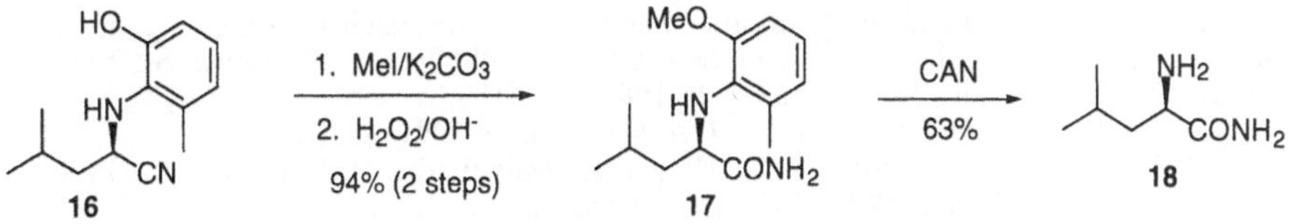

Scheme 7. Synthesis of Leucinamide

α-Amino nitrile **16** was easily converted to leucinamide according to Scheme 7. Thus, after methylation of the phenolic OH of **16** using methyl iodide and potassium bicarbonate, the nitrile group was converted to an amide moiety.[24] Treatment of **17** with CAN gave

189

leucinamide **18**. The absolute configuration assignment (*R*) was made by comparison of its hydrochloride with the authentic sample.

Acknowledgment. The author thanks and expresses his deep gratitude to his coworkers whose names appear in the references. This work was partially supported by CREST, Japan Science and Technology Corporation (JST), and a Grant-in-Aid for Scientific Research from the Ministry of Education, Science, Sports, and Culture, Japan.

References and Notes
1. (a) Narasaka, K. *Synthesis* **1991**, 1. (b) Santelli, M.; Pons, J.-M. *Lewis Acids and Selectivity in Organic Synthesis*; CRC Press: Boca Raton; 1995.
2. Cf. (a) Nugent, W. A. *J. Am. Chem. Soc.* **1992**, *114*, 2768. (b) Bedeschi, P.; Casolari, S.; Costa, A. L.; Tagliavini, E.; Umani-Ronchi, A. *Tetrahedron Lett.* **1995**, *36*, 7897. (c) Hoveyda, A. H.; Morken, J. P. *Angew. Chem., Int. Ed. Engl.* **1996**, *35*, 1262, and references cited therein.
3. Rare earths (Sc, Y, Ln) are also promising candidates for the catalytic activations of aldimines. (a) Kobayashi, S.; Araki, M.; Ishitani, H.; Nagayama, S.; Hachiya, I. *Synlett* **1995**, 233. (b) Kobayashi, S.; Araki, M.; Yasuda, M. *Tetrahedron Lett.* **1995**, *36*, 5773. See also Ref. 4.
4. Ishitani, H.; Kobayashi, S. *Tetrahedron Lett.* **1996**, *37*, 7357.
5. Ishitani, H.; Ueno, M.; Kobayashi, S. *J. Am. Chem. Soc.* **1997**, *119*, 7153.
6. Kronenthal, D. R.; Han, C. Y.; Taylor, M. K. *J. Org. Chem.* **1982**, *47*, 2765.
7. Kunz, H.; Schanzenbach, D. *Angew. Chem., Int. Ed. Engl.* **1989**, *28*, 1068.
8. Kobayashi, S.; Ueno, M.; Ishitani, H. *J. Am. Chem. Soc.* **1998**, *120*, 431.
9. We also observed similar dramatic changes in diastereoselectivities in chiral tin(II) mediated asymmetric aldol reactions. Kobayashi, S.; Horibe, M. *Synlett* **1994**, 147, and references cited therein.
10. (a) Bruncko, M.; Schlingloff, G.; Sharpless, K. B. *Angew. Chem., Int. Ed. Engl.* **1997**, *36*, 1483. (b) Escalante, J.; Juaristi, E. *Tetrahedron Lett.* **1995**, *36*, 4397. (c) Nicolaou, K. C.; Dai, W.-M.; Guy, R. K. *Angew. Chem., Int. Ed. Engl.* **1994**, *33*, 15.
11. (a) Waldmann, H. *Synthesis* **1994**, 535. (b) Waldmann, H. in *Organic Synthesis Highlights II*; Waldmann, H., Ed.; VCH: Weinheim, 1995; pp. 37-48. (c) Weinreb, S. M. In *Comprehensive Organic Synthesis*; Trost, B. M.; Fleming, I., Eds.; Pergamon Press: Oxford, 1991; Vol. 5, pp. 401-449.
12. (a) Hattori, K.; Yamamoto, H. *J. Org. Chem.* **1992**, *57*, 3264. (b) Hattori, K.; Yamamoto, H. *Tetrahedron* **1993**, *49*, 1749. (c) Ishihara, K.; Miyata, M.; Hattori, K.; Tada, T.; Yamamoto, H. *J. Am. Chem. Soc.* **1994**, *116*, 10520.
13. (a) Danishefsky, S.; Kitahara, T. *J. Am. Chem. Soc.* **1974**, *96*, 7807. (b) Kerwin, Jr., J. F.; Danishefsky, S. *Tetrahedron Lett.* **1982**, *23*, 3739.
14. Danishefsky, S.; Bednarski, M.; Izawa, T.; Maring, C. *J. Org. Chem.* **1984**, *49*, 2290.
15. Kobayashi, S.; Komiyama, S.; Ishitani, H. *Angew. Chem., Int. Ed. Engl.* **1998**, *37*, 979.
16. (a) Clive, D. L. J.; Bergstra, R. J. *J. Org. Chem.* **1991**, *56*, 4976. (b) Sugawasa, S.; Yamada, S.-I.; Narahashi, M. *J. Pharm. Soc. Jpn.* **1951**, *71*, 1345.
17. (a) Strecker, A. *Ann. Chem. Pharm.* **1850**, *75*, 27. (b) Shafran, Y. M.; Bakulev, V. A.; Mokrushin, V. S. *Russian Chem. Rev.* **1989**, *58*, 148.
18. (a) Williams, R. M. *Synthesis of Optically Active a-Amino Acids*; Pergamon: Oxford, 1989. (b) Williams, R. M.; Hendrix, J. A. *Chem. Rev.* **1992**, *92*, 889. (c) Duthaler, R. O. *Tetrahedron* **1994**, *50*, 1539.
19. Iyer, M. S.; Gigstad, K. M.; Namdev, N. D.; Lipton, M. *J. Am. Chem. Soc.* **1996**, *118*, 4910.
20. Commercially available (Aldrich, etc.). (a) Luijten, J. G. A.; van der Kerk, G. J. M. *Investigations in the Field of Organotin Chemistry*; Tin Research Institute: Greenford, 1995; p. 106. (b) Tanaka, M. *Tetrahedron Lett.* **1980**, *21*, 2959. (c) Harusawa, S.; Yoneda, R.; Omori, Y.; Kurihara, T. *Tetrahedron Lett.* **1987**, *28*, 4189.
21. Cox, P. J.; Wang, W.; Snieckus, V. *Tetrahedron Lett.* **1992**, *33*, 2253.
22. (a) Brown, J. M.; Chapman, A. C.; Harper, R.; Mowthorpe, D. J.; Davies, A. G.; Smith, P. J. *J. Chem. Soc., Dalton.* **1972**, 338. (b) Davies, A. G.; Kleinschmidt, D. C.; Palan, P. R.; Vasishtha, S. C. *J. Chem. Soc. (C).* **1971**, 3972.
23. Quite recently, we have developed scandium triflate-catalyzed Strecker-type reactions of aldehydes, amines, and Bu3SnCN (achiral reactions). In these reactions, complete recovery of tin compounds towards environmentally-friendly chemical processes has been achieved. Kobayashi, S.; Busujima, T.; Nagayama, S. *J. Chem. Soc., Chem. Commun.* **1998**, 981.
24. Cacchi, S.; Misiti, D.; Torre, F. L. *Synthesis* **1980**, 243.

SULFUR YLIDE MEDIATED CATALYTIC ASYMMETRIC EPOXIDATION AND AZIRIDINATION

Varinder K. Aggarwal,[a*] J. Gair Ford,[a] Alison Thompson,[a] John Studley,[a] Ray V. H. Jones,[b] and Robin Fieldhouse.[b]

(a) Dept. of Chemistry, University of Sheffield, Sheffield S3 7HF, UK.
(b) Zeneca Process Technology Department, Earls Road, Grangemouth, Stirlingshire FK3 8XG, UK.

ASYMMETRIC EPOXIDATION OF ALDEHYDES

The development of catalytic methods for the synthesis of non-racemic epoxides has been a long standing goal in asymmetric synthesis. For epoxide synthesis, attention has largely focused on the asymmetric oxidation of alkenes and good enantioselectivities are now beginning to emerge for an increasing range of substrates.[1-3] Alkenes are themselves commonly obtained by Wittig reaction from the corresponding aldehyde or ketone and so epoxidation is usually a two step process from carbonyl compounds. However, in terms of efficiency and atom economy this overall process is poor compared to direct epoxidation of carbonyl compounds using sulfur ylides. However, in terms of catalysis and asymmetric induction, oxidation of alkenes is still superior as epoxidation of carbonyl compounds using sulfur ylides usually requires stoichiometric amounts of sulfides/sulfur ylides and only gives moderate enantioselectivities.[4] We have described a *catalytic* process for epoxidation involving sulfur ylides which now overcomes this limitation (Scheme 1).[5] Whilst only low levels of enantioselectivity were originally achieved we now report significant improvements and describe chiral sulfides which provide high levels of asymmetric induction.[6]

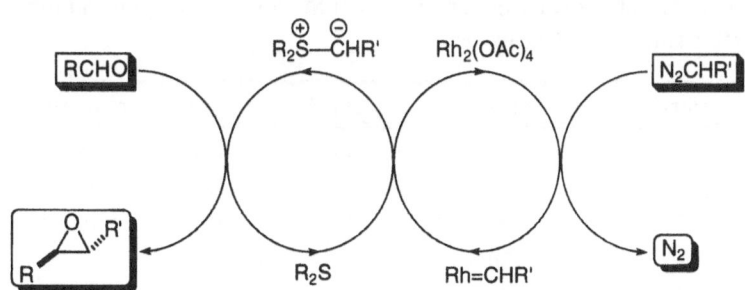

Scheme 1: Catalytic Process for Epoxidation
Reprinted with permission from reference 6. Copyright (1996) American Chemical Society.

In the design of new chiral sulfides it was deemed important to ensure that only one of the two lone pairs reacts with the metallocarbene to give a single sulfur ylide and to avoid formation of diastereomeric sulfur ylides which could react with opposite enantioselectivity. Ideally, we would also want to be able to tune the steric and/or electronic environment of the sulfur to maximise enantioselectivity in a simple way. Sulfide **1** was therefore designed as it possesses only one reactive sulfur lone pair and, being a thioacetal, the R group is readily amenable to 'tuning'.

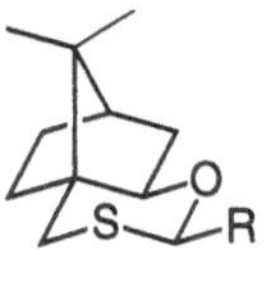

1a	R = H
1b	R = Me
1c	R = *i*-Pr
1d	R = *t*-Bu
1e	R = CH$_2$Ph
1f	R = CH$_2$OPh
1g	R = CH$_2$OMe

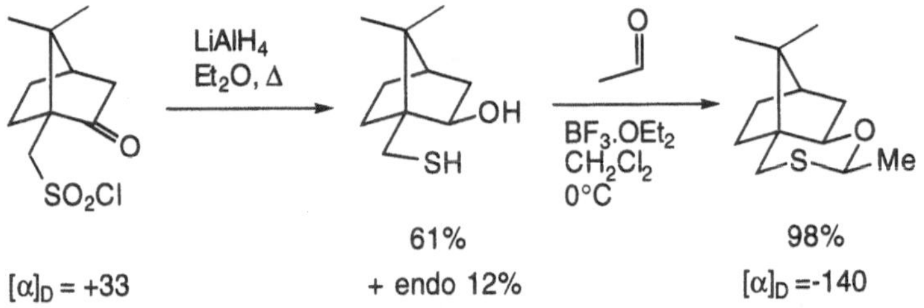

Scheme 2: Preparation of Chiral Sulfides
Reprinted with permission from reference 6. Copyright (1996) American Chemical Society.

Sulfides **1a-g** were prepared as shown in Scheme 3 (**1b** shown) and incorporated in the catalytic cycle with benzaldehyde (Table 1).[6] It was found that high enantio-selectivity could be obtained provided that the thioacetal was substituted at the 2 position (entries 2-7). Sterically hindered (entries 2, 3, 4 and 5) or electron withdrawing groups (entries 6 and 7) resulted in lower yields in the epoxidation process. The optimum sulfide in terms of yield (73%) and enantioselectivity (93%) was **1b** (entry 2).

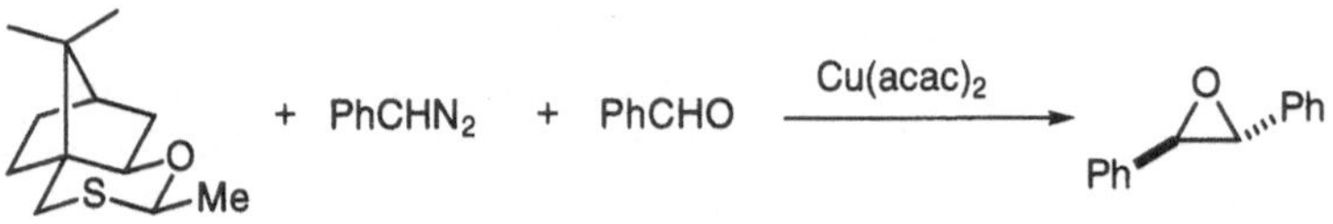

Table 1: Yields, Enantioselectivities and Ratios of Stilbene Oxide Formed From Benzaldehyde Using 0.2 eq Of Sulfides **1a-g**.

entry	sulfide	yield %	ee %[a]	trans: cis
1	3a	83	41 (R,R)	>98:2
2	3b	73	93 (R,R)	>98:2
3	3c	45	93 (R,R)	>98:2
4	3d	0	-	-
5	3e	56	88 (R,R)	>98:2
6	3f	43	83 (R,R)	>98:2
7	3g	70	92 (R,R)	>98:2

(a) Enantiomeric excess determined by chiral HPLC using a Chiralcel OD column.
Reprinted with permission from reference 6. Copyright (1996) American Chemical Society.

Sulfide **1b** was tested with a range of aldehydes and the results are summarised in Table 2. It was found that high enantioselectivity was maintained with both aromatic and aliphatic aldehydes. Aliphatic aldehydes gave lower yields compared to aromatic aldehydes and gave a mixture of *trans* and *cis* epoxides *whereas aromatic aldehydes only gave trans epoxides*. This contrasts with simple sulfides e.g. dimethyl sulfide in which mixtures of *trans:cis* epoxides were obtained with benzaldehyde.[5]

Table 2: Yields, Enantioselectivities and Ratios of Epoxides Formed from Aldehydes Using 0.2 eq. of Sulfide **1b.**

entry	aldehyde	yield %	ee %[a]	trans:cis
1	benzaldehyde	73	93 (R,R)	>98:2
2	p-chlorobenzaldehyde	72	92(R,R)	>98:2
3	p-tolualdehyde	64	92(R,R)	>98:2
4	cinnamaldehyde	73	89	>98:2
5	valeraldehyde	35	68	92:8
6	cyclohexanecarboxaldehyde	32	90	70:30

(a) Enantiomeric excess determined by chiral HPLC using a Chiralcel OD column.
Reprinted with permission from reference 6. Copyright (1996) American Chemical Society.

Our mechanistic rationale for the high asymmetric induction observed is depicted in Scheme 3.[6] The ylide can adopt two conformations **2a** or **2b** but **2a** suffers from 1,3 diaxial interactions of the phenyl group with the axial H's. **2b** May also suffer from 1,3 interactions between the phenyl and methyl groups but as the carbon-bearing ylide is likely to be between sp^2 and sp^3 hybridised, this interaction may be smaller than that encountered in **2a**. The aldehyde can attack either face of ylide **2b** but the equatorial methyl group hinders *Si* face attack and hence *Re* face attack is preferred. We also believe that the oxygen of the oxathiane plays a role in favouring *Re* face attack through a combination of an anomeric and Cieplak effect. Since the *trans* epoxide is obtained this dictates the orientation of the aldehyde as it approaches the *Re* face of the ylide (Scheme 3) and gives the (R,R)-epoxide.

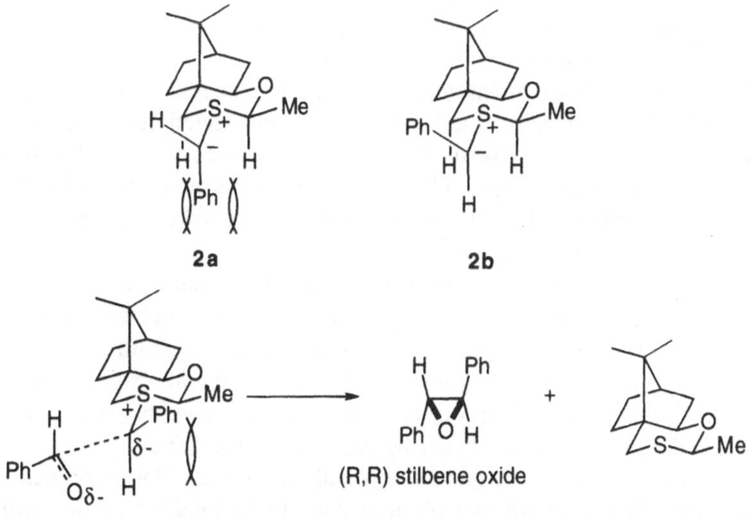

2a　　　　　**2b**

(R,R) stilbene oxide

Scheme 3: Mechanism of Sulfur Ylide Reaction
Reprinted with permission from reference 6. Copyright (1996) American Chemical Society.

The scale up of this reaction however, presents significant operational problems associated with the hazards of handling large quantities of diazocompounds. We therefore considered the possibility of generating the diazocompound *in situ* and extending the cascade process. Diazocompounds have been generated by pyrolysis of the corresponding tosylhydrazone salts[7] and we have discovered that this reaction can be conducted in solution using phase transfer catalysts[8] at much lower temperatures (Scheme 4).

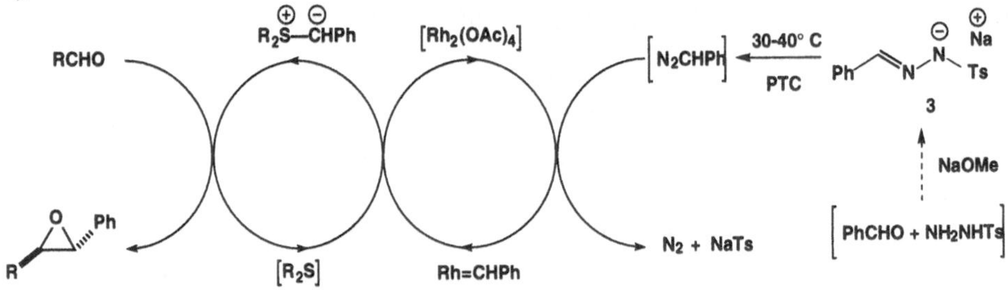

Scheme 4: Catalytic Process for Epoxidation with in situ Generated Diazocompound

Thus heating a rapidly stirred suspension of salt **3** in a solution of the aldehyde, with catalytic quantities of benzyltriethylammonium chloride (PTC), sulfide and rhodium acetate resulted in clean epoxidation, furnishing stilbene oxide in excellent yield and diastereoselectivity (Scheme 5). We are currently preparing chiral sulfides for use in this novel cascade process.

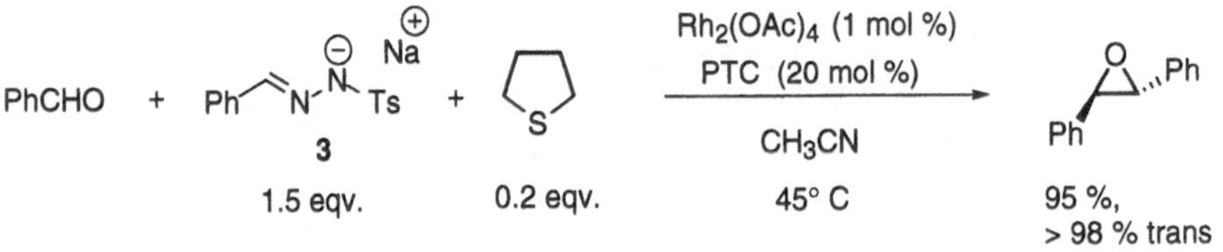

Scheme 5: Catalytic Process for Epoxidation Using Tosylhydrazone Salt

CATALYTIC AND ASYMMETRIC AZIRIDINATION OF IMINES

Like epoxides, aziridines are versatile synthetic intermediates. However, only a limited number of methods for asymmetric aziridination exist: addition of carbenoids to imines[9,10] or copper catalysed additions of nitrenoids to alkenes.[11] Superior results have been obtained by the latter method although high enantioselectivity is still limited to a small subset of alkenes and only *N*-Ts aziridines are accessible. This is a severe limitation of this methodology, especially as harsh conditions are required for cleavage of the strong sulfonamide bond. Despite recent advances, the tosyl group remains a difficult group to remove from sensitive substrates.[12,13]

An alternative strategy for aziridination involves the addition of sulfur ylides to imines and we have considered the application of our novel catalytic cycle for epoxidation to this process. Our proposed catalytic cycle for aziridination is shown in Scheme 6 and involves the slow addition of a diazocompound to a solution of a suitable metal salt, sulfide and imine. From our previous work we knew that sulfur ylides could be generated in the presence of aldehydes by this method and that direct reaction between the diazo-compound or carbenoid and aldehyde did not occur. For aziridination, direct reaction between the carbenoid and imine needed to be avoided as this would give racemic aziridines. Indeed, this reaction is the basis of a method for asymmetric aziridination which was achieved using chiral ligands on the metal.[9,10] It has been found that the efficiency of the direct coupling of metal carbenoids with imines is critically

194

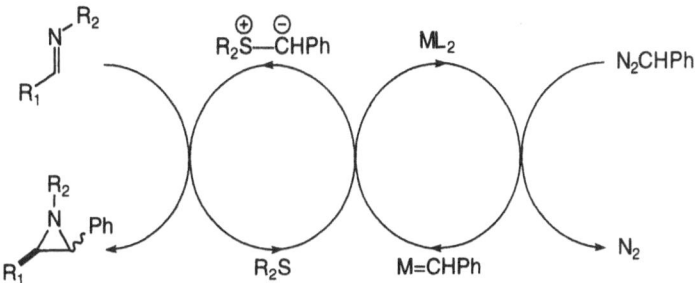

Scheme 6: Catalytic Process for Aziridination of Imines
Reprinted with permission from reference 14. Copyright (1996) American Chemical Society.

dependent on the group on the imine nitrogen: electron donating groups give enhanced yields of azirdines whilst electron withdrawing groups give much reduced yields.[10] Electron withdrawing groups on the imine nitrogen would also enhance the rate of addition of the sulfur ylide to the imine. Thus, we initially chose N-tosyl benzaldimine and carried out the reaction shown below. We were delighted to find that the corresponding aziridine was formed in excellent yield (Table 3, entry 1).[14] Even with catalytic quantities of sulfide (0.2 equivalents), high yields of aziridine were obtained (Table 3, entry 2). We confirmed that the reactions were occurring *via* the sulfur ylide as in the absence of sulfide no aziridine was formed. This experiment confirmed that we had successfully eliminated the direct coupling between the metal carbenoid and imine.

Table 3: Preparation of aziridines from imines and phenyldiazomethane.

Entry	R^1	R^2	Eq. of Me$_2$S	Yield/%[a]	Ratio (trans:cis)
1	Ph	Ts	1.0	90	4:1
2	Ph	Ts	0.2	91	4:1
3	Ph	DPP	0.2	83	3:1
4	Ph	SES	0.2	92	3:1
5	p-Cl-C$_6$H$_4$	SES	0.2	88	3:1
6	p-Me-C$_6$H$_4$	SES	0.2	96	3:1

[a] The yield refers to the total yield of *trans* and *cis* isomers.
Reprinted with permission from reference 14. Copyright (1996) American Chemical Society.

Having successfully shown that N-tosyl imines participated in the aziridination process we wanted to test alternative groups on nitrogen that similarly activated the imine towards nucleophilic attack but were easier to remove. Thus, N-diphenylphosphinyl (DPP)[15] and N-β-trimethylsilylethanesulfonyl (SES)[16] benzaldimines were prepared and subjected to the catalytic cycle. Again, using just catalytic quantities of sulfides, high yields of the corresponding aziridine were obtained (Table 3, entries 3, 4). The SES imines were more easily prepared than the DPP imines and several aryl imines with this group were prepared and tested in the cycle and again, high yields of the corresponding aziridines were obtained (Table 3, entries 5, 6). In all cases a 3:1 mixture of *trans:cis* aziridines was obtained. Deprotection of *trans-N-β*-trimethylsilylethanesulfonyl-2,3-diphenylaziridine using CsF was facile and furnished *trans*-2,3-diphenylaziridine in 89%

yield. This demonstrated the advantage of the current catalytic process over existing methods.

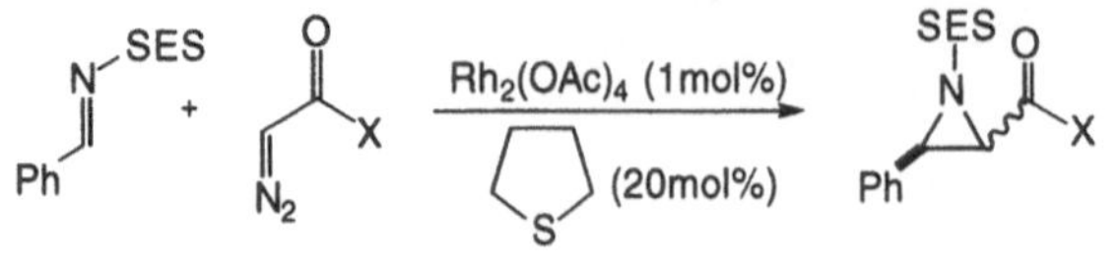

X	Yield (%)	Ratio (trans:cis)
NEt$_2$	91	2:1
OEt	57	1:3

Scheme 7: Aziridination with Functionalised Diazocompounds
Reprinted with permission from reference 14. Copyright (1996) American Chemical Society.

Encouraged by these results we explored alternative diazocompounds (*N,N*-diethyl diazoacetamide and ethyl diazoacetate) in the catalytic cycle. Higher temperatures (60°C) were required for the decomposition of these more stable diazocompounds and so tetrahydrothiophene was used instead of Me$_2$S. As shown in Scheme 7, good to excellent yields of the corresponding aziridines were obtained.[14] Whilst the diazoacetamide gave predominantly the *trans* aziridine, the diazoester gave the *cis* aziridine as the major product. These results show that the new methodology may be applied to the preparation of functionalised aziridines. We believe that the reaction employing the diazoester derived sulfur ylide is under thermodynamic control and that the *cis* isomer is the thermodynamic product. Interestingly, *this stabilised sulfur ylide does not react with aldehydes.*[17-19] As imines are less reactive than aldehydes this may seem contradictory at first. However, the difference in reactivity can be rationalised if the sulfur ylide adds reversibly to C=X in both cases. The bond energy of C=O (191 kcal mol^{-1}) is greater than that of C=N (146 kcal mol^{-1}) and therefore reversion to starting materials should be thermodynamically more favoured in reactions with aldehydes compared to those with imines.

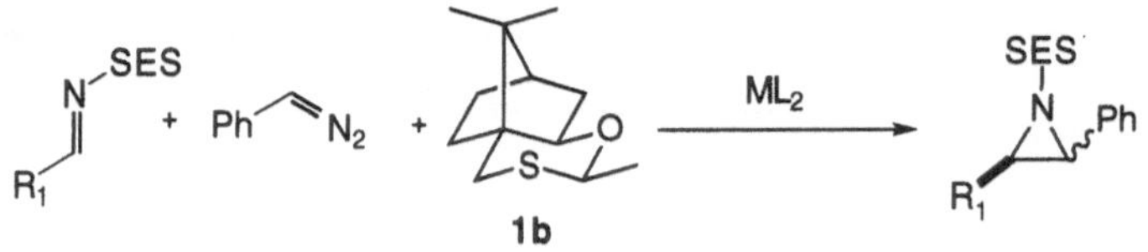

Table 4: Asymmetric aziridination of *N*-SES benzaldimine.

Entry	R^1	Eq. of **1b**	ML$_2$	Yield/%[a]	trans:cis	ee/%
1	Ph	1.0	Rh$_2$(OAc)$_4$	55	3:1	97 (*R,R*)
2	Ph	1.0	Cu(acac)$_2$	83	3:1	95 (*R,R*)
3	Ph	0.2	Cu(acac)$_2$	62	3:1	90 (*R,R*)
4	*p*-Me-C$_6$H$_4$	1.0	Rh$_2$(OAc)$_4$	88	3:1	95 (*R,R*)
5	*p*-Me-C$_6$H$_4$	0.2	Cu(acac)$_2$	50	3:1	88 (*R,R*)
6	*p*-Cl-C$_6$H$_4$	1.0	Rh$_2$(OAc)$_4$	70	3:1	88 (*R,R*)
7	*p*-Cl-C$_6$H$_4$	0.2	Cu(acac)$_2$	44	3:1	85 (*R,R*)

[a] The yield refers to the total yield of *trans* and *cis* isomers.
Reprinted with permission from reference 14. Copyright (1996) American Chemical Society.

In preliminary studies we have tested chiral sulfides derived from (+)-camphorsulfonyl chloride in the catalytic cycle for aziridination (Scheme 7, Table 4).[14] We were delighted to find that the use of sulfide **3b** gave the required aziridine with high enantiomeric excess (97%), although in modest yield (55%) (Table 4, entry 1). In related studies, it was found that using hindered sulfides, Cu(acac)$_2$ gave superior yields of epoxides with less dimerisation of the diazocompound compared to Rh$_2$(OAc)$_4$.[5] Thus, Cu(acac)$_2$ was used in the catalytic cycle in place of Rh$_2$(OAc)$_4$ and a higher yield (83%) of the corresponding aziridine was obtained (Table 4, entry 2).[20] Use of a catalytic amount of sulfide **1** (20 mol%) also furnished the corresponding aziridine but in slightly lower yield (Table 5, entry 3). Other aldimines were tested in the aziridination process using both rhodium and copper salts and high enantioselectivity was maintained (Table 4, entries 4-7). The small reduction in the enantiomeric excess observed when Cu(acac)$_2$ was used in place of Rh$_2$(OAc)$_4$ is presumably due to limited, but competing, direct reaction between the imine and copper carbenoid, a pathway that became more pronounced when catalytic amounts of sulfides were used.

Conclusion

In summary we have developed a new method for the preparation of epoxides and aziridines from aldehydes and imines which requires catalytic amounts of metal salts and sulfides and operates under neutral conditions. Using enantiomerically pure sulfides, high enantioselectivity has been achieved in these processes.

References

1   T. Katsuki and V. S. Martin, *Org. React. (N.Y.)*, 1996, **48**, 1.
2   E. N. Jacobson, *Catalytic Asymmetric Synthesis*; I. Ojima, Ed., VCH, New York, 1993, pp. 159.
3   Z.-X. Wang, Y. Tu, M. Frohn, J.-R. Zhang, and Y. Shi, *J. Am. Chem. Soc.*, 1997, **119**, 11224 .
4   A. H. Li, L. X. Dai, and V. K. Aggarwal, *Chem. Rev.*, 1997, **97**, 2341.
5   V. K. Aggarwal, H. Abdel-Rahman, F. Li, R. V. H. Jones, and M. Standen, *Chem. Eur. Jn.*, 1996, **2**, 212.
6   V. K. Aggarwal, J. G. Ford, A. Thompson, R. V. H. Jones, and M. Standen, *J. Am. Chem. Soc.*, 1996, **118**, 7004.
7   X. Creary, *Org. Synth*, 1986, **64**, 207.
8   D. S. Wulfman, S. Yousefian, and J. M. White, *Synth. Comm.*, 1988, **18**, 2349.
9   K. B. Hansen, N. S. Finney, and E. N. Jacobsen, *Angew. Chem., Int. Ed. Engl.*, 1995, **34**, 676.
10   K. G. Rasmussen and K. A. Jorgensen, *J. Chem. Soc., Chem. Commun.*, 1995, 1401.
11   D. A. Evans, M. M. Faul, M. T. Bilodeau, B. A. Anderson, and D. M. Barnes, *J. Am. Chem. Soc.*, 1993, **115**, 5328.
12   E. Vedejs and S. Z. Lin, *J. Org. Chem.*, 1994, **59**, 1602.
13   T. Fukuyama, C. K. Jow, and M. Cheung, *Tetrahedron Lett.*, 1995, **36**, 6373.
14   V. K. Aggarwal, A. Thompson, R. V. H. Jones, and M. C. H. Standen, *J. Org. Chem.*, 1996, **61**, 8368.
15   W. B. Jennings and C. J. Lovely, *Tetrahedron*, 1991, **47**, 5561.
16   S. M. Weinreb, D. M. Demko, and T. A. Lessen, *Tet. Lett.*, 1986, **27**, 2099.
17   A. J. Speziale, C. C. Tung, K. W. Ratts, and A. Yao, *J. Am. Chem. Soc.*, 1965, **87**, 3460.
18   K. W. Ratts and A. N. Yao, *J. Org. Chem.*, 1966, **31**, 1689.
19   C. R. Johnson and D. McCants, *J. Am. Chem. Soc.*, 1965, **87**, 1109.
20   The quality of the Cu(acac)$_2$ is critical to the success of the reaction. Use of commercial Cu(acac)$_2$ was ineffective but use of Cu(acac)$_2$ prepared by the method of Cervelló followed by sublimation was successful; Cervelló, J.; Marquet, J.; Moreno-Mañas, M. *Synth. Commun.* **1990**, *20*, 1931-1941.

STEREOSELECTIVE ALKYLATION OF CHIRAL GLYCINE AND CHIRAL β-AMINOPROPIONIC ACID DERIVATIVES IN THE PREPARATION OF ENANTIOPURE α- AND β-AMINO ACIDS

Eusebio Juaristi*, Margarita Balderas, José Luis León-Romo,
Heraclio López-Ruiz, and Adelfo Reyes

Departamento de Química, Centro de Investigación y de Estudios Avanzados del Instituto
Politécnico Nacional, Apartado Postal 14-740, 07000 México, D.F. (MEXICO)

INTRODUCTION

As a result of the wide spectrum of applications of α- and β-amino acids, an unprecedented degree of activity has been recorded in the field of enantioselective synthesis of chiral amino acids.[1,2]

Among the various methods available for the preparation of enantioenriched α-amino acids, those employing chiral glycine derivatives[3] have been particularly successful. By the same token, chiral derivatives of 3-aminopropionic acid have afforded efficient enantioselective synthesis of substituted, chiral β-amino acids.[4] The present report summarizes our work in the area, with particular emphasis on developments being disclosed at the 12th International Conference on Organic Synthesis, on July 2, 1998.

RESULTS AND DISCUSSION

Part 1. Highly Diastereoselective Addition of a Racemic β-Alanine Enolate Derivative to Electrophiles.

Some year ago, Seebach and coworkers[3b,5] demonstrated the potential of imidazolidinones (R)- and (S)-1 (chiral derivatives of glycine) as useful precursors of enantiopure α-amino acids (Scheme 1).

Motivated by these reports, β-alanine (an achiral β-amino acid) was converted into racemic 2-tert-butyl-perhydropyrimidinone, rac-2, which was alkylated with high diastereoselectivity via its corresponding enolate (Table 1).[6]

Current Trends in Organic Synthesis
Edited by Scolastico and Nicotra, Kluwer Academic/Plenum Publishers, 1999

Scheme 1

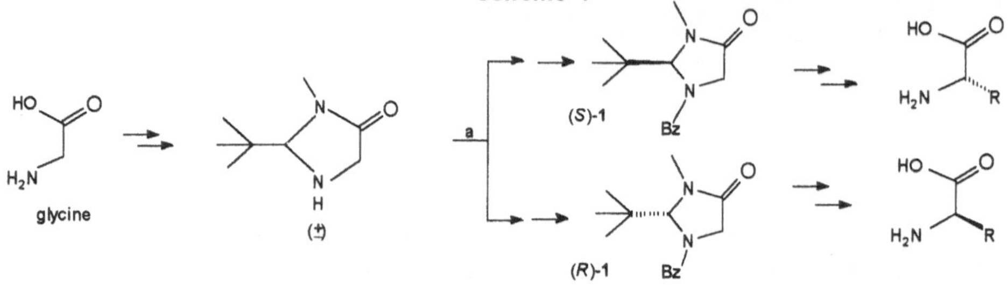

a Resolution by fractional crystallization with mandelic acid.

Table 1. Diastereoselectivity of the Alkylation of rac-2-Li.[6]

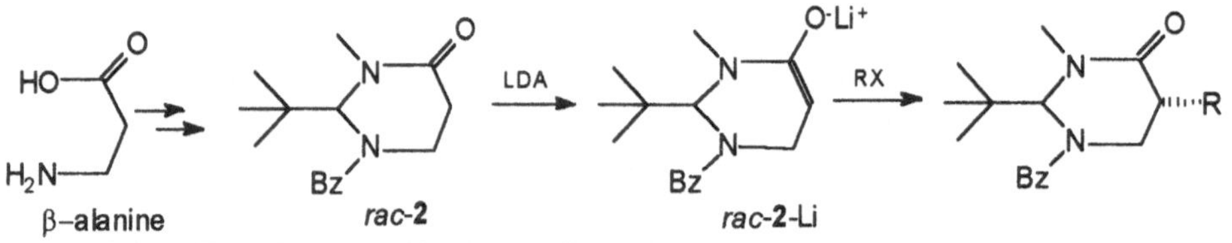

RX	Isolated yield (%)	Diastereoselectivity trans (%)	m.p. (°C)
CH_3I	77	96.7	98-99
$C_6H_5CH_2Br$	75	95.5	144-145
n-BuI	76	>96.0	89-90
n-$C_6H_{13}I$	76	>96.0	49-50
H_2CO	72	86.0	134-135
CH_2=$CHCH_2Cl$	78	86.0	80-81
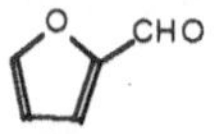	78	80.0	144-145

The high stereoselectivity observed in the alkylation of rac-2-Li is a consequence of the axial disposition of the tert-butyl group at the N,N-acetal carbon, which directs reaction on the enolate face opposite to this group.[4a,6] Hydrolysis of the alkylated products provided the desired α-substituted β-amino acids in good yields (Table 2).

Table 2. Hydrolysis of the Alkylated Pyrimidinones to Afford α-Substituted β-Amino Acids.

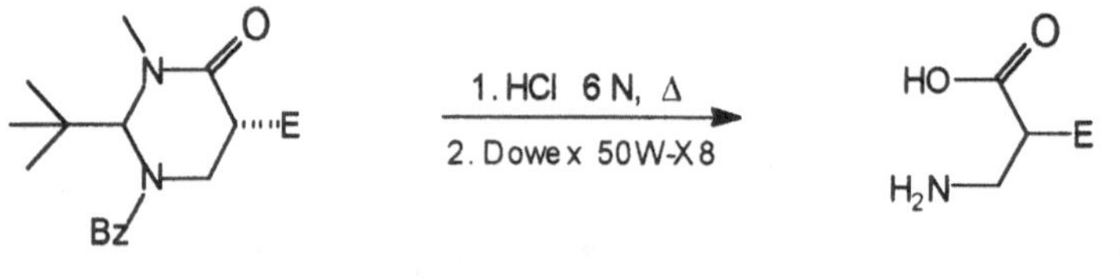

E	Isolated yield (%)	m.p. (°C)
CH_3	82	170-171
CH_2Ph	84	240-241
n-C_4H_9	80	230-231
n-C_6H_{13}	79	215-216

Part 2. Enantioselective Synthesis of α-Substituted β-Amino Acids from (*S*)-2.

While the results described in the previous section paved the road for the development of novel asymmetric methodology to chiral β-amino acids, a procedure was required for the efficient preparation of enantiomerically pure starting pyrimidinone **2**. Accordingly, (*S*)-asparagine was condensed with pivalaldehyde according to the procedure described by Konopelski, et at.[7] to give heterocycle *cis*-**3**, which was decarboxylated, *N*-methylated, and hydrogenated to afford enantiopure (*S*)-**2** (Scheme 2).[8,9]

Scheme 2

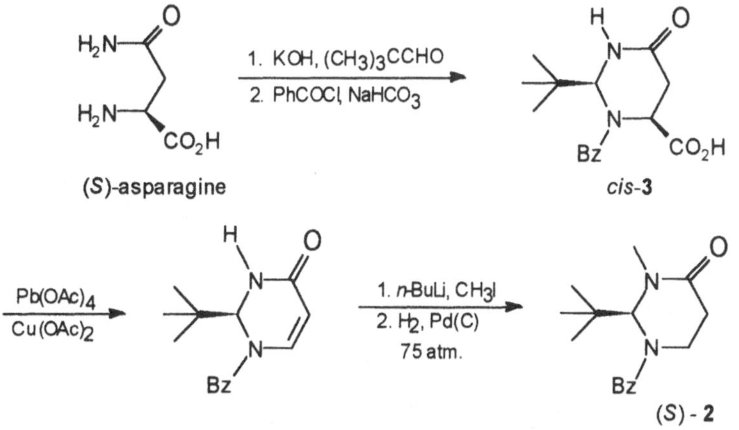

Enantiopure pyrimidinone (*S*)-**2** was then successfully used as a convenient substrate for the enantioselective synthesis of α-substituted β-aminopropionic acids.[10] To this end, enolate (*S*)-**2**-Li was treated with several electrophiles to produce the *trans*-alkylated products (2*S*,5*R*)-**4** with high diastereoselectivity and good yields (Table 3).

Table 3. Diastereoselectivity of Enolate (*S*)-**2**-Li Alkylations.[10]

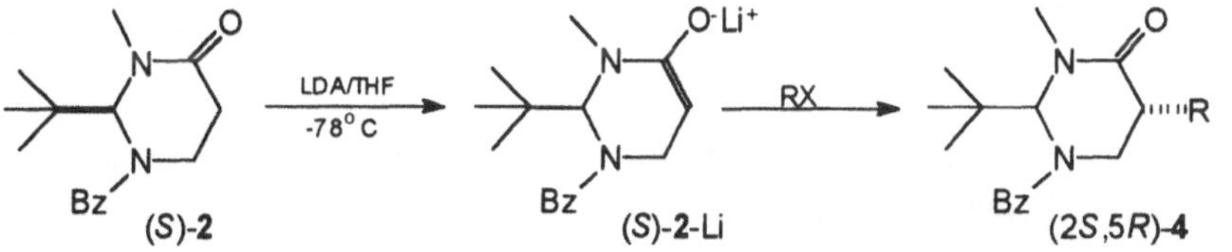

RX	$[\alpha]_D^{28}$	m.p. (°C)	ds, %	yield %
CH$_3$I	+39.5	121-122	>95	77
n-C$_4$H$_9$I	+26.7	80-81	95	75
n-C$_6$H$_{13}$I	+31.2	70-71	95	80
PhCH$_2$Br	-64.0	173-174	>95	80

The last step in the overall conversion of β-alanine to 2-alkyl-3-aminopropionic acid, namely the hydrolysis of the alkylated pyrimidinones, was achieved by acid hydrolysis (6 N HCl, 90-100°C) followed by purification on an ion-exchange column (Table 4).[10]

Table 4. Hydrolysis and Isolation of (*R*)-α-Substituted β-Amino Acids.[10]

E	$[\alpha]_D^{28}$	m.p. (°C)	yield, %
CH_3	-11.8	185-186	80
n-C_4H_9	+5.3	170-171	80
n-C_6H_{13}	+6.6	219-220	80
$PhCH_2$	+11.3	225-226	85

Epimerization of trans adducts (2*S*,5*R*)-**4** afforded the cis diastereoisomers (2*S*,5*S*)-**5** (Table 5), whose hydrolysis provided then the enantiomeric α-alkylated β-amino acids (Table 6).[10]

Table 5. Epimerization of Pyrimidinone Adducts (2*S*,5*R*)-**4**.[10]

E	$[\alpha]_D^{28}$	m.p. (°C)	yield, %
CH_3	+33.5	108-109	85
n-C_4H_9	+18.7	64-65	85
n-C_6H_{13}	+20.7	94-95	90
$PhCH_2$	-6.4	98-98.5	88

Table 6. Hydrolysis of Epimerized Adducts (2S,5S)-**5** to Give (*S*)-α-Substituted β-Amino Acids.[10]

E	$[\alpha]_D^{28}$	m.p. (°C)	yield, %
CH_3	+11.6	184-185	80
n-C_4H_9	- 5.0	169-170	80
n-C_6H_{13}	- 6.3	219-220	85
$PhCH_2$	- 12.0	224-225	85

202

The epimerization described above (Table 5) constitutes clear evidence that protonation (aqueous NH_4Cl) of enolates generated from (2S,5R)-4 takes place on the face opposite to the *tert*-butyl group. This result was useful in the highly stereoselective preparation of α,α-disubstituted β-amino acids. (Next section).

Part 3. Enantioselective Synthesis of α,α-Disubstituted β-Amino Acids.

In recent years, the preparation of enantiopure α,α-dialkylated α-amino acids has attracted considerable attention in view of the useful chemical and biological properties exhibited by these compounds.[11] Not surprisingly, analogous α,α-dialkylated β-amino acids are now attracting the attention of synthetic organic chemists.[12]

As summarized in Table 7, alkylation of pyrimidinones (2S,5R)-4 takes place with very high diastereoselectivity, to give the dialkylated adducts **6**. Subsequent hydrolysis afforded the desired β-amino acids **7** in enantiopure form. (Table 7).[13]

Table 7. Enantioselective Synthesis of α,α-Dialkylated β-Amino Acids.[13]

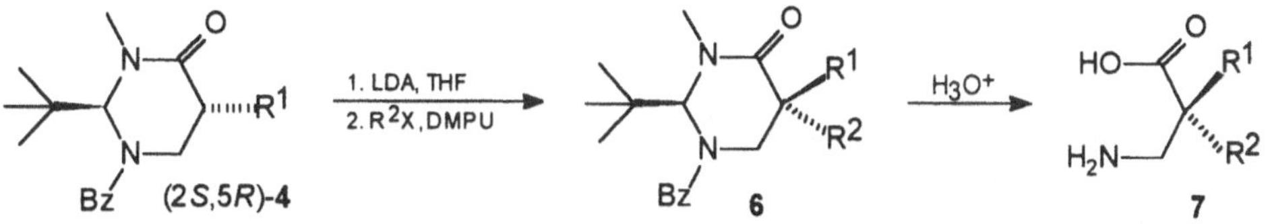

R^1	R^2	m.p.(6),°C	$[\alpha]_D^{25}$ (6)	m.p. (7)	$[\alpha]_D^{25}$ (7)
CH_3	$PhCH_2$	209-10	-48.8	205-6	-17.2
$PhCH_2$	CH_3	129-30	-38.7	205-6	+17.8
CH_3	n-Bu	184-5	+30.3	187-8	- 6.8
n-Bu	CH_3	147-8	+12.6	187-8	+ 7.0

Part 4. Enantioselective Synthesis of α-Substituted Aspartic Acids.

Because of their relevant role in physiological events, enantiopure α-alkylated aspartic acids are especially interesting synthetic targets.[14] Condensation of (S)-asparagine with isobutyraldehyde followed by in-situ benzoylation afforded (2S,6S)-7, whose relative configuration was assigned by X-ray crystallographic analysis. Treatment with two equivalents of silver oxide and three equivalents of methyl iodide provided imino ether (2S,6S)-8, which proved to be useful precursor of α-alkylated aspartic acids via alkylation at C(6), followed by hydrolysis.[15] (Scheme 3 and Table 8).

Scheme 3

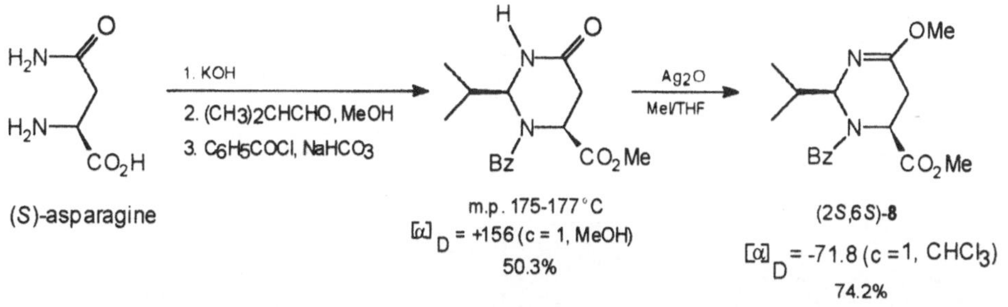

203

Table 8. Enantioselective Synthesis of α-Alkylated Aspartic Acids.[15]

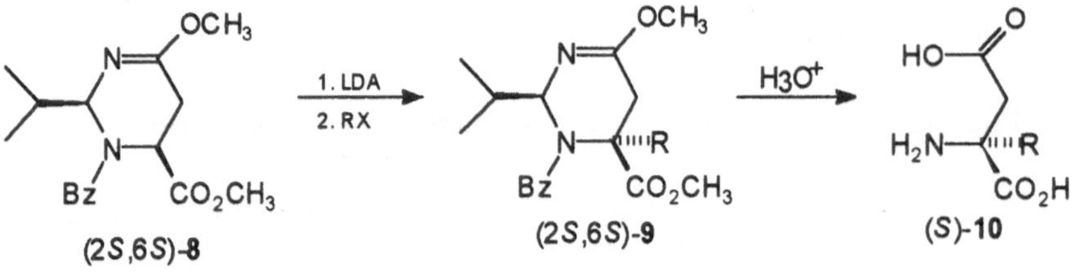

RX	m.p. (9) (°C)	$[\alpha]_D^{28}$ (9)	$[\alpha]_D^{28}$ (10)
CH$_3$I	---	-47.0	+43.0
CH$_3$CH$_2$I	158-9	-67.0	+35.1
n-C$_4$H$_9$I	---	-20.0	+26.7
PhCH$_2$Br	122-4	+90.0	+50.0

Part 5. Enantioselective Synthesis of α-Amino Acids from Chiral Glycine Derivatives Containing the α-Phenylethyl Group.

(R)- and (S)-α-Phenylethylamine are simple, yet efficient chiral auxiliaries in asymmetric synthesis.[16] Very recently, incorporation of (S)-α-phenylethylamine into 1,4-benzodiazepin-2,5-dione, (S)-**11**, provided a novel chiral glycine derivative (Scheme 4), which was alkylated with high diastereoselectivity to give alkylated products **12**. These adducts were then hydrolyzed with 57% HI to produce enantiopure α-amino acids **13** in excellent yields. (Table 9).[17]

Scheme 4

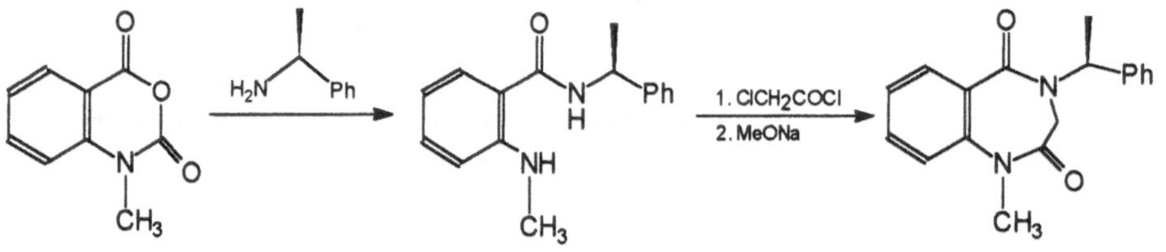

(S)-**11**

(57% overall yield)

Table 9. Diastereoselectivity of Enolate (*S*)-**11**-Li Alkylation and Hydrolysis to α-Substituted Amino Acids.[17]

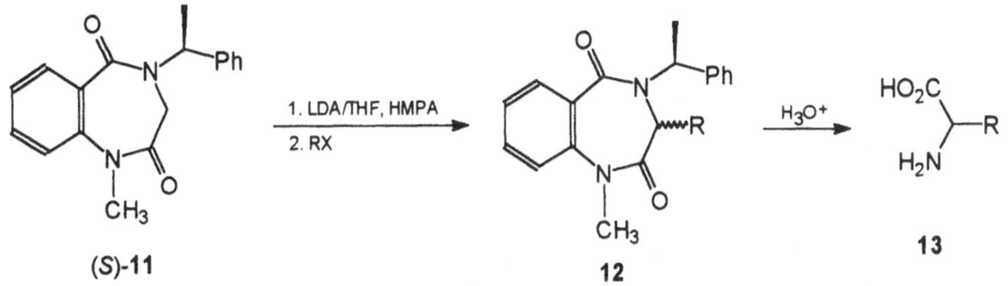

RX	ds(**12**) %	configuration of main product	yield (**13**) %, overall
CH$_3$I	90	(*R*)	79
PhCH$_2$Br	91	(*S*)	91
CH$_3$CH$_2$I	83	(*R*)	72
n-BuI	81	(*R*)	71

On the other hand, C_2-symmetric chiral glycinamide (*S,S*)-**14** was alkylated with good to excellent diastereoselectivity (Table 10). Isolation of the corresponding α-substituted α-amino acid was accomplished as depicted in Scheme 5.[18]

Table 10. Diastereoselectivity of Enolate (*S,S*)-**14**-Li Alkylation.[18]

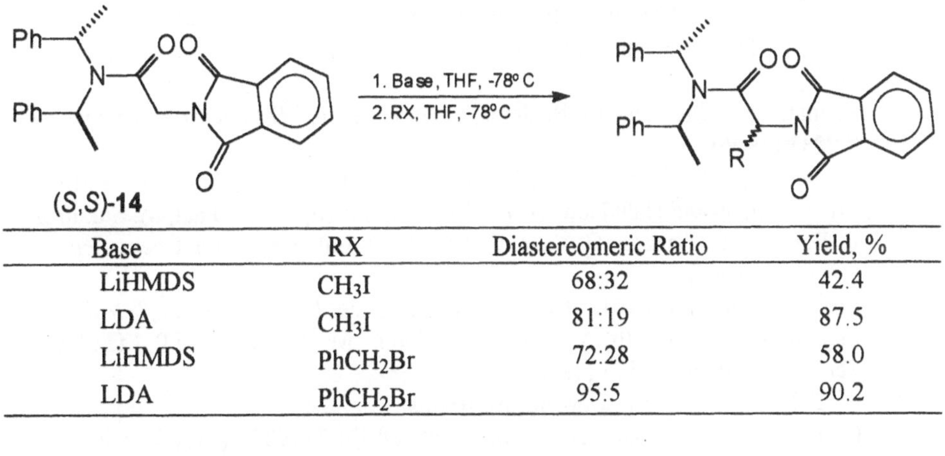

Base	RX	Diastereomeric Ratio	Yield, %
LiHMDS	CH$_3$I	68:32	42.4
LDA	CH$_3$I	81:19	87.5
LiHMDS	PhCH$_2$Br	72:28	58.0
LDA	PhCH$_2$Br	95:5	90.2

Scheme 5

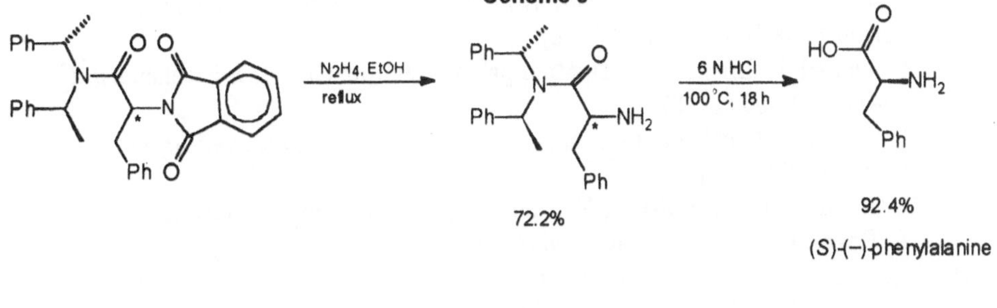

205

REFERENCES AND NOTES

1. (a) M.J. O'Donnell, Ed., *α-Amino Acid Synthesis* (*Tetrahedron* Symposium-in-Print), *Tetrahedron* 44:5253 (1988). (b) R.M. Williams, *Synthesis of Optically Active α-Amino Acids*, Pergamon Press: Oxford (1989). (c) R.O. Duthaler, *Tetrahedron* 50:1539 (1994).

2. (a) E. Juaristi, D. Quintana, J. Escalante, *Aldrichimica Acta* 27:3 (1994). (b) D.C. Cole, *Tetrahedron* 50:9517 (1994). (c) G. Cardillo, C. Tomasini, *Chem. Soc. Rev.* 25:117 (1996). (d) E. Juaristi, Ed., *Enantioselective Synthesis of β-Amino Acids*, Wiley-VCH: New York (1997).

3. (a) U. Schoellkopf, *Tetrahedron* 39:2085 (1983). (b) D. Seebach, E. Juaristi, D.D. Miller, C. Schickli, T. Weber, *Helv. Chim. Acta* 70:237 (1987). (c) R.M. Williams, P.J. Sinclair, D. Zhai, D. Chen, *J. Am. Chem. Soc.* 110:1547 (1988). (d) Y.N. Belokon, A.S. Sagyan, S.M. Djamgaryan, V.I. Bakhmutov, V.M. Belikov, *Tetrahedron* 44:5507 (1988).

4. (a) E. Juaristi, D. Seebach, in ref. 2d, Chapter 13, pp. 261-277. (b) G. Cardillo, C. Tomasini, in ref. 2d, Chapter 11, pp. 211-248. (c) J.P. Konopelski, in ref. 2d, Chapter 12, pp. 249-259.

5. (a) R. Fitzi, D. Seebach, *Tetrahedron* 44:5277 (1988). (b) D. Seebach, E. Dziadulewicz, L. Behrendt, S. Cantoreggi, R. Fitzi, *Liebigs Ann. Chem.* 1215 (1989).

6. E. Juaristi, D. Quintana, B. Lamatsch, D. Seebach, *J. Org. Chem.* 56:2553 (1991).

7. K.S. Chu, G.R. Negrete, J.P. Konopelski, F.J. Lakner, N.T. Woo, M.M. Olmstead, *J. Am. Chem. Soc.*, 114:1800 (1992).

8. E. Juaristi, D. Quintana, *Tetrahedron: Asymmetry* 3:723 (1992).

9. The assignment of the absolute configuration of dextrorotatory (*S*)-**2** was accomplished by chemical correlation to levorotatory α-methyl-β-aminopropionic acid.

10. E. Juaristi, D. Quintana, M. Balderas, E. García-Pérez, *Tetrahedron: Asymmetry* 7:2233 (1996).

11. (a) M.J. Jung, In *Chemistry and Biochemistry of the Amino Acids*, G.C. Barrett, Ed., Chapman and Hall: New York (1985), p. 227. (b) P.K.C. Paul, M. Sukumar, R. Bardi, A.M. Piazzesi, G. Valle, C. Toniolo, P. Balaram, *J. Am. Chem. Soc.* 108:6363 (1986).

12. See, for example: (a) W.L. Scott, C. Zhou, Z. Fang, M.J. O'Donnell, *Tetrahedron Lett.* 38:3695 (1997). (b) A. Gaucher, F. Bintein, M. Wakselman, J.P. Mazaleyrat, *Tetrahedron Lett.* 39:575 (1998).

13. M. Balderas, E. Juaristi, Unpublished results.

14. (a) J.D. Aebi, D. Seebach, *Helv. Chim. Acta* 68:1507 (1985). (b) A. Fadel, J. Salaün, *Tetrahedron Lett.* 28:2243 (1987). (c) G.I. Georg, X. Guan, J. Kant, *Tetrahedron Lett.* 29:403 (1988). (d) C.O. Chan, D. Crich, S. Natarajan, *Tetrahedron Lett.* 33:3405 (1992).

15. E. Juaristi, H. López-Ruiz, D. Madrigal, Y. Ramírez-Quirós, J. Escalante, *J. Org. Chem.* 63:0000 (1998).

16. E. Juaristi, J. Escalante, J.L. León-Romo, A. Reyes, *Tetrahedron: Asymmetry* 9:715 (1998).

17. E. Juaristi, J.L. León-Romo, Y. Ramírez-Quirós, Unpublished results.

18. A. Reyes, E. Juaristi, Unpublished results.

DEVELOPMENT AND APPLICATION OF CATALYTIC HIGHLY ENANTIOSELECTIVE HETERO-DIELS-ALDER REACTIONS OF ALDEHYDES AND KETONES

Karl Anker Jørgensen
Center for Metal Catalyzed Reactions, Department of Chemistry
Aarhus University, DK-8000 Aarhus C, Denmark

INTRODUCTION

The hetero-Diels-Alder reaction of carbonyl compounds with conjugated dienes [eq (1)] constitutes one of the cornerstone reactions in organic chemistry.[1] The products formed in this reaction are of fundamental importance in chemistry, and numerous examples of the use of these compounds in academia and industry have appeared.

$$\text{(1)}$$

The new dimension of hetero-Diels-Alder reaction is to catalyze the reaction applying Leiws acids as catalysts, and especially enantioselective catalysis is a field in rapid development.[2] A variety of different Lewis acid complexes have been shown to be able to catalyze the hetero-Diels-Alder reaction of aldehydes, and both activated and non-activated aldehydes are found to react with various types of dienes, whereas for ketones, only very few examples of catalytic enantioselective hetero-Diels-Alder reactions are known.

This paper presents the development of highly enantioselective hetero-Diels-Alder reactions of both activated aldehydes and ketones using a common catalytic complex, the copper(II) bisoxazolines (S)-**1a** and (R)-**1b**. The aldehydes and ketones which are substrates for these highly enantioselective reactions are α-dicarbonyl compounds.

(S)-**1a**   (R)-**1b**

RESULTS AND DISCUSSION

Reactions of Glyoxylates (Aldehydes)

The Lewis acid catalyzed hetero-Diels-Alder (HDA) reaction of α-dicarbonyl compounds, such as ethyl glyoxylate **2** with conjugated dienes can take two different reaction paths, if the diene contains an allylic carbon-hydrogen bond [eq (2)].[3,4] Under thermal conditions the hetero-Diels-Alder product **4** is exclusively formed, whereas using BINOL-TiX$_2$ complexes as the catalyst, the éne product **5** is the major product with a HDA:éne ration of up to 1:9.[3] Application of the copper(II) bisoxazolines (*S*)-**1a** and (*R*)-**1b** (X = OTf) as catalysts for the reaction of **2** with 2,3-dimethyl-1,3-butadiene **3** [eq (2)] changes the chemoselectivity and a HDA:éne (**4:5**) ratio of 1:1.8 is found with an ee of 85% of **4** when applying (*S*)-**1a** as the catalyst and 83% ee using (*R*)-**1b**.[4] It is notable that the (*S*)-enantiomer of **4** is formed as the major enantiomer when using catalysts with opposite absolute stereochemistry.

2	**3**	**4**	**5**	(2)
		(*S*)-**1a**: 85% ee	(*S*)-**1a**: 83% ee	
		(*R*)-**1b**: 83% ee	(*R*)-**1b**: 88% ee	

The HDA reaction also proceeds for other conjugated dienes; isoprene reacts also with ethyl glyoxylate **2** in the presence of (*R*)-**1b** as the catalyst giving a 1:1 ratio of the HDA:éne products with 33% isolated yield of the HDA product having an ee of 80%, while 1,3-butadiene reacts with the same glyoxylate giving the HDA product in 55% isolated yield with an ee of 87%.[4a]

The reaction of ethyl glyoxylate **2** with 1,3-cyclohexadiene **6** in the presence of (*S*)-**1a** (X = OTf) as the catalyst proceeds well in CH$_3$NO$_2$ giving the HDA product **7** in 93% isolated yield with an ee >97%.[4]

2	**6**	**7**	(3)
		93% yield	
		>97% ee	

The catalytic enantioselective HDA reaction in eq (3) can be used for the total synthesis of various natural products such as the synthesis of the monoterpenes dihydroactinidiolide and actinidiolide as outlined in the reterosynthetic approach in Scheme 1.[5]

Scheme 1

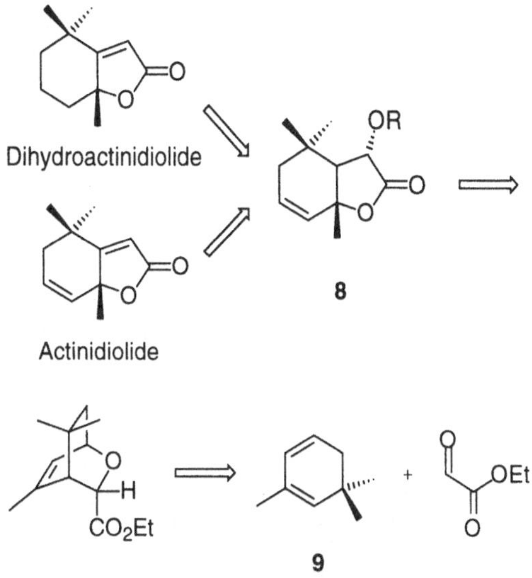

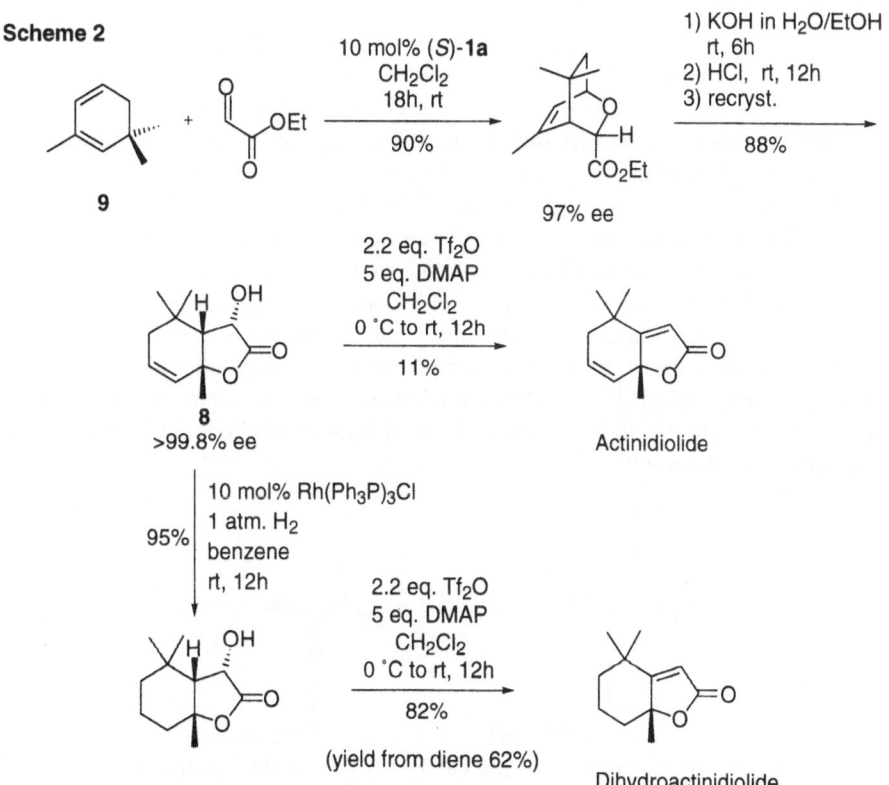

Dihydroactinidiolide

Actinidiolide

8

9

The crucial reaction for the synthesis of dihydroactinidiolide and actinidiolide is the HDA reaction of 2,6,6-trimethyl-1,3-cyclohexadiene **9** with ethyl glyoxylate followed by an intramolecular rearrangement leading to the monoterpene **8**.[5]

The actual synthesis of dihydroactinidiolide and actinidiolide is presented in Scheme 2.[5]

Scheme 2

9

10 mol% (S)-**1a**
CH₂Cl₂
18h, rt

90%

97% ee

1) KOH in H₂O/EtOH
rt, 6h
2) HCl, rt, 12h
3) recryst.

88%

8
>99.8% ee

2.2 eq. Tf₂O
5 eq. DMAP
CH₂Cl₂
0 °C to rt, 12h

11%

Actinidiolide

95%

10 mol% Rh(Ph₃P)₃Cl
1 atm. H₂
benzene
rt, 12h

2.2 eq. Tf₂O
5 eq. DMAP
CH₂Cl₂
0 °C to rt, 12h

82%

(yield from diene 62%)

Dihydroactinidiolide

The HDA methodology proceeds well for the synthesis of dihydroactinidiolide and actinidiolide with high yield, diastereo- and enantioselectivity leading to the HDA product, which by treatment base followed by acid gives the monoterpene from which the natural products are formed.[5]

Reactions of Ketones

The catalytic approach for the reaction of the glyoxylates can also be applied for the HDA reaction of ketones. The reaction proceeds with very high enantioselective using (S)-1a (10 mol%) as the catalyst for the reaction of alkyl- and aryl α-keto esters, and α-dicarbonyls 10 with activated dienes 11 [eq (3)].[6]

$$R^1 = alkyl; R^2 = O\text{-}alkyl: ee\ up\ to\ 99\%$$
$$R^1 = aryl; R^2 = O\text{-}alkyl: ee\ up\ to\ 99\%$$
$$R^1 = R^2 = alkyl: ee\ up\ to\ 98\%$$
$$R^1 = aryl; R^2 = O\text{-}alkyl: ee\ up\ to\ 99\%$$

The catalyst (S)-1a is a general catalyst for the HDA reaction in eq (3) showing large substrate tolerance allowing a variety of different substituted alkyl- and aryl α-keto esters, and α-dicarbonyls 10 to react with the activated dienes 11. The HDA product 12 is formed with very high enantioselectivity - up to 99%. This reaction allows in a simple manner, the formation of quaternary carbon atoms, which are of utmost importance in organic synthesis,[7] from simple non-optically active substrates.

Careful investigation of the HDA reaction of the α-dicarbonyls 10 with the activated dienes 11 revealed that the reaction can take place with a very low loading of catalyst (S)-1a. It was found that only 0.05 mol% of the catalyst (S)-1a is sufficient to achieve a reaction leading to both high yield and enantioselectivity of the HDA products. In some cases the ee was slightly improved compared with the reactions using 10 mol% catalyst. The scope of the HDA reaction is presented below where the regio- and enantioselectivity of the reaction of 2,3-pentanedione and 3-phenyl-2,3-propanedione is shown.[6b]

10 (97.8% ee)   98 (96.4% ee)

The turn over numbers for the catalyst (S)-1a in the reaction of the ketones with the activated dienes using only 0.05 mol% of the catalyst shows, that for the present HDA reaction the catalytic effect is beginning to be enzyme-like - *chemzymatic reactions*.[6b]

Based on the absolute stereochemistry of the HDA products in the reactions of the aldehydes and ketones with conjugated dienes, an intermediate in which the two carbonyl oxygens of the substrate are coordinated to the copper(II) bisoxazoline catalyst (*S*)-**1a** has been proposed as outlined in Figure 1. The geometry at the copper(II) center is proposed to planar which leads to a shielding of the *re*-face of the carbonyl functionality allowing the diene to approch the carbonyl from the *si*-face.[4,6]

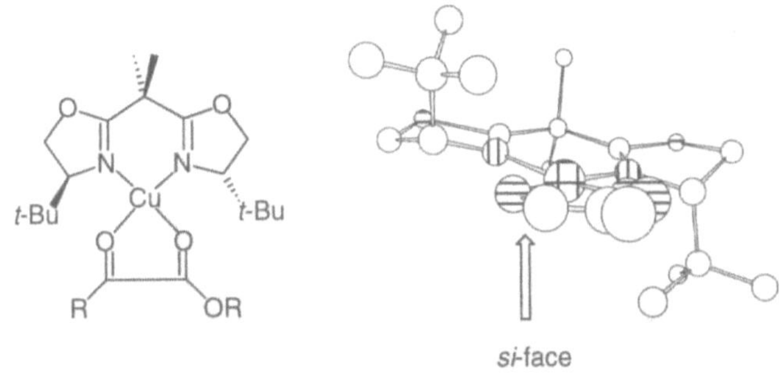

si-face

Figure 1. The proposed intermediate in the HDA reaction of α-dicarbonyl compounds with dienes catalyzed by the copper(II) bisoxazoine complex (*S*)-**1a**.

The HDA reactions discussed in the previous part are all normal electron demand reactions. However, the copper(II) bisoxazolines can also be used as catalysts for the HDA reaction of α,β-unsaturated carbonyl compounds with electron rich alkenes, giving excellent access to substituted 3,4-dihydro-2*H*-pyrans which are very useful precursors for the synthesis of carbonhydrates and numerous natural products.[1] This reaction is an inverse electron demand reaction, and is controlled by the LUMO of the α,β-unsaturated carbonyl compound and the HOMO of the dienophile.

The copper(II) bisoxazoline (*S*)-**1a** can catalyze the [4+2] cycloaddition of β,γ-unsaturated-α-ketoesters, substituted the γ-position with both alkyl, aryl and alkoxy substituents **13** with electron-rich alkenes as ethyl vinyl ether **14** [eq (4)].[8]

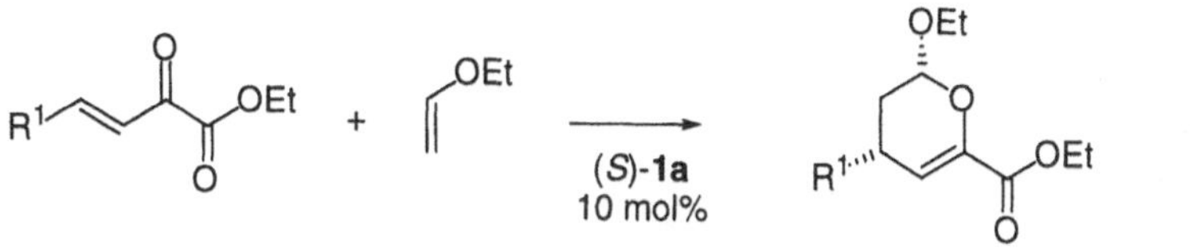

$$(4)$$

13a: R^1 = Me   14   15a: R^1 = Me, ee >99.5%
13b: R^1 = Ph   15b: R^1 = Ph, ee 99.5%
13c: R^1 = OEt   15c: R^1 = OEt, ee >99.5%

The reaction of the α,β-unsaturated carbonyl compounds **13** with ethyl vinyl ether **14** catalyzed by (*S*)-**1a** gives the [4+2] cycloaddition product **15** in high yield and with very high enantioselectivity (ee's >99.5%).[8] The reaction proceeds also well for other electron rich alkenes and similar results as those presented in eq (4) have been found for the cyclic alkene 2,3-dihydrofurane.[8]

In relation to the reaction presented in eq (4) it should be noted that Evans *et al.* have developed a related approach for the copper(II) bisoxazoline catalyzed HDA reaction of α,β-unsaturated acyl phosphonates with vinyl ethers.[9]

The present work has shown that the copper(II) bisoxazolines are excellent catalysts for the HDA reaction of α-dicarbonyl compounds with alkenes leading to highly valuable products in high yield and enantioselectivity.

Acknowledgements

This work was made possible by a grant from The Danish National Science Foundation.

References

1.   See *e. g.*: (a) G. Desimoni, G. Tacconi, *Chem. Rev.* **1975**, *75*, 651. (b) D. L. Boger; S. M. Weinreb, *Hetero Diels-Alder Methodology in Organic Synthesis*; Academic Press New York, 1987. (c) L. F. Tieize; G. Kettschau, *Stereoselective Heterocyclic Synthesis I*, P. Metz, Ed. (Springer-Verlag, Berlin, 1997), vol. 189, pp. 1-120.

2.   See *e.g.*: H. B. Kagan, O. Riant, *Chem. Rev.* **1992**, *92*, 1007.

3.   K. Mikami, M. Shimizu, *Chem. Rev.* **1992**, *92*, 1021.

4.   (a) M. Johannsen, K. A. Jørgensen, *J. Org. Chem.* **1995**, *60*, 5757. (b) *Tetrahedron* **1996**, *52*, 7321. (c) *J. Chem. Soc., Perkin Trans. 2* **1997**, 1183. (d) S. Yao, M. Johannsen, K. A. Jørgensen, *J. Chem. Soc., Perkin Trans. 1* **1997**, 2345.

5.   S. Yao, M. Johannsen, K. A. Jørgensen, *J. Org. Chem.* **1998**, *63*, 118.

6.   (a) M. Johannsen, S. Yao, K. A. Jørgensen, *J. Chem. Soc., Chem. Commun.* **1997**, 2169. (b) S. Yao, M. Johannsen, H. Audrain, R. Hazell, K. A. Jørgensen, *J. Am. Chem. Soc.* **1998** in press.

7.   E. J. Corey, A. Guzman-Perez, *Angew. Chem. Int. Ed. Engl.* **1998**, *37* 388.

8.   J. Thorhauge, M. Johannsen, K. A. Jørgensen, *Angew. Chem. Int. Ed. Engl.* **1998** in press.

9.   D. A. Evans, J. S. Johnson, *J. Am. Chem. Soc.* **1998**, *120*, 4895.

ENANTIOPURE PYRROLINE-N-OXIDES FOR THE SYNTHESIS OF PYRROLIZINE AND INDOLIZINE ALKALOIDS

Alberto Brandi,* Francesca Cardona, Stefano Cicchi, Franca M. Cordero, Andrea Goti

Dipartimento di Chimica Organica "Ugo Schiff" and Centro di Studio sulla Chimica e la Struttura dei Composti Eterociclici e loro Applicazioni, C.N.R., Università degli Studi di Firenze, via G. Capponi 9, 50121 Firenze

The total synthesis of natural products requires new and selective methods to address the structural complexity that the nature is able to introduce in natural compounds. This is the challenge that gives the impetus to organic chemists to find new reactions, new reagents and new catalysts to run reactions in a chemoselective, regioselective and stereoselective way. Recently, the horizon has moved to explore the development of new synthetic strategies, like domino or multicomponent,[1] combinatorial[2] or enzymatic[3] processes, that have expanded the synthetic tools available to researchers. On the other hand, the concern of chemists has shifted to modern topics like atom economy[4] and environmentally friendly syntheses. In the light of all these points we approached the synthesis of natural products belonging to the classes of indolizine[5] and pyrrolizine[6] alkaloids which are attracting growing interest from synthetic chemists for their various and essential biological activities. Among these are glycosidase inhibitors (with indolizidine or pyrrolizidine skeletons)[7] and necine bases (with pyrrolizidine skeletons)[6] (Chart 1).

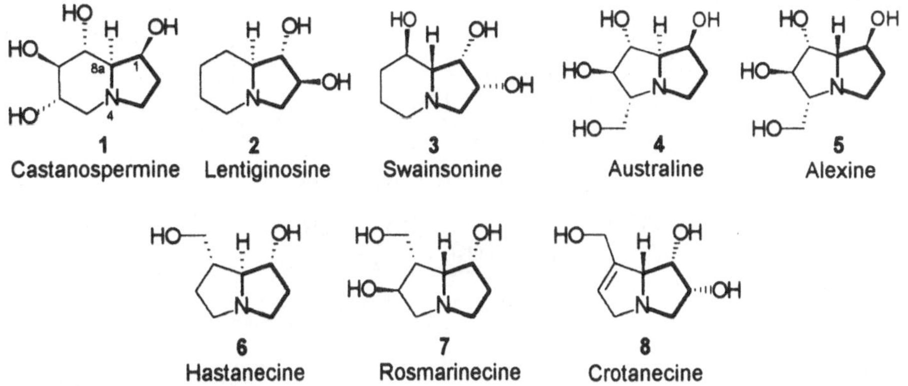

1	2	3	4	5
Castanospermine	Lentiginosine	Swainsonine	Australine	Alexine

6	7	8
Hastanecine	Rosmarinecine	Crotanecine

Chart 1

A structural analysis of these polyhydroxylated structures showed that they might be synthesized in a general way starting from a common five membered pyrroline synthon **9** bearing one or two hydroxy functions on the 3 and 4 position of the ring (Scheme 1).

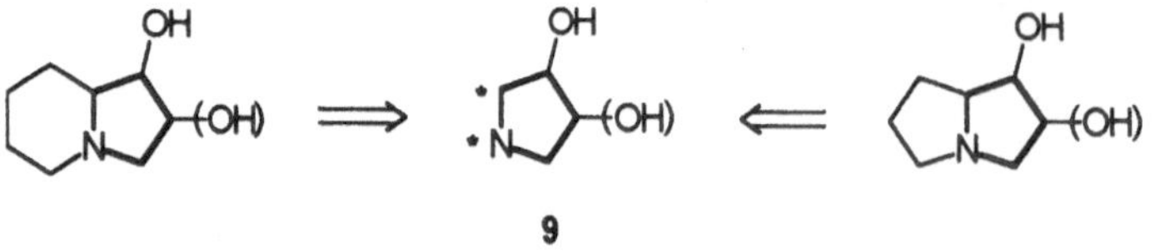

Scheme 1

The choice of the appropriate reaction partner should provide the direct entry to indolizidine or pyrrolizidine skeletons. Pyrroline-N-oxides appeared to us excellent synthons for this purpose, as they can afford these ring systems by 1,3-dipolar cycloadditions processes followed by simple elaboration of the cycloadducts. The whole process can fulfill all the requirements of a modern synthetic methodology like chemoselectivity, stereoselectivity and high atom economy. Many strategies were described in the literature for the synthesis of these ring skeletons[8,9] starting from pyrroline-N-oxide **10**, the most common pathways being summarized in Scheme 2.

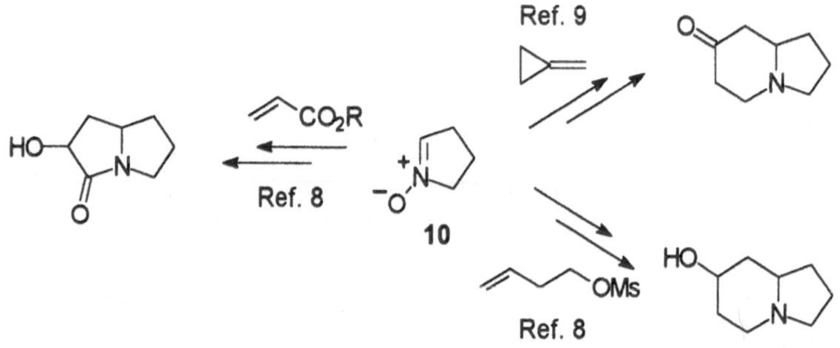

Scheme 2

Hydroxylated pyrroline-N-oxides **11** in both enantiomeric forms can be conveniently synthesized from malic and tartaric acids. Complementary syntheses were developed by us to allow the synthesis of alkyl and silyl protected 3-hydroxy- and 3,4-dihydroxy pyrroline-N-oxides. For alkyl protected ones a strategy that makes use of a cyclization of 1,4-(bis)tosylated tetrols (or 1,4-(bis)mesylated triols) by NH_2OH is able to afford the nitrones **11** in 40-50% overall yield.[10]

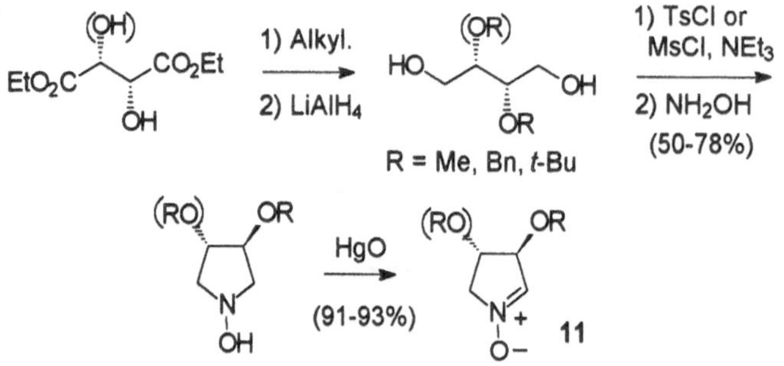

Scheme 3

For silylated derivatives **13** a precedent procedure that makes use of a *trans*-3,4-dihydroxy-N-benzylpyrrolidine from tartaric acid is required.[11]

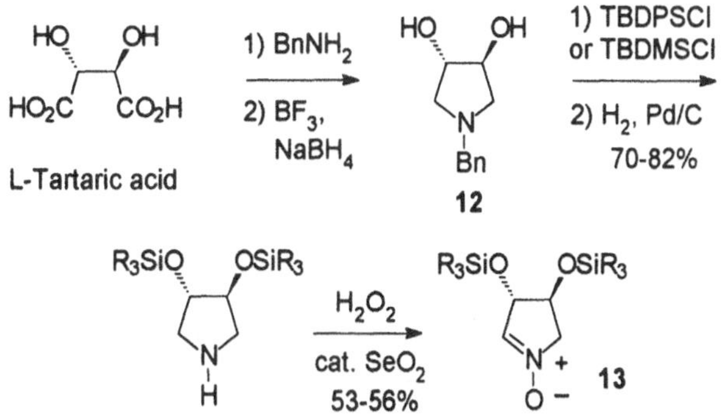

HO OH   1) BnNH₂   HO OH   1) TBDPSCl
                                          or TBDMSCl
HO₂C CO₂H 2) BF₃,   ───────────→
            NaBH₄   N   2) H₂, Pd/C
L-Tartaric acid   Bn   70-82%
                        12

R₃SiO OSiR₃   R₃SiO OSiR₃

        N   H₂O₂   N +
        H   ─────────→ | 13
              cat. SeO₂   O −
              53-56%

Scheme 4

The use of TIPS as a O-substituent in the malic acid series allows the use of the procedure of Scheme 3 for a silyl protected nitrone 17 (Scheme 5).

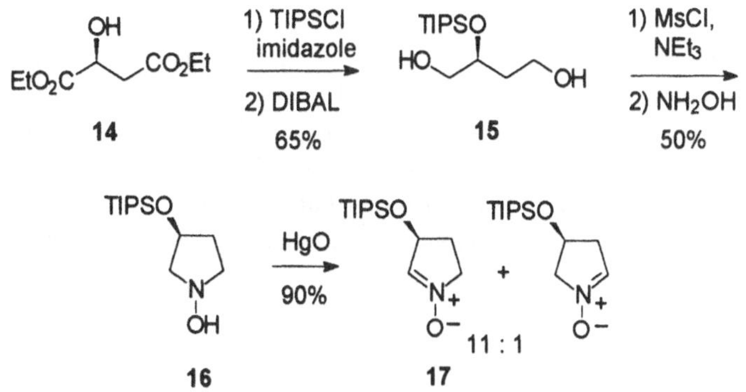

        OH   1) TIPSCl   TIPSO   1) MsCl,
                        imidazole   HO   NEt₃
EtO₂C   CO₂Et   ─────────→   OH   ─────────→
                        2) DIBAL   2) NH₂OH
        14   65%   15   50%

TIPSO   TIPSO   TIPSO

      N   HgO   N +   +   N +
      | ─────────→ | |
      OH   90%   O −   O −
                                     11 : 1
      16   17

Scheme 5

The observation of the multiform diversity offered by nature showed that a *cis*-OH relationship on the pyrroline oxide was necessary to address the synthesis of swainsonine 3, crotanecine 8 or their analogues. In order to obtain enantiopure *cis*-OH nitrones applying the methodologies either of scheme 3 or 4, it was necessary to differentiate the two hydroxy groups before oxidation, a process that is generally carried out by the use of enzymes ("*meso* trick"). We devised a new methodology that is formally able to desymmetrize a *meso* diol, like the desired *cis*-dihydroxypyrrolidine, starting from the enantiopure *trans*-dihydroxypyrrolidine. The MEMCl or TBDMSCl monoprotected pyrrolidines 18, treated in standard Mitsunobu conditions with benzoic acid, gave stereoselectively the diprotected *cis*-dihydroxypyrrolidines 19 in good yields (Scheme 6).[12]

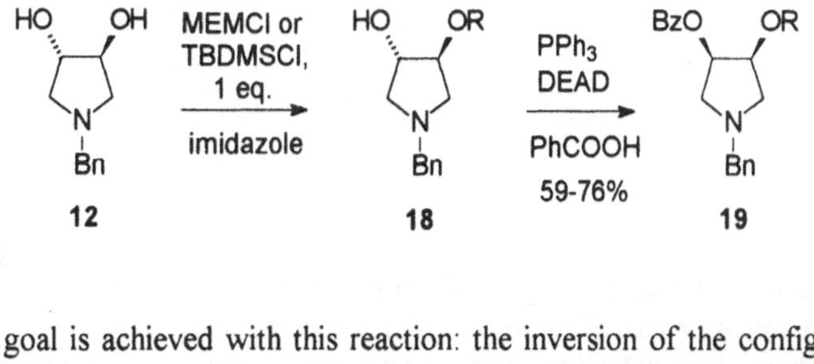

HO OH   MEMCl or   HO OR   BzO OR
        TBDMSCl,   PPh₃
         1 eq.   DEAD
   N   ─────────→ N   ─────────→   N
   | imidazole | PhCOOH |
   Bn   Bn   59-76%   Bn
   12   18   19

Scheme 6

A double goal is achieved with this reaction: the inversion of the configuration of the free hydroxy group, and the orthogonal protection of the second hydroxy group, that can be of high utility in further synthetic applications. This process, which has been named

"Mitsunobu trick", represents a formal desymmetrization of a *meso*-diol (Scheme 7), and promises useful applications especially when substrates cannot be sufficiently differentiated by an enzyme cavity.[3]

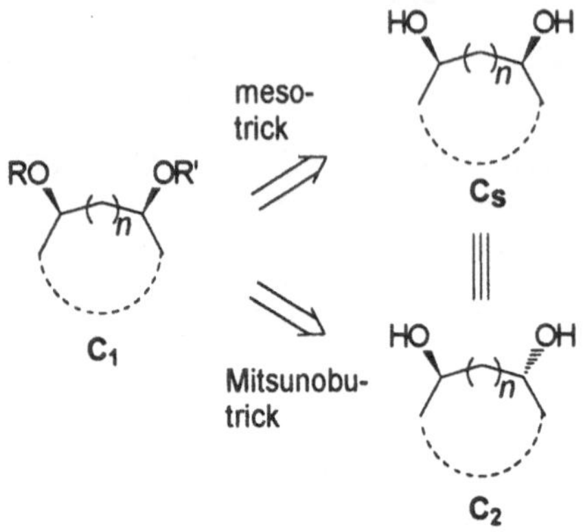

Scheme 7

Debenzylation and oxidation with oxaziridine **20** affords an almost equimolar mixture of the two regioisomeric nitrones **21** and **22**, which can be separated by chromatography (Scheme 8).[12] The slight preference for the oxidation vicinal to the benzoyloxy group was expected on the basis of previous observations.[10c] The final outcome of the process is the production of regioisomeric *cis*-3,4-dihydroxypyrroline-*N*-oxides **21** and **22** which afford the two enantiomeric 1,2-*cis*-trihydroxyindolizidines **23** and **24** in an enantiodivergent synthesis from the common precursor **19** (Scheme 8).

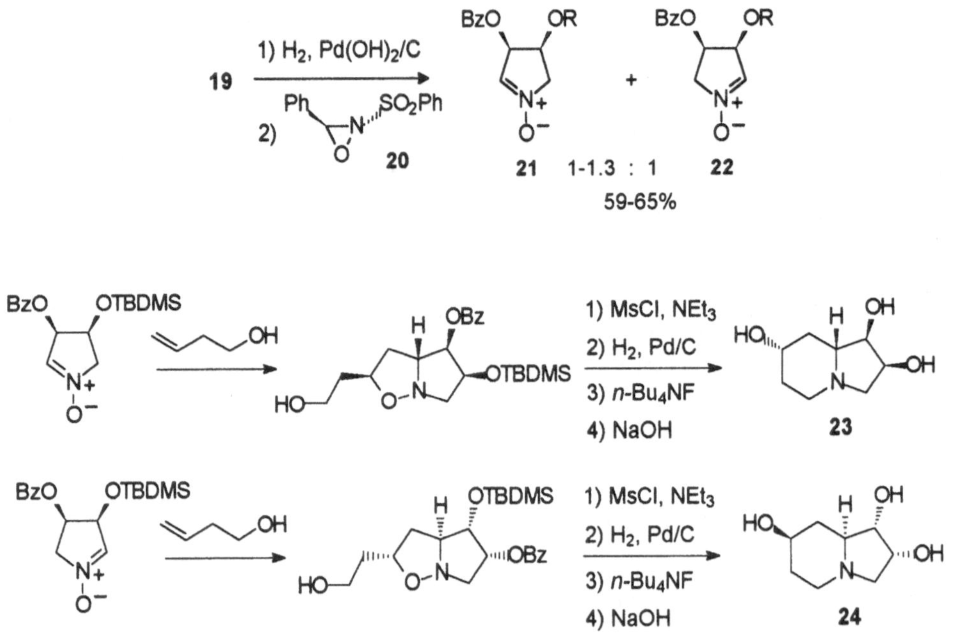

Scheme 8

The big success of a synthetic methodology consists of the possibility to fulfill different stereochemical requirements of the products starting from the same precursors. A general outlook of glycosidase inhibitors and necine bases structures shows that one main stereochemical feature is represented by the relative stereochemistry between the hydroxy group at C1 and the bridgehead proton, which is relevant to the biological activity of these compounds. As shown for the synthesis of compounds **23, 24** (Scheme 8) a *cis*-{OH,H} relationship originates from a highly favoured *anti* transition state approach of the reactants in an intermolecular cycloaddition (Scheme 9). To induce the inversion of the stereochemistry and a *trans*-{OH,H} relationship like in castanospermine (**1**), swainsonine (**3**), australine (**4**) and rosmarinecine (**7**), the dipolarophile has to approach *syn* to the 3-alkoxy substituent on the nitrone, and this is possible quantitatively only if the dipolarophile is directly linked to the substituent in an intramolecular process.

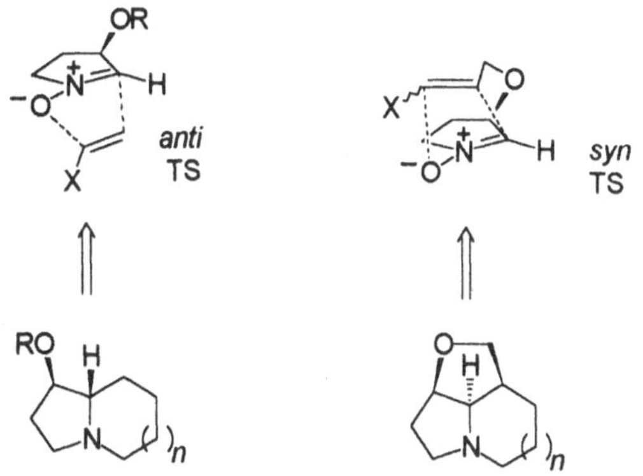

Scheme 9

The successful application of these concepts has been shown in the synthesis of alkaloids belonging to both classes.

(+)-Lentiginosine (**2**), the most potent indolizidine inhibitor of amyloglucosidase (K_i = 2.0 µM)[11a] was obtained from L-tartaric acid derived nitrones either by cycloaddition to methylenecyclopropane[11a] or to butenol[13] in 2.4% or 25% yield, respectively. In both cases a good diastereoselectivity of the cycloaddition step (12:1 to 5:1, respectively) warranties the required *cis*-{1-OH, 8a-H} relationship (Scheme 10).

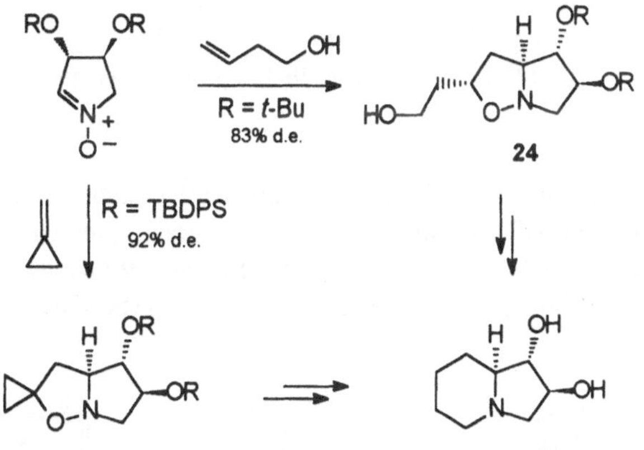

Scheme 10

The synthesis of (−)-hastanecine exhibited the same result in the pyrrolizine series (Scheme 11).[14] (−)-Hastanecine, the necine base of hastacine,[15] requires the *R* configuration of the C1 carbon that can be furnished by D-malic acid. This chiral precursor, however, is thirty times more expensive than its L enantiomer and less optically pure. For the sake of economical syntheses we planned a strategy starting from L-malic acid that in one single step was able to reach several goals: i) the use of a single precursor for the synthesis of different final products; ii) the protection of the OH group on nitrone; iii) the introduction of a group that is able to pilot the regioselectivity in the oxidation step to the nitrone. A Mitsunobu reaction on dimesylate **27** which afforded the benzoyl protected dimesylate **28** with the correct configuration for hastanecine was the solution (Scheme 11).[14]

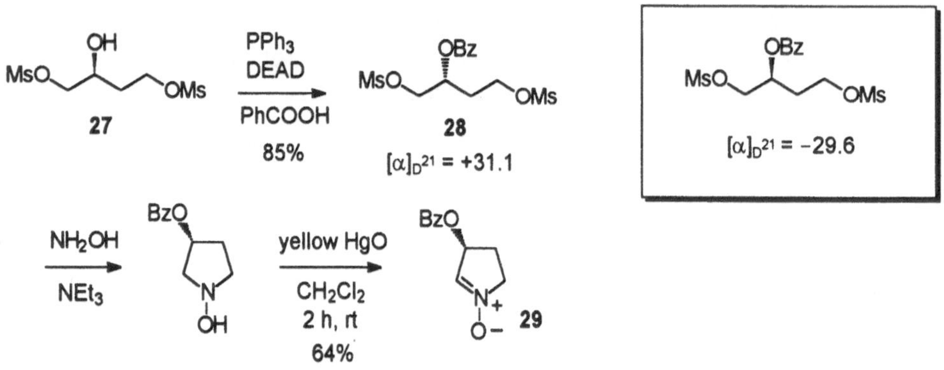

Scheme 11

The cycloaddition of **29** to dimethylmaleate gave the adduct **30** in 56% yield as the result of a preferred (4:1.5:1) *exo-anti* approach. Ring opening and reclosure gave pyrrolizidinone **31** identical to the intermediate used by Denmark[16] for the synthesis of hastanecine (Scheme 12).

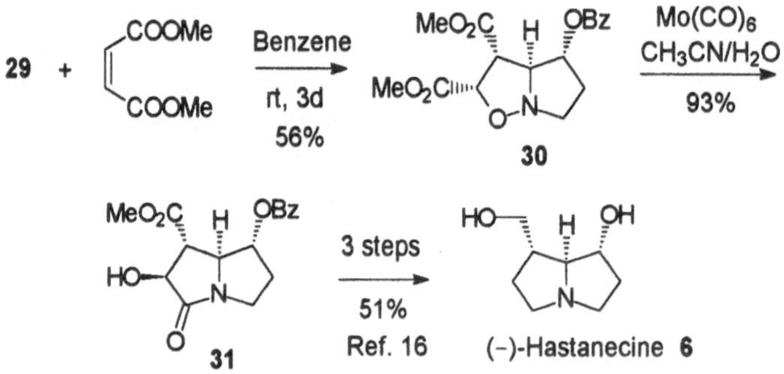

Scheme 12

(−)-Rosmarinecine, the necine base of rosmarinine,[17] was chosen as a target to prove the general application of an intramolecular 1,3-dipolar cycloaddition process from hydroxylated pyrroline-*N*-oxides. As a *trans*-{OH,H} relative stereochemistry is required the nitrone **33** bearing also the dipolarophile unit has to be set up (Scheme 13).

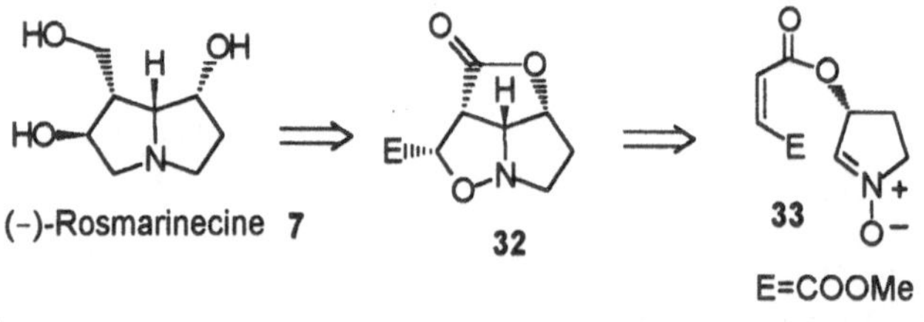

Scheme 13

Nitrone **33** with R configuration of the stereocentre could be synthesized from the S dimesylate by a Mitsunobu reaction with maleic acid monomethylester followed by standard

218

procedures. This route proved to be impracticable, as the cyclization with hydroxylamine was accompanied by aza-Michael addition to the maleic acid moiety. The solution was found carrying out the same reaction on the preformed 3-hydroxypyrroline-*N*-oxide 34, which could be directly synthesized from the TIPS protected nitrone 17. The reaction with maleic acid monomethyl ester, PPh₃ and DEAD gave directly the tricyclic isoxazolidine 32 in 70% yield (Scheme 14).

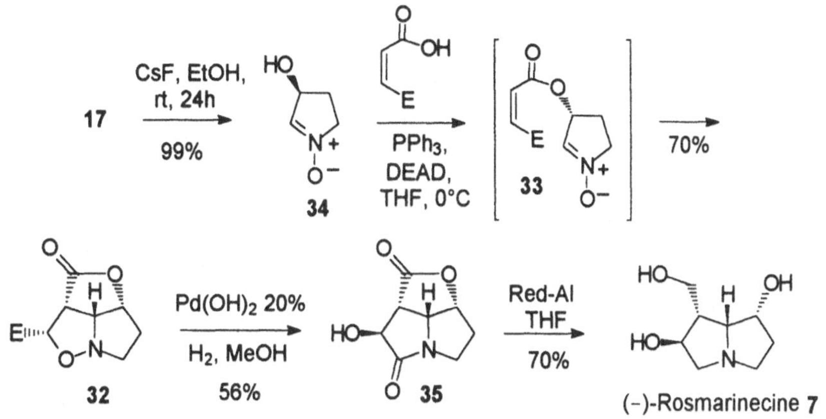

Scheme 14

It is worth to be noted that this method represents a novel example of a domino process involving a Mitsunobu reaction followed by an intramolecular nitrone cycloaddition. The complete inversion of configuration in the Mitsunobu reaction has been proved by the synthesis also of the enantiomeric product. The reductive cleavage of the N-O bond followed by reclosure and final reduction steps affords a new convenient synthesis of rosmarinecine and its enantiomer.

CONCLUSIONS

Hydroxylated pyrroline-*N*-oxides provide a general entry to the synthesis of polyhydroxylated pyrrolizidine and indolizidine alkaloids. Their use appears attractive for many aspects that are a major topic in modern organic synthesis. They allow a high "atom economy" in the overall process as all the skeleton atoms are kept intact from reagents to products. All the key steps occur by simple thermal induction without any need of added catalyst. This feature allow to carry out several steps in one pot as a domino process, another appealing aspect for environmentally friendly syntheses. Finally, the described methodology allows the synthesis of stereodiverse products, starting from the same precursor, by simple changes of reagents and conditions in the reaction sequence. This aspect, being so similar to what nature does in a larger extent, might suggest to assign the term of biomimetic to the methodology hitherto reported.

REFERENCES

1) L.F. Tietze, *Chem. Rev.* 96:115 (1996). L.F. Tietze, and U. Beifuss, *Angew. Chem.* 105: 137 (1993). *Angew. Chem. Int. Ed. Eng.* 32:131 (1993).

2) L.A. Thompson and J.A. Ellman, *Chem. Rev.* 96:555 115 (1996). R.W. Armstrong, A.P. Combs, P.A. Tempest, S.D. Brown, and T.A. Keating *Acc. Chem. Res.* 29:123 (1996).

3)   For a recent review, see: E. Schoffers, A. Golebiowski, and C.R. Johnson, *Tetrahedron* 52:3769 (1996).

4)   B.M. Trost, *Angew. Chem., Int. Ed. Engl.* 34:259 (1995).

5)   A.D. Elbein, and R.J. Molyneux, In *Alkaloids: Chemical and Biological Perspectives* S.W. Pelletier, ed., Wiley-Interscience New York, Vol. 5 (1987). Howard, A. S. Michael, J. P. In *The Alkaloids*, Brossi, A., Ed., Academic Press: New York, **1986**, Vol. 28, Ch 3.

6)   T. Hartmann and L. Witte, "Chemistry, Biology and Chemoecology of the Pyrrolizidines Alkaloids" in *Alkaloids: chemical and biological perspectives*, S.W. Pelletier, ed., Pergamon, Oxford, Vol. 9, pp.155-233 (1995). D.J. Robins, *Nat. Prod. Rep.*, 12:413 (1995), and references cited therein.

7)   A.D. Elbein, *Annu. Rev. Biochem.* 56:497 (1987). P. Vogel, *Chimica Oggi* 9 (1992). B. Winchester and G.W.J. Fleet, *Glycobiology* 2:199 (1992). Legler, G. *Adv. Carbohydr. Chem. Biochem.* 48:319 (1990). C.-H. Wong, R. L. Halcomb, Y. Ichibaka, and T. Kajimoto, *Angew. Chem., Int. Ed. Engl.* 34:521 (1995).

8)   For indolizidine and pyrrolizidine skeleton: J.J. Tufariello, *Acc. Chem. Res.* 12:396 (1979). J.J. Tufariello and J.P. Tette, *J. Org. Chem.* 40:3866 (1975). J.J. Tufariello and J.J. Tegeler, *Tetrahedron Lett.* 45:4037 (1976).

9)   For indolizidine skeleton: A. Brandi, F.M. Cordero, F. De Sarlo, A. Goti, and A. Guarna, *Synlett* 1 (1993). A. Goti, F.M. Cordero, and A. Brandi, *Top. Curr. Chem.* 178:1 (1996).

10) S. Cicchi, I. Höld, and A. Brandi, *J. Org. Chem.* 58:5274 (1993). S. Cicchi, A. Goti, and A. Brandi, *J. Org. Chem.* 60:4743 (1995). A. Goti, S. Cicchi, V. Fedi, L. Nannelli, and A. Brandi, *J. Org. Chem.* 62:3119 (1997).

11) A. Brandi, S. Cicchi, F.M. Cordero, R. Frignoli, A. Goti, S. Picasso, and P. Vogel, *J. Org. Chem.* 60:6806 (1995). A. Goti, F. Cardona, A. Brandi, S. Picasso , and P. Vogel, *Tetrahedron: Asymmetry* 7:1659 (1996).

12) S. Cicchi, J. Nunes Jr., A. Goti, and A. Brandi, *Eur. J. Org. Chem.* 419 (1998).

13) A. Goti, F. Cardona, and A. Brandi, *Synlett* 761 (1996).

14) A. Goti, V. Fedi, L. Nannelli, F. De Sarlo, and A. Brandi, *Synlett* 577 (1997).

15) V.S. Konovalov and G.P. Menshikov, *Zh. Obshch. Khim.* 15:328 (1945). D.J. Robins, *Prog. Chem. Org. Nat. Prog.* 41:115 (1982).

16) S.E. Denmark and A. Thorarensen, *J. Org. Chem.* 59:5672 (1994).

17) K. Tatsuta, H. Takahashi, Y. Ameniya, M. Kinoshita, *J. Am. Chem. Soc.* 105:4096 (1983). S.E. Denmark, A. Thorarensen, and D.S. Middleton, *J. Am. Chem. Soc.* 118:8266 (1996).

NOVEL TYPES OF CHIRAL CATALYSTS FOR ASYMMETRIC C-C BOND FORMATION REACTIONS

Yuri N. Belokon,[1] Michael North,[2] Henri B. Kagan[3]

[1]A.N.Nesmeyanov Institute of Organoelement Compounds, Russian Academy of Sciences, 117813, Vavilov 28, Moscow, Russia
[2]Department of Chemistry, University of Wales, Bangor, Gwynedd, UK LL57 2UW
[3]Univerite de Paris Sud. 91405, Orsay Cedex, Institut de Chimie Moleculaire d'Orsay, Laboratoire de Synthese Asymetrique, URA CNRS 1497, France

INTRODUCTION

The development of asymmetric catalytic reactions is a major challenge and a subject of intense activities of the chemical community. In particular, the elaboration of simple, cheap and efficient asymmetric catalysts for the formation of C-C bonds is urgently needed.[1] We describe herein two new chiral catalytic systems for promoting two important asymmetric organic reactions of C-C bond formation.

The first one is cyanosilylation of aldehydes and ketones, which makes available protected cyanohydrins, useful for the synthesis of pyrethroids and serving as starting materials for the preparation of several classes of enantiomerically enriched compounds.[2] These include α-hydroxy acids, β-hydroxy amines, α-amino acid derivatives, α-hydroxy ketones, etc.[2]

The second type of reaction is asymmetric catalytic phase-transfer (PTC) C-alkylation of an alanine synthon with the formation of enantiomerically enriched α-alkyl-α-amino acids, an important class of compounds. This approach opens a new perspective for the synthesis of complex molecules, containing quaternary (fully substituted) stereocenters. The progress in this area was recently reviewed.[3]

ASYMMETRIC CYANOSILYLATION OF ALDEHYDES AND KETONES PROMOTED BY CHIRAL SALEN-Ti(IV) COMPLEXES

Catalytic asymmetric addition of HCN or $(CH_3)_3SiCN$ onto aldehydes can be achieved in several ways.[2] Amongst these, the catalysis by chiral metal complexes feature prominently.[1,4,5] Recently, N.Oguni and co-workers developed a new catalytic system based on a Ti(IV)-tridentate Schiff's base complexes, derived from N-(2-hydroxy-3-t-butyl-bezylidene)-(S or R)-valinol (or other aminoalcohols or peptides) and Ti(O-i-Pr)$_4$, for trimethylsilylcyanation of aldehydes. E.e. of the corresponding reaction products up to 96% at a ratio substrate / catalyst equal to 5 were reported by the group.[5] Some experimental evidence was presented to support the reaction mechanism, including as a key step the substitution by the aldehyde carbonyl group of an isopropanol molecule in the co-ordination sphere of the catalytic complex.[5]

Jacobsen's group reported the highly enantioselective asymmetric epoxidation of simple olefins and the ring opening of epoxides with Me_3SiN_3 catalysed by complexes derived from (1R,2R)-[N,N'-bis(2'-hydroxy-3'-t-butyl-benzylidene)]-1,2-disubstituted-ethylenediamine and Mn(III) and Cr(III) ions respectively.[6,7] A possible mechanism, involving the nucleophile delivery to the epoxide via co-ordination of the nucleophile with the central metal ion was tentatively suggested by the authors.[7]

We supposed that chiral (salen)Ti(IV) catalysts prepared in situ from titanium tetraisopropoxide or $TiCl_4$ and (1R,2R)-[N,N'-bis(2'-hydroxy-3'(and/or-5')-alkyl- benzylidene)]-1,2-diaminocyclohexane might become efficient in the asymmetric trimethylsilylcyanation of aldehydes.

The reaction (see Scheme 1) was conducted in CH_2Cl_2 at -78 at a catalyst / substrate ratio of 0.1. The reaction was terminated after 120 h and the e.e. of the reaction products were monitored by chiral GLC.[8] The results are summarized in Table 1. As can be seen from the data, the complexes proved to be effective catalysts of the reaction (Table 1, runs 1-12). Invariably whichever substrate was chosen the catalysts induced the (S)-configuration to the silyl ethers formed. The introduction of substituents (R_1 and R_2) into the salen moieties increased the enantiomeric purity of the products (Table 1, compare runs 1-4 with runs 5-8 and 9-12). It also seemed evident that the nature of the axial coordinating ligands had no influence on the e.e. of the final product (Table 1, runs 9 and 12) which could indicate that the real catalytic species had the same structure whatever the initial axial ligand.

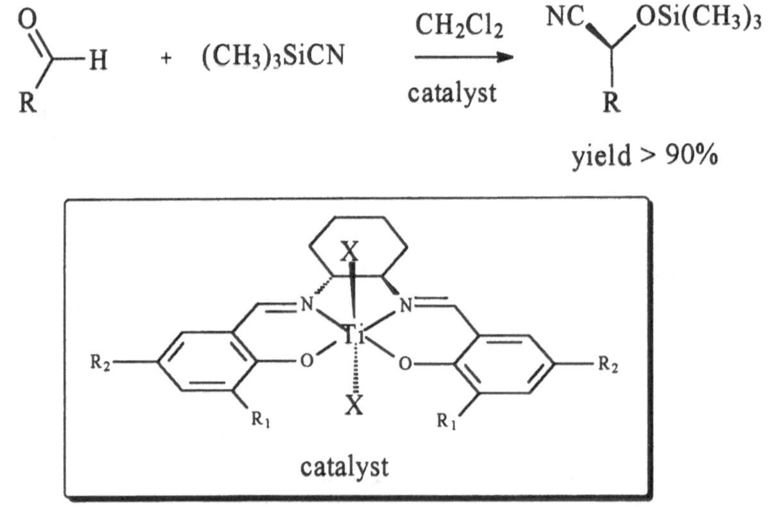

Scheme 1 Cyanosilylation of aldehydes catalyzed by chiral salen-Ti(IV) complexes

The structure of of $TiCl_2L$ [H_2L = (R,R)-N.N'-bis(3,5-di-t-butylsalicylidene)hexane-1,2-diamine] was established by X-ray crystallography.[9] and was found to be a monomeric compound with the salen ligand, adopting equatorial positions around the titanium with the chloride ligands situated trans to each other.

A semi quantitative kinetic analysis of the disappearance of benzaldehyde in the presence of $(CH_3)_3SiCN$ at room temperature catalyzed by salen-Ti(IV) complexes and monitored by spectrophotometry at 243 nm revealed an unexpected dependence of the rate of the reaction on the amount of water present in solution (see Fig.1).

Table 1 Enantioselective Trimethylsilylcyanation of Aldehydes Catalysed by Chiral Schiff's Base-Titanium Complexes, TiX₂L [H₂L= (*1R,2R*)-N.N'-bis[salicylidene)hexane-1,2-diamine or other Schiff's bases derived from (*1R,2R*)-hexane-1,2-diamine and substituted salicylaldehyde].[1]

run	R	R₁	R₂	e.e.%[2]
1	C₆H₅	H	H	60
2	p-CH₃OC₆H₄	H	H	60
3	(E)-C₆H₅CH=CH	H	H	70
4	(CH₃)₃C	H	H	16
5	C₆H₅	(CH₃)₃C	H	75
6	p-CH₃OC₆H₄	(CH₃)₃C	H	62
7	(E)-C₆H₅CH=CH	(CH₃)₃C	H	78
8	(CH₃)₃C	(CH₃)₃C	H	77
9	C₆H₅	(CH₃)₃C	(CH₃)₃C	90
10	(E)-C₆H₅CH=CH	(CH₃)₃C	(CH₃)₃C	70
11	(CH₃)₃C	(CH₃)₃C	(CH₃)₃C	40
12 [3]	C₆H₅	(CH₃)₃C	(CH₃)₃C	86

[1] Catalysts were obtained *in situ* from the corresponding Schiff's bases and Ti(OiPr)₄, the only exception was TiCl₂L (run 12) which was used as a pure compound. The reaction conditions were described in the text. See Scheme 1 for the designations of R, R₁ and R₂

[2] In all the cases the configuration of the final compounds was (*S*).

[3] TiCl₂L was used as a catalyst at a catalyst / substrate ratio of 0.1.

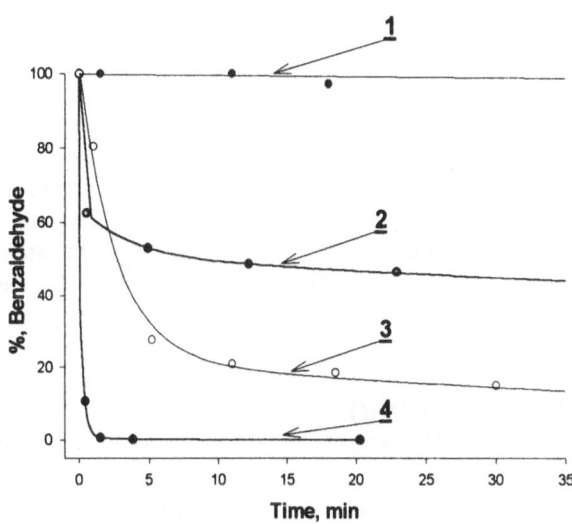

Fig 1 The disappearance of benzaldehyde (0.1 M initial concentration) in the presence of (CH₃)₃SiCN (mol ratio 1/1) catalyzed in CH₂Cl₂ at 25°C by: TiX₂L, H₂L=(*1R,2R*)-N.N'-bis(3-butylsalicylidene)hexane-1,2-diamine. 1) X= Cl, ratio substrate/catalyst = 100 water content in the solvent was less than 0.02%. 2) X= OiPr, substrate/catalyst ratio = 80 water content in the solvent was less than 0.02%. 3) X= Cl, substrate/catalyst ratio = 450, water content was 0.1% (CH₂Cl₂. saturated with water), two eq. of Et₃N were added. 4) X= OiPr, substrate/catalyst ratio = 100 water content was 0.1% (CH₂Cl₂. saturated with water).

As can be seen from Fig.1, both TiCl₂L or Ti(OiPr)₂L seemed to be unable to produce effective catalysts in CH₂Cl₂ with low concentration of water (Fig. 1, curve 1 and 2). The addition of water to Ti(OiPr)₂L produced very efficient catalyst (Fig. 1, curve 4). TiCl₂L was also activated by the addition of water and Et₃N (to neutralize generated HCl). The enantioselectivity of the reaction at 25°C was the same for both initial complexes (e.e. 80%).

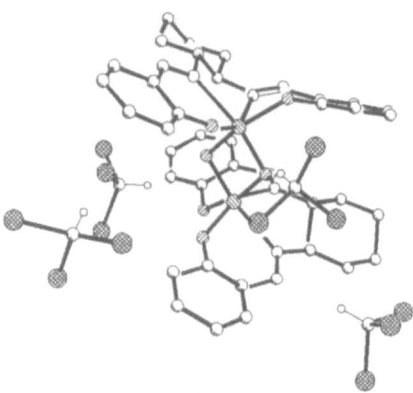

Fig.2 Solid state structure of a dimer derived from Ti(OiPr)₄ and (*1R,2R*)-N.N'-bis(salicylidene)hexane-1,2-diamine, according to an X-ray single crystal structure analysis.

The same procedure produced effective catalysts for all the other complexes described above. The modification of other Ti complexes with water was already disclosed.[10] Analysis of the structures of the water activated Ti(salen) complexes indicated that under the experimental conditions dimers were formed with two Ti(IV) atoms linked by two oxygen bridges. The dimers were stable both in aprotic solvents and crystals. The X-ray structure of the dimer generated from Ti(OiPr)₄ and (*1R,2R*)-N.N'-bis-(salicylidene)hexane-1,2-diamine is shown (Fig. 2). The salient features of the structure are the great distortions of the ligands and a non-planar cis-β arrangement the salen ligands with Δ-configuration of both Ti complexes. A similar dimeric structure derived from a salenophos ligand has been reported in the literature.[11] All the dimers were found to be highly efficient catalysts, operating at ratios of substrate / catalysts 400-2000. The e.e of the reactions were very similar to those collected in Table 1. It seems that the real catalytic particles responsible for the catalysis and the asymmetric induction (Table 1) is the dimeric one, being either an intermediate on the reaction coordinate or the real catalytic entity. Further work on the mechanism and synthetic applications of this chemistry is underway and will be reported in due course.

ASYMMETRIC PTC *C*-ALKYLATION MEDIATED BY TADDOL - NOVEL ROUTE TO ENANTIOMERICALLY ENRICHED α-ALKYL-α-AMINO ACIDS

Asymmetric reactions of C-H acids carried out under the effects of chiral bases constitute an important class of organic transformations.[12] Use of chiral ligands for the purpose of asymmetric catalysis of the *C*-alkylation reaction of achiral Li-enolates represents another important development.[12b] Recently, pioneering work of L. Duhamel reported use of chiral alkoxides (derived from chiral β-amino alcohols) either as stoichiometric bases or basic catalysts[13] to effect an asymmetric dehydrobromination reaction.

We believed that another important reactions of organic synthesis, alkyltion of CH-acids could be performed, using chiral alkoxides. We have chosen (*4R,5R*)-2,2-dimethyl-α,α,α',α'-tetraphenyl-1,3-dioxolane-4,5-dimethanol (TADDOL,[14] see chart 1) for the purpose. We report the results of the application of the idea to the asymmetric synthesis of α-methyl substituted α-amino acids which represent an important class of nonproteinogenic amino acids.[15]

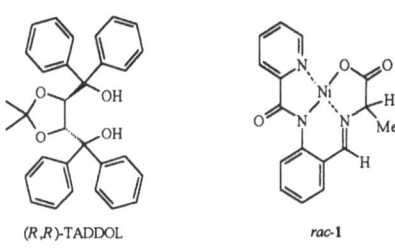

(R,R)-TADDOL   rac-1

Chart 1.

For this purpose four C-H acids were chosen: a Ni(II) complex of the racemic alanine Schiff's base, **1** (see the chart); the Schiff's base derivatives of benzaldehyde and racemic alanine methyl ester, **2a**; isopropyl ester, **2b**; and *(S)*-alanine *t*-butyl ester, **2c**. The alkylations of the substrates with benzyl bromide were conducted in toluene (carefully dried before use) at the ambient temperature (15-20°C) using solid MOH (ground under Ar) as bases (M= Li, K, Na) and TADDOL (either stoichiometric or catalytic amounts) as a chiral promoter (see Scheme 2, illustrating the alkylation of class-2 substrates). The experimental results are summarized in Table 2.

The alkylation of **2** with other activated alkyl halides (substituted benzyl halides, allyl bromide, and so on) proceeded very effectively, whereas the reaction with ethyl bromide under the experimental conditions proceeded very slowly.

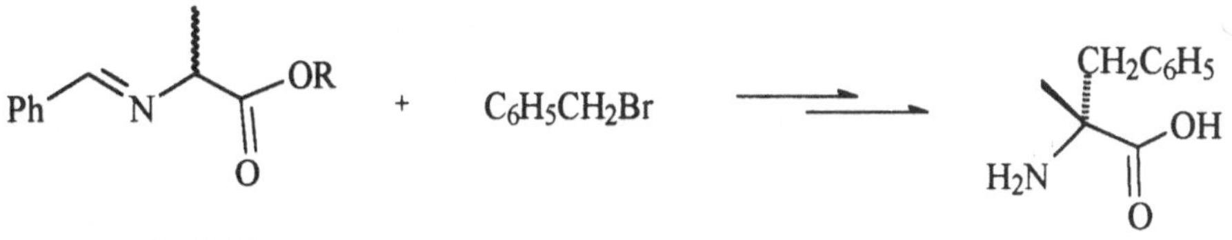

rac-**2a**: R=Me
rac-**2b**: R=iPr
(S)-**2c**: R=t-Bu

Scheme 2 Synthesis of α-methyl-α-phenylalanine by alkylation of Schiff's bases **2 a-c**

As can be seen from the data, both (R,R)- and (S,S)-TADDOL were efficient asymmetric promoters of the alkylation reactions (e.e. in the range of 20-82%) with the active alkylation agent (see Table 2, runs 1,3-5,7,8). (R,R)-TADDOL furnished (R)-α-methylphenylalanine whereas (S,S)-TADDOL gave (S)-α-methylphenylalanine (run 4).

Any possible enrichment of the final alkylation product by TADDOL during the product isolation was excluded by a control experiment where TADDOL and the final racemic product were mixed and treated in the conditions of the work up and no e.e. were found for the final α-methylphenylalanine. The O-monobenzyl derivative of TADDOL was also a promoter of the reaction but the asymmetric induction was very low.

We believe that under the experimental conditions TADDOL functioned as a base (see Fig. 3). To prove this a mixture of TADDOL (1 eq.) and solid NaOH (4 eq.) were stirred for several hours under Ar at ambient temperature in toluene. The mixture was filtered and the filtrate was added to the solution of **2b** (1eq.) and benzyl bromide (1 eq.). After 24 hours the reaction was terminated and the product analyzed to indicate that α-methylphenylalanine was formed in 35% yield and had 80% e.e. Evidently, TADDOL was ionized by NaOH in toluene, although the ion pair formed in this way might have to contain one or two other molecules of TADDOL, serving to stabilize it. The rigid structure of TADDOL could provide the necessary features to make it a hydrophobic complexing agent of cations in the manner indicated by Fig.3. In fact, separate experiments proved that TADDOL was able to extract some sodium picrate (solid) to toluene. Finally, an intermolecular hydrogen bond between the ionized substrate and TADDOL might stabilize the complex between the enolate ion pair and TADDOL. In fact, the structure of the complex would be the same whether TADDOL sodium salt served initially as a hydrophobic base or the neutral TADDOL functioned in the same manner as disclosed for another chiral catalyst of Li-enolate alkylations.[12]

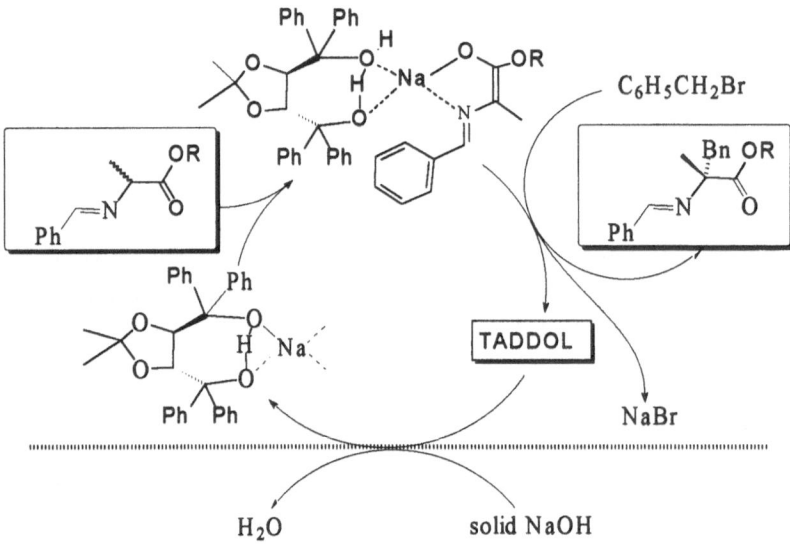

Fig.3 Possible mechanism of TADDOL mediated asymmetric PTC alkylation of CH-acids as illustrated by **2 a-c** case

Table 2 The asymmetric alkylation of different alanine derivatives mediated by (R,R)-TADDOL in toluene at 15-20°C [1]

Runs	Substrate	R	Base	TADDOL (equiv)	yield.% [2]	e.e. (R)-aa [3]
1	**1**		NaOH	1	40	20
2	**2a**	Me	NaOH	1.0	0	-
3	**2b**	i-Pr	NaOH	1.0	80	82
4	**2b**	i-Pr	NaOH	0.1	81	82 [4,5]
5	**2b**	i-Pr	KOH	1.0	31	24
6	**2b**	i-Pr	LiOH	1.0	0	-
7 [6]	**2c**	t-Bu	NaOH	1.0	45	38
8 [6]	**2c**	t-Bu	NaOH	0.1	93	22

[1] The concentration of the substrates was 0.2 M unless indicated otherwise; the reactions were conducted for 15-24 h with a ratio of alanine derivatives **1** or **2** : BnBr : TADDOL : NaOH = 1.0 : 1.2 : 0.1-1.0 : 4.0-5.0 (2.0-4.0.

[2] Determined by ^{1}H NMR, using leucine as an internal standard.

[3] Chiral GLC analysis (Chirasil-L-Val type phase) on crude products.

[4] After crystallization it was recovered in 40% yield with greater than 99% e.e.

[5] (S,S)-TADDOL gave (S)-α-methylphenylalanine with the same e.e.

[6] The alanine moiety was of (S) configuration; the reaction was continued for 168 h..

The e.e. of the reaction depended strongly on the structure of the substrate. The lowest e.e.s were observed for the alkylation of **1** with benzyl bromide (20% e.e., Table 2, run 1). The substrates of type **2** gave invariably better chemical yields and enantiomeric purities (see Table 2, runs 3-5, 7,8). The alkylation of **2a,** using NaOH as a base, was the only exception (Table 2, run 2) due to the methyl ester group hydrolysis of the substrate, efficiently removing it from the reaction media. The obvious reason for the better performance of the substrates **2b-c**, as compared to **1**, lies in the ability of their carbanions to chelate sodium ion in a mixed chiral complex with TADDOL (see Fig. 3), providing a rigid structure where the *Si*-side of the carbanions was effectively shielded from the electrophilic attack (the mixed complex might have a much more complicated structure. Obviously, the carbanion of **1** was functioning as a monodentate ligand with too many degrees of rotation freedom in the chiral ion pair.

The cation of the base was important. For example, LiOH was completely inactive (Table 2, run 6) whereas use of KOH furnished the product with disappointingly low e.e. 24% (Table 2, run 5). The structure of the class **2** substrates also influenced the e.e of the alkylation, as the increase of the size of the ester group from i-Pr to t-Bu was accompanied by the decrease of the alkylation e.e. from 70% to 60% (Table 2, compare runs 3,4 with 7,8). **2c** proved to be a slow reacting substrate, giving low e.e. of the final product (Table 2, runs 7,8).

The enantiomeric purity of (R)-α-methyl-α-phenylalanine could be greatly increased by crystallization (Table 1, run 4), as had already been disclosed.[16] In summary, our results offer a novel approach to the creation of the new generations of efficient chiral catalysts of asymmetric C-alkylation of C-H-acids, including PTC conditions. The conditions of the alkylation were not optimized and higher e.e. of the alkylation could be expected by modifying the reaction conditions. Our results[17] compare favorably with other methods of asymmetric PTC alkylations, employing chiral derivatives of alkaloids in terms of both stability of the catalyst and e.e. of the final products.[16,18] TADDOL and its derivatives are easily available and can be readily modified.[14] Thus a host of new asymmetric alkylations of different C-H-acids can be envisaged, using our approach with different TADDOL modifications tailored for each particular applications.

Acknowledgment

The work was supported by EU, INCO-Copernicus grant IC15-CT96-0722

References

1. B.M. Trost, Atom economy- a challenge for organic synthesis: homogeneous catalysis leads the way, *Angew.Chem.Int.Ed.Ingl.*,34: 259 (1995); B.Bosnich, *Asymmetric Catalysis*, Martinus Nijhoff Publishers, Dordrecht, The Netherlands, (1986); R. Noyori, *Asymmetric Catalysis in Organic Synthesis*, John Wiley and Sons, New York, (1994).
2. M. North, Catalytic asymmetric cyanohydrin synthesis, *Synlett*, 807 (1993); C.G. Kruse, Chiral cyanohydrins-their manufacture and utility as chiral building blocks, in: *Chirality in Industry*, A.N.Collins, G.N.Sheldrake and J.Crosby, ed., John Wiley and Sons, New York, (1992)
3. E.J.Corey, and A. Guzman-Perez, The catalytic enantioselective construction of molecules with quaternary carbon stereocenters, *Angew. Chem. Int. Ed.*, 37: 389 (1998)
4. W. Yang, and J. Fang, Asymmetric addition of trimethylsilyl cyanide to benzaldehyde catalyzed by samarium(III) chloride and chiral phosphorus(v) reagents, *J.Org.Chem.*, 63:1356 (1998), and references cited therein.
5. M. Hayashi, Y. Miyamoto, T. Inoue, and N. Oguni, Enantioselective trimethylsilylcyanation of some aldehydes catalyzed by chiral Schiff's base-titanium alkoxide complexes, *J. Org. Chem.*,58: 1515 (1993)

6. N. Finney, P. Pospisil, s. Chang, M. Palucki, R. Konsler, K. Hansen, and E. Jacobsen, On the validity of oxametallacyclic intermediates in the (salen)Mn-catalyzed asymmetric epoxidation, *Angew. Chem. Int. Ed. Engl.*, 36: 1720 (1997) and references cited therein.

7. K. Hanson, J. Leighton, and E.Jacobsen, On the mechanism of asymmetric nucleophilic ring-opening of epoxides catalyzed by (salen)Cr(III) complexes, *J.Am.Chem.Soc.*, 118: 10924 (1996)

8. Y. Belokon, N. Ikonnikov, M. Moscalenko, M. North, S. Orlova, V. Tararov, and L.Yashkina, Asymmetric trimethylsilylcyanation of aldehydes catalyzed by chiral (salen)Ti(IV) complexes, *Tetrahedron Asymmetry*, 7: 851 (1996)

9. V. Tararov, D.E. Hibbs, M.B. Hursthouse, N. Ikonnikov, K.M. Abdul Malik, M. North, C. Orizu, and Y. Belokon, First structurally defined catalyst for the asymmetric addition of trimethylsilyl cyanide to benzaldehyde, *Chem. Commun.*, 387 (1998)

10. P.Pitchen, E.Dunach, N.Desmukh, and H.B.Kagan, An efficient asymmetric oxidation of sulfides to sulfoxides, *J. Am. Chem. Soc.*, 106: 8188 (1984); M. Terada, Y. Matsumoto, Y. Nakamura, and K. Mikami, Anomalous role of molecular sieves 4A in the preparation of a binaphthol-derived μ_3-oxo titanium catalyst, *Chem. Commun.*, 281 (1997); D. Kitamoto, H. Imma, and T. Nakai, Asymmetric catalysis by a new type of chiral binaphthol-titanium complex, *Tetrahedron Letters*, 36: 1861 (1995)

11. A. Kless, C. Lefeber, A. Spannenberg, R. Kempe, W. Baumann, J. Holz, and A. Borner, The first chiral early-late heterobimetallic complex-a titanium(IV)-palladium(II) complex based on salenophos, *Tetrahedron*, 52: 14599 (1996)

12. M. Imai, A. Hagihara, H. Kawasaki, K. Manabe, and K. Koga, Catalytic asymmetric benzylation of achiral lithium enolates using a chiral ligand for lithium in the presence of an achiral ligand, *J.Am. Chem.Soc.*, 116: 8829 (1994); H. Fujieda, M. Kanai, T. Kambara, A. Iida, and K. Tomioka, A ternary complex reagent for an asymmetric reaction of lithium ester enolates with imines, *J. Am. Chem. Soc.*, 119: 2060, (1997) and references cited therein.

13. M. Amadji, J. Vadecard, J. C. Plaquevent, L. Duhamel, P. Duhamel, First catalytic enantioselective proton abstraction using chiral alkoxides, *J. Am. Chem. Soc.*, 118: 12483 (1996)

14. D. Seebach, A. Beck, M. Hayakawa, G. Jaeschke, F. Kuhnle, I. Nageli, A. Pinkerton, P. Beat Rheiner, R. Duthaler, P. Rothe, W. Weigand, R. Wunsch, S. Dick, R. Nesper, M. Worle, V. Gramlich, TADDOLs on their way to late transition metal complexes-synthesis and crystal structure of N- and S-containing TADDOL-derived compounds, *Bull. Soc.Chim. Fr.*, 134: 315 (1997) and references cited therein.

15. For recent references see: T. Wirth, New strategies to α-alkylated α-amino acids, *Angew. Chem. Int. Ed. Engl.*, 36:, 225 (1997)

16. M. O'Donnell, S. Wu, and J. Huffman, Anew active catalyst species for enantioselective alkylation by phase-transfer catalysis, *Tetrahedron: Asymmetry*, 50: 4507 (1994)

17. Y. Belokon, K. Kochetkov, T. Churkina, N. Ikonnikov, A. Chesnokov, O. Larionov, V. Parmar, R. Kumar, and H. Kagan, Asymmetric PTC C-alkylation mediated by TADDOL-novel route to enantiomerically enriched α-alkyl-α-amino acids, *Tetrahedron Asymmetry*, 9: 851 (1998)

18. After the completion of our work recently appeared a significant improvement of asymmetric alkylation of a glycine Schiff's base by PTC using modified alkaloids as catalysts: E. Corey, F. Xu, and M. Noe, A rational approach to catalytic enantioselective enolate alkylation using a structurally rigidified and defined chiral quaternary ammonium salt under phase transfer conditions, *J.Am.Chem.Soc.*, 119: 12414 (1997); B. Lygo, and P. Wainwright, A new class of asymmetirc phase-transfer catalysts derived from Cinhona alkaloids- application in the enantioselective synthesis of α-amino acids, *Tetrahedron Letters*, 38: 8595 (1997)

RECENT DEVELOPMENTS IN ASYMMETRIC HYDROGENATION WITH CHIRAL Ru(II) CATALYSTS AND SYNTHETIC APPLICATIONS TO BIOLOGICALLY ACTIVE MOLECULES

Jean Pierre GENET

Ecole Nationale Supérieure de Chimie de Paris
Laboratoire de Synthèse Sélective Organique et Produits Naturels
Associées CNRS (UMR 7573)
11 rue Pierre et Marie Curie
75231 PARIS Cedex 05, France

INTRODUCTION

Among all the possible methods of creating optically active compounds, enantioselective catalysis using chiral transition metal complexes is the most effective methods since a small amount of chiral material can, in principle, produce a large amount of optically active product. The enantioselective hydrogenation of prochiral C=C and C=X (X=N,O) double bonds constitutes a powerful technology to establish chirality at stereogenic carbon atoms. This tool has enabled the chemist to compete successfully with nature in the creation of enantiomerically pure compounds. The Rh(I) complexes bearing chiral tertiary phosphines are remarkable catalysts in the hydrogenation of dehydroamino acids[1,2,3,4]. Optimisation of the conditions made it possible for the industrial production of several amino acids[5]. In spite of such impressive achievments and improvments in the design of new ligands for rhodium(I) catalysts[6], the scope of asymmetric hydrogenation is not so wide. A breakthrough came by the discovery of hexacoordinated ruthenium catalysts containing the highly effective axially symmetric BINAP ligand[7,8]. A disguishing feature of theses catalysts is their universality, they are suitable for enantioselective hydrogenation of wide range of substrates[9]. In the application of such technology to the synthesis optically active compounds in perfumery, food industry and drugs research, chemists are facing the task of designing broad libraries of chiral catalysts. Therefore developments of chemistry which allows preparation of wide range of chiral ruthenium catalysts is highly desirable for a rapid screening. As part of our research programm we have initiated and developed a novel and general route for the preparation of chiral ruthenium (II) catalysts. Such chemistry and the usefulness of these catalysts will be briefly presented with the successful development of enantioselective hydrogenation of prochiral substrates including olefins and functionalized ketones. Moreover, several applications in the synthesis of various complex biologically active molecules will be reported.

Current Trends in Organic Synthesis
Edited by Scolastico and Nicotra, Kluwer Academic/Plenum Publishers, 1999

DIVERSITY IN CATALYSTS PREPARATION

By treatment of [RuCl$_2$(COD)]n **1** with BINAP in the presence of triethylamine, in toluene at reflux, a dinuclear complex formulated as Ru$_2$Cl$_4$(BINAP)$_2$NEt$_3$, **2** was isolated by Ikariya, Saburi et al., its structure not being elucidated. However, very recently

such dinuclear complex bearing p-MeO-BINAP (p-MeO-BINAP = 2,2'-bis-(bis p-methoxyphenyl)phosphino)-1,1'-binaphtyl instead of BINAP has been characterized[10] (X-ray analysis) as an unique anionic dinuclear structure 4. The Ikariya-Saburi complex has been used by Noyori for the preparation of mononuclear dicarboxylate complex **3** by treatment with sodium acetate in t-butanol at 80°C [11] (fig.1)

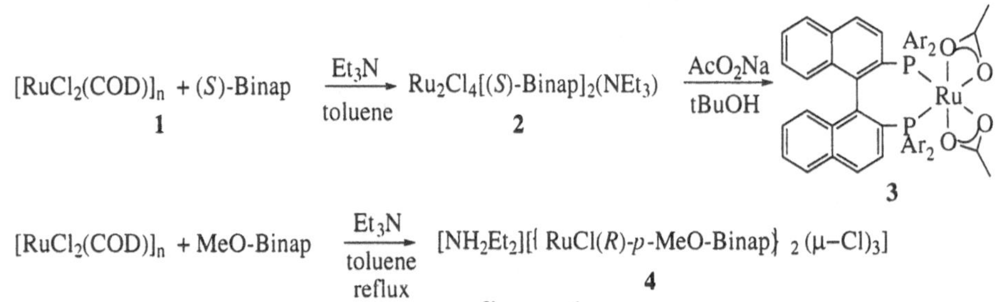

$$[RuCl_2(COD)]_n + (S)\text{-Binap} \xrightarrow[\text{toluene}]{Et_3N} Ru_2Cl_4[(S)\text{-Binap}]_2(NEt_3) \xrightarrow[\text{tBuOH}]{AcO_2Na}$$

1   **2**

3

$$[RuCl_2(COD)]_n + \text{MeO-Binap} \xrightarrow[\substack{\text{toluene} \\ \text{reflux}}]{Et_3N} [NH_2Et_2][\{RuCl(R)\text{-}p\text{-MeO-Binap}\}_2(\mu\text{-Cl})_3]$$

4

figure 1

A number of monomeric BINAP-Ru(II) complexes have been prepared and are effective catalysts for the asymmetric hydrogenation of variety of prochiral olefins and ketones[2,3,4]

Our catalysts preparation (fig. 2) starts with commercially available (COD)Ru(2-methylallyl)$_2$ **5** which is transformed into a wide range of chiral ruthenium(II) complexes **6** by displacement of COD-ligand by the appropriate chiral phosphine[12,13] including DIPAMP[14]. A subsequent protonation by HX (X=Br, Cl, BF$_4$, PF$_6$) **7** produces the corresponding catalysts P*PRuX$_2$[15]. Interestingly an improved and rapid in situ preparation of this type of ruthenium-catalysts has been developed. In one step from (COD)Ru(2-methylallyl)$_2$ by protonation 1.5 to 2 eq. of HX in acetone, dichloromethane or ethylacetate with the appropropriate chiral phosphine DUPHOS, SKEWPHOS, BINAP, MeO-BIPHEP the catalysts defined by the empirical formula RuX$_2$P*P, which are excellent catalysts for the asymmetric hydrogenation of ketones and olefins.

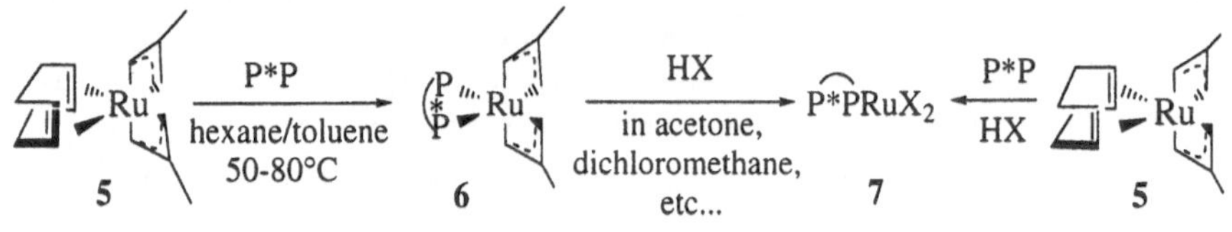

P*P= CHIRAPHOS, BINAP, MeO-BIPHEMP, DUPHOS, DIOP,DIPAMP, SKEWPHOS, etc...

figure 2

With respect to the preparation of chiral ruthenium(II) catalysts, the established route has some advantages. Specifically, the rapid screening of ligands (fig. 3). Therefore in addition to Ru-catalysts bearing atropisomeric ligands (e.g. BINAP, BIPHEMP, MeO-BIPHEP)[12], new dibromides Ru(II)-catalyst containing C$_2$-symmetric (bisphopholane) (e.g. MeDUPHOS[16], BPE[17] (1,2-bis-phospholano ethane) and SKEWPHOS[18] (fig.3) are emerging. Moreover in the case of molecules which can be difficult to reduce, individual procedure (e.g. screening of solvents) are easy to carry out[19]. The usefulness of such chiral ruthenium(II) catalysts will now be demonstrated with some enantioselective hydrogenation examples and applications to the synthesis of molecule of biological interest.

(R,R)-DIPAMP   (S,S)-CHIRAPHOS   (S,S)-SKEWPHOS   (R)-BINAP

(R)-MeO-BIPHEP   (S)-BIPHEMP   (R,R)-Me-DUPHOS   (R,R)-iPr-BPE

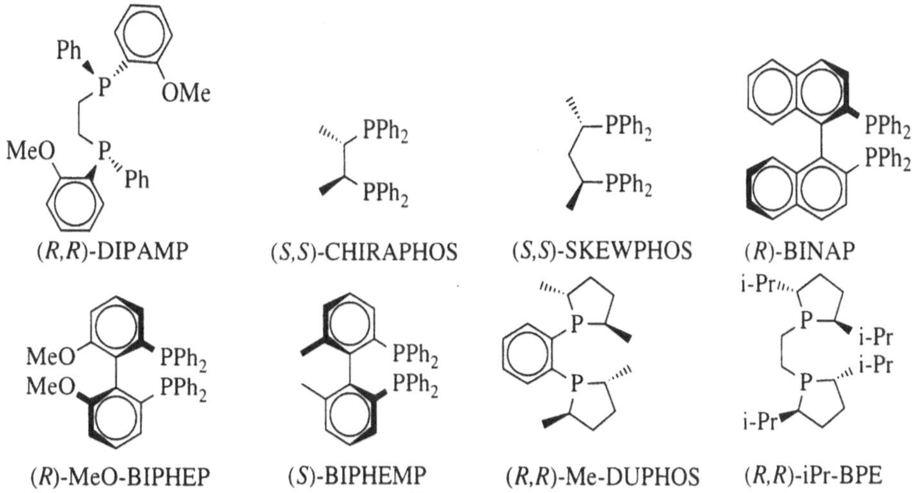

figure 3

ENANTIOSELECTIVE HYDROGENATION

C=C Bonds Hydrogenations

The P*P Ru (2-methylallyl)$_2$ catalysts **6** displayed good efficiency giving the saturated acids with high selectivity[16]. An important application of this technique is the enantioselective hydrogenation of **9** catalyzed by (S)-BINAPRu (2-methylallyl)$_2$ affording naproxen **10**[16] an useful anti-inflammatory agent. More recently the α, β-unsaturated vinyl phosphonic acids **11** were hydrogenated to the corresponding arylethylphosphonic acids **12** with good selectivity up to 86% using MeO-BIPHEP ruthenium dibromide complex at 80°C, 80 bars[20] (fig. 4).

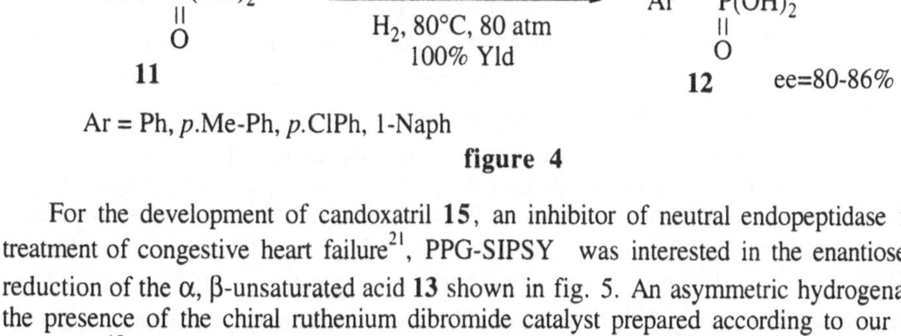

Ar = Ph, p.Me-Ph, p.ClPh, 1-Naph

figure 4

For the development of candoxatril **15**, an inhibitor of neutral endopeptidase for the treatment of congestive heart failure[21], PPG-SIPSY was interested in the enantioselective reduction of the α, β-unsaturated acid **13** shown in fig. 5. An asymmetric hydrogenation in the presence of the chiral ruthenium dibromide catalyst prepared according to our *in situ* procedure[15] produced the desired compound **14** in 98% enantioselectivity and quantitative yield. The asymmetric hydrogenation process after optimisation (substrate catalyst, ratio, solvent, etc..) was used on a 200 kg scale for a total production of 2 t.[22].

231

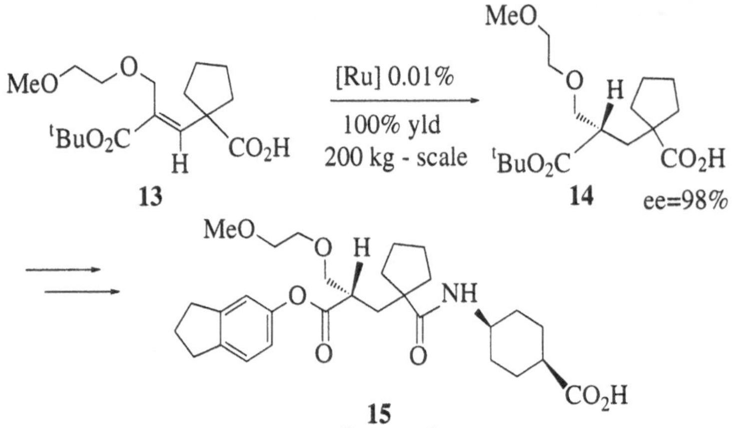

figure 5

The asymmetric hydrogenation of prochiral allylic and homoallylic alcohols catalyzed by the Ru(II) catalyst produced the corresponding chiral products with exceptionally high selectivities[3,12]. The enantiomerically pure alcohol **17** was prepared using asymmetric hydrogenation of **16** catalyzed by (R)-MeO-BiphepRu(II)Br$_2$[16]. The 1β-methyl carbapenem prepared above was used in the synthesis of several bicyclic derivatives bicyclic carbapenem **18** via π-allyl palladium cyclization[23,24].

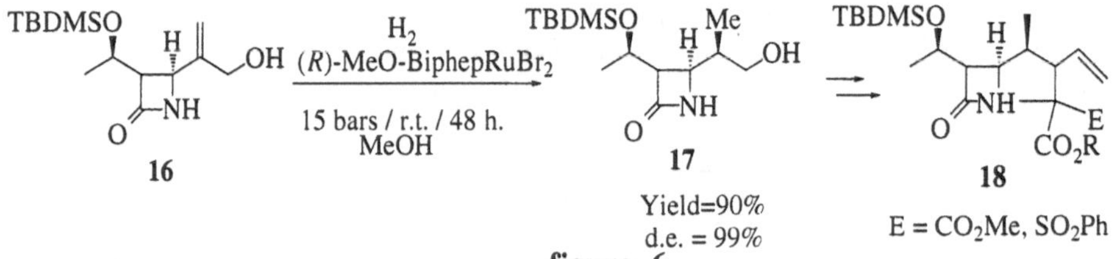

figure 6

C=O Bond Hydrogenations.

It is well established now that the reduction of carbonyl groups with ruthenium diphosphines[3,25] is competitive to chiral hydrides. We have also demonstrated that our *in situ* prepared (P*P)RuX$_2$ catalysts are extremely efficient for the asymmetric hydrogenation of a wide range of functionalized ketones including α and β-keto esters[16], β-keto phophonates[26], β-thiophosphonates, thioketones[27] with excellent enantioselectivities. The catalytic behavior of complexes **7** has been extensively investigated. These *in situ* prepared catalysts exhibit high catalytic activity and enantioselectivity in the hydrogenation of keto groups, ruthenium complexes bearing atropisomeric ligands MeO-BIPHEP, BINAP[16] as well as SKEWPHOS[18] were efficient for the asymmetric hydrogenation under or atmospheric pressure of β-keto esters (fig. 7)[28]. Interestingly the β-keto esters having an unsaturated alkyl chain were chemoselectively hydrogenated to optically pure unsaturated β−hydroxyesters as shown in fig. 7.

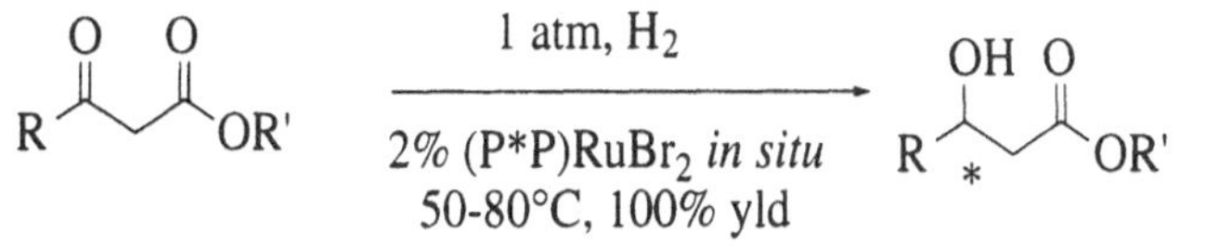

1 atm, H$_2$

2% (P*P)RuBr$_2$ *in situ*
50-80°C, 100% yld

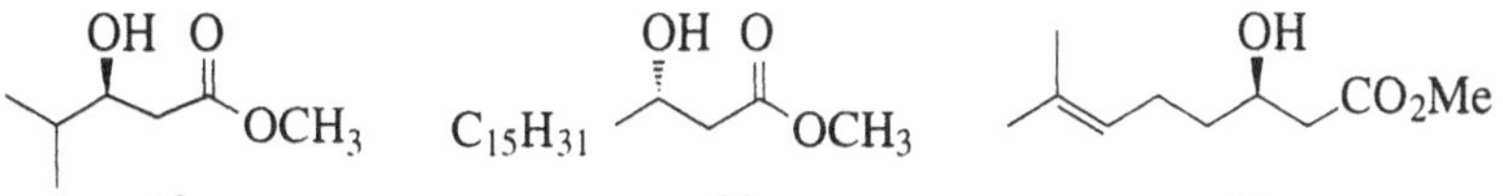

19	20	21
(R)-MeO-BiphepRuBr$_2$	(S)-BinapRuBr$_2$	(R)-MeOBiphepRuBr$_2$
ee = 99%	ee = 96%	ee = 98%

figure 7

Asymmetric Hydrogenation of Chiral Substrates.

Double stereodifferentiations that uses the chirality of a ketonic substrate and chiral ligand allows increased selectivity. An application is shown in the asymmetric hydrogenation of ethyl (2'S)-3-N-Boc-2'-pyrrolidinyl-3-oxopropanoate 22 with the ru-catalysts[29]. In this case, both the efficiency of the catalyst-substrate chirality transfer (catalyst control) and intramolecular control (substrate control) are important as shown in table 1.

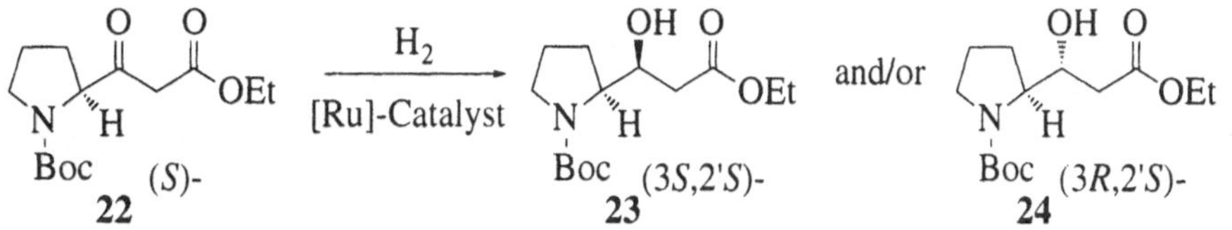

Table 1 : Double stereo differentiation with (S)-proline derivative

Catalyst	Pressure (bars)	Temp. (°C)	Time (h)	Yields %	Diastereoselectivity (3S,2'S)- : (3R, 2'S)-
2% (R)-BinapRuBr$_2$	1	50	48	90	>99 : 1
1%(R)-MeO-BiphepRuBr$_2$	10	50	48	100	>99 : 1
1%(S)-BinapRuBr$_2$	10	50	48	100	7.5 : 92.5
1%(S)-MeO-BiphepRuBr$_2$	10	50	24	100	6 : 94

The preparation of chiral building blocks 23, 24 provide a new and versatile entry to pyrrolizidine alkaloids[29].

Dynamic Kinetic Resolution of α-Substituted β-Keto Esters and Synthetic Applications.

A racemic substrate β-keto ester containing an α subtituent should in principle provide four stereoisomers (fig. 8). However such racemic compounds with a chirally labile stereogenic center may, under certain conditions, be converted to one major stereoisomer. Noyori and our group successfully discovered the first example of stereo and enantioselective hydrogenation through dynamic kinetic resolution using ruthenium catalysts of racemic α-acetamido-β-ketoesters (R = NHCOR'). The reaction provides a efficient route to *syn* β-hydroxy α-amino acids[3,12,13b].

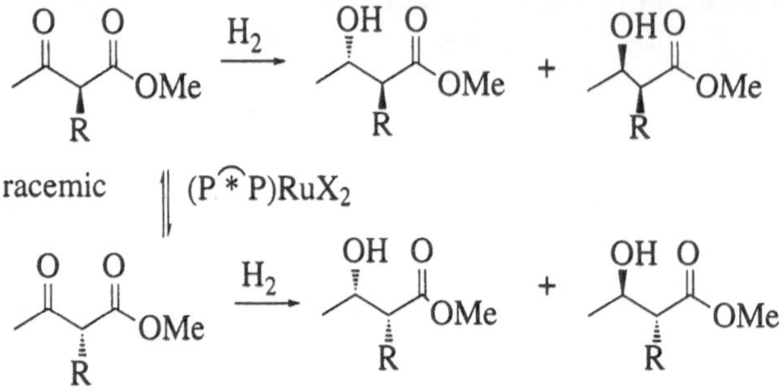

R = Heteroatom : (halogen, nitrogen)

figure 8

More recently using this Ru-catalyzed hydrogenation of **25** we have prepared efficiently the (2S,3R)-3-hydroxylysine **26** (fig. 9) an intermediate for the preparation of azepane moiety of balanol **27** (proteine kinase inhibitor)[30].

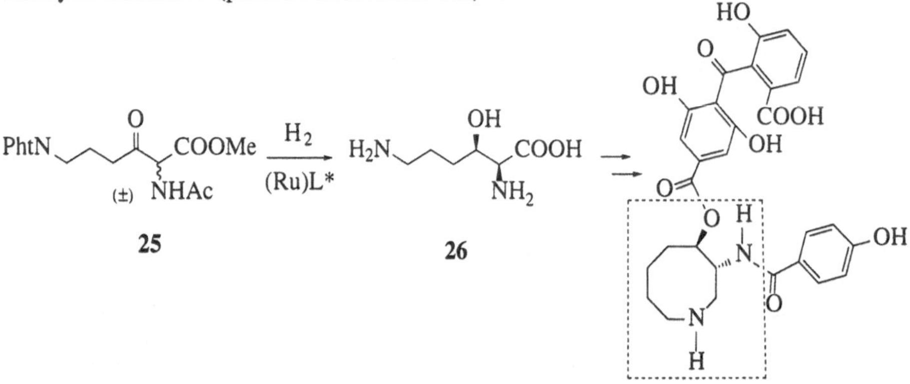

figure 9

We have found that hydrogenation of racemic α-chloro β-ketoesters **28** under optimized conditions with MeOBiphepRu-catalysts proceeds with high *anti* diastereoselectivity to give the corresponding α-chloro β-hydroxy esters **29, 30** in excellent ee[31] (fig. 10).

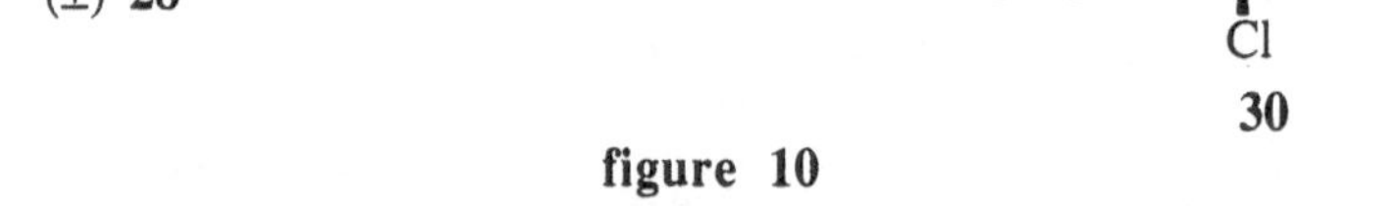

figure 10

This technology offers a new route to optically active lycidates **31** and **33** key intermediates for the the synthesis of C-13 side chain of taxol, taxotere[32] **34** and diltiazem **32** (a *potent* channel blocker used for the treatment of hypertension) have been efficeintly prepared (fig. 11).

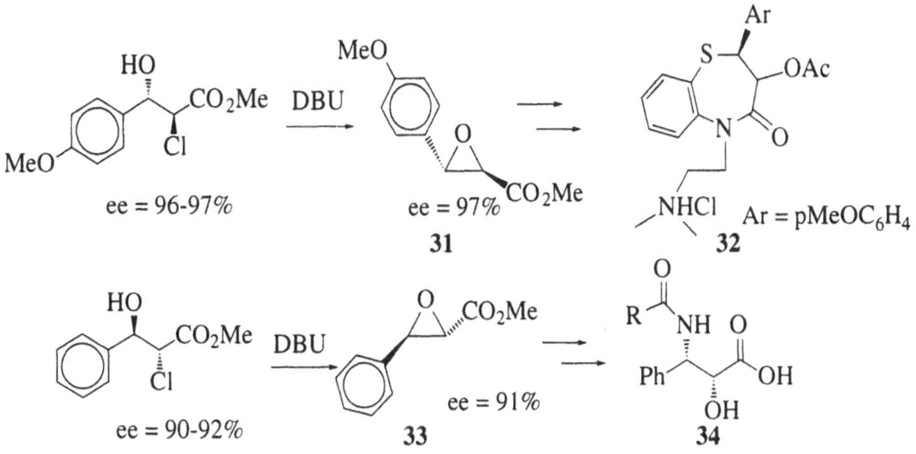

R = Ph (taxol C-13 side chain) ; R = O*t*Bu (taxotere C-13 side chain)
figure 11

We also developed an efficient synthesis (fig. 12) of enantiomerically pure *anti* β-hydroxy α-amino acids by sequential ruthenium catalyzed hydrogenation and electrophilic amination[33].

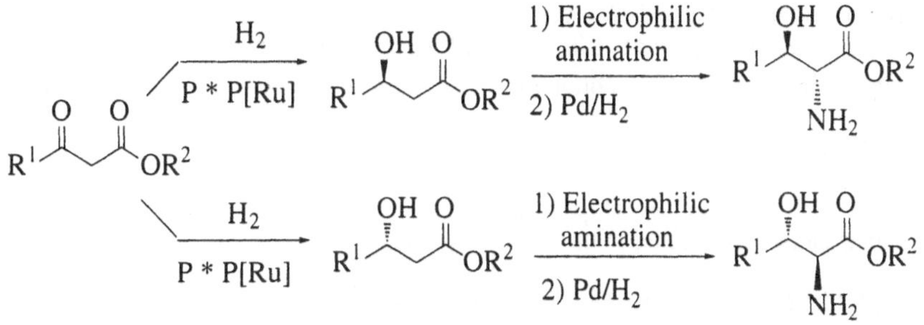

figure 12

Having the two technologies[13b,33] in hand for the preparation of *syn* and *anti* β-hydroxy α-amino acids we prepared efficiently both β-hydroxy α-amino acids key components of vancomycin[34].

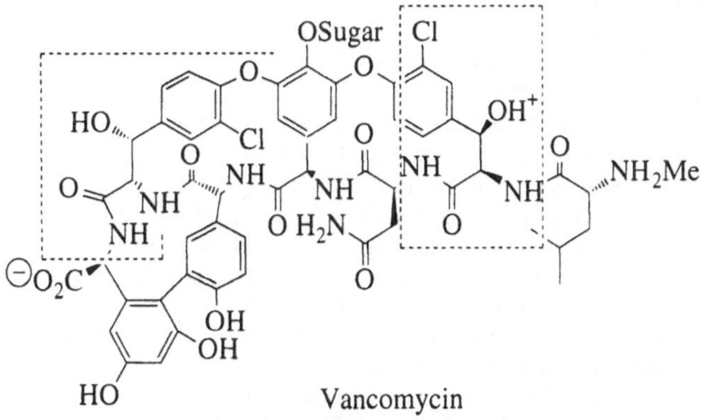

Vancomycin

Heterocycles of biological interest have been synthetized using ruthenium based asymmetric hydrogenation of β-keto esters containing unsaturated chain **35** and **37** (fig. 13).

For instance, (3S,4S)-4-hydroxy-2,3,4,5-tetrahydropyridazine-3-carboxylic acid, component of luzopeptine A[35] **36**, an antitumor antiobiotic and (-) swainsonine[36] **39** having high α-D-mannosidase inhibitory and antimetastatic activities have been prepared efficiently.

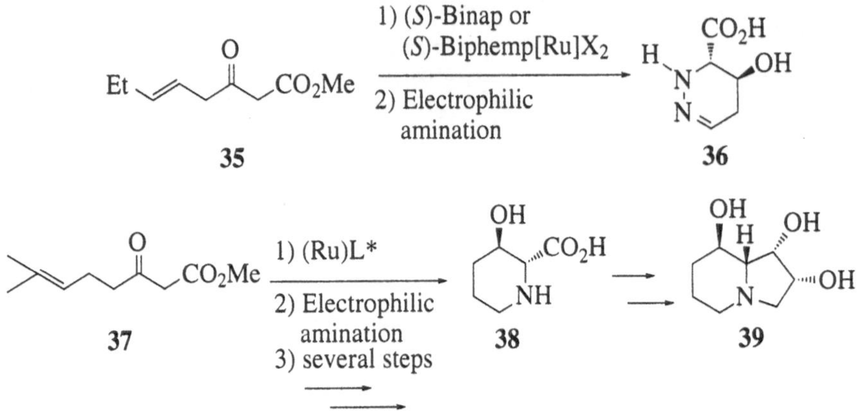

figure 13

CONCLUSION

Homogeneous asymmetric hydrogenation using chiral ruthenium(II) catalysts produced in situ from CODRu(2-methylallyl)$_2$ is a powerful tool that can provide entry to the preparation of highly functionalized optically enriched building blocks. This asymmetric catalysis is clean, economical versatile and often proved to be superior to any other method. Of importance, is the fact that this technology is also amenable to large scale production[37].

ACKNOWLEDGEMENTS

I would like to thank the co-workers who were contributors of the results presented in this lecture. Their names are quoted in the list of references.

REFERENCES

1. K.E. Koenig in *Asymmetric Catalysis*, J.E. Morrison, Ed. Academic Press, New York, Vol.5 (1985).
2. H. Takaya, T. Ohta and R. Noyori *Catalytic Asymmetric Synthesis*, I. Ojima, ed., Chapter 1, VCH publisher, New York (1993).
3. R. Noyori *Asymmetric Catalysis in Organic Synthesis*, Chapter 2, p.16, J. Wiley, New York (1994).
4. J.P. Genet *Advanced Asymmetric Synthesis*, Chapter 8, G.R. Stephenson, Ed., p.146, Chapman and Hall, London (1996).
5. H.B. Kagan, *Bull. Soc. Chim. Fr.*, 846, (1988).
6. K. Inogushi, S. Sakuraba and K. Achiwa *Synlett*, 169, (1992).
7. T. Ikariya, I. Ishii, H. Kawano, T. Arai, M. Saburi, S. Yoshikawa and S. Akutagawa *J. Chem. Soc. Chem. Commun.*, 922, (1985).
8. R. Noyori, M. Ohta, Y. Hsiao, M. Kitamura, H. Takaya *J. Am. Chem. Soc.*, 108, 7117, (1986).
9. Reviews : a) R. Noyori *Chem. Soc. Review*, 18, 2, 187, (1989) ; b) R. Noyori *Science*, 248, 1194, (1990) ; c) R. Noyori and H. Takaya *Acc. Chem. Res.*, 23, 345, (1990) ; d) H. Takaya, T. Ohta, K. Mashima and R. Noyori *Pure Appl. Chem.*, 62, 1135, (1990) ; e) R. Noyori *Tetrahedron*, 50, 4259, (1994) ; f) R. Noyori and S. Hashiguchi *Acc. Chem. Res.*, 30, 97, (1997).
10. T. Ohta, Y. Tonomura, K. Nozaki, H. Takaya, K. Mashima *Organometallics*, 15, 6, 1521, (1996).
11. T. Ohta, H. Takaya, R. Noyori *Inorg. Chem.* 27, 566, (1988).
12. Review : J.P. Genet in *Reduction Organic Synthesis, Am. Chem. Soc. Symposium Series* 641, Abdel F. Magid, ed., Chapter 2, 31, (1996).
13. a) J.P. Genet, S. Mallart, C. Pinel, S. Jugé and J.A. Laffitte *Tetrahedron : Asymmetry*, 2, 43, (1991) ; b) J.P. Genet, S. Mallart, C. Pinel, S. Thorimbert, S. Jugé, J.A. Laffitte *Tetrahedron : Asymmetry*, 555, (1991).

14. J.P. Genet, C. Pinel, S. Mallart, S. Jugé, N. Cailhol and J.A. Laffitte *Tetrahedron Lett.*, 33, 5343, (1992).

15. J.P. Genet, C. Pinel, V. Ratovelomanana-Vidal, S. Mallart, X. Pfister, M.C. Caño de Andrade, J.A. Laffitte *Tetrahedron : Asymmetry*, 5, 665, (1994).

16. J.P. Genet, V. Ratovelomanana-Vidal, X. Pfister, M.C. Caño de Andrade, J.A. Laffitte, S. Darses, C. Pinel, L. Bischoff, C. Galopin *Tetrahedron : Asymmetry*, 5, 675, (1994).

17. M. Burk, T. P. Harper and C.S. Kalberg *J. Am. Chem. Soc.*, 117, 4423, (1995).

18. D. Blanc, J.C. Henry, V. Ratovelomanana-Vidal and J.P. Genet *Tetrahedron Lett.*, 38, 6603, (1997).

19. J.P. Genet, V. Ratovelomanana-Vidal unpublished results.

20. J.C. Henry, D. Lavergne, V. Ratovelomanana-Vidal, J.P. Genet, T.M. Dolgina, I.P. Beletskaya *Tetrahedron Lett.*, 39, 3473 (1998).

21. Pfizer, Sandwich Drug Discoveries (October 1997)

22. J. Lastenet, S. Coulon, S. Roussiasse, M. Bulliard, B. Laboue PPG SIPSY, (1997).

23. S. Roland, Ph.D. Thesis Pierre and Marie Curie University(1995)

24. J.P. Genet, J.C. Galland, unpublished results.

25. H. Takaya, T. Ohta, N. Sayo, H. Kumobayashi, S. Akutagaria, S. Inoue, I. Kasahara, R. Noyori, *J. Am. Chem. Soc.*, 109, 1596 (1987).

26. I. Gautier, V. Ratovelomanana-Vidal, P. Savignac, J.P. Genet *Tetrahedron Lett.*, 37, 7721, (1996).

27. J.P. Tranchier, V. Ratovelomanana-Vidal, J.P. Genet, S. Tong, T. Cohen *Tetrahedron Lett.*, 38, 2951, (1997).

28. J.P. Genet, V. Ratovelomanana-Vidal, M.C. Caño de Andrade, X. Pfister, P. Guerreiro, J.Y. Lenoir *Tetrahedron Lett.*, 36, 4801, (1995).

29. P. Guerreiro, P. Bertus, V. Ratovelomanana-Vidal, J.P. Genet unpublished results.

30. E. Coulon, S. Duprat de Paule, V. Ratovelomanana-Vidal, M.C. Caño de Andrade, J.P. Genet unpublished results.

31. J.P. Genet, M.C. Caño de Andrade, V. Ratovelomanana-Vidal *Tetrahedron Lett.*, 36, 2663, (1995).

32. M.C. Caño de Andrade, V. Ratovelomanana-Vidal, unpublished results.

33. A. Girard, C. Greck, D. Ferroud, J.P. Genet *Tetrahedron Lett.*, 37, 7967, (1997).

34. Review : C. Greck, J.P. Genet *Synlett*, 741, (1997).

35. C. Greck, L. Bischoff, J.P. Genet *Tetrahedron : Asymmetry*, 6, 1989, (1995).

36. a) C. Greck, F. Ferreira, J.P. Genet *Tetrahedron Lett.*, 37, 203, (1996) ; b) F. Ferreira, C. Greck, J.P. Genet *Bull. Soc. Chim. Fr.*, 134, 615, (1997).

37. For some other developments of this chemistry with industrial perspectives see review : R. Schmidt, E.A. Broger, M. Cereghetti, Y. Crameri, J. Foricher, M. Lalonde, R.K. Müller, M. Scalone, G. Schoettlel, U. Zutter *Pure & Appl. Chem.*, 68, 1, 131 (1996).

ENANTIO AND DIASTEREOSELECTIVE ADDITION OF ORGANOMETALLIC REAGENTS TO ALDEHYDES AND IMINES

Pier Giorgio Cozzi, Emilio Tagliavini, and Achille Umani-Ronchi

Dipartimento di Chimica "G. Ciamician"
Università di Bologna
Via Selmi 2, 40126-Bologna, Italy

INTRODUCTION

Catalytic enantioselective addition of organometallic reagents to aldehydes and imines represents an important methodology for the preparation of optically and biologycally active compounds. In the last few years we have developed a general and reproducible methodology for the high enantioselective addition of allyltin reagents to aldehydes based on Lewis acid prepared *in situ* from optically active 1,1'-binaphtalene-2,2'-diol (BINOL) or bis(oxazolines) (Box) and metal salts. Moreover, we have recently exploited a new strategy for the reduction of prochiral ketones (Scheme **1**).

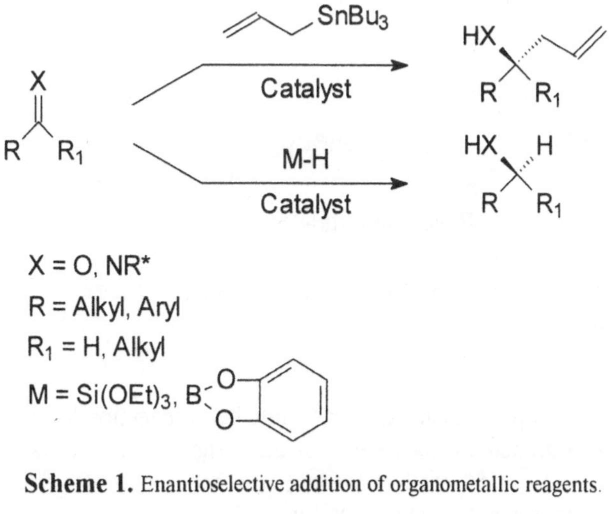
SnBu3
Catalyst
M-H
Catalyst
X
R R1
HX
HX H
X = O, NR*
R = Alkyl, Aryl
R1 = H, Alkyl
M = Si(OEt)3, B

Scheme 1. Enantioselective addition of organometallic reagents.

BINOL-TITANIUM LEWIS ACID CATALYSTS

The enantioselective formation of a new C-C bond can be accomplished by a facial discrimination in the Lewis acid-substrate complex . Futhermore, the coordination with the Lewis acid enhances the reactivity of the substrate.

In the last few years powerful and effective Lewis acids have been introduced to promote Diels Alder, hetero-Diels-Alder, Mukaiyama, ene, and other reactions. In the case of allylation reaction the addition of allylsilanes and allylstannanes to aldehydes promoted by catalytic amount of chiral boron Lewis acid were reported by Yamamoto[1] and Marshall.[2] A more effective catalytic system based on BINOL and $TiCl_2(OiPr)_2$ (catalyst **1**) was introduced by us.[3] Independently, Keck reported the use of a BINOL / $Ti(OiPr)_4$.[4] Recently, we have improved the preparation of our catalyst system, that can be simply obtained by treating the mixture of BINOL (1 equiv) and $TiCl_2(OiPr)_2$ (1 equiv.), with allyltributyltin (2 equiv), avoiding the use of activated molecular sieves (MS) (catalyst **2**, Scheme **2**).[5]

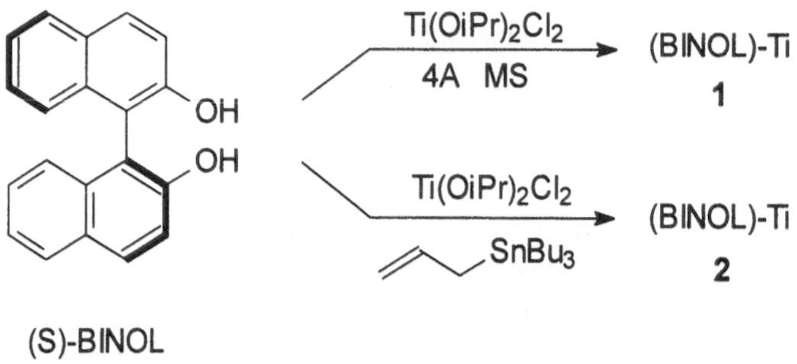

Scheme 2. Preparation of the BINOL-titanium catalysts (**1** and **2**).

The two catalysts were used for the enantioselective allylation of aldehydes with allyltributyltin (Scheme **3**) to afford homoallylic alcohols in good yields and excellent e.e.. Some selected results are reported in Table **1**.

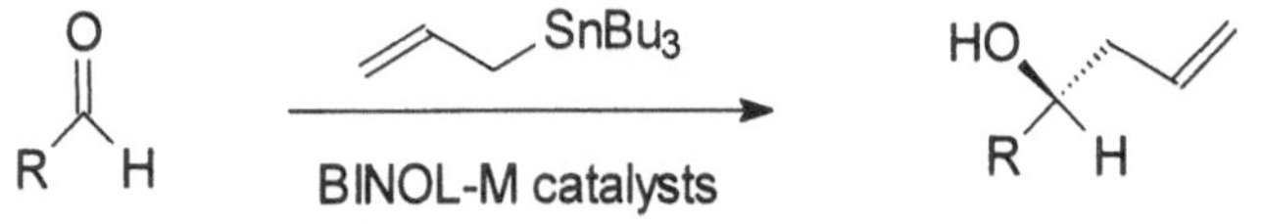

Scheme 3. Enantioselective addition of allyltributyltin to aldehydes catalyzed by BINOL-Metal complexes.

Although similar in performances, only slightly better enantioselectivity can be achieved with the catalyst **1**, a dramatic enhancement of the turnover number was obtained with the catalyst **2**. In fact, a 2 to 5% catalyst loading was enough to afford high level of facial discrimination also operating at room temperature.

Table 1. Enantioselective allylation of aldehydes with the titanium catalysts **1** and **2** [a]

R	catalyst,	%	T. °C	Yields % [b]	e.e %[c]
Ph	1	20	RT	96	82
n-C$_7$H$_{15}$	1	20	-20	75	98
(E)-PhCH=CH	1	20	RT	75	93
Ph	2	20	RT	74	86
Ph	2	5	0	74	74
n-C$_7$H$_{15}$	2	20	RT	80	94
n-C$_7$H$_{15}$	2	5	RT	80	94
n-C$_7$H$_{15}$	2	2	RT	50	89

a) The reactions were carried out in CH$_2$Cl$_2$ by addition of allyltributyltin to the catalyst prepared *in situ*
. followed by the aldehyde. b) yields of purified products. c) The e.e. was determined by chiral GC
analysis.

BINOL-ZIRCONIUM LEWIS ACID CATALYSTS

In serching for a more reactive catalyst we moved to a more oxophilic metal by adding
BINOL to Zr(OiPr)$_4$·iPrOH we obtained the catalyst **3** (Scheme **4**) showing an increased
reactivity with respect to the catalyst **1** (Scheme **3**, Table **2**).[6] Analougously to the
preparation of the catalyst **2**, the treatment of ZrCl$_4$(thf)$_2$ (1 equiv.) and BINOL (2 equiv.)
with allyltributyltin (4 equiv.) led to a new catalyst (catalyst **4**). The use of uncomplexed
ZrCl$_4$ afforded less reproducible results.[7]

ACTIVATION OF BINOL-ZIRCONIUM CATALYST BY CALIX[n]ARENES

Recently Mikami [8] has reported that certain BINOL-titanium catalysts are activated by
the addition of chiral phenols increasing the e.es of ene and Diels Alder reactions. On the
basis of these results we have tried to prepare new mixed ligand complexes using a
combination of optically pure BINOL and achiral ligand as activator. We have investigated a
series of macrocyclic phenols and we have found that calix[4]arene is the best activator of our
allylation reaction (Scheme **5**, catalyst **5**).[7] The remarkable aspect of this novel procedure is
the low amount of catalyst that can be used without decreasing of the enantioselectivity
(Table **3**).

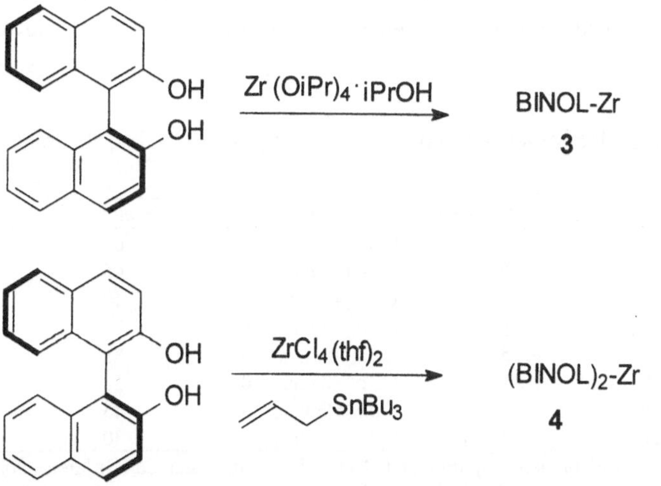

Scheme 4. Preparation of the BINOL-zirconium catalysts (**3** and **4**).

Table 2. Enantioselective allylation of aldehydes with the zirconium catalysts 3^a and 4^b

R	catalyst	solvent	T, °C	Yields %	e.e %
Ph	3	CH_2Cl_2	-40	79	93
n-C_7H_{15}	3	CH_2Cl_2	-20	61	88
(E)-PhCH=CH	3	CH_2Cl_2	-20	89	91
Ph^c	4	CH_2Cl_2	-20	64	79
Ph	4	Et_2O	0	74	74
n-C_7H_{15}	4	CH_2Cl_2	-20	78	73
n-C_7H_{15}	4	Et_2O	-20	65	85

a) 20% mol % of the catalyst 3 was used in the presence of 4 Å MS. b) The catalyst 4 was prepared by adding 10% mol BINOL and $ZrCl_4(thf)_2$, followed by the addition of 200% mol of allyltributyltin, in the solvent at room temperature. c) In this case was used uncomplexed $ZrCl_4$.

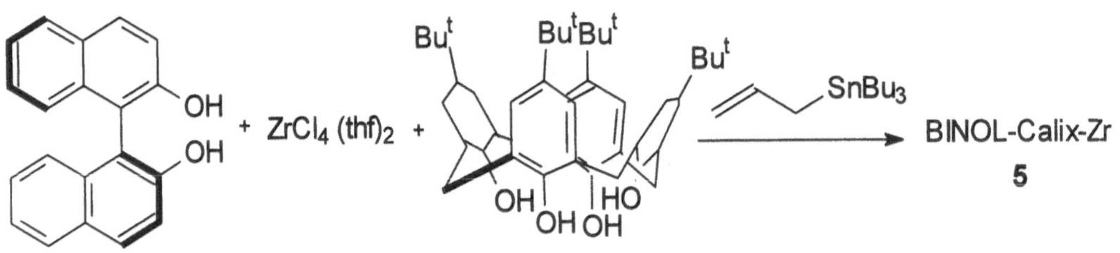

Scheme 5. Preparation of the BINOL-zirconium catalyst (5)

INVESTIGATIONS ON THE NATURE OF THE CHIRAL CATALYSTS

With the aim of improving the catalytic and stereoselective efficiency of our systems, we have undertaken a more systematic study on the various stages of the reaction process. The formation of the active catalyst has been investigated through NMR spectroscopy. For both the Ti and Zr derivatives, it clearly emergs that the rate of catalyst formation from the BINOL ligand is determined by the number of chlorine atoms in the $MCl_n(Oi\text{-}Pr)_{4-n}$ starting salts. Fast reactions occur for n = 0, while for n = 2 only little amount of BINOL-M complexes are formed within 2 h, and for n = 4 no complexation was observed in the reaction mixture also after long time. $TiCl_2(Oi\text{-}Pr)_2$ gives a very slow (CH_2Cl_2, RT, 15 d) or fast (toluene, reflux, 2h) conversion to a single symmetric species (cat. 6) as showed by single ^{13}C

Table 3. Enantioselective allylation of aldehydes with the BINOL-Calixarene-zirconium 5^a

R	$ZrCl_4(thf)_2$ (%)	Calix[n]arene, %	Yields %	e.e %
n-C_5H_{15}	10	Calix[4]arene, 10	68	93
n-C_5H_{15}	5	Calix[4]arene, 5	65	96
n-C_5H_{15}	2	Calix[4]arene, 1	40	92
Ph^b	6	Calix[4]arene, 6	78	78
Ph	5	Calix[4]arene, 5	85	85
(E)-PhCH=CH	10	Calix[4]arene, 10	38	77
(E)-PhCH=CH	4	Calix[6]arene, 3	43	62
(E)-PhCH=CH	4	Calix[8]arene, 3	30	70

a) The catalyst 5 was prepared by mixing BINOL, $ZrCl_4(thf)_2$ and Calix[n]arene, followed by the addition of allyltributyltin (200% mol) in Et_2O at room temperature for 1 hour. The aldehyde was added at -20°C and the reaction was carried out at the same temperature. The equiv of BINOL used were the same as $ZrCl_4(thf)_2$ b) The reaction was performed at 0 °C.

resonances (C-O-Ti)   for the complexed BINOL. The catalyst **6** gives inferior enantioselectivity then cat. **1** and **2**. However, the mixture of BINOL with $TiCl_2(Oi\text{-}Pr)_2$ or $ZrCl_4$, either in the presence or absence of MS, react with allyltributyltin by proton exchange to give propene, chlorotributyltin, and M-BINOL complexes, catalyst **2** and catalyst **4**, respectively. (Scheme **6**) All the active catalysts, with the exception of **6**, but including the Keck's catalyst, are mixtures of oligomeric species: four or more C-O-Ti BINOL resonances are recorded in the ^{13}C NMR spectra .

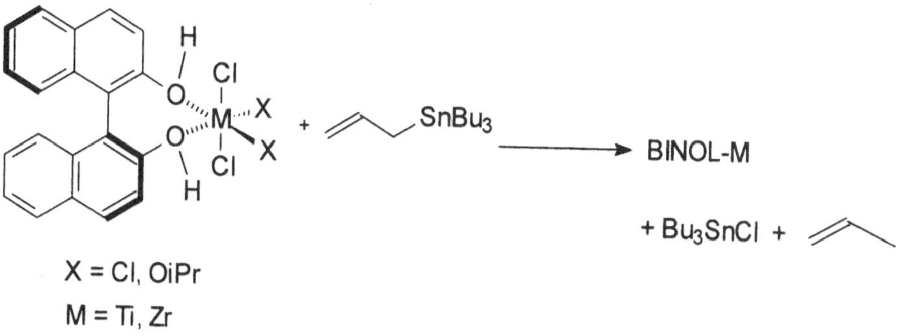

X = Cl, OiPr

M = Ti, Zr

Scheme 6. Formation of the BINOL-M complexes by deprotonation of BINOL determined by allyltributyltin

DIASTEREO- AND ENANTIOSELECTIVE ALLYLATION OF IMINES

The use of Lewis acids in diastereoselective allylation of imines was also investigated by our group. We discovered that using catalytic amount of Lewis acid allyltributyltin reacts with imines to afford homoallylic amines. In particular lanthanide triflates (Yb, La) or $Sc(OTf)_3$ give high diastereoselective addition to chiral imines derived from (S)-valine methyl ester (Scheme 7).[9]

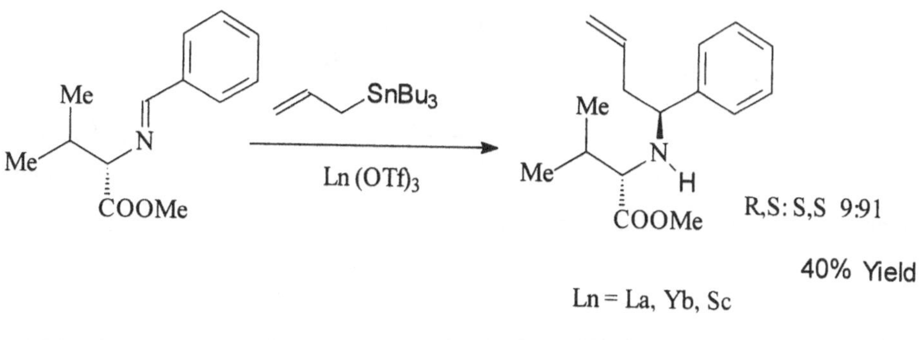

R,S:S,S 9:91

40% Yield

Ln = La, Yb, Sc

Scheme 7. Diastereoselective addition of allyltributyltin to chiral imine.

However the drawback of this methodology was the type of chiral auxiliary that was removed with a selective oxidative cleavage after the formation of the desired homoallylic amine. We are now exploring the possibility of using a chiral catalist for the enantioselective allylation of imines. The preliminary result obtained with the catalyst **4** is reported in the Scheme **8**.[10]

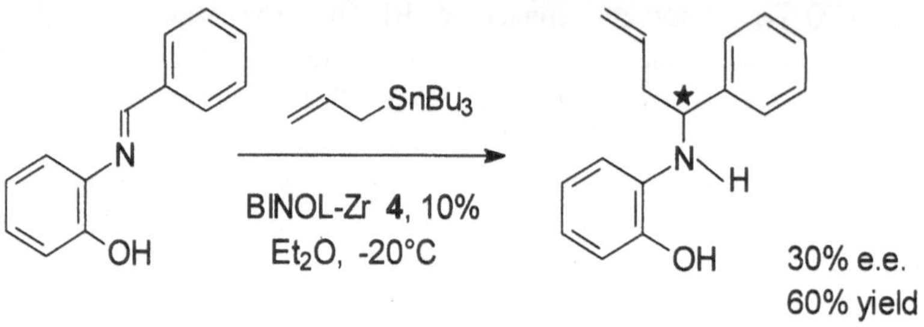

Scheme 8. Enantioselective addition of allyltributyltin to imine.

BIS(OXAZOLINE)-ZINC CHIRAL CATALYST

The application to asymmetric synthesis of chiral bis(oxazoline) metal complexes[11] as Lewis acid was developed to promote Diels Alder reactions, hetero Diels Alder reactions, [2+3] dipolar cycloadditions, addition of Me_3SiCN to aldehydes, Mukaiyama and Mukaiyama-Michael reactions. However, in almost all the cases, the enantioselectivity of bis(oxazoline) metal mediated reactions is high only if the electrophilic substrate is chelated to the metal center in a two-site interaction. In searching for other complexes able to promote the allylation reactions, we screened a large numbers of metals and ligands combination and found interesting catalytic properties of the bis(oxazoline)-zinc complexes 7, prepared *in situ* by reacting the bis(oxazoline) with $Zn(OTf)_2$ in CH_2Cl_2 at room temperature.[12] The catalyst can afford homoallylic alcohols from aldehydes and allyltributyltin in moderate e.e. (Scheme 10).

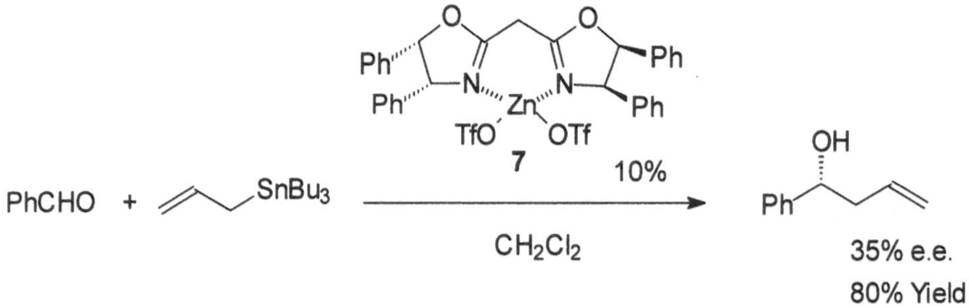

Scheme 9. Enantioselective allylation of aldehydes promoted by bis(oxazoline)-zinc complex.

CATALYTIC ENANTIOSELECTIVE REDUCTION OF KETONES PROMOTED BY BIS(OXAZOLINE)-TITANIUM COMPLEXES

Enantioselective reduction of ketones is an important and widely utilized methodology for the preparation of key intermediates for the synthesis of drugs and biologically active compounds. We have extended the use of chiral bis(oxazoline) ligands (Box) to prepare bis(oxazoline)-titanium complexes and to apply them in asymmetric catalysis. These complexes, treated with silyl hydrides, or catecholborane, have shown to generate active catalysts capable to reduce ketones and α-halogeno ketones, in good yield and excellent e.e. (Scheme 10).[13]

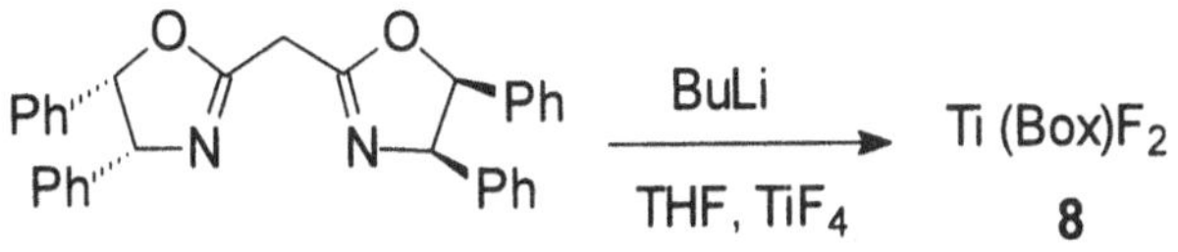

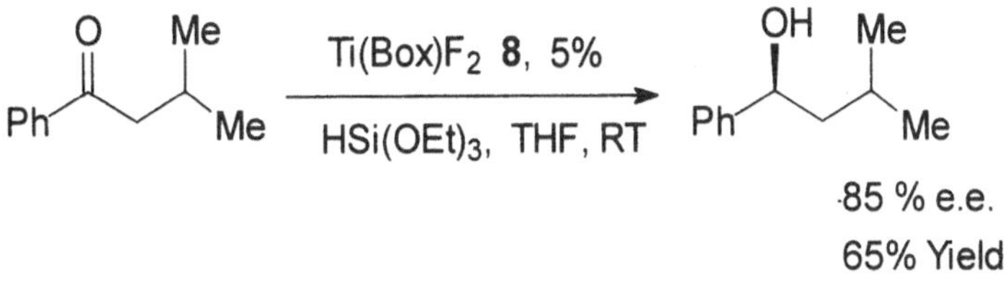

Scheme 10. Enantioselective reduction of ketones promoted by Box-Ti **8**.

Further application of titanium- and zirconium-Box complexes to asymmetric catalysis are under investigation in our laboratory.

ACKNOWLEDGEMENTS.

We thank the CNR (Rome), the University of Bologna (funds for selected research topics) and the M.U.R.S.T -Rome (National Project "Stereoselezione in Sintesi Organica. Metodologie ed Applicazioni") for financial support. We thank also the Ph.D. and undergraduate students involved in this research.

REFERENCES

1. K. Furuta, M. Mouri and H. Yamamoto, Chiral (acyloxy)borane catalyzed asymmetric allylation of aldehydes, *Synlett.* 561 (1991).
2. J.A. Marshall and Y. Tang, Catalyzed asymmetric $S_{E'}$ addition of allylstannanes to aldehydes, *Synlett*, 653 (1992).
3. A.L.Costa, M.G.Piazza, E. Tagliavini, C. Trombini and A. Umani-Ronchi, Catalytic asymmetric synthesis of homoallylic alcohols, *J. Am. Chem. Soc.* 115: 7001 (1993).
4. G.E. Keck, K.H.Tarbet and L.S. Geraci, Catalytic asymmetric allylation of aldehydes, *J. Am. Chem. Soc.* 115: 8467 (1993).
5. S. Casolari, P.G.; Cozzi, E. Tagliavini, A. Umani-Ronchi, unpublished results.
6. P. Bedeschi, S. Casolari, A.L. Costa, E. Tagliavini, A. Umani-Ronchi, Catalytic asymmetric synthesis promoted by a chiral zirconate: highly enantioselective allylation of aldehydes, *Tetrahedron Lett.* 36: 7897 (1995).
7. S. Casolari, P.G. Cozzi, P. Orioli, E. Tagliavini and A. Umani-Ronchi, Chiral-achiral ligand synergy: activation of a zirconium-BINOL Lewis acid complex by the addition of 4-*tert*-butylcalix[4]arene, *Chem. Commun.* 2123 (1997).
8. K. Mikami, S. Matsukawa, Asymmetric synthesis by enantiomer-selective activation of racemic catalysts, *Nature* 385: 613 (1997).
9. C. Bellucci, P.G. Cozzi and A. Umani-Ronchi, Catalytic allylation of imines promoted by lanthanide triflates, *Tetrahedron Lett.*, 36: 7289 (1995).
10. Bandini, M.; Cozzi, P.G.; Tagliavini, E.; Umani-Ronchi, A. unpublished results.
11. A.K. Gosh, P. Mathivanan, K. Cappiello, C_2-symmetric chiral bis(oxazoline)-metal complexes in catalytic asymmetric synthesis, *Tetrahedron:Asymmetry*, 7: 2165 (1996).

12. P.G. Cozzi, P. Orioli, E. Tagliavini and A. Umani-Ronchi, Enantioselective allylation of aldehydes promoted by chiral zinc bis(oxazoline) complexes, *Tetrahedron Lett.*, 38: 145 (1997).
13. M. Bandini, P.G. Cozzi, L. Negro, A. Umani-Ronchi, Enantioselective reduction with Triethoxysilane catalyzed by titanium complexes with C_2-chiral bis(oxazolines), submitted for publication.

SOME RECENT STUDIES ON THE DEVELOPMENT OF NEW METAL MEDIATED REACTIONS FOR ORGANIC SYNTHESIS

William B. Motherwell

Department of Chemistry
Christopher Ingold Laboratories
University College London
20 Gordon Street
London WC1H OAJ , UK

INTRODUCTION

The increasingly sophisticated array of reagents and methods which require a metal mediated transformation have revolutionised Organic Synthesis over the past 50 years, and the practitioner of this art now has an enormous "tool kit" from which he can make a judicious selection, based on his own particular requirements of chemo-, regio- and stereoselectivity. While it could therefore be argued that there is now little need for further additions to this armamentarium, the environmental and economic constraints imposed by the modern chemical industry provide a considerable challenge and stimulus for further improvement; particularly in terms of catalytic processes, atom efficiency and the ability to carry out controlled reactions leading to the creation of several new bonds in a single step.

The purpose of the present chapter is to illustrate this theme by describing our recent work on a new catalytic reaction for which we have coined the term "Cascade Polyhydroxylation of Arenes".

As is so often the case, the rich and wonderful world of Biological Chemistry and Natural Products provided both the inspiration and the challenge for our foray into this area. Thus, as encapsulated in Figure 1, the remarkable ability of the enzyme system present in strains of *Pseudomonas Putida* to achieve stereospecific vicinal *cis* dihydroxylation of benzene and related monosubstituted arenes[1] has provided synthetic organic chemists with a group of *cis* diene diols which have been extensively used as starting materials for the preparation of biologically important conduritol and inositol derivatives.[2,3,4] Accordingly, our own objective was to develop a chemical method, which, in contrast to reagents such as ruthenium tetroxide and ozone, would achieve controlled dihydroxylation of the aromatic ring without concomitant destruction of the carbocyclic ring.

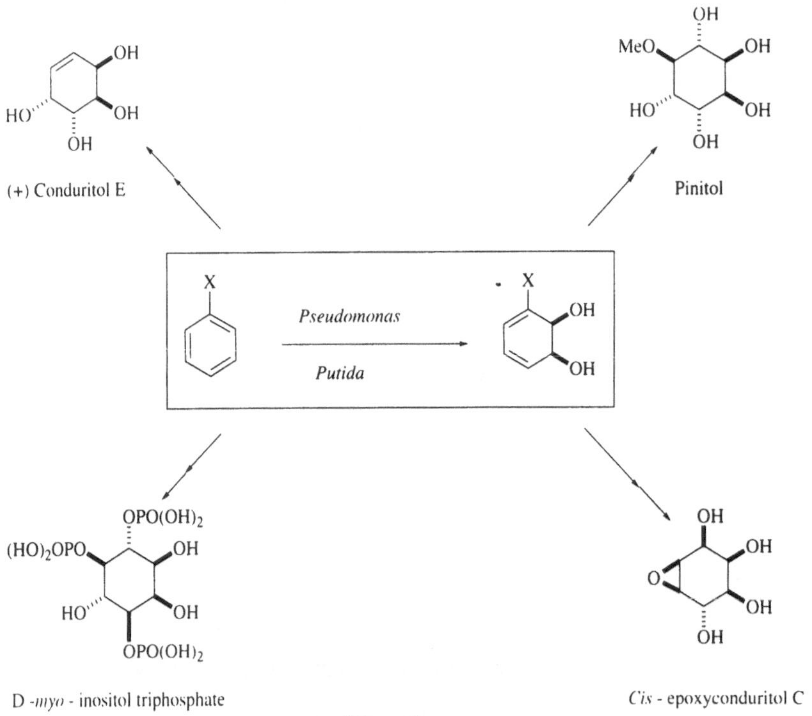

Figure 1.

RESULTS AND DISCUSSION

We were particularly attracted to an elegant series of studies by Kochi[5], who demonstrated that the Electron Donor Acceptor (EDA) complex **1** responsible for the yellow colouration produced when osmium tetroxide is dissolved in benzene, could lead, on actinic irradiation at a frequency equal to or greater than the absorption maximum, to charge transfer from **1**, with the resultant formation of a tight ion pair **2**. The fate of **2** was either to collapse to starting materials or to form the osmate ester of benzene diol **3**. Under the conditions studied by Kochi in the presence of an excess of osmium tetroxide, a second thermal *anti* osmylation occurred to give a polymeric material **4**, which was isolated as its pyridine adduct **5**.

The above study clearly established the principle of controlled dihydroxylation of benzene and related arenes, even although the yields based on stoichiometric osmium

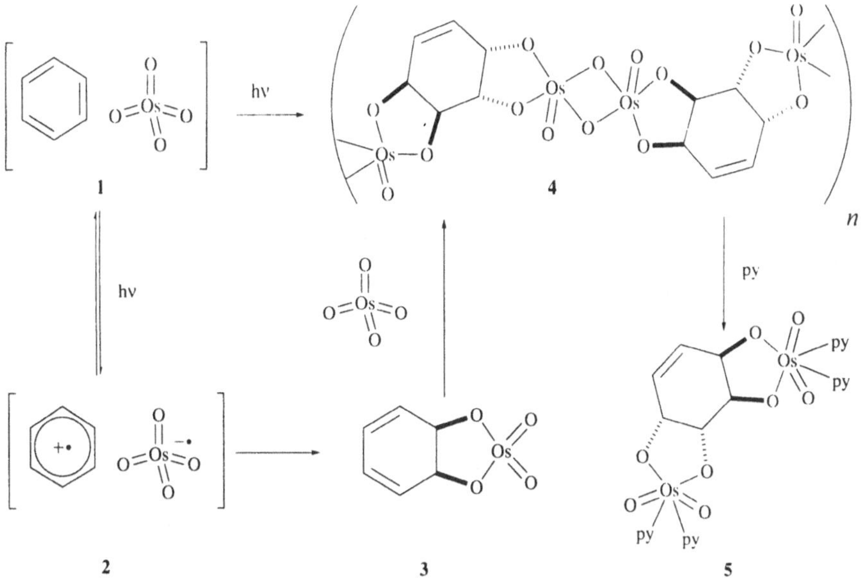

Figure 2.

tetroxide were disappointingly low. We therefore reasoned that the development of a *catalytic* photoinduced charge transfer osmylation sequence should be possible if a suitable solvent system and oxygen atom transfer reagent could be found. While recent debate in the literature has focussed on the mechanism of formation of osmate esters, it should also be recognised that the details of the oxidative hydrolysis step required to recycle an osmium catalyst also require clarification, particularly in terms of understanding such facets as the timing of the hydrolysis as a function of the substitution pattern around the double bond or the role of tetraalkylammonium acetates or the sodium salt of N-methylsulfonamide as catalysts to accelerate osmate ester hydrolysis. In our own case however, the fragility of the EDA complex **1**, whose heat of formation has been estimated[6] to be around 0.5kcal mol^{-1}, posed the most immediate problem, inasmuch as it was vital to avoid the use or generation of any molecule such as pyridine, N-methylmorpholine or even acetone or *tert*-butanol whose donor ability for osmium tetroxide would give a more stable complex than that with the arene.

In any event, one of the more traditional combinations;- the use of an aqueous solution of barium chlorate in a biphasic system which was introduced by Hoffman[7] in 1912, proved to provide a suitable environment which did not disrupt the necessary photoinduced charge transfer step. Our early experiments in this area would certainly have shocked a photochemical purist, since they involved vigorous stirring of the two phase system in a simple Pyrex florentine flask fitted with a Teflon tap with external irradiation from a 400W medium pressure Hg bulb.

The results of some preliminary experiments using benzene are shown in Figure 3 and the isolation of the *meso* derivative, *allo*-inositol hexaacetate **6**, confirmed that cascade polyhydroxylation was indeed a viable catalytic pathway. It was also of interest to note that

249

at lower operating temperatures, the tertiary osmylation sequence for the third double bond was sufficiently slow to permit isolation of significant amounts of conduritol E tetraacetate **7** which is derived from the major but less reactive *anti* isomer at the tetrol stage. The all *syn* tetrol was not detected, but of course would pose no steric problem for a final tertiary osmylation step leading once again to *allo* inositol.

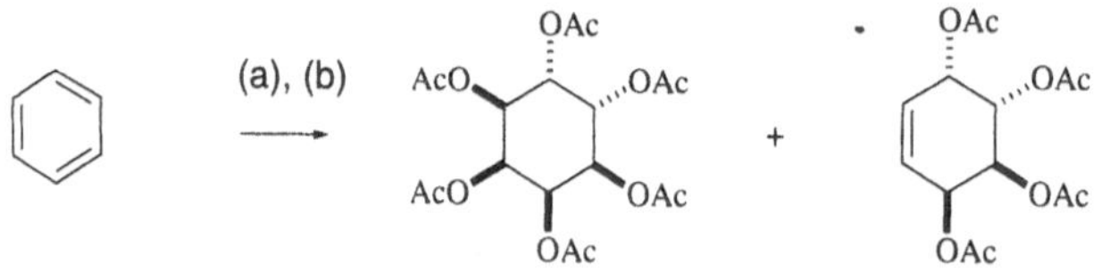

Reagents. a) OsO$_4$ (cat.). hv, Ba(ClO$_3$)$_2$ (0.22M); b) Ac$_2$O, Et$_3$N, DMAP;
6 : 7 (6.2:1) 36%.

Figure 3.

We then elected to carry out a preliminary screening of some simple monosubstituted arenes. Toluene proved to be a substrate of particular stereochemical interest, as shown by the results in Figure 4 , which reveal that the major reaction pathway proceeds *via* the C-methyl conduritol E isomer **8** to the isomeric mixture of C-methyl inositols **9** and **10**, with the third isomer **11** being a minor constituent of this mixture.

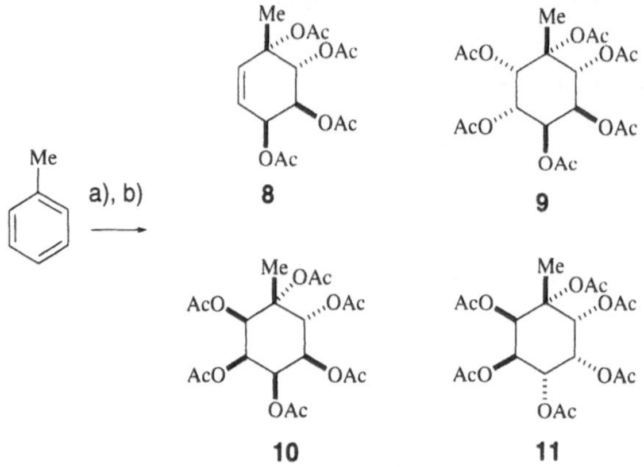

Reagents. a) OsO$_4$ (cat.), hv, Ba(ClO$_3$)$_2$ (0.22M); b) Ac$_2$O, Et$_3$N, DMAP; **8** 2.2%, **9 : 10 : 11** (2 : 1 : 1) 10.8%

Figure 4.

Since the initial dihydroxylation of toluene by *Pseudomonas Putida* occurs at the 2,3 position, the present method clearly provides a complementary pathway to "unnatural" C-methyl inositol derivatives, with the major initial photoosmylation step occuring either at the 1,2 or 3,4 positions. It was also of great interest that only trace amounts of benzaldehyde were formed in this reaction, particularly since the Etard oxidation[8] of toluene using chromyl chloride yields only benzylic oxidation products and also involves a metal mediated electron transfer step. The nature and subsequent reactivity of the ion pairs formed in these two reactions are significantly different, although both formally involve the same radical cation, and this aspect is certainly worthy of further study.

Our attention then turned to the halogenated arenes, which on the basis of their higher ionisation potentials alone might well be expected to be more recalcitrant substrates. Indeed, under our standard screening conditions, catalytic "turnover" is practically non existent. The nature and relative ratios of the diastereoisomeric products formed, however, was intriguing(Figure 5).

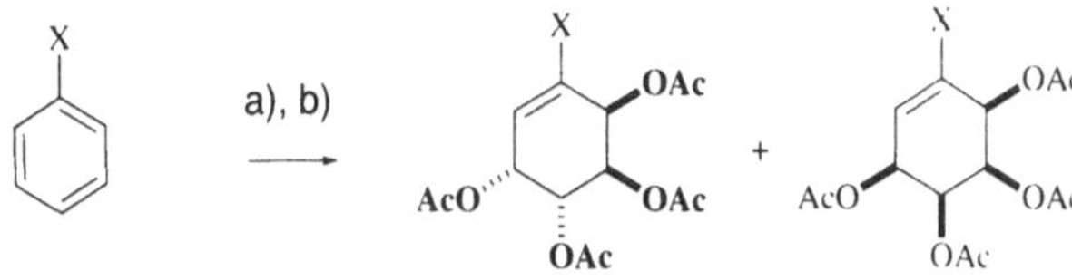

12	X = F	12a	4.4 : 1	12b
13	X = Cl	13a	5.0 : 1	13b
14	X = Br	14a	5.5 : 1	14b

Reagents. a) OsO_4 (cat.), hv, $Ba(Cl)_3)_2$(0.22M); b) Ac_2O, Et_3N, DMAP ; 12 0.6%; 13 3.4%; 14 2.6%.

Figure 5.

Thus, no inositol derivatives (or their derived penta-hydroxy cyclohexanones) were detected in the reactions, indicating that the remaining vinylic halide in the conduritol products 12, 13 and 14 was resistant to further osmylation. The stereochemical preference for the formation of *anti*-tetrols in these reactions is also much higher than in the osmium tetroxide catalysed dihydroxylation of the 2,3 diol, derived from microbial oxidation of bromobenzene by *Pseudomonas Putida*,[9,3c] which displays almost no stereochemical preference (Ratio 14a : 14b is 1.1 : 1.0). These observations may be explained by saying that secondary osmylation is much faster than oxidative hydrolysis of the initial diene osmate ester photoproduct, even when only a catalytic amount of osmium tetroxide is used, irrespective of whether the initial photoosmylation step occurs at the 2,3(ortho-meta) or 3,4(meta-para) positions.

While our studies of the monosubstituted arenes were in hand, I had also asked my postgraduate colleague Mr. Alvin Williams to convince me of the practical utility of the reaction which he was developing, by producing at least one gram of *allo*-inositol hexaacetate from benzene. It was at this stage that Mr. Williams betrayed his own Swiss background as well as the fact that he was acutely aware of his Scottish supervisor's cost concious personality. Thus, rather than ask the glassblower to make a larger florentine flask and carry out a simple scale up, he decided to carry out an experiment using his usual apparatus but at a much higher chlorate anion concentration (1.10M) and a reduced water : arene ratio (1 : 1.5). To our mutual delight, this process produced 1.12g of *allo*-inositol hexaacetate **6** from 15ml of benzene with a catalytic turnover number of 108 for osmium tetroxide. We were even more excited however by the fact that this reaction also yielded a significant proportion (0.6g) of the three deoxychloroinositols **15, 16** and **17** shown in Figure 6.

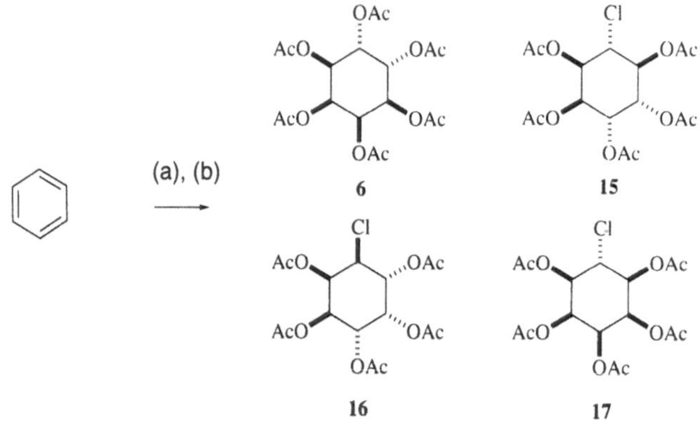

Reagents. a) OsO_4 (cat.). hv, $Ba(ClO_3)_2$ (1.10M); b) Ac_2O, Et_3N, DMAP;
15 : 16 : 17 (2:2:1) .

Figure 6.

From this chance observation, we immediately recognised that selection of the bromate anion as the oxygen atom transfer reagent should favour the formation of the corresponding deoxybromoinositols to an even greater extent, simply by virtue of the weaker bromine-oxygen bond. In the event, this proved to be the case, and, as shown in Figure 7 the stereoselective monobromopentahydroxylation of benzene proved to be a preparatively useful reaction with effective suppression of the tertiary osmylation sequence leading to *allo*-inositol hexaacetate **6** and routine isolation of the two deoxybromoinositol **18** and **19** in a combined yield of 18.8%.

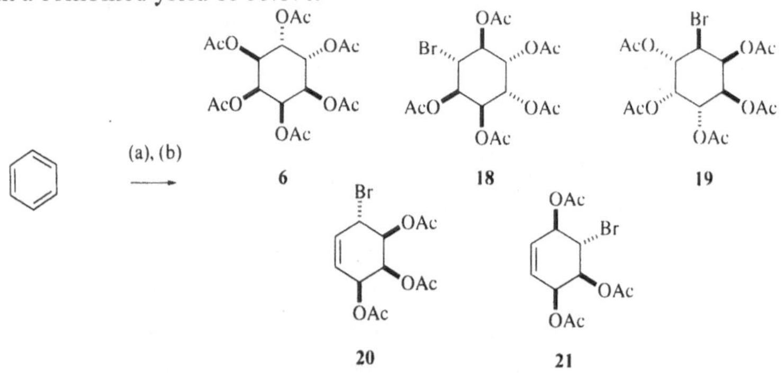

Reagents. (a) OsO_4 (cat.), hv, $NaBrO_3$(0.22M) (b) AcOH, Ac_2O. **6** 7.7%; **18 : 19** (5.1 : 1), 18.8%; **20: 21** (1 : 5), 2.3%

Figure 7

Although a casual inspection of the isolated products clearly suggests that they may be formally derived by the *trans* addition of hypobromous acid at some intermediate stage after the initial photoosmylation step, we were particularly curious to try and understand the relative timing of the incorporation of the bromine atom and the remaining hydroxyl groups. The isolation of bromoconduritols C **20** and F **21** as minor by-products indicated that addition at the diene stage was a possibility, and this in turn was indicative of the presence of hypobromous acid during the reaction. A systematic study of the influence of temperature on the reaction also provided a valuable mechanistic clue, by revealing that it was possible to control the relative ratios of the two diastereoisomers **18** and **19** in a highly significant way. (Table and Figure 8).

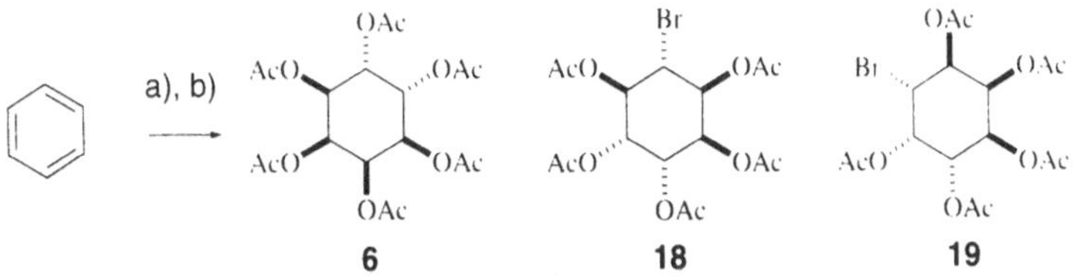

Temperature T/^{0}C	Products (Yield %)		Relative ratio 18 : 19
	6	18 + 19	
2	5.1	12.6	2 : 1
15	7.4	16.3	1 : 1
30	7.7	18.8	1 : 5
45	5.7	16.4	1 : 5

Reagents. a) OsO_4 (cat. 1.3 mol%), hv, $NaBrO_3$(0.36M), b) AcOH, Ac_2O

Figure 8

We therefore believe, as shown in Figure 9, that a bifurcated reaction pathway is operating, in which addition at the diene stage becomes progressively more important with increase in temperature, while addition of hypobromous acid to conduritol E is favoured at lower temperatures and occurs in fact during the reductive metabisulfite work up. This suggestion is supported by the fact that addition of a vast excess of cyclohexene as a sacrifical olefin following a photoosmylation at 2^0C and prior to the addition of metabisulfite led to the isolation of conduritol E as its tetraacetate (8.6%) and a concomitant reduction in the isolated yield of the pentaacetoxy bromides **18** and **19** (combined yield 6.2%).

From a purely practical standpoint, the above study demonstrated that the monobromopentahydroxylation of benzene could be considered as a viable sequence and that a useful degree of control could be exercised over the production of either the *chiro-*

diastereoisomer **18** or the *neo*-derivative **19**.The ready availability of such usefully fuctionalised building blocks in a one-pot operation naturally proved to be an irresistable temptation for synthetic studies in the cyclitol area.

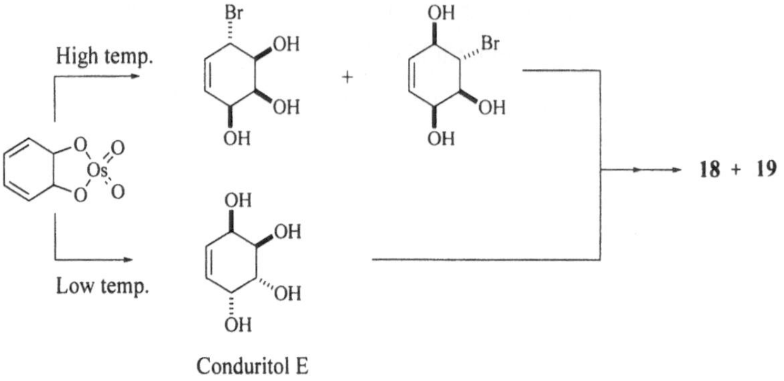

Figure 9.

(±)Pinitol **24**, the 3-O-methyl ether of *chiro*-inositol was selected as a suitable target. Both enantiomers have been found in various plant sources and the dextrorotatory form has been shown to posses significant hypoglycemic and antidiabetic activity in diabetic albino mice[10]. The *cis*- dienediols available by microbial oxidation of arenes using *Pseudomonas Putida*, have of course, featured prominently in previous syntheses either of the racemic form, or of either antipode.[2a,11,12,13]

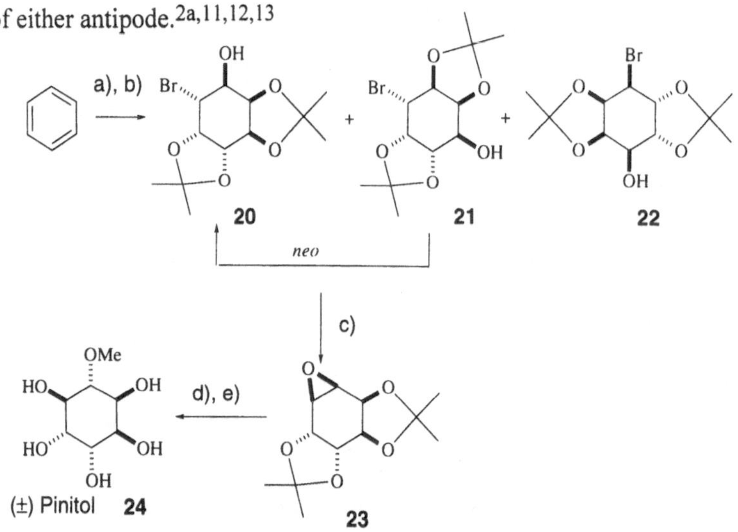

Reagents. a) OsO$_4$ (cat.), hv, NaBrO$_3$(0.22M), 15^{0}C ; b) Ac$_2$O,
TsOH, **20** 8%, **21** 6.7% , **22** 2.5%; c) K$_2$CO$_3$, MeOH, heat 2h, 83%;
d) Al$_2$O$_3$, MeOH, neat, 24h; e) H$_2$O, THF HCl(cat.), 60% from **23**.
Figure 10

The overall "synthetic" sequence, which requires only three operations. and five discrete steps from benzene, is set out in Figure 10 and begins with a bromate driven photoosmylation of benzene at 15^{0}C followed by isolation of the products as their

254

diisopropylidene derivatives **20, 21** and **22** by simple treatment of the crude reaction mixture with acetone and *para*-toluenesulfonic acid. Under these conditions, the required *neo* diastereoisomers **20** and **21** are favoured and can be isolated in 14.7% yield. The unwanted regioisomer **21**, can of course be recycled to provide additional quanities of **20**. Treatment of the *trans* bromohydrin **20** with potassium carbonate in methanol proceeded "smoothly and in high yield"[14] to furnish the epoxide **23**, which was then treated with alumina in methanol followed by hydrolytic work up to achieve *in situ* deprotection of the isopropylidene groups and afford racemic pinitol **24** in 60% isolated yield from **23**. The regiospecificity of the key epoxide ring opening is determined by the necessity for a *trans* opening *anti* to the isopropylidene groups and is precedented in Hudlicky's elegant enantiodivergent strategy for the synthesis of both enantiomers.[13]

CONCLUSIONS

The foregoing tale on our introductory steps[15,16] to discover catalytic photoinduced charge-transfer osmylation of arenes has hopefully indicated, once again, the power of metal-mediated reactions to achieve useful transformations for Organic Synthesis. In the present instance, for any single substrate, given the number of kinetically discrete steps, both thermal and photochemical, which involve both the formation and oxidative hydrolysis of osmate esters, it is somewhat surprising in retrospect, that some measure of regio- and stereocontrol over product formation can be achieved. Our belief, as always however, is that new reactions should be simple in concept, even if complex in mechanistic detail.

ACKNOWLEDGMENTS

First and foremost, it has been my pleasure and privilge to be associated with Mr. (now Dr.) Alvin Williams and Dr. Pierre Jung who have initiated our studies in this area. I owe both of them an enormous debt of gratitude for their insight and dedication. Finally, I wish to acknowledge British Petroleum for the provision of a studentship (to ASW) and the EPSRC for the award of a postdoctoral fellowship (to PMJJ).

REFERENCES

1. D.T.Gibson, J.R.Koch and R.E.Kallio, Oxidative degradation of aromatic hydrocarbons by microorganisms.I. Enzymatic formation of catecol from benzene, *Biochemistry*, 7:2653(1968).

2. a).S.V.Ley,F.Sternfeld and S.Taylor, Microbal oxidation in synthesis: a six step preparation of (±) pinitol from benzene,*Tetrahedron lett.* 28:225(1987); b). S.V.Ley and F.Sternfeld, Microbal oxidation in synthesis: preparation from benzene of the cellular secondary messenger *myo*- inositol-1,4,5-triphosphate (IP3) and related derivatives, *ibid*,29:5305(1988); c) S.V.Ley and A.J.Redgrave, Microbal oxidation in synthesis: concise preparation of (+)-conduritol F from benzene, *Synlett* 393(1990).

3. a). H.A.J.Carless and O.Z.Oak, Short synthesis of conduritols A and D, and dehydroconduritols, from benzene: the photo-oxidation of *cis*- cyclohexa-3,5-diene-1,2-diol,*Tetrahedron lett.* 30:1719(1989); b) H.A.J.Carless and K. Busia, Total synthesis of *chiro*- inositol-2,3,5-triphosphate: a *myo*- inositol-1,4,5-triphosphate analogue from benzene *via* photo-oxidation, *ibid,* 31:1617(1990); c). H.A.J.Carless, K.Busia, Y.Dove and S.S.Malik, Syntheses of conduritol-d derivatives from aromatic-compounds. *J. Chem. Soc. Perkin Trans. 1,* 2505(1993).

4. a). T.Hudlicky, M.Mandel, J.Rouden, R.S.Lee, B.Bachmann, T.Dudding, K.J.Yost and J.S.Merola, Microbal oxidation of aromatics in enantiocontrolled synthesis 1. Expedient and general asymmetric synthesis of inositols and carbohydrates *via* an unusual oxidation of a polarized diene with potassium permanganate, *J. Chem. Soc. Perkin Trans.1,*1553(1994); b). T.Hudlicky, J.Rouden, H.Luna and S.Allen, Microbal oxidation of aromatics in enantiocontrolled synthesis 2. Rational design of aza sugars (*endo* -nitrogenous). Total synthesis of (+)-kifunensine, mannojirimycin, and other glycosidase inhibitors, *J. Am. Chem. Soc.*116:5099(1994); c). T.Hudlicky, H.F.Olivo and B.McKibben, Microbial oxidation of aromatics in enanticontrolled synthesis 3. Design of amino cyclitols (*exo* -nitrogenous) and total synthesis of (+)-lycoricidine *via* acylnitrosyl cycloaddition to polarized 1-halo-1,3-cyclohexadienes, *ibid,* 116:5108(1994).

5. J.M.Wallis and J.K.Kochi, Direct osmylation of benzenoid hydrocarbons. Charge-transfer photochemistry of osmium tetraoxide, *J. Org. Chem.* 53:1679(1988); Electron-transfer activation in the thermal and photochemical osmylations of aromatic EDA complexes with osmium(VIII) tetroxide, *J. Am. Chem. Soc.*110:8207(1988).

6. P.R.Hammond and R.R.Lake, Electron acceptor- electron donor interactions Part XXII. Charge-transfer interactions of some of the highest-valency halides, oxyhalides, and oxides with aromatic hydrocarbons and fluorocarbons. Ball-plane interactions. Some remaining elements, *J. Chem. Soc. (A)* 3819(1971).

7. K.A.Hoffman, Sauerstoff-ubertragung durch osmiumtetroxyd und aktivierung von chlorat losungen. *Chem. Ber. Dtsh.Chem.Ges.*45:3329(1912).

8. W.H.Hartford and M.Darin, The chemistry of chromyl compounds, *Chem. Rev.*58:1(1958).

9. For an analysis of this selectivity see H.C.Kolb, M.S.VanNieuwenhze and K.B.Sharpless, Catalytic asymmetric dihydroxylation, *Chem. Rev.* 94:2483(1994).

10. C.R.Narayanan, D.D.Joshi, A.M.Mujumdar and V.V.Dhekne, Pinitol-a new anti-diabetic compound from the leaves of *Bougainvillea-spectabilis*, *Curr. Sci.* 56:139(1987).

11. H.A.J.Carless, J.R.Billinge and O.Z.Oak, Photochemical routes from arenes to inositol intermediates: the photo-oxidation of substituted *cis*- cyclohexa-3,5-diene-1,2-diols, *Tetrahedron Lett.*30:3113(1989).

12. S.V.Ley, F. Sternfeld and S.Taylor, Microbial oxidation in synthesis: preparation of (+) - and -(-) pinitol from benzene, *Tetrahedron* 45:3463(1989).

13. T.Hudlicky, J.D.Price, F.Rulin and T.Tsunoda, Efficient and enantiodivergent synthesis of (+) - and - (-) pinitol, *J. Am. Chem. Soc.* 112:9439(1990).

14. R.A.Raphael, *ipse dixit.*

15. W.B.Motherwell and A.Williams, Catalytic photoinduced charge-transfer osmylation: a novel pathway from arenes to cyclitol derivatives, *Angew. Chem. Int. Ed.* 34:2031(1995).

16. P.M.J.Jung, W.B.Motherwell and A.S.Williams, Stereochemical observations on the bromate induced monobromopentahydroxylation of benzene by catalytic photoinduced charge transfer osmylation. A concise synthesis of (±)-pinitol, *J. Chem. Soc. Chem. Commun.* 1283(1997).

STEREOSELECTIVE ROUTES TO CONJUGATED POLYENES

Francesco Naso

Centro CNR di Studio sulle Metodologie Innovative di Sintesi Organiche,
Dipartimento di Chimica, Università degli Studi di Bari,
Via Amendola 173
70126 Bari, Italy

INTRODUCTION

Our work on sequential cross-coupling reactions has shown that 1-bromo-2-phenylthioethenes **1** and **2** represent very useful building blocks for the construction of stereodefined 1,2-disubstituted ethenes.[1] The methodology is based upon the selective substitution of the halogen atom in the first step by reaction with Grignard reagents in the presence of Ni or Pd complexes followed by the substitution of the phenylthio group. Good yields are obtained. The stereospecificity is higher than or equal to 99% in the case of E-isomers and slightly lower in the case of Z-isomers.

The process, which is described in Figure 1, represents a conceptually simple and experimentally easy route to olefins.

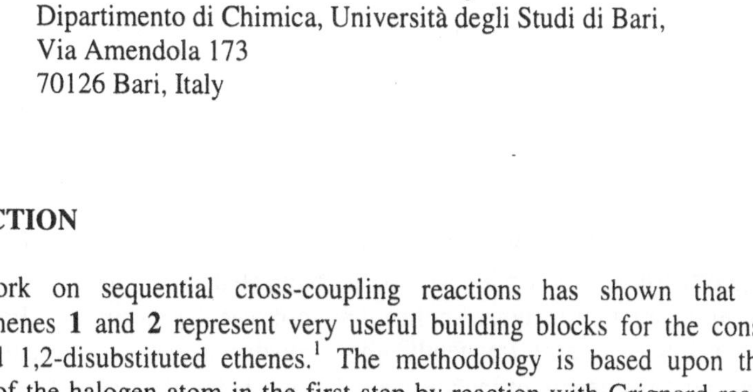

R_1, R_2 = alkyl, aryl, vinyl, alkynyl

Figure 1. Sequential cross-coupling reactions of Grignard reagents with 1-bromo-2-phenylthioethenes

Dienes can be also obtained provided that a vinyl Grignard reagent is used. However, due to the importance of this type of compounds, we developed additional stereoselective routes.[2,3,4]

Actually, one of these routes represents a variation of the scheme reported in Figure 1. In fact, an elimination leading to phenylthioacetylene is performed starting from **2**. The elimination is followed by an addition of cuprates and the final products, Z,E-dienes, are easily obtained in the cross-coupling step.[2]

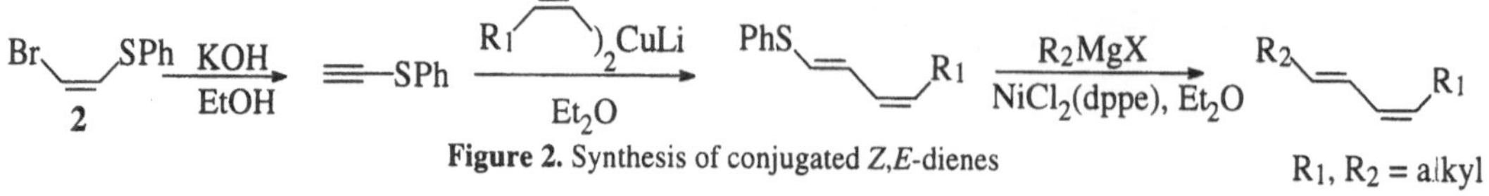

Figure 2. Synthesis of conjugated Z,E-dienes

Another methodology involves the elongation of the chain of the starting building block from one to two double bonds. The dienic counterpart of **1**, *i. e.* compound **3**, is then easily transformed into a series of E,E-dienes according to the usual sequential cross-coupling strategy, as shown in Figure 3.[3]

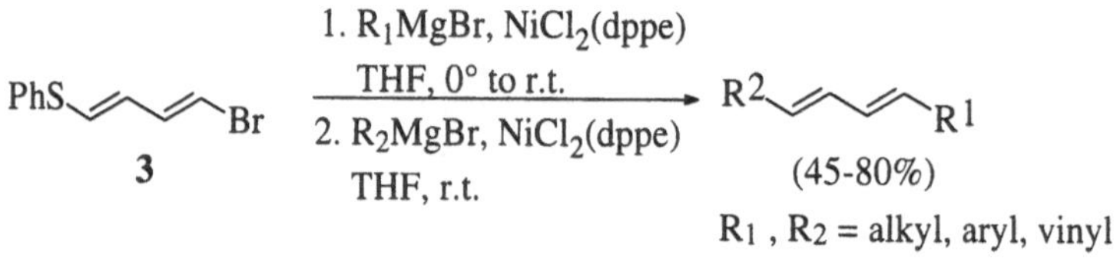

Figure 3. Sequential cross-coupling reactions between (1E,3E)-1-bromo-4-phenylthio-1,3-butadiene and Grignard reagents

Conjugated polyenes are well known for their importance in the field of natural compounds. Furthermore, molecules with long conjugated π systems are envisaged as the most promising materials for future applications in molecular electronics. For this reason, the attention is focused most frequently upon polydisperse macromolecules. However, polyenes with a defined length of the π system, besides being interesting by themselves, are also taken as discrete models of polymeric conjugated materials. This background has led us to strengthen our efforts in the synthesis of π conjugated systems and, with this aim, several approaches were introduced in our work.

SYNTHESIS OF POLYENES FROM UNSATURATED SILYL DERIVATIVES

Two novel building blocks, 1,4-bis(trimethylsilyl)-1,3-butadiene **4**[5] and 1,6-bis(trimethylsilyl)-1,3,5-hexatriene **5**,[6] were prepared as described in Figure 4. It is worth noting that for the synthesis of the triene **5**, 1-chloro-2-phenylthioethene **6** was used as a starting material.

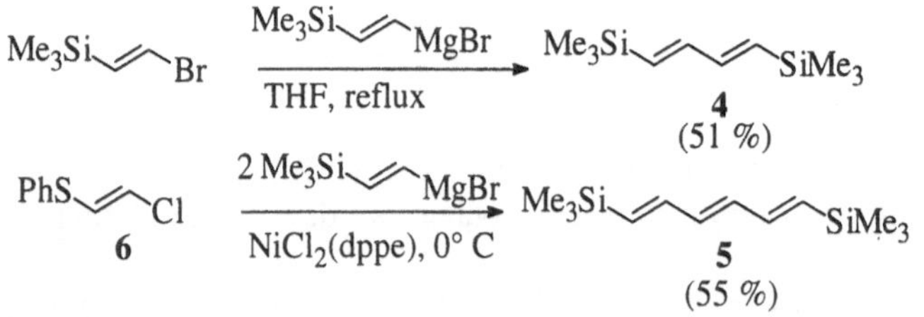

Figure 4. Synthesis of bis(trimethylsilyl)-1,3-butadiene **4** and 1,3,5-hexatriene **5**

When reacted with electrophiles, *e. g.* acyl chlorides in the presence of a Lewis acid, the new building blocks **4** and **5** showed a chemoselective behaviour. In fact, it was possible to replace only one of the two silyl groups in a first reaction, whereas the substitution of the second silyl group occurred in a slower step[7] (Figure 5).

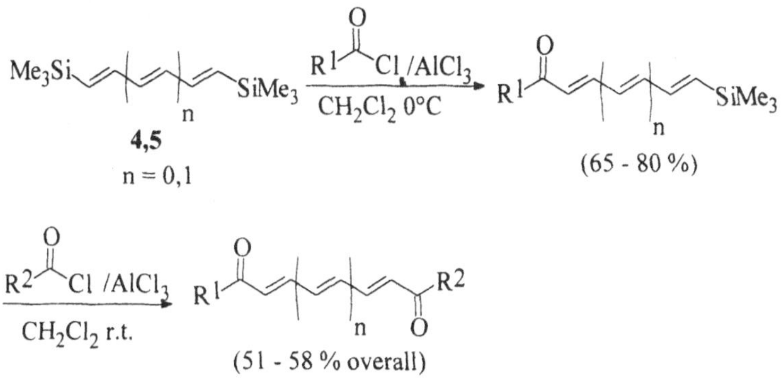

$R^1 = Ph, MeO_2C(CH_2)_3, MeO_2C(CH_2)_8, n\text{-}C_3H_7, n\text{-}C_7H_{15}$

$R^2 = CH_3, n\text{-}C_3H_7, n\text{-}C_7H_{15}, Ph, PhCH_2$

Figure 5. Synthesis of diketo-dienes and trienes

It is worth noting that the change from the halogeno-phenylthio building blocks to the bis-silylated counterparts permits a useful switch of the strategy from a "nucleophilic" to an "electrophilic" coupling process.

The chemoselectivity of the electrophilic acylation was the key feature for the elaboration of straigthforward routes to several types of compounds such as ostopanic acid 7, a cytotoxic E,E fatty acid,[8] the tetraene 8,[9] methyl ester of β-parinaric acid, which is used as a fluorescent probe for biological membranes, and 6E-LTB₃ 9.[8] Ostopanic acid was obtained in a straigthforward manner by sequential acylation reactions of the diene 4 with two acyl halides. In a similar manner, a diketone was prepared for the synthesis of β-parinaric acid methyl ester. Reduction of the diketone and dehydration gave the tetraenic ester 8. Two consecutive acylation reactions of the building block 5 led to a trienic diketone which was reduced to give (6E)-LTB₃ 9.

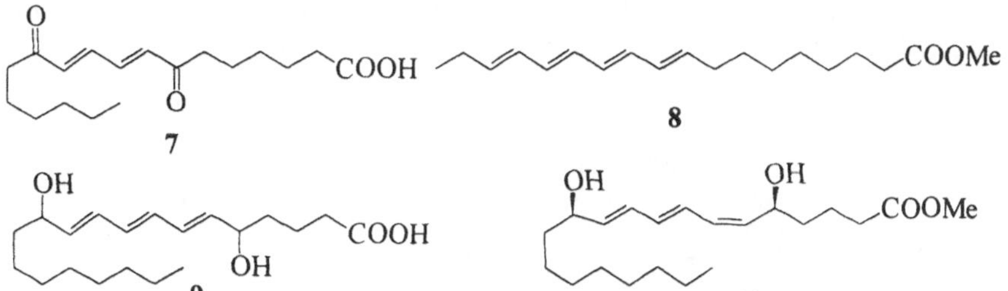

Figure 6. Ostopanic acid 7, β–parinaric acid methyl ester 8, (6E)-LTB₃ 9, and LTB₁ methyl ester 10

Chiral non racemic products were also prepared. Although we were able to obtain hydroxypolyenes by enzymatic kinetic resolution,[10] for the synthesis of the optically active LTB₃ methyl ester 10, we preferred to use a different strategy[11]. The strategy, which still involves the chemoselective electrophilic substitution on unsaturated bis-silylated compounds 11 and 12, is described in Figure 7.

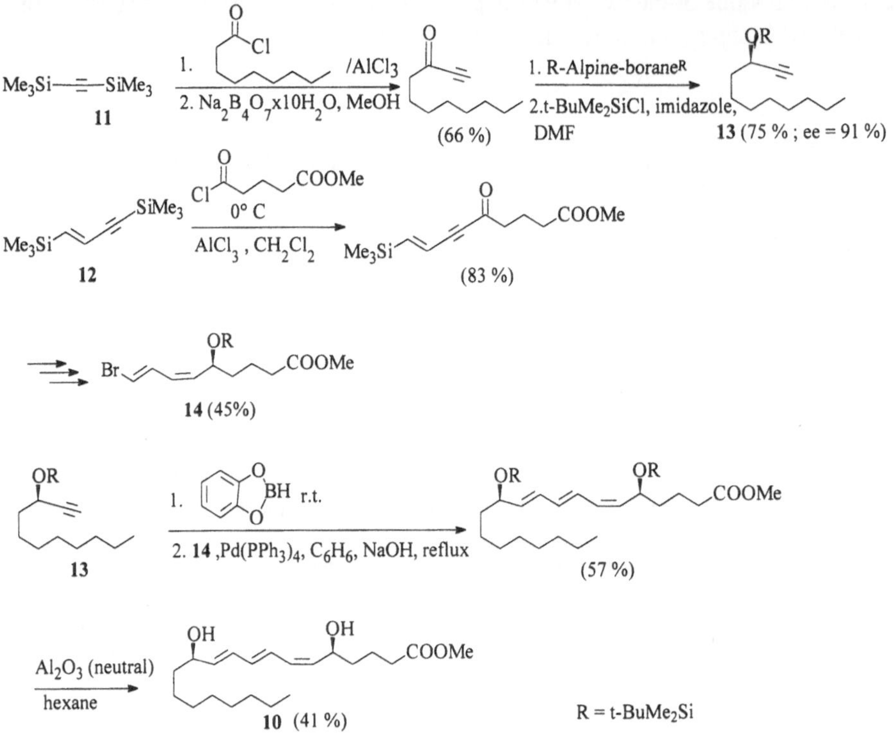

Figure 7. Synthesis of LTB$_3$ methyl ester **10**

Two unsaturated protected optically active alcohols, **13** and **14** were prepared. Transformation of **13** into a vinylic boron derivative, followed by a Suzuki-Miyaura cross-coupling process and deprotection, led to LTB$_3$ methyl ester **10**.

THE SELF-COUPLING OF MONOSILYLATED POLYENES

An interesting possibility of increasing the number of the conjugated double bonds was offered by the monosilylated polyenes. In fact, these compounds were easily subjected to a self-coupling reaction promoted by a PdCl$_2$-CuCl$_2$-LiCl system. Polyenes with four, six or eight double bonds were obtained[12], according to Figure 8.

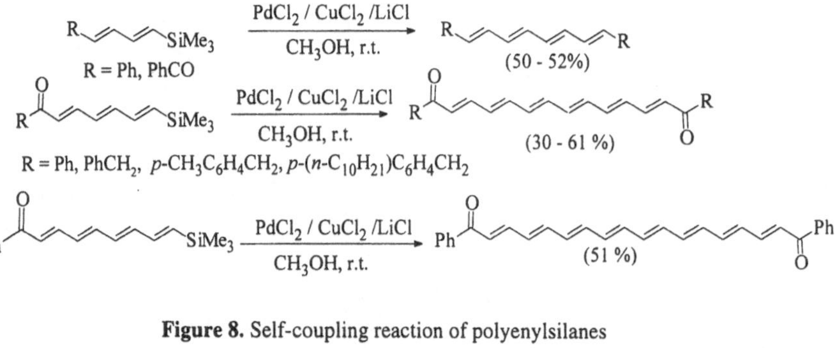

Figure 8. Self-coupling reaction of polyenylsilanes

A change of the solvent from methanol to acetonitrile was found to switch the reaction from the self-coupling process to a halogeno-desilylation.

A FORMAL SUZUKI-MIYAURA CROSS-COUPLING PROCESS STARTING WITH VINYLSILANES

The versatility of the polyenylsilanes in the construction of longer chains was expanded by our results concerning the possibility of using alkenylsilanes as starting materials in a formal Suzuki-Miyaura cross-coupling reaction. The unsaturated silanes were transformed *in situ* in boron derivatives which were subsequently coupled with halides in the presence of Pd catalysts (Figure 9).[13]

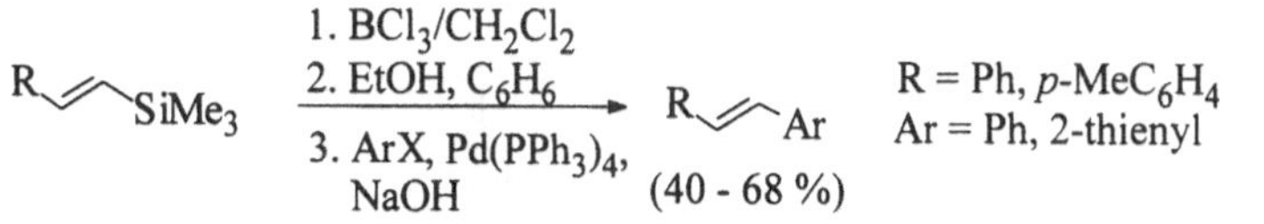

Figure 9. Formal Suzuki-Miyaura cross-coupling reaction starting with vinylsilanes

The formal Suzuki-Miyaura cross-coupling process described for alkenylsilanes was also performed starting from bis(trimethylsilyl)diene **4** or triene **5** and, owing to the chemoselectivity of the silicon-boron substitution, various polyenylsilanes were prepared as shown in Figure 10.[14]

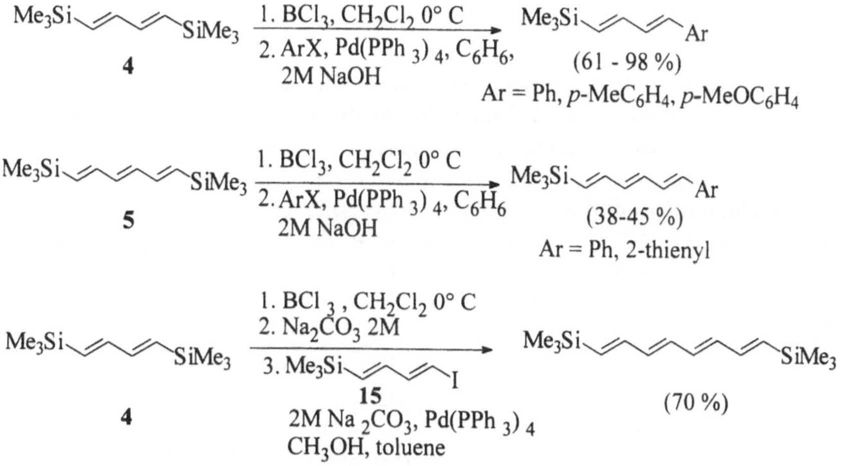

Figure 10 Synthesis of trimethylsilyl-polyenes *via* formal Suzuki-Miyaura cross-coupling reaction

It is worth noting that the organic halide **15** necessary for the cross-coupling process could be obtained, with retention of configuration, by halogenation of the boron derivative prepared from the bis-silylated compound **4**.

The use of this approach in the synthesis of natural compounds is best exemplified by the procedure shown in Figure 11 and leading to navenone B **19**, an alarm pheromone of the mollusk *Navanax inermis*.[15]

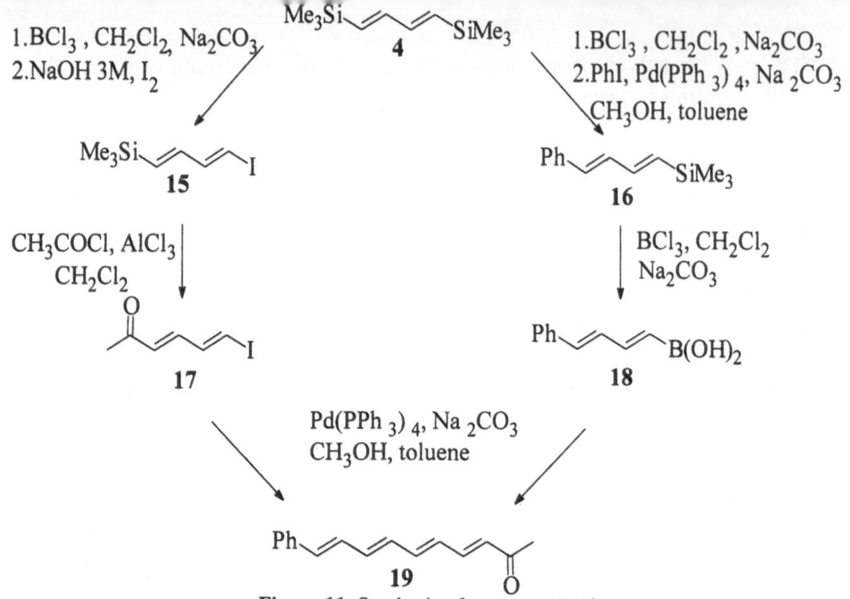

Figure 11. Synthesis of navenone B **19**

The all-*E* tetraenic structure was build up by a Suzuki-Miyaura cross-coupling reaction between two dienic fragments **17** and **18**. The iodo-dienyl fragment **17** was prepared from 1,4-bis(trimethylsilyl)diene **4** by a borodesilylation-iodination procedure leading to **15**, followed by an acylation reaction. The boro-desilylation methodology was used for the preparation of the intermediate **16** and repeated to obtain the boronic fragment **18**.

SYNTHESIS OF CONJUGATED POLYMERS

As stated above, polyenes are of special interest in the field of materials. Moreover, the actual systems which are valid candidates for the use in electrical or electronic devices are represented by polymers. Among these materials an important position is occupied by poly(*p*-phenylenevinylene) (PPV).[16] This polymer is most commonly prepared by the precursor route, *i. e.* by elimination from a soluble polymeric sulphonium salt.

By resorting to an organometallic route (*i. e.* the Stille coupling reaction), we have prepared a soluble dialkoxy-substituted PPV **20** (Figure 12).[17]

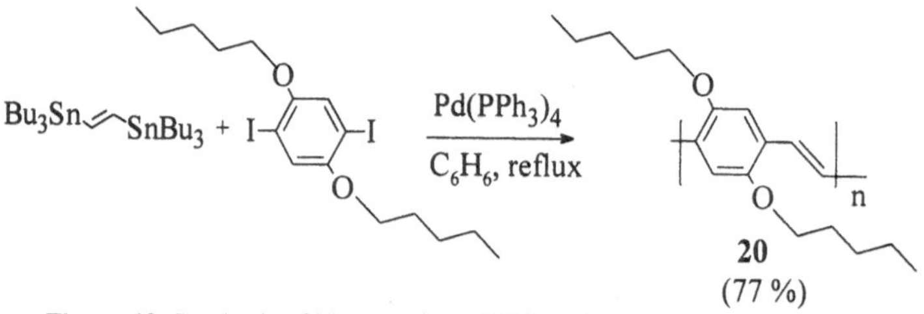

Figure 12. Synthesis of bis(pentyloxy)PPV *via* Stille coupling reaction

The material, which has weight-average molecular weights M_w ranging from 3347 to 3878, depending on the reaction conditions, was successfully used for the preparation of a light emitting diode (LED).[15]

In view of the versatility of the Stille coupling, a variety of soluble PPV materials were prepared[15] with the aim of establishing a structure-properties relationship. The series of compounds prepared includes an optically active PPV **21**, a silyloxy substituted PPV **22**, an *ansa*-compound **23**, a *para*-crown ether system **24** and a fluorinated PPV **25** (Figure 13).

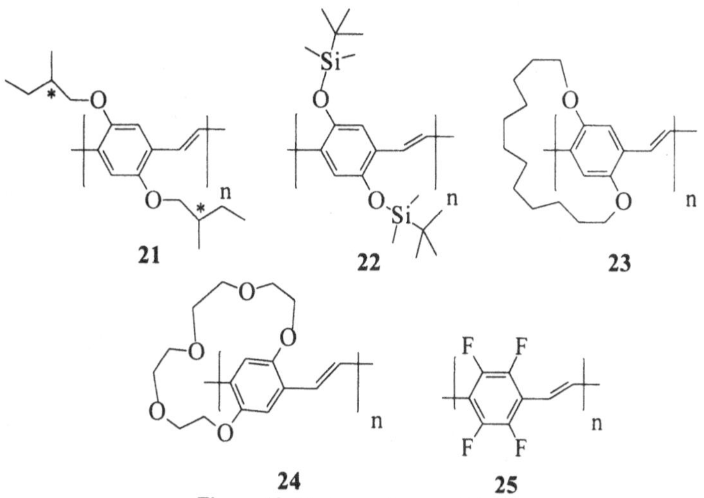

Figure 13. Various substituted PPVs

Various physical properties can be fine-tuned by simply changing the groups present in the conjugated backbone[16]. Besides the solubility, other characteristics, more strictly related to the semiconducting behaviour of this class of organic molecules, can be controlled.[18] In particular, the modulation of bandgap and the consequent control of the color of the photo- or electro-emitted radiation can be achieved. The search for blue light-emitting polymers is particularly challenging because this property is seldom shown by inorganic semiconductors. In the case of our polymers, photoemission of the linear dialkoxy PPV **20** shows a maximum in the yellow-orange region and a slight blue-shift is observed for the silyloxy PPV **22** and the *ansa*-dialkoxy PPV **23**. Preliminary experiments have shown that a more pronounced blue-shift of the photoemission arises in the fluoro-substituted PPV **25**.

Another interesting property exhibited by some of our polymers is the stimulated emission that could find application in the fabrication of organic polymer-based lasers.

CONCLUSION AND OUTLOOK

We have shown that a variety of procedures can be used in order to prepare stereodefined systems presenting a conjugated backbone. Straigthforward routes to several natural compounds have been set up. However, we believe that the design and synthesis of monodisperse or polydisperse systems showing interesting electro-optical properties represent a field which at the present has been explored only to a sligth extent. Indeed, the development of synthetic metodologies could offer a powerful tool for a fine control of the properties of this class of organic semiconductors. Such control appears of special importance when one considers that it cannot be so easily achieved in the case of classical inorganic semiconductors.

ACKNOWLEDGEMENT

The author wishes to express his appreciation to his coworkers and students who have contributed their enthusiasm and hard work to the investigations described above. Partial support of this work by Ministero dell'Università e della Ricerca Scientifica e Tecnologica, Rome, Consiglio Nazionale delle Ricerche (Progetto finalizzato MSTA II), and Università di Bari is gratefully acknowledged.

REFERENCES

1. Naso, F., Stereospecific Synthesis of Olefins through Sequential Cross-coupling Reactions, *Pure Appl. Chem.*, 60:79 (1988).
2. Fiandanese, V., Marchese, G., Naso, F., Ronzini, L., Rotunno, D., An Easy Route to Insect Pheromones with a *E-Z* or a *Z-E* Conjugated Diene Structure, *Tetrahedron Lett.*, 30:243 (1989).
3. Babudri, F., Fiandanese, V., Mazzone, L., Naso, F., A General Approach to Conjugated (*E, E*)-Dienes through Sequential Coupling Reactions, *Tetrahedron Lett.*, 35:8847 (1994).
4. Babudri, F., Fiandanese, V., Naso, F., Punzi, A., A New Straightforward and General Approach to Dienamide Natural Products 35: 2067 (1994).
5. Naso, F., Fiandanese, V., Babudri, F., (1*E*, 3*E*)-1,4-Bis(trimethylsilyl)-1,3-butadiene, in: *Encyclopedia of Reagents for Organic Synthesis,* John Wiley and Sons, ed, L.A. Paquette, Columbus, OH, USA (1995).
6. Naso, F., Fiandanese, V., Babudri, F., (1*E*, 3*E*, 5*E*)-1,6-Bis(trimethylsilyl)-1,3,5-hexatriene, in: *Encyclopedia of Reagents for Organic Synthesis,* John Wiley and Sons, ed, L.A. Paquette, Columbus, OH, USA (1995).
7. Babudri, F., Fiandanese, V., Marchese, G., Naso, F., A New Approach to Silylated Ketones and Dicarbonyl Compounds with Conjugated (all-*E*) Diene or Triene Structure, *J. Chem. Soc., Chem. Commun.*, 237 (1991).
8. Babudri, F., Fiandanese, V., Naso, F., Conjugated Polyene Synthesis via Disilyl Derivatives: a Direct Access to Ostopanic Acid, a Plant Anticancer Agent, and to (6*E*)-LTB$_3$ Leukotriene, *J. Org. Chem.*, 56: 6245 (1991).
9. Babudri, F., Fiandanese, V., Naso, F., Punzi, A., An Easy Route to Conjugated (all *E*) Tetraene Compounds via Disilyl Derivatives Exemplified by β-Parinaric Acid Methyl Ester, *Synlett*, 221 (1992).
10. Fiandanese, V., Hassan, O., Naso, F., Scilimati, A., A Highly Efficient Kinetic Resolution of ω-Trimethylsilyl Polyunsaturated Secondary Alcohols by Lipase-Catalyzed Transesterification, *Synlett*, 491 (1993).
11. Babudri, F., Fiandanese, V., Hassan, O., Punzi, A., Naso, F., New Synthesis of Leukotriene B$_3$ Methyl Ester from bis(Trimethylsilyl) Unsaturated Derivatives, *Tetrahedron*, 54: 4327 (1998).
12. Babudri, F., Cicciomessere, A.R., Farinola G.M., Fiandanese, V., Marchese, G., Musio, R., Naso, F., Sciacovelli, O., Highly Stereoselective Synthesis of Conjugated Polyenes via a Homocoupling Reaction of Unsaturated Silanes, *J. Org. Chem.*, 62: 3291 (1997).
13. Farinola, G.M., Fiandanese, V., Mazzone, L., Naso, F., A Novel and Efficient Route to (*E*)-Alk-1-enyl Boronic Acid Derivatives from (*E*)-1-(Trimethylsilyl)alk-1-enes and a Formal Suzuki-Miyaura Cross-coupling Reaction Starting with Vinylsilanes, *J. Chem. Soc., Chem. Commun.*, 2523 (1995).
14. Babudri, F., Farinola, G.M., Fiandanese, V., Mazzone, L., Naso, F., A Straightforward Route to Polyenylsilanes by Palladium or Nickel-Catalyzed Cross-Coupling Reactions, *Tetrahedron*, 54: 1085 (1998).
15. Unpublished results.
16. Kraft., A., Grismade A. C., Holmes, A. B., Electroluminescent Conjugated Polymers-Seeing Polymers in a New Light, *Angew. Chem. Int. Ed.*, 37: 402 (1998).
17. Babudri, F., Cicco, R.S., Farinola G.M., Naso, Bolognesi A., Porzio, W., F., Synthesis, Characterization and Properties of a Soluble Polymer with a Poly(phenylenevinylene) Structure, *Macromol. Rapid Commun.*, 905 (1996).
18. Roncali, J., Synthetic Principles for Bandgap Control in Linear π Conjugated Systems, *Chem. Rev.*, 97: 173 (1997).

ALLENES IN NOVEL PALLADIUM-CATALYZED AND ACID-MEDIATED CYCLIZATION PROCESSES

Henk Hiemstra*, Winfred G. Beyersbergen van Henegouwen,
Willem F. J. Karstens, Marinus J. Moolenaar, and Floris P. J. T. Rutjes

Laboratory of Organic Chemistry, Institute of Molecular Chemistry,
University of Amsterdam, Nieuwe Achtergracht 129,
1018 WS Amsterdam, The Netherlands

INTRODUCTION

The unique location of two orthogonal carbon carbon double bonds in the allene moiety makes this functional group suitable for consecutive (tandem) reactions and the rapid build-up of molecular complexity.[1] An addition reaction to one double bond usually leads either to an allylic or a vinylic functional group, or to both, which in several cases leads to enhanced reactivity of the molecule for further processes.

This article describes our recent investigations into two different types of allene reactions. The first part deals with palladium-catalyzed cyclization reactions of allenes with the lactam nitrogen atom functioning as the nucleophile.[2] The allene moiety is made electrophilic through palladium(II) complexation. The second part of this article discusses cyclization reactions in which the allene serves as π-nucleophile to react with N-acyliminium intermediates. The utility of this process is illustrated in a synthetic approach towards the oxindole alkaloid gelsedine.[3] A major point of interest in these intramolecular allene reactions is the regiochemistry, which is determined by structural effects like the length of the tether and the substitution pattern of the allene.

PALLADIUM-CATALYZED ALLENE CYCLIZATIONS

The reaction of a palladium(II) species with an allene is a well-studied process which usually leads to a η^3 π-allylpalladium intermediate after a highly regioselective olefin insertion of an incipient η^2 π-complex.[4] Most relevant for synthetic applications are palladium(II) species containing an aryl or vinyl group which originate from oxidative addition of Pd(0) into the corresponding halides or triflates. Carbon,[5] oxygen[6,7] and nitrogen[8-10] nucleophiles have been used to trap the π-allylpalladium intermediate both in inter- and intramolecular fashion. Eq 1 and 2 show two examples of intramolecular reactions with nitrogen as nucleophile, including the most reasonable reaction intermediates. Eq 1 illustrates the most usual course of allene cyclization, featuring formation of a π-allyl intermediate, which is trapped by the nitrogen nucleophile at the proximate carbon to give a pyrrolidine.[8] In eq 2 a somewhat different process takes place initiated by palladium chloride which probably activates the terminal allene double bond for attack by nitrogen at the distant carbon atom. The resulting σ-vinylpalladium species reacts with allyl bromide to the interesting carbapenem structure.[9]

We were interested in the synthetic use of ω-allenyl-substituted lactams such as the starting material in eq 2. Such compounds are readily accessible by N-acyliminium ion chemistry.[11] We investigated the palladium-catalyzed ring closure of the corresponding pyrrolidinone **1** (eq 3),[11] which much to our surprise was fully unsuccessful by using either the conditions of eq 1 or 2. However, the homologue **2** made by an acyliminium coupling and the Crabbé reaction[12] did cyclize by using the conditions of eq 1.

As shown in eq 4 the product of the cyclization of **2** was not the six-membered ring product **3** but the pyrrolizidine skeleton **4**. Although the ¹H NMR data provided some evidence (the vinylic hydrogen gave a broad singlet at 4.79 ppm, pointing to an enamide) the structure of **4** was proven by acid-induced reduction to crystalline **5**, the identity of which was assured by X-ray crystallography. Remarkably, the nitrogen nucleophile is connected to the central allene carbon as a result of a novel type of allene cyclization.

A proposal for the mechanism of the process is depicted in eq 5 for the homologous lactam **6**. In addition to the major product **7** a small amount of byproduct **8** (single isomer) was isolated in this case (ratio 88:12). After activation of one of the allene double bonds by the *in situ* formed η^2 palladium(II) complex, nitrogen attack on the central carbon is apparently faster than transfer of the phenyl group. The favourable geometry for 5-membered ring formation is probably an important factor. The resultant π-allyl complex **9** then suffers reductive elimination via the two σ-complexes to give the products.

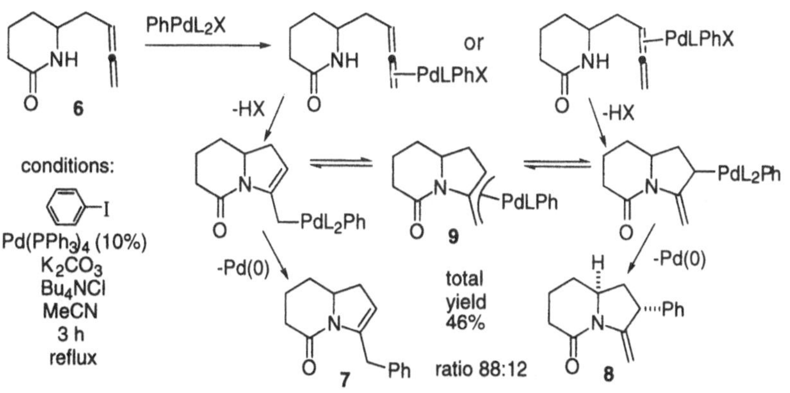

(5)

The intermediacy of a π-allyl complex was confirmed by treatment of allene **2** with PdCl$_2$ and triphenylphosphine, respectively, to give the palladium complex **10**, identified by X-ray crystallography (eq 6). However, a π-allyl palladium intermediate is not required, because acetylene **11** (prepared from **2** by strong base treatment)[13] cyclized under identical conditions to enamide **12** (eq 7). The remarkable double bond geometry in **12** is probably the result of a favourable coordination of the carbonyl to palladium in the intermediate **13**. The generality of the allene cyclization process is indicated by the cyclizations of **14** and enantiopure **15** in ca. 75% yield to tricycle **16** and carbapenem **17**, respectively.

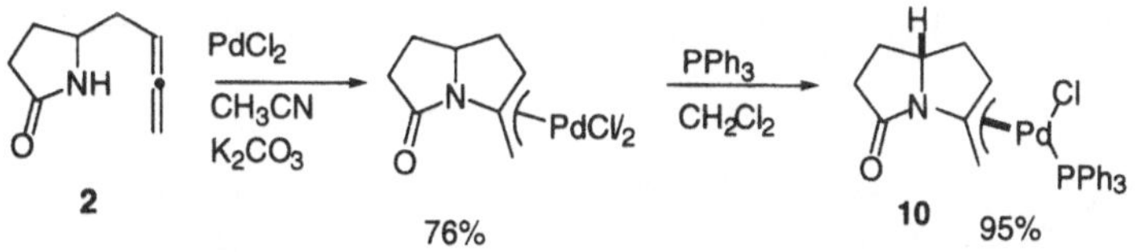

(6)

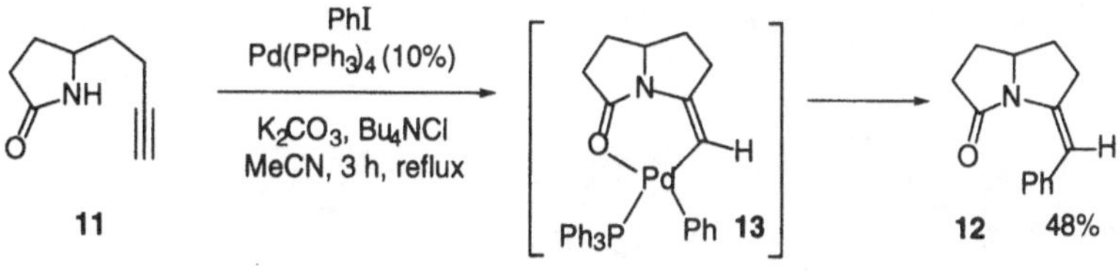

(7)

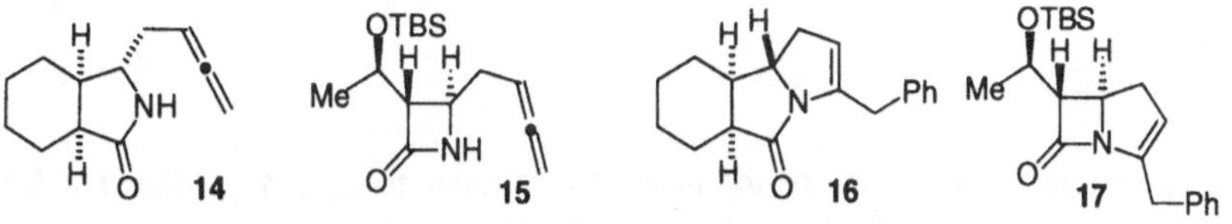

Our synthetic approach to the allenes (eq 3) was not fully satisfactory. We discovered a more flexible route to enantiopure allenes from 5-iodomethylpyrrolidinone (**18**), readily prepared from cheap (*S*)-pyroglutamic acid.[14] Thus, **18** was transformed into the zinc reagent **19** by using activated zinc in DMF. This solvent was essential, because in THF complete β-elimination was observed.[15] Zinc compound **19** reacted in the presence of a catalytic amount of CuBr·Me$_2$S with several propargylic tosylates or mesylates and with bromoallene to enantiopure allenes **20** - **24** and acetylene **25**, respectively.[16]

The Pd-catalyzed coupling and cyclization of the substituted allenes **21** and **23** with iodobenzene proceeded smoothly (conditions as in eq 5) to give **26** and **27**, both in 64% yield. However, allene **22** gave only 16% of the usual cyclization product and 60% of a 7:3 mixture of dienes **28** and **29**. These latter products show that the attack of nitrogen is hindered in this case and that the alternative acyclic phenyl-substituted π-allylpalladium complex does not cyclize but loses palladium hydride. Allene **24** gave under the usual conditions only 7% of the normal product and 50% of diene **30**. This simple, yet chiral and enantiopure diene readily underwent a Diels-Alder reaction with *N*-methylmaleimide and will be the subject of further investigations with respect to its applications in synthesis.

More insight in the course of the process as shown in eq 5 was obtained from our study of allenic α-amino acid derivatives (eq 9). Cyclization of **31** with iodobenzene in

DMF gave a mixture of the 4- and 6-membered ring products **32** and **33** without a trace of a 5-membered ring. Thus, with an acyclic nitrogen nucleophile (similar selectivity was found for a *tert*-butoxycarbonyl and a benzyl substituent on nitrogen), cyclization takes place after formation of a π-allylpalladium intermediate. The nature of the nucleophile clearly has a prominent influence on the course of these allene cyclizations. It was further found that the kinetic product **32** is slowly converted into **33** in a Pd(0)-catalyzed process.

$$(10)$$

Zinc reagent **19** appeared also very useful for the synthesis of enantiopure lactam **34** (eq 10). We had already discovered some years ago that **34** is an excellent intermediate in the synthesis of enantiopure epibatidine, a simple alkaloid with highly interesting physiological activities.[17] Our first approach to **34** involved the coupling of a organometallic derivative of propargylsilane with iodide **18** but suffered from low yields. Our new approach (eq 10) features a copper-mediated coupling of zinc compound **19** with 3-trimethylsilyl-1-iodo-1-propyne and gave the product in an unoptimized yield of 52%. After *N*-functionalization of **34**, carbonyl reduction and *N*-acyliminium ion cyclization allene **35** was obtained. Ozonolysis gave enantiopure ketone **36** which is a known intermediate in the synthesis of the alkaloid.[18]

ALLENE-TERMINATED *N*-ACYLIMINIUM ION CYCLIZATIONS

Although *N*-acyliminium ion cyclizations constitute a well-known, versatile method for the synthesis of cyclic nitrogen compounds,[19] the use of an allene as π-nucleophile has received only little attention. Several years ago it was observed qualitatively that in the formation of the indolizidine **37** allene **38** cyclized faster than alkyne **39** (eq 11).[20] In both reactions a vinyl formate is the initial product, but the two different vinyl formates are hydrolyzed to the same ketone with a trace of water which is present in formic acid.

$$(11)$$

We then set out to investigate allenic lactams in *N*-acyliminium chemistry. γ-Lactams **1** (racemic) and **21** (enantiopure) were converted into cation precursors **40** and **43** via well-known methodology.[21] On stirring for 17 h in formic acid the indolizidine skeleton **41** was obtained from **40** in excellent yield as a single isomer (eq 12). The stereochemistry can be explained by invoking the *N*-acyliminium ion **42** with the more stable trans geometry as intermediate. Only 6-membered ring formation was observed due to regioselective reaction of the terminal double bond. Stirring of **43** in formic acid gave only reaction of the internal double bond followed by trapping of the intermediate allylic cation at the exocyclic carbon atom to lead to indolizidine **44** in 44% yield (eq 13). The methyl group was essential for a good result as the precursor without the methyl group gave a complex mixture.

A different type of cyclization was studied with precursor **45** readily obtained by alkylation of 1-benzyl-5-ethoxy-2-pyrrolidinone[22] with 1-bromo-2,3-butadiene (eq 14). A very clean cyclization to the bridged bicyclic lactam **46** took place in neat formic acid at 85 °C. Interestingly, in the presence of sodium iodide the vinyl iodide **47** was the major product. The latter result was exploited as one of the key steps in a synthetic approach to the oxindole alkaloid gelsedine (**48**). Gelsedine, a minor but highly toxic alkaloid from *Gelsemium sempervirens*,[23,24] has not succumbed to total synthesis, despite some interesting attempts.[25-27] A partial synthesis from koumidine was recently published.[28]

Our synthetic route (eq 15) aimed at *ent*-gelsedine started from cheap (*S*)-malic acid. The crystalline cyclic imide **49** was regioselectively reduced to a hydroxylactam which on ethanolysis and deacetylation produced **50**. The alkylation of **50** proceeded via the dianion and showed complete stereoselectity to give *trans*-product **51**.[29] The key *N*-acyliminium cyclization occurred as in the model system (eq 14) to furnish vinyl iodide **52**, formylated at the bridge hydroxyl group.

(16)

The vinyl iodide is an ideal functional group to introduce the spiro oxindole moiety. To this end **52** was first converted to anilide **53** via Pd-catalyzed aminocarbonylation and removal of the formyl group. To make conditions for a stereochemically correct Heck cyclization as favourable as possible the bridge atom was turned into an sp^2 carbon via oxidation and Wittig reaction. The top face of the endocyclic double bond in **54** should now be accessible to allow the desired Heck cyclization in the required stereochemical sense. In the event *N*-methylation followed by Heck cyclization in a sealed tube at 120 °C gave the desired oxindole **55** in high yield.[30,31] Chemoselective hydroboration of the exocyclic double bond also proceeded with high stereoselectivity. The resulting primary alcohol was then cyclized by using mercuric trifluoroacetate. While this cyclization proceeded well the reduction of the organomercurial gave only a moderate yield of **56** despite the use of triethylborane, which was recently published to improve the yield of such reactions.[32] Nevertheless, the construction of the skeleton of gelsedine had now been achieved.

The remaining steps were aimed at the introduction of the ethyl and the *N*-methoxy group. The deprotection of the oxindole nitrogen occurred with remarkable ease by using benzoyl peroxide in a radical reaction followed by aminolysis of the intermediate benzoate.[33] The benzyl group was removed by lithium in ammonia to give bislactam **57** as a crystalline solid. Both nitrogen atoms were then reprotected with a *tert*-butoxycarbonyl (Boc) group to activate the pertinent carbonyl for reaction with a Grignard reagent. Treatment with excess ethylmagnesium bromide at -78 °C followed by work-up and immediate reduction with triethylsilane in trifluoroacetic acid gave the desired secondary amine **58** as its trifluoroacetate salt in a remarkable 44% yield.[3] In this process the Grignard reagent also attacked the Boc group of the oxindole which led to simple removal of that Boc group. The

273

other Boc group was removed by TFA after stereoselective acid-induced hydride reduction of the intermediate N,O-hemiacetal via an iminium ion. The structure of **58** was confirmed by extensive NMR studies, in particular ^{1}H NMR NOE measurements. The introduction of the oxindole N-methoxy group awaits further study.

ACKNOWLEDGEMENTS

We thank Rob H. Balk, Rutger M. Fieseler, dr Urszula Grabowska, Gertjan Mentink and Marianne Stol for their synthetic contributions and Professor W. Nico Speckamp for his continuous support and useful discussions. These investigations were supported (in part) by DSM Research in Geleen, Organon Akzo Nobel in Oss, and the Netherlands Foundation for Chemical Research (SON) with financial aid from the Netherlands Organization for Scientific Research (NWO).

REFERENCES

1. H.F. Schuster and G.M. Coppola. *Allenes in Organic Synthesis*, Wiley, New York (1984).
2. W.F.J. Karstens, F.P.J.T. Rutjes, and H. Hiemstra, *Tetrahedron Lett.* 38:6275 (1997).
3. W.G. Beyersbergen van Henegouwen and H. Hiemstra, *J. Org. Chem.* 62:8862 (1997).
4. J. Tsuji. *Palladium Reagents and Catalysts*, Wiley, New York (1995).
5. B. Cazes, *Pure Appl. Chem.* 62:1867 (1990).
6. R.D. Walkup, L. Guan, M.D. Mosher, S.W. Kim, and Y.S. Kim, *Synlett.* 88 (1993).
7. C. Jonasson and J.E. Bäckvall, *Tetrahedron Lett.* 39:3601 (1998).
8. J.S. Prasad and L.S. Liebeskind, *Tetrahedron Lett.* 29:4257 (1988).
9. I.W. Davies, D.I.C. Scopes, and T. Gallagher, *Synlett.* 85 (1993).
10. V.M. Arredondo, F.E. McDonald, and T.J. Marks, *J. Am. Chem. Soc.* 120:4871 (1998).
11. H. Hiemstra, H.P. Fortgens, and W.N. Speckamp, *Tetrahedron Lett.* 25:3115 (1984).
12. S. Searles, Y. Li, B. Nassim, M.T.R. Lopes, P.T. Tran, and P. Crabbé, *J. Chem. Soc., Perkin Trans. 1* 747 (1984).
13. S.R. Macaulay, *J. Org. Chem.* 45:734 (1980).
14. O. Provot, J.P. Célérier, H. Petit, and G. Lhommet, *J. Org. Chem.* 57:2163 (1992).
15. C. S. Dexter and R.F.W. Jackson, *J. Chem. Soc., Chem. Commun.* 75 (1998).
16. R. Duddu, M. Eckhardt, M. Furlong, H.P. Knoess, S. Berger, and P. Knochel, *Tetrahedron* 50:2415 (1994).
17. Z. Chen and M.L. Trudell, *Chem. Rev.* 96:1179 (1996).
18. S.R. Fletcher, R. Baker, M.S. Chambers, R.H. Herbert, S.C. Hobbs, S.R. Thomas, H.M. Verrier, A.P. Watt, and R.G. Ball, *J. Org. Chem.* 59:1771 (1994).
19. H. Hiemstra and W.N. Speckamp, Additions to N-Acyliminium Ions, in *Comprehensive Organic Synthesis*, B.M. Trost and I. Fleming, eds., Pergamon: Oxford, (1991); Vol. 2, pp. 1047.
20. P.M.M. Nossin and W.N. Speckamp, *Tetrahedron Lett.* 22:3289 (1981).
21. P.M. Esch, I.M. Boska, H. Hiemstra, R.F. de Boer, and W.N. Speckamp, *Tetrahedron* 47:4039 (1991).
22. W.J. Klaver, M.J. Moolenaar, H. Hiemstra, and W.N. Speckamp, *Tetrahedron* 44:3805 (1988).
23. H. Schwarz and L. Marion, *Can. J. Chem.* 31:958 (1953)
24. E. Wenkert, J.C. Orr, S. Garrett, J.H. Hansen, B. Wickberg, and C.L. Leicht, *J. Org. Chem.* 27:4123 (1962).
25. S.W. Baldwin and R.J. Doll, *Tetrahedron Lett.* 3275 (1979).
26. A.S. Kende, M.J. Luzzio, and J.S. Mendoza, *J. Org. Chem.* 55:918 (1990).
27. N.K. Hamer, *J. Chem. Soc., Chem. Commun.* 102 (1990).
28. H. Takayama, Y. Tominaga, M. Kitajima, N. Aimi, and S. Sakai, *J. Org. Chem.* 59:4381 (1994).
29. W.J. Klaver, H. Hiemstra, and W.N. Speckamp, *J. Am. Chem. Soc.* 111:2588 (1989).
30. M.M. Abelman, T. Oh, and L.E. Overman, *J. Org. Chem.* 52:4130 (1987).
31. N.J. Newcombe, Y. Fang, R.J. Vijn, H. Hiemstra, and W.N. Speckamp, *J. Chem. Soc., Chem. Commun.* 767 (1994).
32. S.H. Kang, J.H. Lee, and S.B. Lee, *Tetrahedron Lett.* 39:59 (1998).
33. S. Nakatsuka, O. Asano, and T. Goto, *Heterocycles* 24:2791 (1986).

A CATALYTIC TANDEM ADDITION ROUTE
TO γ,δ-UNSATURATED CARBONYLS

Xiyan Lu*, Zhong Wang, Aiwen Lei

Shanghai Institute of Organic Chemistry, Chinese Academy of Sciences
354 Fenglin Lu, Shanghai 200032, China

INTRODUCTION

The conjugate addition of organometallics to α,β-unsaturated carbonyl compounds constitutes one of the most important carbon-carbon bond formation reactions.[1] Among them the base-promoted addition of stabilized carbanions and organocopper reagents mediated additions have been extensively studied,[1,2] and many successful total syntheses of natural products have hinged on such conjugate additions.[3] More recently, addition reactions of nucleophiles including silyl enol ethers and organosilanes under Lewis acid catalysis have been developed, complementing the conventional basic conditions.[4] However, most of these methods require the preparation of stoichiometric organometallic reagents such as organolithium reagents, Grignard reagents and other highly reactive intermediates which precludes the incorporation of sensitive functional groups. Tedious protection-deprotection sequences were often used to circumvent the difficulty in preparing multifunctional compounds needed for sophisticated synthesis.[5]

Recently, coupling of alkynes and allylic alcohols catalysed by a ruthenium complex was reported to give γ,δ-enones and enals,[6,7] but harsh conditions were required and the regioselectivity was only moderate. In a catalytic precedent, transmetallated organopalladium species mediated 1,4-addition of organomercurials to α,β-unsaturated carbonyls.[8] On the other hand, Alkenylpalladium intermediates are easily obtained from the halopalladation and oxypalladation of alkynes.[9-13] In our ongoing studies of divalent palladium catalyzed reactions, we have established a series of efficient methods for synthesizing α-methylene-γ-butyrolactone derivatives using halopalladation-cyclization-deheteropalladation as the key step.[14-16] While studying the scope and synthetic application of this protocol, we serendipitously discovered an intramolecular Pd-catalyzed conjugate addition reaction, which was applied to the formal synthesis of (+)-pilocarpine.[17] In this reaction, an aldehydic enyne ester 1 was cyclized under the catalysis of Pd(II) to give the aldehydic γ-lactones 2. Based on this intramolecular version, we developed the tandem halopalladation-conjugate addition[18] and tandem oxypalladation-conjugate addition reactions[19] which allow efficient and highly stereoselective synthesis of multifunctional carbonyl compounds.

CATALYTIC TANDEM ADDITION ROUTE TO γ,δ-UNSATURATED CARBONYLS[18]

When methyl propynoate **3a** (1.0 mmol) was treated with acrolein (5.0 mmol) in the presence of Pd(OAc)$_2$ (0.02 mmol) and LiBr (4.0 mmol) in acetic acid , two products **4a** and **5a** were isolated.

Compound **4a** was the 1:1 codimerization product and **5a** was the 2:1 cotrimerization product. The formation of these two products was rationalised by the mechanism shown in the following. Vinylpalladium intermediate **6** was first formed by *trans*-halopalladation of **3a** in HOAc, followed by acrolein insertion (path A) or subsequent alkyne and acrolein insertion (path B). The competing alkyne insertion may be ascribed to the higher reactivity of acetylenes than olefins. Both reaction paths involved the protonolysis of the carbon-palladium bond to regenerate the catalytic Pd(II) species, giving **4a** and **5a** respectively.

The yield of **4** could be increased by slow addition of **3** to a solution of the other reactants including acrolein, LiBr and Pd(OAc)$_2$ in HOAc. Electron-deficient alkynes and phenyl acetylene all gave good yields of the 1:1 codimerization product with acrolein and methylvinyl ketone. The regiochemistry was affected by the electronic properties of the substituent on the triple bond, in accordance with the results reported by Kaneda et al.:[11] the halide attacks the more electropositive carbon atom. Higher stereoselectivity was observed for the propynoates: only Z-isomers were produced. With substituted alkynoates, a mixture of Z/E isomers were obtained, which is determined by the stereoselectivity of halopalladation of alkynes.[14-16]

EFFICIENT PREPARATION OF FUNCTIONALIZED (E,Z)-DIENES[20]

The characteristic (E,Z) double bond configuration of the molecules which is usually responsible for the special functions meanwhile poses great challenges for their stereoselective synthesis. A naive retrosynthetic analysis suggests that two acetylene molecules can be added together to form the (E,Z)-double bonds, provided the addition reaction occur in a stereoselective manner. Indeed, from the well-documented organometallic elementary reactions, we could devise that a tandem addition incorporating two molecules of acetylene may lead to (E,Z)-conjugated diene structure.

In this sequence, the stereochemical requirements of trans-addition and cis-insertion (carbometallation) helped to establish the otherwise hard-to-access conjugate (E,Z) double bond configuration. Despite the extreme efficiency this sequence may bring about, there was only one precedent in literature realizing such a concept: The palladium catalyzed cotrimerization reaction of acetylene and allyl chloride developed by Kaneda et al.,[11] but the reaction gave in low yield a mixture of codimer and cotrimer and the stereochemistry of the cotrimer was not established.

From the recently developed synthesis of γ,δ-unsaturated carbonyl compounds in which a halide-assisted protonolysis efficiently recycles Pd(II) catalytic species[18,19] thus effecting the tandem addition reaction of halide, an alkyne and an α,β-unsaturated carbonyl. In this context, we attempted the reaction of acetylene with α,β-unsaturated electron-deficient alkenes in the presence of a palladium catalyst to explore the possibility of developing new methods for (E,Z)-diene synthesis.

We first use acrolein as solvent to avoid the excessive polymerization of acetylene. When acetylene was passed through a mixture of acrolein, HOAc, Pd(OAc)$_2$ and LiBr at 15°C for 2h, rapid formation of palladium black was observed. The reaction afforded the expected dienal **10** together with the α,β-unsaturated trienal **11** in 220% and 80% yield, respectively (yields are calculated on the basis of Pd(OAc)$_2$). Compound **10** could be purified carefully by column chromatography and its NMR data unambiguously show the (E,Z)-configuration of the molecule.

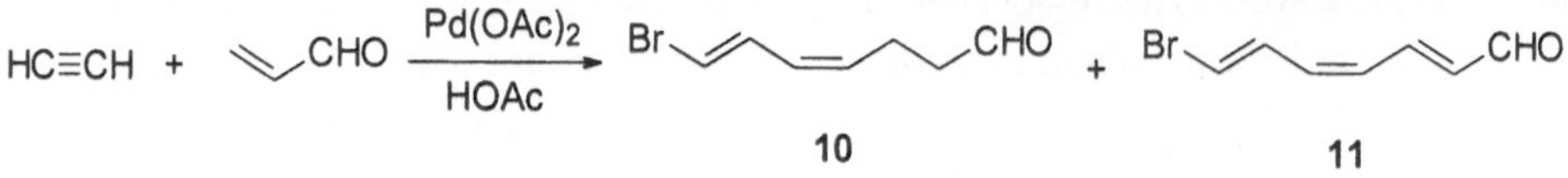

The formation of **10** and **11** could be explained with the following mechanism which also rationalizes the stereoselectivity:

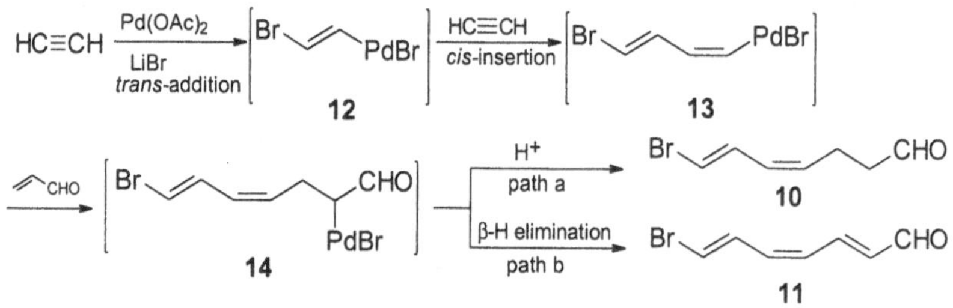

Preliminary results showed that polar solvents, high acetylene concentration, high halide concentration and low temperature favor the formation of **10**. In most cases, using HOAc as a solvent and rapidly passing acetylene into the reaction mixture, **10** was isolated as the main product after chromatography.

The clean reaction, high catalytic efficiency and high stereoselectivity encouraged us to explore the acetylene-alkene tandem addition reactions of other electron-deficient alkenes in the presence of lithium halides. Preliminary results showed that different α,β-unsaturated carbonyl compounds all gave (*E,Z*) cotrimerization products with high stereoselectivity. Relative lower yields were obtained for sterically hindered substrates. Lithium chloride and bromide gave comparable yields of halogenodienic carbonyl compounds, while iodide showed much lower reactivity and iodosubstituted analogue was not isolated in its pure form.

Alkenes with weaker electron-withdrawing groups such as acrylonitrile and methyl acrylate, failed to give cotrimerization product. However, nitroethylene was found to be an effective electrophile to afford the corresponding nitro-substituted diene. Thus, with the abundant transformations related to vinyl halide, carbonyl and nitro groups, the present reaction provides sound precursors to a wide range of unsaturated compounds with (*E,Z*) diene structures.

The utility of this method can be demonstrated by the concise synthesis of spilanthol, also known as afinin, which was isolated from *Spilanthes oleraceae* Jacq. and is the most insecticidally active and stable of the natural isobutylamides thus far isolated and identified.[21] The synthesis of spilanthol was reported by several groups in multi-steps and low overall yields.[21,22] Wittig olefination of the (*E,Z*)-dienal **10** prepared from acetylene-acrolein cotrimerization,[20] followed by the palladium catalyzed coupling with MeZnCl afforded spilanthol **16** in 55% overall yield.

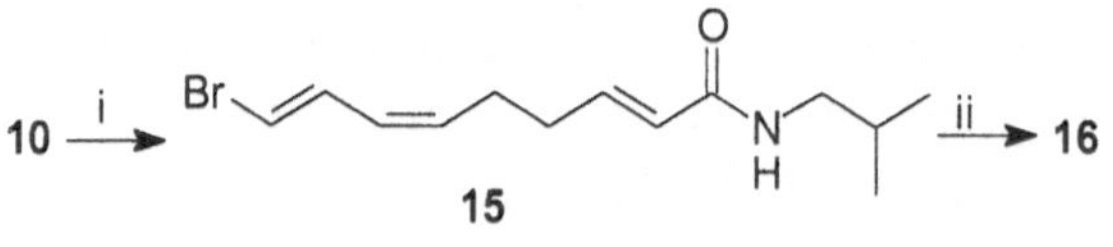

Reaction condition: i) Ph₃PCH₂CONHCH₂CHMe₂⁺Br⁻, LDA, THF, 62%.
ii) MeZnCl, Pd(OAc)₂, THF, 89%.

NUCLEOPHILE-ALKYNE-α,β-UNSATURATED CARBONYL COUPLING THROUGH TANDEM NUCLEOPALLADATION AND CONJUGATE ADDITION[19]

Oxypalladation of triple bonds is a process analogous to halopalladation.[12,13] With the hope that oxypalladation of alkynes followed by insertion of α,β-unsaturated carbonyls may lead to 1,5-dicarbonyl compounds, we examined the reaction using lithium acetate and methyl propynoate in the presence of catalytic amount of Pd(OAc)$_2$. Only halogen incorporated product **4a** was formed in the presence of LiBr, and the reaction without LiBr resulted in fast decomposition of Pd(OAc)$_2$ to Pd metal with no identifiable products from the reaction of **3a**.

However, an intramolecular carboxyl did effect this oxypalladation-acrolein coupling sequence. Using Pd(OAc)$_2$ as the catalyst, 3-heptynoic acid (**17a**) and 5 equiv of acrolein in the presence of excess LiBr (2 equiv) in HOAc gave the cyclized product **18a** in high yield.[19] Other 4-substituted 3-butynoic acid reacted smoothly to give cyclized acrolein coupling products. It was noteworthy that only intramolecular oxypalladation products were obtained although halopalladation was in competition. The reaction of 4-alkynoic acids with acrolein afforded γ-alkylidene-γ-butyrolactone **19** derivatives in good yield and stereoselectivity. Only (*E*)-**19** are produced from **17**, conforming to *trans*-oxypalladation of the triple bonds. Methyl vinyl ketone as the olefin partner also gave good yields of lactonic ketones, revealing the potential of the present reaction for easy access to a number of polyfunctionalized lactones.

A general mechanism similar to the formation of γ,δ-unsaturated carbonyls discussed above is proposed. Oxypalladation of the coordinated triple bond of **17** gives the vinylpalladium intermediate **20** which on acrolein insertion forms the (2-oxoalkyl)palladium intermediate **21**. The key step to form the products **18** and regenerate the Pd(II) catalytic species is again the protonolysis of the carbon-palladium bond in **21**.

ROLE OF HALIDE IONS IN THE REACTION

The key step to form the products and regenerate the Pd(II) catalytic species in above reactions is the protonolysis of the carbon-palladium bond in **7**, **9**, **14** and **21**. Vinylpalladium species usually react with α,β-unsaturated carbonyls to give vinylation products through β-hydride elimination of the insertion intermediate rather than addition products through protonolysis.[23,24] There are a few reports concerning the quenching of carbon-palladium bonds by acid hydrolysis,[25-27] but the formation of addition products was influenced by many factors.[28]

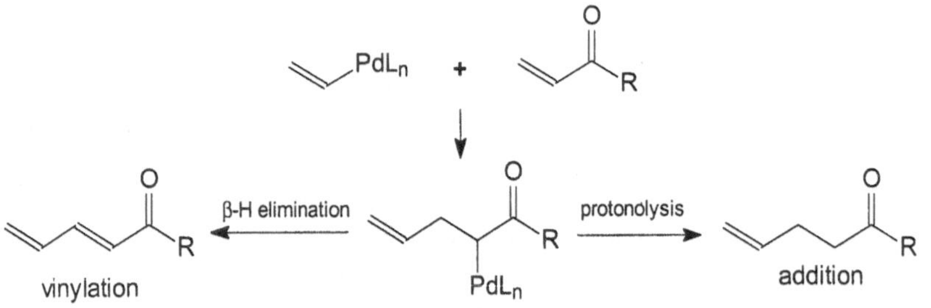

22

Although protonolysis of the carbon-palladium bond is, in principle, also an important reaction of organopalladium compounds, its study and application are far less reported in the literature than the β-hydride elimination and other elementary reactions.[25,27]

Inhibition of the normal β-hydride elimination pathway was achieved by addition of excess phosphine,[29,30] but this inibited the insertion step as well when we attempted the reaction of acrolein with the vinylpalladium generated by halopalladation of an alkyne. With the thought that an increase in electron density at the palladium center might also diminish cis-hydride elimination[30] and that excess halide ligand might serve this purpose, we found that the addition of the vinylpalladium intermediate to acrolein did occur in the presence of excess LiBr in HOAc.[18] Thus, a divalent palladium catalyzed halide-alkyne-α,β-unsaturated carbonyl coupling reaction was developed to form the γ,δ-unsaturated carbonyl compounds.[18-20] In this case, the protonolysis of (2-oxoalkyl)palladium intermediate effectively recycles the catalytic species.

Our studies reveal that excess coordinating halide inhibits β-H elimination and facilitates protonolysis in acidic conditions.[19,20] In the reaction of preparation of (E,Z)-dienes,[20] while pure **10** could be obtained in high concentration of LiBr, the reaction afforded the expected dienal **10** together with the trienal **11** in 220% and 80% yield (based on Pd), respectively, with a low concentration of LiBr. We found that the LiBr:Pd ratio has a great impact on the reaction: when the LiBr:Pd ratio was increased from 10:1 to 200:1, the catalyst activity was dramatically enhanced and the reaction yield went up from 550% to 3980% (based on Pd), respectively. This can be accounted for by the predominant complexation of halide ion on palladium which favors protonolysis over β-H elimination.

While high yield of lactone **18** was obtained by oxypalladation procedure in high concentration of LiBr, in the presence of less amount of LiBr, **23**, produced by β-hydride elimination, was obtained in 94% yield (based on Pd) together with precipitated palladium black.[19] This, again, demonstrated the significant role of halide in the present reaction.

The ease of this process in our case might be due to several factors: (1) The large excess of halide ion may make the β-hydride elimination less feasible through electron donation to palladium and occupancy of the coordination site needed for *cis*-hydride elimination; (2) intermediate **22** which is actually a mesomeric palladium enolate may readily undergo a heterolytic Pd-O fission by nucleophilic attack of halide ion on the palladium center.[8]

In conclusion, we have developed a high-yielding coupling reaction assembling a nucleophile, an alkyne, and an α,β-unsaturated carbonyl or nitroolefins in one step. This method permits facile entry to γ,δ-unsaturated carbonyl compounds and lactonic aldehydes (ketones) from halide, and carboxyl nucleophiles, respectively, and may well be extended to other nucleophiles giving carbocyclic or different hetercyclic structures. Protonolysis of the carbon-palladium bond was developed as the key step to form the products and regenerate the Pd(II) catalytic species. The high regio-/stereoselectivity, simplicity of operation, and mild condition ensuring compatibility for senseitive functional groups and possible application to large scale preparation suggest the method an attractive tool in synthesis.

ACKNOWLEDGMENTS

Financial support from National Natural Science Foundation of China and Chinese Academy of Sciences is gratefully acknowledged.

REFERENCES

1. P. Perlmutter. *Conjugate Addition Reactions in Organic Synthesis*, Pergamon, Oxford (1992).
2. J.A. Kozlowski. Organocuprates in the conjugate addition reaction in: *Comprehensive Organic Synthesis*, B.M. Trost and I. Fleming, eds., Pergamon, Oxford, Vol. 4, p.169 (1991).
3. N. Anand, J.S. Bindra, and S. Ranganathan. *Art in Organic Synthesis*, 2nd Ed., John Wiley & Sons, New York (1988).
4. V.J. Lee. Conjugate addition of carbon ligands to activated alkenes and alkynes mediated by Lewis acids in: *Comprehdnsive Organic Synthesis*, B.M. Trost and I. Fleming, eds., Pergamon, Oxford, vol. 4, p. 139 (1991).
5. K. Nonoshita, H. Banno, K. Maruoka, and H. Yamamoto, *J. Am. Chem. Soc.* 112:316 (1990) and references cited therein.
6. B.M. Trost, J.A. Martinez, R.J. Kulawiec and A.F. Indolese, *J. Am. Chem. Soc.* 115: 10402 (1993).

7. S. Derien and P.H. Dixneuf, *J. Chem. Soc. Chem. Commun.* 2551 (1994).

8. S. Cacchi, F. La Torre, D. Misiti, *Tetrahedron Lett.* 25:4591 (1979).

9. H. Dietl, H, Reinheimer, J. Moffat, P.M. Maitlis, *J. Am. Chem. Soc.* 92:2276 (1970).

10. P.M. Maitlis, *The Organic Chemistry of Palladium*, Academic Press, New York, Vol. 2, p. 47 (1971).

11. K. Kaneda, T. Uchiyama, Y. Fujiwara, T. Imanaka and S. Teranishi, *J. Org. Chem.* 44:55 (1979).

12. C. Lambert, K. Utimoto, H. Nozaki, *Tetrahedron Lett.* 25:5323 (1984).

13. N, Yanagihara, C. Lambert, K. Iritani, K. Utimoto, H, Nozaki, *J. Am. Chem. Soc.* 108:2753 (1986).

14. S. Ma, G. Zhu, and X. Lu, *J. Org. Chem.* 58:3692 (1993).

15. X. Lu, G. Zhu, and Z. Wang, *Synlett* 115 (1998).

16. X. Lu, G. Zhu, Z. Wang, S. Ma, J. Ji and Z. Zhang, *Pure Appl. Chem.* 69:553 (1997).

17. Z. Wang and X. Lu, *Tetrahedron Lett.* 38:5213 (1997).

18. Z. Wang and X. Lu, *Chem. Commun.* 535 (1996).

19. Z. Wang and X. Lu, *J. Org. Chem.* 61:2254 (1996).

20. Z. Wang, X. Lu, A. Lei and Z. Zhang, *J. Org. Chem.* in the press.

21. M. Jacobson, in: *Naturally Occurring Insecticides*, M. Jacobson, D.G. Crosby, Eds., Marcel Dekker, New York, pp.137-176 (1971).

22. Y. Ikeda, J. Ukai, N. Ikeda, H. Yamamoto, *Tetrahedron* 43:731 (1987).

23. R.F. Heck, *Org. Reactions* 27:345 (1982).

24. W. Cabri and I. Candiani, *Acc. Chem. Res.* 28:2 (1995).

25. S. Cacchi, *Pure Appl. Chem.* 62:713 (1990).

26. S. Cacchi, *Pure Appl. Chem.* 68:45, (1996).

27. C. Coperet, P. Sugihara, G. Wu, I. Shimovama, E.-I. Negishi, *J. Am. Chem. Soc.* 117:3422 (1995).

28. A. Arcadi, S. Cacchi, G. Fabrizi, F. Marinelli, P. Pace, *Synlett* 1996; 568.

29. G. Yagupsky, W. Mowat, A. Shortland, G. Wilkinson, *J. Chem. Soc. Chem. Commun.* 1369 (1970).

30. R. J. Cross in: *The Chemistry of the Metal-Carbon Bond*, F. R. Hartley, S. Patai, Eds., Wiley, Chichester, Vol. 2, p.559 (1985).

PALLADIUM CATALYZED NEW BENZANNULATION *VIA* ENYNES

Yoshinori Yamamoto

Department of Chemistry
Graduate School of Science
Tohoku Unviersity
Sendai 980-8578, Japan

INTRODUCTION

Since the earliest example of thermal trimerization of acetylene to benzene reported by Berthelot in 1866, and the first transition-metal-catalyzed version of this reaction demonstrated by Reppe in 1948, the [2 + 2 + 2] cycloaddition of acetylenes was extensively studied by a large number of research groups and became a vast field three decades ago. A large number of transition metal catalysts, as well as Ziegler-type catalysts, give rise to this reaction. Although this approach becomes one of the most powerful methods to assemble a benzene ring, it suffers from serious chemo- and regioselectivity problems which normally lead to complex mixtures of products, thus severely limiting the synthetic utility of this reaction. We communicated two complementary methods of construction of benzene skeleton under the palladium catalysis: the formation of 1,3,5-unsymmetrical benzenes via trimerization of terminal diynes,[1] and synthesis of 1,4-disubstituted benzenes *via* dimerization of conjugated enynes.[2]

HOMO-BENZANNULATION OF CONJUGATED ENYNES[2]

We reported that the palladium ($Pd_2dba_3 \cdot CHCl_3$-dppf) catalyzed reaction of conjugated

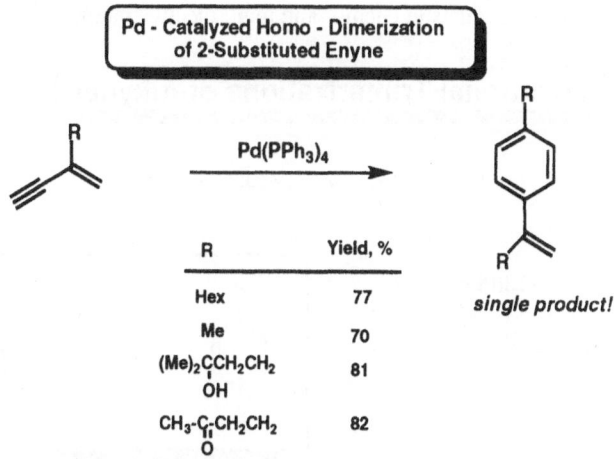

R	Yield, %
Hex	77
Me	70
(Me)$_2$CCH$_2$CH$_2$ OH	81
CH$_3$-C-CH$_2$CH$_2$ O	82

enynes with pronucleophiles affords the corresponding 1,4-addition products (allenes) in high yields.[3] We also examined the reaction between conjugated enynes and pronucleophiles by using the Trost catalyst system (bis π-allylpalladium chloride dimer-KO-^{t}Bu-dppf). Very interestingly, an aromatic compound derived from dimerization of enyne was obtained as a minor product along with the major product (allenes)! Accordingly, we investigated the reaction of 2-hexyl-1-buten-3-yne under various catalyst systems to find the optimum condition for obtaining this unprecedented reaction product in higher yields. Among the catalyst systems we examined, Pd(PPh$_3$)$_4$/toluene or Pd$_2$dba$_3$•CHCl$_3$-tris(2,6-dimethoxyphenyl)phosphine/toluene gave better results. The best solvent for the reaction was toluene, benzene, or acetonitrile; the rate of the reaction was slower in THF, DMSO, and DMF.

To help clarify the mechanism of this unprecedented cyclodimerization, the reaction of enyne in which the deuterium content was 90% at the 4-position was carried out. Disubstituted benzene, which contained deuterium mainly at the 2- and 6-positions (83% deuterium content at each position), was obtained in high yield. Deuterium was not distributed to other positions except for the protons attached to the benzene ring. The formation of the deuterated benzene indicated that the acetylenic C-H(D) bond is not cleaved in the present benzannulation. Transition metal catalysts effective for the cyclotrimerization of acetylenes (CpCo(Co)$_2$ or RhCl(PPh$_3$)$_2$) were not effective for the cyclodimerization of enynes. Taken together, it seems difficult to explain the unprecedented cyclodimerization by the ordinary accepted mechanisms.

(10% H, 90 % D)

R = n-C$_6$H$_{13}$

PALLADIUM-CATALYZED ENYNE-DIYNE [4+2] CROSS-BENZANNULATION REACTION

The well known [2 + 2 + 2] cycloaddition of acetylenes suffers from serious chemo- and regioselectivity problems which normally lead to complex mixtures of products, thus severely limiting the synthetic utility of this reaction[4]. Vollhardt succeeded to solve these problems for several types of *intramolecular* or *partially intramolecular* modes of cyclotrimerization: three new bonds were formed under the cobalt catalysis affording cyclophane-type aromatic products in chemo- and regioselective manner.

TRANSITION METAL-CATALYZED:

not regioselective

Why is the traditional trimerization not regioselective? Various different ways of the orientation of acetylenes are possible for assembling an intermediate **i**, since three new bonds are formed in [2+2+2] cyclotrimerization. It occurred to us that in the case where a conjugated enyne would react with alkyne in a [4+2] cycloadditon manner, this reaction could be more regioselective than the [2+2+2] mode of cycloaddition since only the regioselectivity of two bond formation remains questionable. After trying a number of alkynes in a role of enyne partner in the [4+2] cycloaddition, we discovered that conjugated diynes underwent regiospecific cross-cycloaddition with enynes substituted at the C-2 position, as shown below.

Why Traditional Trimerization is not Regioselective

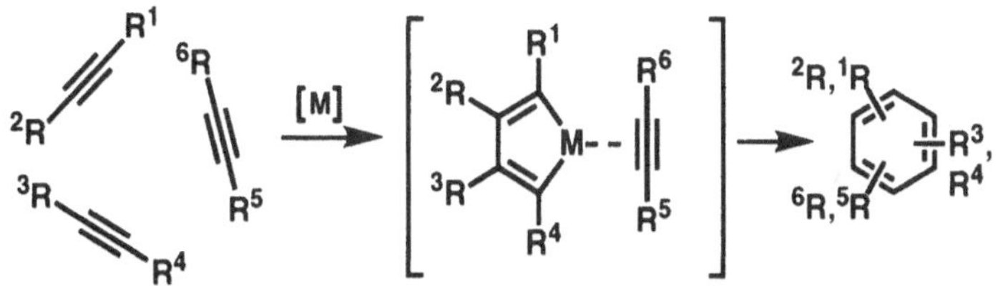

Are Alternative Approaches Toward Benzene Ring Possible?

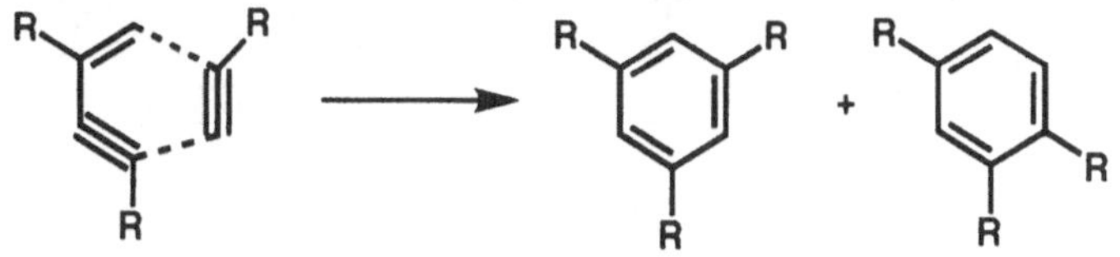

We discovered the first example of *intermolecular* enyne-diyne [4+2] cross-benzannulation reaction: the reaction of enynes with diynes affording regioselectively 1,2,4-trisubstituted benzenes in high to quantitative yields with none of the regioisomers being produced.[5]

Enyne-Diyne Cross-Benzannulation

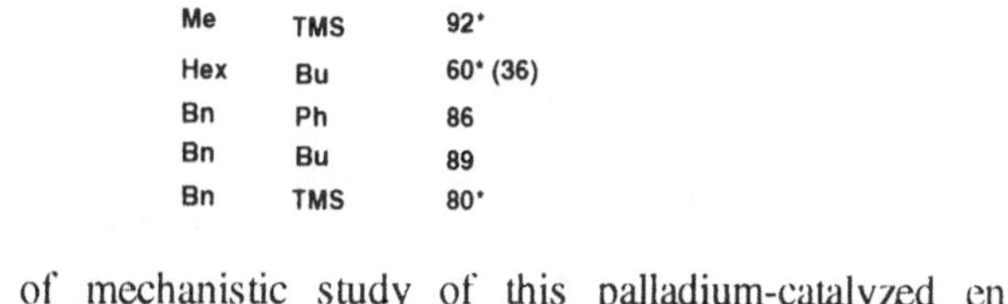

R	R'	Yield (recovery, %)
Me	Ph	100
Me	Bu	89 (10)
Me	TMS	92*
Hex	Bu	60* (36)
Bn	Ph	86
Bn	Bu	89
Bn	TMS	80*

As a part of mechanistic study of this palladium-catalyzed enyne-diyne cross-benzannulatoin, the deuterium analogues were employed in the reaction with dodeca-5,7-diyne. Accordingly, the cross-benzannulation of the mono-deuterated enyne afforded the corresponding benzene derivative in 64% yield with a deuterium atom attached to the C-6 position of benzene ring, and cycloaddition of the di-deuterated enyne gave the di-deuterated benzene in 61 % yield with deuterium atoms at C-3 and C-5 positions of ring, exclusively.

The results on deuterium labeling experiments, taken together with the fact of *regiospecific* formation of *para*-oriented regioisomer, encouraged us to propose the following mechanistic rationale for this reaction. The reversible coordination of palladium with enyne and diyne would produce palladacycle, stabilized by coordination of Pd atom with neighboring η^3-propargyl moiety. The reductive elimination of palladium would form strained cyclic cumulene, which *via* sigmatropic rearrangement would be transformed into cross-annulation product. Alternatively, the 1,3-shift of deuterium would give the triene-π-allylpalladium complex, which would produce the benzene derivative upon reductive coupling.

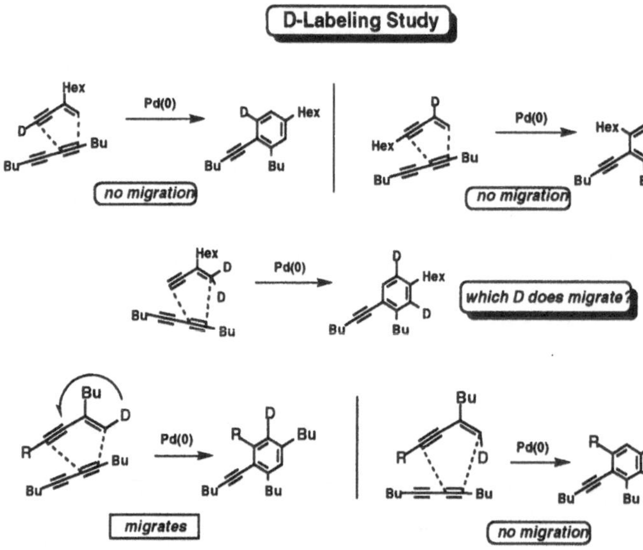

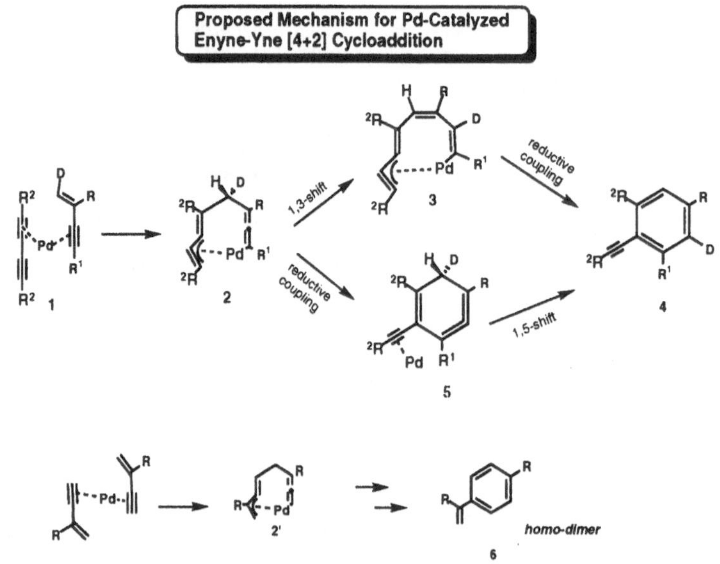

D-Labeling Study

Pd(0) no migration

Pd(0) no migration

Pd(0) which D does migrate?

Pd(0) migrates

Pd(0) no migration

Proposed Mechanism for Pd-Catalyzed Enyne-Yne [4+2] Cycloaddition

1 → 2 → 1,3-shift → 3 → reductive coupling → 4
2 → reductive coupling → 5 → 1,5-shift → 4

homo-dimer 2' 6

Even 1,2,4-trisubstituted enynes underwent the cross-benzannulation with diynes to produce regioselectively penta-substituted benzenes in high to allowable yields.[6] Interestingly, carborane substituted enynes gave carboranyl benzenes in good yields, in which *o*, *m*, and *p*-carboranes could be introduced.

Z reacts faster than E

43-95%

R - R³ = alkyl, aryl, TMS, CO₂R

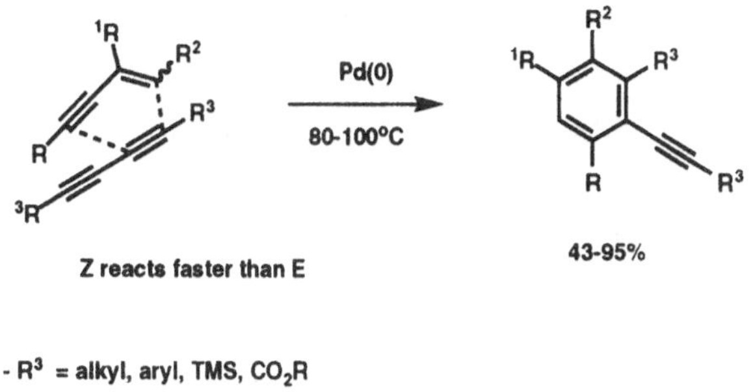

Synthesis of Carboranyl Ethynyl Benzenes

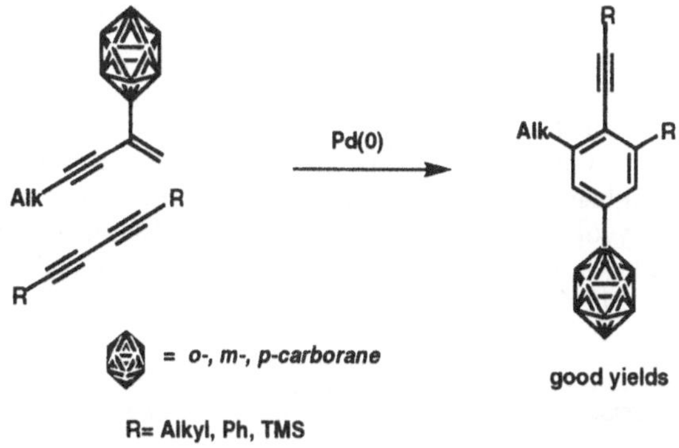

= o-, m-, p-carborane

good yields

R= Alkyl, Ph, TMS

FIRST SYNTHESIS OF EXOMETHYLENE PARACYCLOPHANES

A series of novel exomethylene paracyclophanes was synthesized by the intramolecular benzannulation of conjugated enynes in the presence of tetrakis (triphenylphosphine) palladium(0).[7] Detailed spectral studies revealed that the conformation of the exomethylene paracyclophanes depends on the size of the ring; the alkene moiety does not conjugate with the adjacent phenyl group in the smaller exomethylene paracyclophanes.

Synthesis of [15] - Paracyclophane

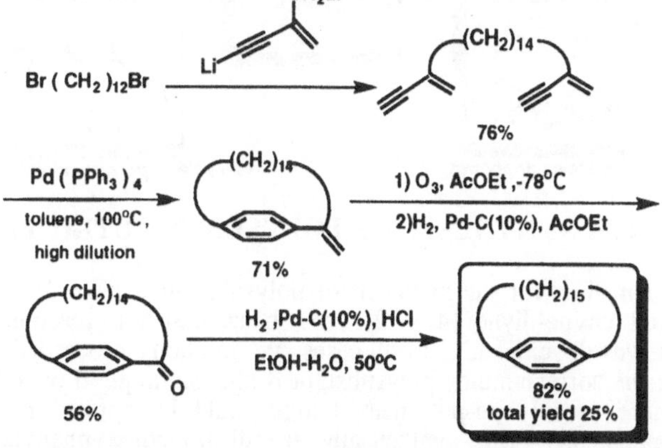

Not only *p*-cyclophanes, but also *m*- and *o*- cyclophanes were synthesized by using cyclic enynes and cyclic diynes, respectively.

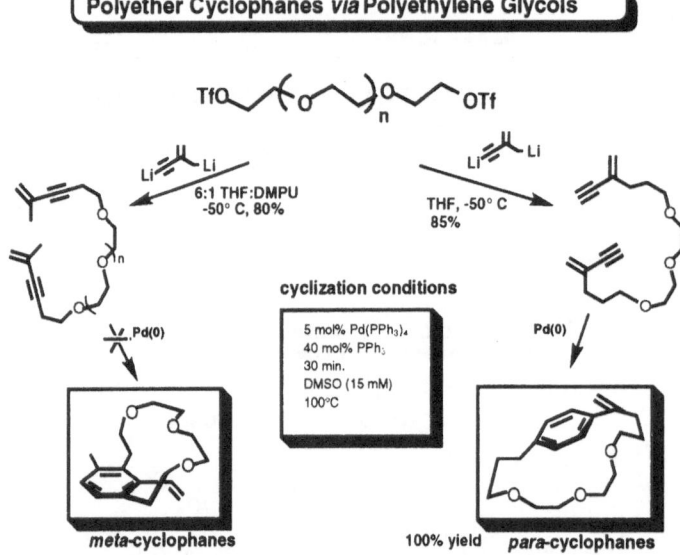

SYNTHESIS OF POLYETHER EXOMETHYLENE PARACYCLOPHANES

Several novel crownlike exomethylene paracyclophanes were efficiently synthesized via an intramolecular palladium-catalyzed benzannulation of conjugated bis-enynes.[8] A method for the efficient synthesis of the bis-enynes, bearing an oligo(oxyethylene)linkage, was developed by utilizing a two-step procedure from commercially available and inexpensive ethylene glycols.   A remarkable complete reversal of reactivity of the ambident dilithiated nucleophile toward the triflate in the presence of 1,3-dimethyl-3,4,5,6-tetrahydro-2(1H)-pyrimidinone (DMPU)was found.

REGIOSELECTIVE SYNTHESIS OF POLYSUBSTITUTED PHENOLS

An efficient method for the synthesis of polysubstituted phenols via the consecutive palladium-catalyzed enyne-diyne [4 + 2] cross-benzannulation reaction and subsequent deprotection step was developed.[9]   In all cases, the reactions proceeded in a regiospecific manner affording the corresponding polysubstituted phenols in good overall yields.   It was shown that a more useful one-pot methodology could be applied to the synthesis of polysubstituted phenols.   The synthetically useful p-methoxyphenylacetylene and its

288

monosilylated derivative were smoothly prepared via exhaustive or partial desilylation of bis-silylated aromatic adduct, respectively.

Polysubstituted Phenols *via* **Pd-Catalyzed Enyne-Diyne Cross- Benzannulation**

R-R^2 = H, Alkyl, Aryl

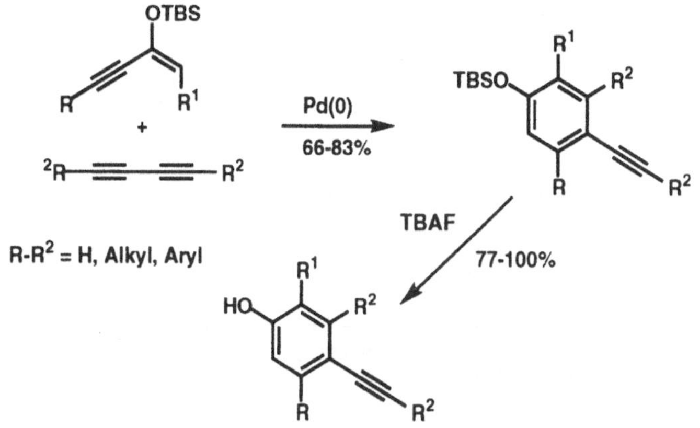

REGIOSELECTIVE SYNTHESIS OF ANILINES

The enynes, substituted with nitrogen atom at the C-2 position, also underwent the [4 + 2] cross benzannulation reaction to give the corresponding aniline derivatives in good to allowable yields. As a protective group of the nitrogen atom, Boc(t-butoxycarbonyl) group is essential to obtain the cross-benzannulation products in good yields.

Polysubstituted Anilines *via* **Pd-Catalyzed Enyne-Diyne Cross- Benzannulation**

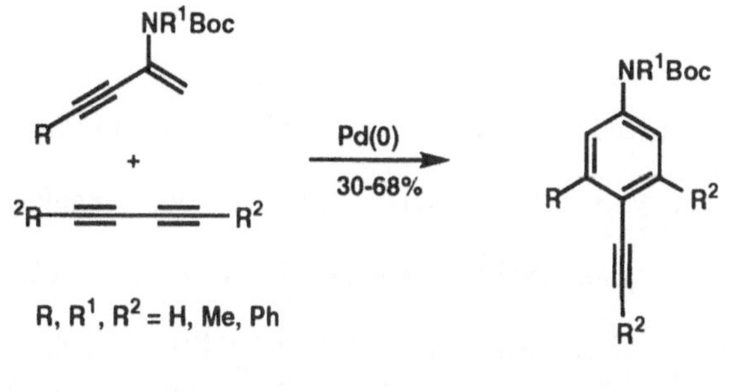

R, R^1, R^2 = H, Me, Ph

SEQUENTIAL TRIMERIZATION OF ALKYNES

We tried to carry out sequential trimerization of different three alkynes, based upon our finding on the enyne-diyne cross-benzannulation. The basic idea is following. As shown above, benzene skeleton is derived from the [4 + 2] cycloaddition between enynes and certain acetylenes having an activating group, such as triple or double bond. It occurred to us that the enynes may be synthesized in situ from a terminal acetylene and another acetylene.

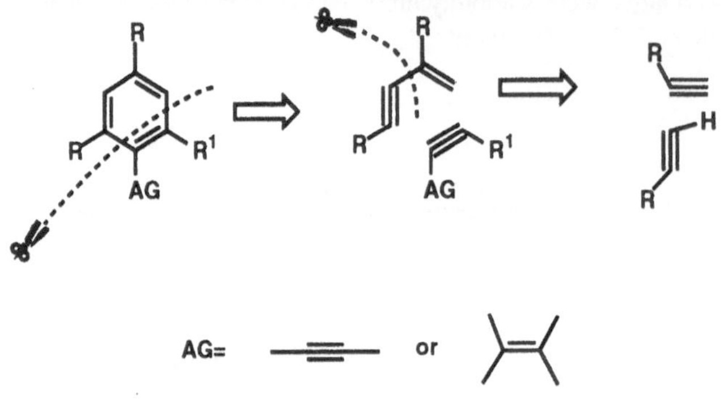

AG= ——≡—— or >=<

Actually, the one pot reaction of terminal acetylenes, acetylenes having an electron withdrawing group, and diynes gave the sequential trimerization products regioselectively in good to allowable yields.

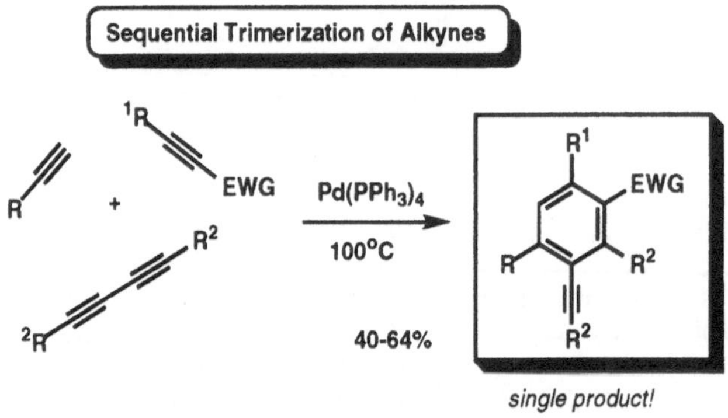

Sequential Trimerization of Alkynes

single product!

R= alkyl, Ph, CH_2NEt_2;

R^1= alkyl, Ph;

EWG= COR, CO_2R;

R^2= alkyl, Ph

Reference

1. A.Takeda, A.Ohno, I.Kadota, V.Gevorgyan, and Y.Yamamoto, *J.Am.Chem.Soc.*, 119, 4547 (1997).
2. S.Saito, M.M.Salter, V.Gevorgyan, N.Tsuboya, K.Tando, Y.Yamamoto, *J.Am.Chem.Soc.*, 117, 3970 (1996).
3. M.M.Salter, V.Gevorgyan, S.Saito, Y.Yamamoto, *J.Chem.Soc., Chem.Commun.*, 17 (1996).
4. W.Reppe,*et al. Justus Liebigs Ann.Chem.*, 560, 1 (1948). K.P.C.Vollhardt, *Angew.Chem.*, Int.Ed.Engl., 23, 539 (1984). N.E.Schore, *Chem.Rev.*, 88, 1081 (1988).
5. V.Gevorgyan, A.Takeda, Y.Yamamoto, *J.Am.Chem.Soc.*, 119, 11313 (1997).
6. V.Gevorgyan, N.Sadayori, Y.Yamamoto, *Tetrahedron Lett.*, 38, 8603 (1997).
7. S.Saito, N.Tsuboya, and Y.Yamamoto, *J.Org.Chem.*, 62, 5042 (1997).
8. D.Weibel, V.Gevorgyan, and Y.Yamamoto, *J.Org.Chem.*, 63, 1217 (1998).
9. V.Gevorgyan, L.G.Quan, and Y.Yamamoto, *J.Org.Chem.*, 63, 1244,(1998).

CHEMISTRY AND BIOLOGY OF CALYCULIN C

Ari M.P. Koskinen,* and Petri M. Pihko

Department of Chemistry
University of Oulu
Linnanmaa
FI-90570 Oulu, Finland

INTRODUCTION

Calyculins A and C, two primary members of a polyketide antibiotic family now comprising 11 members, were originally isolated from *Discodermia calyx*, a marine sponge living in the secluded waters of Edo Bay, Japan.[1] Calyculins exhibit strong serine/threonine protein phosphatase inhibitory activity. Their target enzymes belong to the PP1/PP2A/PP2B subfamily, which form a phylogenetically and structurally homogenous subset of protein phosphatases. By dephosphorylating serine and threonine residues, these enzymes act as important "on/off" switches, and modulate the activity of numerous other proteins, including many important signal transduction messengers. In contrast to most enzymes, members of the PP1/PP2A/PP2B subfamily exhibit broad substrate specificity and they (especially PP1 and PP2A) are involved in almost every part and function of the cell.[2] Thus, discerning the phosphatases responsible for a particular effect or pathway has been possible largely with the help of natural toxins inhibiting these enzymes. Calyculins have been especially fruitful in these studies.

Fusetani and his group first solved the intriguing structure of calyculins by X-ray crystallography in 1986.[1] At that time, the absolute stereochemistry remained unknown. However, the combination of a challenging, novel structure and the increasing importance for cell biological studies led to several total synthesis approaches.[3] Initially, these studies were all aimed at the enantiomer of the natural product. Later, in 1991, both Fusetani and Shioiri ascertained the absolute stereochemistry to that shown in the structures in this review.[4]

THE BINDING MODE OF CALYCULINS TO PP1[5]

With the publication of the crystal structure of PP1 complexed with the cyanobacterial toxin microcystin in 1995,[6] the binding interactions of other toxins, including the calyculins, with the enzyme could be explored by means of molecular modeling. We initially began our modeling efforts with the published crystal structure of calyculin A (inverted to represent the natural enantiomer). However, we soon found that the pseudo-cyclic structure of the crystal structure, held together by hydrogen bonds between the phosphate and the Me_2N, oxazole, amide and hydroxyl groups, simply cannot fit efficiently into the pocket at the active site and the more shallow grooves extending from it. Opening up of the crystal structure to an extended conformation allowed crucial contacts between the acidic residues in the acidic groove (Glu-256, Asp-208) and the basic Me_2N-containing chain. The spiroketal-phosphate unit then lies at the active site, with the hydrophobic tetraene chain extending into the hydrophobic groove (see Figure 1).[5]

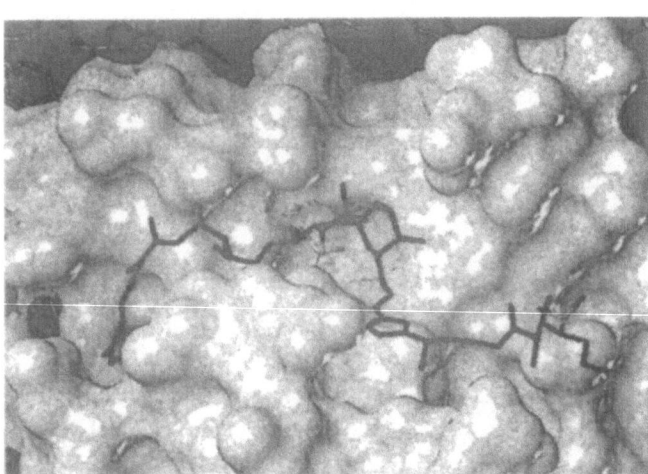

Independently from us, Chamberlin and co-workers have also proposed a binding model, which is very similar to ours.[7] Two other approaches to the binding have also been proposed, both only with the unnatural enantiomer.[8]

Figure 1. Proposed binding mode of calyculin A to PP1. Calyculin A is shown in black and the Connolly surface of the enzyme in grey background. The hydrophobic groove is to the left, the active site pocket is in the center and the acidic groove is to the right.

SYNTHETIC ANALYSIS

Our own interest in the synthesis of calyculins had a somewhat different origin. Having only recently started my own independent academic career (AMK), and being limited in funding, I was looking for prospective candidates for total synthesis for my new students. The structure published for Calyculin A was definitely challenging enough, and the tail portion (A, C_{33}-C_{39}, Scheme 1) of the molecule could naively be seen as originating from L-serine (albeit in enantiomeric form to that then believed to be the absolute configuration of the molecule). Bearing this in mind, the synthetic analysis straightforwardly unraveled as shown in Scheme 1: the obvious disconnections involve a late stage Wittig-type olefination to give the Northern and Southern parts of the molecule. The Northern part displays a trivial amide-disconnection between N-32 and C-33. The Southern part provides a little more challenge for synthetic chemists: construction of the spiroketal and accumulation of the tetraene conjugation was anticipated to provide provocation during the synthesis.

In this paper, we describe our approach to the syntheses of the Northern half segments A and B, and will also illustrate our initial efforts towards the synthesis of the Southern hemisphere, in particular towards the synthesis of the tetraene unit ($D+D'$).

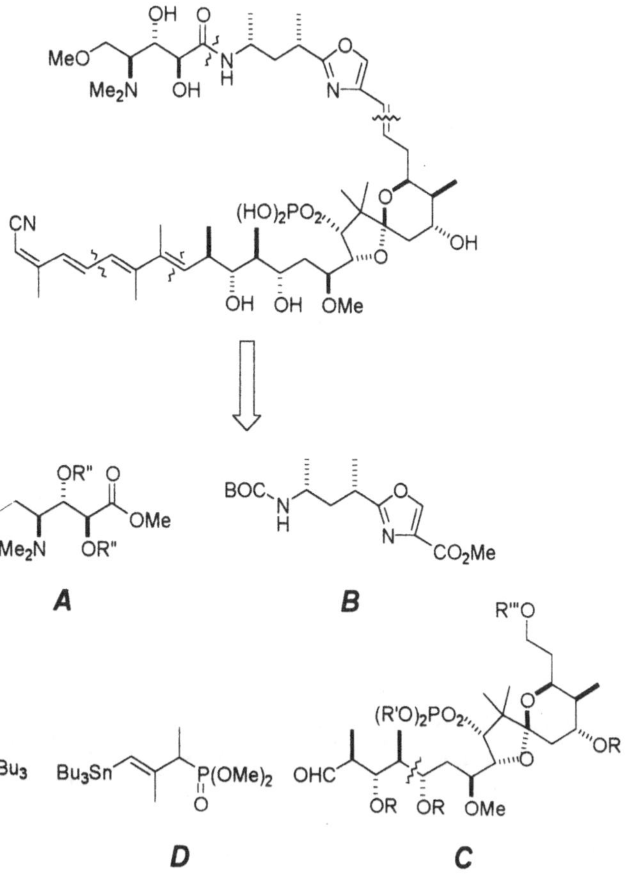

SCHEME 1. Retrosynthetic analysis of Calyculin C.

AMINO ACID SEGMENT C$_{33}$-C$_{38}$

Our first retrosynthetic analysis of the C$_{33}$-C$_{38}$ segment started, rather naively in retrospect, from the simple assumption that a Kishi-selective dihydroxylation of a suitably substituted Z-enoate as shown in Scheme 2 would lead into the desired stereochemistry.

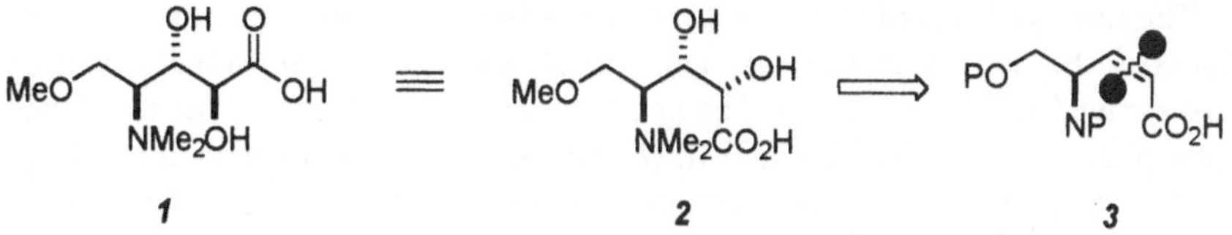

SCHEME 2. Synthetic analysis of the C33-C38 fragment.

Luckily for us, the crucial work by Shioiri and Yokokawa,[9] had not been published by then. They had conducted nearly identical synthetic steps on an *open chain protected serine analogue, which led to a slight predominance for the undesired anti-Kishi dihydroxylation*. In our case, the use of *cyclic* protection (originally derived from the Garner aldehyde) proved totally satisfactory: dihydroxylation of the Z-enoate gave *better than 50:1 Kishi-selectivity* (as evidenced by 500 MHz NMR).

Learning later about this discrepancy with the Shioiri work, we wanted to know the reasons why the selectivity in our case is so much higher. We were lucky to obtain crystals of the Z-enoate derived from L-threoninal. This compound gave dihydroxylation products much *faster, and with no trace of anti-Kishi products detectable*. After considerable

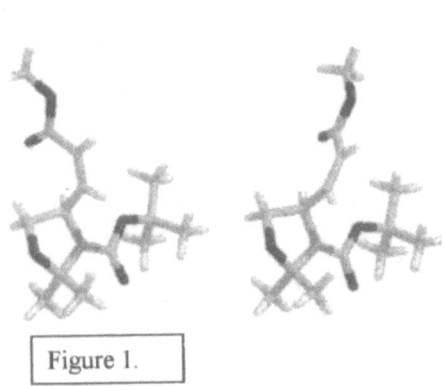

experimentation, we also managed to obtain diffraction quality crystals of the serine derived Z-enoate. This information proved very informative: the crystal contains two molecules (Figure 1), where the orientation of the side chain is different. This obviously leads to different facial selectivities in comparison to the threonine-derived enoate, and thus also explains the erosion in the dihydroxylation selectivities. Based on crude MacroModel calculations (version 6.0, MMFF) the differences in populations of the two lowest energy conformations are approx. 4

Figure 1.

kcal/mol, which is also consistent with experimental results.

The diol was finally converted into the amino acid fragment, a known hydrolysis product from natural Calyculin A.[9]

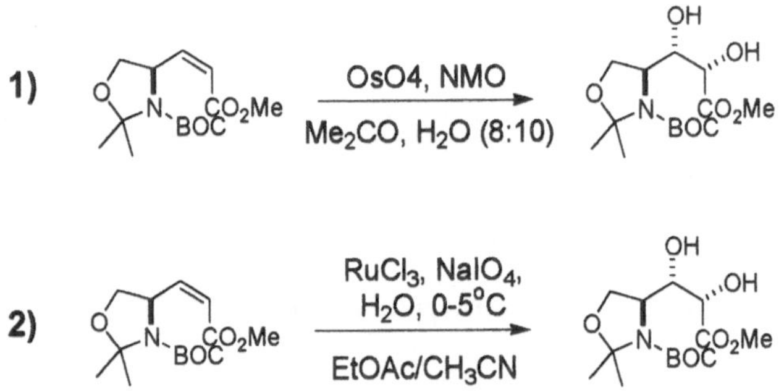

1) $\xrightarrow[\text{Me}_2\text{CO, H}_2\text{O (8:10)}]{\text{OsO}_4, \text{NMO}}$

2) $\xrightarrow[\text{EtOAc/CH}_3\text{CN}]{\text{RuCl}_3, \text{NaIO}_4, \text{H}_2\text{O, 0-5}^\circ\text{C}}$

3   4

SCHEME 3. Dihydroxylation of the enoate.

OXAZOLE FRAGMENT C$_{26}$-C$_{32}$[10]

The retrosynthesis of the oxazole fragment began by noting that the stereochemistry at C$_{32}$ could be derived from D-alanine. To test our strategy, we initially focused on L-alanine-derived starting materials. The known L-alaninal derivative **5** was olefinated with the phosphorane **6** to afford the known (*E*)-enoate **7** in 95 % yield with an (*E*):(*Z*) ratio 18:1. Attempts at hydrogenating the double bond with the Pfaltz-type bisoxazoline, semicorrin or pyridyloxazoline ligands and Co(II)/NaBH$_4$ proved unsuccessful. Direct hydrogenation over Pd/C in EtOH, however, cleanly afforded a 2:1 diastereomeric mixture of the *anti-* and *syn-*isomers **8a** and **8b** in a quantitative yield.

The predominance of the *anti*-isomer could be rationalized on the basis of 1,3-allylic strain: the (*E*)-enoate adopts a conformation where the *si* face is hindered by the Boc group (see inset).

Since the undesired *anti* diastereomer was the major product obtained with the (*E*)-enoate, the synthesis was then attempted *via* the corresponding (*Z*)-enoate **12**, prepared from Boc-L-alaninal through the Still-Gennari modification of Horner-Emmons-Wadsworth reaction using the phosphonate **11**. To our surprise, hydrogenation of **12** over Pd/C gave the esters **8a** and **8b** in *nearly the same diastereomer ratio (5:3), with the undesired* anti-*isomer predominating*.

In this case, a γ-turn type conformation, stabilized by the dipolar interaction between the NH and the ester groups, could be invoked to explain the poor selectivity. Allylic isomerization on the surface of the catalyst can also lead to similar results.[11]

Saponification of the mixture **8a/8b** with aqueous NaOH afforded a mixture of the acids **9a** and **9b**, from which the pure *anti* isomer **9a** was readily obtained by fractional recrystallization in 61% yield. Single crystal X-ray diffraction of **9a** provided a proof of the relative stereochemistry.

295

A successful route to the desired *syn* isomer was then found which involves *cyclic stereocontrol*, thereby imposing better control over the stereochemistry of hydrogenation. The sequence shown below begins with the enantiomeric D-alaninal derivative **10**. Cyclization of the (*Z*)-enoate **12** to the corresponding lactam **13** with Boc$_2$O and DMAP followed by hydrogenation over Pd/C, afforded a 10:1 mixture of the *syn* and *anti* pyrrolidinones **14a** and **14b**. Subsequent hydrolysis with lithium hydroperoxide[30] gave the open-chain acids **15a** and **15b**, which were directly carried over to the coupling with L-serine methyl ester under mixed anhydride conditions

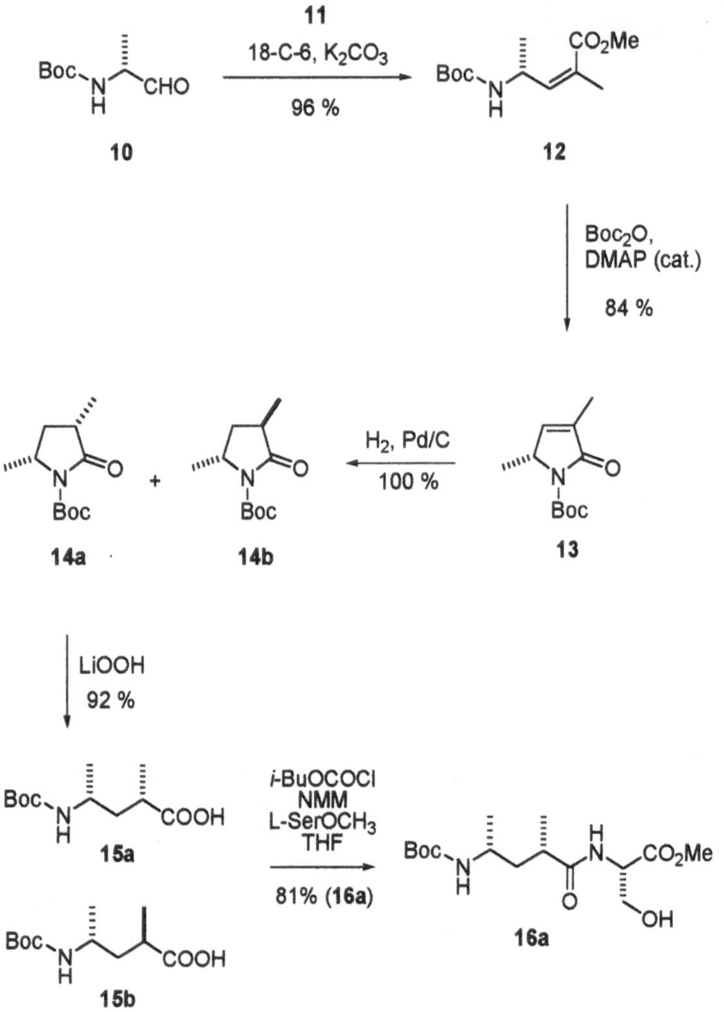

11

10 → [18-C-6, K$_2$CO$_3$, 96 %] → **12**

12 → [Boc$_2$O, DMAP (cat.), 84 %]

13 → [H$_2$, Pd/C, 100 %] → **14a** + **14b**

14a → [LiOOH, 92 %] → **15a**

15a / **15b**

15a → [*i*-BuOCOCl, NMM, L-SerOCH$_3$, THF, 81% (**16a**)] → **16a**

to afford the dipeptide 16a in 81 % yield after separation of the minor diastereomer by chromatography.

After some experimentation, we found that the conversion of **16a** to the oxazoline **18** was best accomplished with the Burgess reagent in 83 % yield. The final step, oxidative aromatization to the oxazole, proved particularly challenging. Several novel methods for oxidizing oxazolines to the oxazoles have been disclosed in recent years, and we began exploring them with high hopes. Of the more than a dozen methods tried, only two gave satisfactory and reproducible results. The CuBr$_2$/HMTA/DBU oxidation (developed at Bristol Myers-Squibb) proved particularly clean and reproducible with different substrates, but gave only 29% yield with the *syn* oxazoline **18**. Our own method involving temporary TMS protection of the carbamate nitrogen, deprotonation of the oxazoline with KHMDS and oxidation of the intermediate enolate directly with iodine gave a satisfactory 42 % yield of the desired oxazole **20**.

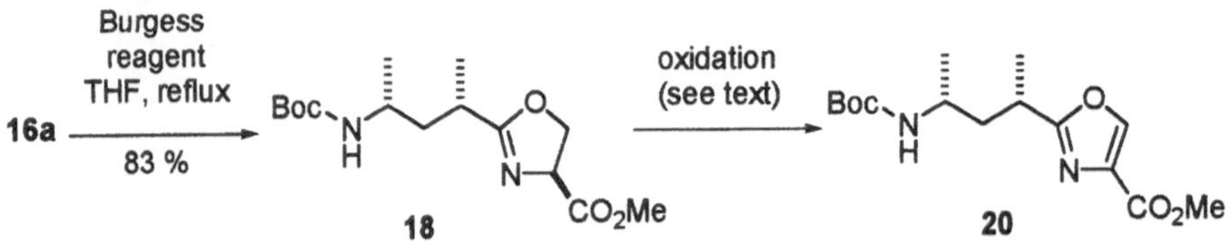

16a → [Burgess reagent, THF, reflux, 83 %] → **18** → [oxidation (see text)] → **20**

In all oxidations, side reactions were a problem—the major side product, the spirocyclic orthoester aminal **17** could usually also be isolated in up to 52 % yield.

17

TETRAENE FRAGMENT

The synthetic strategy for a particularly unstable, nitrile-capped tetraene fragment must involve mild coupling conditions to preserve the delicate olefin geometry. Sp2-sp2 – couplings with Pd(0) catalysts have served particularly well in many syntheses of very complex molecules, and have also found use in the published strategies towards the calyculins.

Bearing in mind the instability brought in by the nitrile group, the most efficient strategy would be to install as much of the tetraene as possible *via* a Horner-Emmons-type coupling and finally add the nitrile-containing "cap" *via* a Pd(0)-catalyzed coupling. This leads to the following fragments, of which **21** is a known compound, but **22** is more arduously accessible.

Disconnect at C$_3$-C$_4$

21 **22** + RCHO

After considerable experimentation with Stille-type couplings towards the synthesis of **22**, we finally found that the ready exchange of one of the tins to lithium in the bis-stannylethene **22** actually opens a way to exchange the tin to zinc, thereby giving access to a Negishi-type reagent which are known to be more reactive than the corresponding organotin reagents in Pd(0) catalyzed couplings.

i) *n*-BuLi
ii) ZnCl$_2$
23: M = SnBu$_3$
24: M = ZnCl

Br CO$_2$Me **25**
Pd(PPh$_3$)$_4$
(81 % crude)

26

63 % (two steps) DIBAL-H

22 **27**

In our case, the coupling between the bismetallic reagent **24** and the bromide **25** proceeded extremely rapidly, giving the diene in 81% crude yield. Reduction with DIBAL-H gave the alcohol **27** in 63 % yield over two steps.

ACKNOWLEDGEMENTS

This work has been financially supported by the Ministry of Education, Finland, Graduate School on Bioorganic Chemistry. Support from the Academy of Finland and TEKES (Technology Development Centre, Finland) is also gratefully acknowledged.

Skillful technical assistance by Mr. Vesa Rauhala, M.Sc is gratefully acknowledged. The assistance of Mrs. Leena Otsomaa, Lic. Phil. (University of Oulu), and Ms. Maija Nissinen, M.Sc. and Prof. Kari Rissanen (both of University of Jyväskylä, Finland) was imperative for obtaining the X-ray structures.

REFERENCES

1. (a) Kato, Y.; Fusetani, N.; Matsunaga, S.; Hashimoto, K.; Fujita, S.; Furuya, T. *J. Am. Chem. Soc.* **1986**, *108*, 2780-2781. (b) Kato, Y.; Fusetani, N.; Matsunaga, S.; Hashimoto, K.; Koseki, K. *J. Org. Chem.* **1988**, *53*, 3930-3932. (c) Matsunaga, S.; Fujiki, H.; Sakata, D.; Fusetani, N. *Tetrahedron* **1991**, *47*, 2999-3006. (d) Matsunaga, S.; Wakimoto, T.; Fusetani, N.; Suganuma, M. *Tetrahedron Lett.* **1997**, *38*, 3763-3764. (e) Matsunaga, S.; Wakimoto, T.; Fusetani, N. *J. Org. Chem.* **1997**, *62*, 2640-2642. (f) Dumdei, E.; Blunt, J.W.; Munro, M.H.G.; Pannell, L.K. *J. Org. Chem.* **1997**, *62*, 2636-2639.

2. For reviews of protein phosphatases 1 and 2A, see: (a) Wera, S.; Hemmings, B. A. *Biochem. J.* **1995**, 311, 17–29. (b) Barford, D. *Trends Biochem. Sci.* **1996**, *21*, 407-412. (c) Widlanski, T.S.; Myers, J.K.; Stec, B.; Holtz, K.M.; Kantrowitz, E.R. *Chem. Biol.* **1997**, *4*, 489-492.

3. For total syntheses of calyculins, see: (a) Evans, D. A.; Gage, J. R.; Leighton, J. L. *J. Am. Chem. Soc.* **1992**, *114*, 9434-9453.(b) Tanimoto, N.; Gerritz, S.W.; Sawabe, A.; Noda, T.; Filla, S.A.; Masamune, S. *Angew. Chem. Int. Ed. Engl.* **1994**, *33*, 673-675. (c) Yokokawa, F.; Hamada, Y.; Shioiri, T. *Chem. Commun.* **1996**, 871-872. For leading references to other synthetic approaches to the calyculins, see: (d) Scarlato, G. R.; DeMattei, J. A.; Chong, L. S.; Ogawa, A. K.; Lin, M. R.; Armstrong, R. W. *J. Org. Chem.* **1996**, *61*, 6139-6152. (e) Ogawa, A. K.; DeMattei, J. A.; Scarlato, G. R.; Tellew, J. E.; Chong, L. S.; Armstrong, R. W. *J. Org. Chem.* **1996**, *61*, 6153-6161. (f) Trost, B. M.; Flygare, J. A. *Tetrahedron Lett.* **1994**, *35*, 4059-4062. (g) Smith, A. B., III; Salvatore, B. A. *Tetrahedron Lett.* **1994**, *32*, 1329-1333. (h) Barrett, A. G. M.; Edmunds, J. J.; Horita, K.; Parkinson, C. J. *J. Chem. Soc., Chem. Commun.* **1992**, 1236-1238.

4. (a) Matsunaga, S.; Fusetani, N. *Tetrahedron Lett.* **1991**, *32*, 5605-5606. (b) Hamada, Y.; Tanada, Y.; Yokokawa, F.; Shioiri, T. *Tetrahedron Lett.* **1991**, *32*, 5983-5986.

5. Lindvall, M. K.; Pihko, P. M.; Koskinen, A. M. P. *J. Biol. Chem.* **1997**, *272*, 23312-23316.

6. Goldberg, J.; Huang, H.; Kwon, Y.; Greengard, P.; Nairn, A. C.; Kuriyan, J. *Nature* **1995**, *376*, 745-753.

7. Gauss, C-M.; Sheppeck, J. E.; Nairn, A. C.; Chamberlin, R. *Bioorg. Med. Chem.* **1997**, *5*, 1751-1773.

8. (a) Bagu, J. R., Sykes, B. D, Craig, M. M. and Holmes, C. F. B. (1997) *J. Biol. Chem.* **272**, 5087-5097. (b) Gupta, V.; Ogawa, A. K.; Du, X.; Houk, K. N.; Armstrong, R. W. *J. Med. Chem.* **1997**, *40*, 3199-3206.

9. Yokokawa, F.; Hamada, Y.; Shioiri, T. *Synlett* **1992**, 703-705.

10. Koskinen, A.M.P.; Chen. J. *Tetrahedron Lett.* **1991**, *32*, 6977-6980.

11. Kauppinen, P.M.; Koskinen, A.M.P. *Tetrahedron Lett.* **1997**, *38*, 3103-3106.

SYNTHESIS OF CYTOTOXIC MARINE MACROLIDES: AN ALDOL-BASED APPROACH TO SPONGISTATIN 1 (ALTOHYRTIN A)

Ian Paterson,* Karl R. Gibson, Linda E. Keown, Roger D. Norcross, Renata M. Oballa and Debra J. Wallace

University Chemical Laboratory
Lensfield Road
Cambridge CB2 1EW
UK

INTRODUCTION

Marine organisms provide an important source of natural product diversity with an associated range of significant biological activities, which may provide new leads for drug discovery. For example, some of the most primitive multicellular invertebrates are sponges, which may be associated with cytotoxic metabolites having potential for development as anti-cancer drugs (*e.g.*, the halichondrins and discodermolide). In this regard, the spongistatins [1,2] (altohyrtins [3]) are a novel group of potent cytotoxic macrolides,[4] which have recently attracted considerable interest and excitement from marine natural product and synthetic organic chemists.

spongistatin 1 (1, X = Cl) = *altohyrtin A*
spongistatin 2 (2, X = H) = *altohyrtin C*

- -

- *Extremely potent cytotoxic agents*
- *Antimitotic action due to inhibition of microtubule assembly*
- *Isolated from sponges in only ~ 10^{-7}% yield*
- *42-membered macrolide containing 6 pyran rings, including 2 spiroacetal systems, with 24 stereocentres*

Pettit *et al.* isolated the spongistatins from sponges of the genus *Spongia*[1a,b] and *Spirastrella*,[1c,d] while the altohyrtins were obtained by the Kobayashi/Kitagawa group from *Hyrtios altum*.[3] As antimitotic agents, these compounds show powerful growth inhibitory activity against multi-drug resistant cancer cells and they may function by inhibiting microtubule assembly by binding to the vinca alkaloid domain of tubulin.[2] For example, spongistatin 1 is reported to be among the most potent cytotoxic compounds tested in the US National Cancer Institute's panel of 60 human cancer cell lines (mean $GI_{50} = 10^{-11}$ *M*).

Their highly complex, polyketide structures, *e.g.* **1** for spongistatin 1 (altohyrtin A[3a,b]), and potent antimitotic action, combined with an extremely meagre natural supply, has provided the impetus for a growing number of synthetic efforts.[5-9] Notably, the first total syntheses of altohyrtins A (**1**) and C (**2**) (structurally identical to spongistatins 1 and 2) have recently been achieved independently by the groups of Kishi[6] and Evans,[7] respectively. This confirmed the full stereochemical assignment made by the Kobayashi/Kitagawa group for the altohyrtins,[3a,b] which was based on NOESY and Mosher ester ^{1}H NMR experiments.

Herein, we review our own progress[5] towards the stereocontrolled synthesis of spongistatin 1 (altohyrtin A), following a novel aldol strategy. In particular, we demonstrate the use of Ipc-substituted, boron enolates derived from methyl ketones, as developed earlier in our laboratory,[10] for the generation of much of the 1,3-polyol framework of the spongistatins, achieving high levels of 1,3-, 1,4-, and 1,5-stereocontrol.

SYNTHETIC STRATEGY FOR THE SPONGISTATINS

As outlined in Scheme 1, our retrosynthetic analysis for spongistatin 1 (**1**) is based on a 3-fold disconnection of the 42-membered macrolide ring across the lactone, the C_{28} *cis*-alkene and the C_{15}–C_{16} bond linking the two spiroacetal ring systems. We planned a highly convergent synthetic route based on the controlled coupling of the resulting three subunits **3** (containing the AB spiroacetal), **4** (containing the CD spiroacetal) and **5** (containing the E and F pyran rings, together with the chlorinated, unsaturated side-chain). Due to the highly functionalised nature of the target macrolide and its likely indiscriminate lability to many reagents, it was highly desirable to minimise the number of steps conducted after the 3-component assembly – at best, this would be 2 steps, a regioselective macrolactonisation (at the C_{41}-OH) followed by global deprotection.

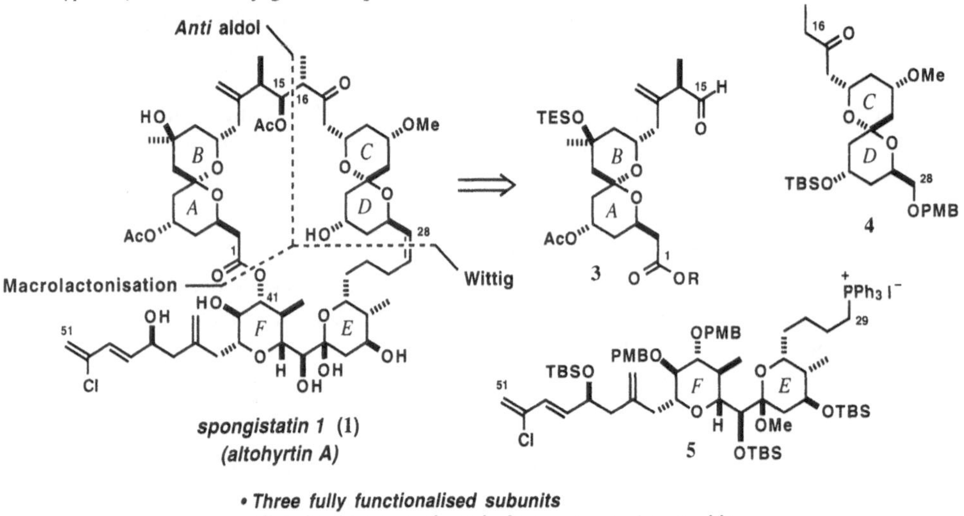

- Three fully functionalised subunits
- Minimise steps conducted after component assembly

Scheme 1: 3-Fold disconnection at C_1–O, C_{15}–C_{16} and C_{28}–C_{29}

Altogether, there are some 24 stereogenic centres to be considered in the synthetic plan – these should be introduced using a combination of suitable chiral building blocks with substrate- and reagent-controlled reactions. The high level of oxygenation calls for a suitable selection of protecting groups, particularly for the subunit **5** containing the E and F rings. Our strategy for assembling most of the carbon and oxygen skeleton in the subunits **3–5**, together with the associated stereochemistry, is centred around the use of a variety of boron-mediated, asymmetric aldol reactions conducted between ketones and aldehydes. We planned to introduce the bridging chain between the AB and CD spiroacetal ring systems by a suitable (C_{15}–C_{16}) *anti* aldol coupling, which provides a uniquely powerful method for the AB + CD component assembly.

1,3-POLYOL SYNTHESIS USING BORON ALDOL REACTIONS

As shown in Schemes 2 and 3, high levels of diastereoselectivity can be achieved in the boron aldol reactions of the appropriate methyl ketones with aldehydes,[10a] leading to the required acyclic precursors, **6** and **7**, for the AB- and CD-spiroacetal containing subunits, respectively. This involves the use of Ipc chiral ligands (from (+)-α-pinene) on boron matched with substrate-based induction from the aldehyde and/or enolate component – corresponding to double and triple asymmetric induction. For these boron-mediated aldol reactions, enolisation of the methyl ketone is best performed using (–)-Ipc$_2$BCl/Et$_3$N in Et$_2$O and addition to the *re*-face of the aldehyde is preferred *via* a twist-boat TS.[10b] Despite the steric demands of the components, high yields are generally obtained.

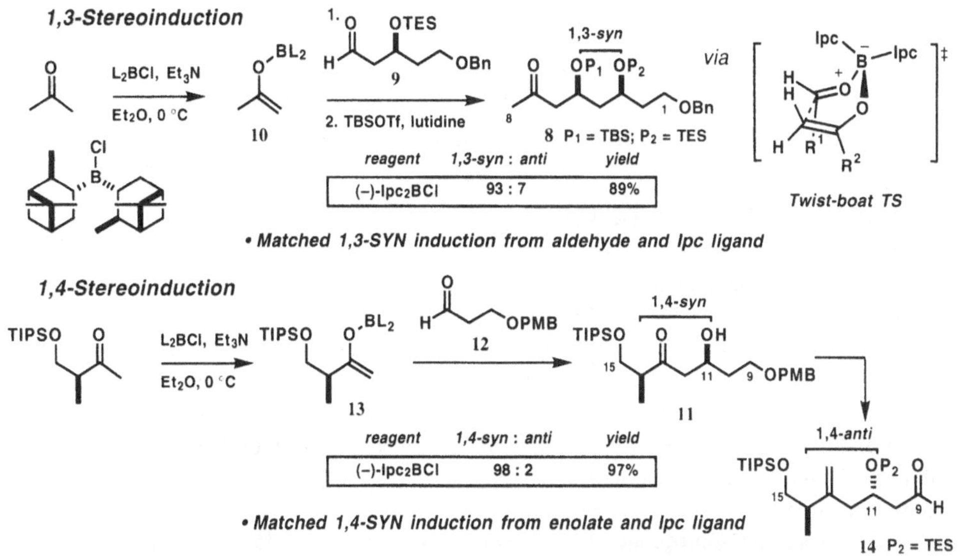

1,3-Stereoinduction

reagent	1,3-*syn* : anti	yield
(–)-Ipc$_2$BCl	93 : 7	89%

• *Matched 1,3-SYN induction from aldehyde and Ipc ligand*

1,4-Stereoinduction

reagent	1,4-*syn* : anti	yield
(–)-Ipc$_2$BCl	98 : 2	97%

• *Matched 1,4-SYN induction from enolate and Ipc ligand*

Scheme 2: Synthesis of the C_1–C_8 and C_9–C_{15} acyclic precursors using boron aldol reactions

In Scheme 2, the methyl ketone **8** (C_1–C_8) was obtained[5a] by matching the modest level of 1,3-*syn* induction from the chiral aldehyde **9** with the sense of Ipc ligand asymmetric induction on the enolate of acetone **10**. The aldol adduct **11** (C_9–C_{15}) was obtained[5d] from aldehyde **12** by matching 1,4-*syn* induction in the enolate **13** with that from the Ipc ligand on boron, followed by a series of steps including a Mitsunobu inversion to give the aldehyde **14**.

In Scheme 3, the ketone **8** was coupled with the aldehyde **14** to provide the aldol adduct **6** with ≥97% diastereoselectivity, which is the open-chain precursor to the AB-spiroacetal subunit.[5d] This reaction is an example of triple asymmetric induction, where there is complete matching of the chiral influence from all three components – the aldehyde, enolate **15** and the associated Ipc ligands. This led to the remarkable discovery of high levels of substrate-based, 1,5-*anti* stereoinduction in the boron aldol reactions of certain chiral β-alkoxy methyl ketones with prochiral aldehydes.[5b,11] This reaction has great potential for achieving remote asymmetric induction in the synthesis of other 1,3-polyol systems (*e.g.* the polyene macrolides). Similarly, the methyl ketone **16** (C_{17}–C_{24}) was enolised regio-selectively to give the boron enolate **17**, which was added to the aldehyde **18** to generate the ketone **7**, as required for formation of the CD-spiroacetal subunit.[5e] Again, high levels of diastereoselectivity were obtained due to matching of the 1,3-*syn* and 1,5-*anti* induction from the aldehyde and enolate components, respectively, with the influence of the boron reagent.

301

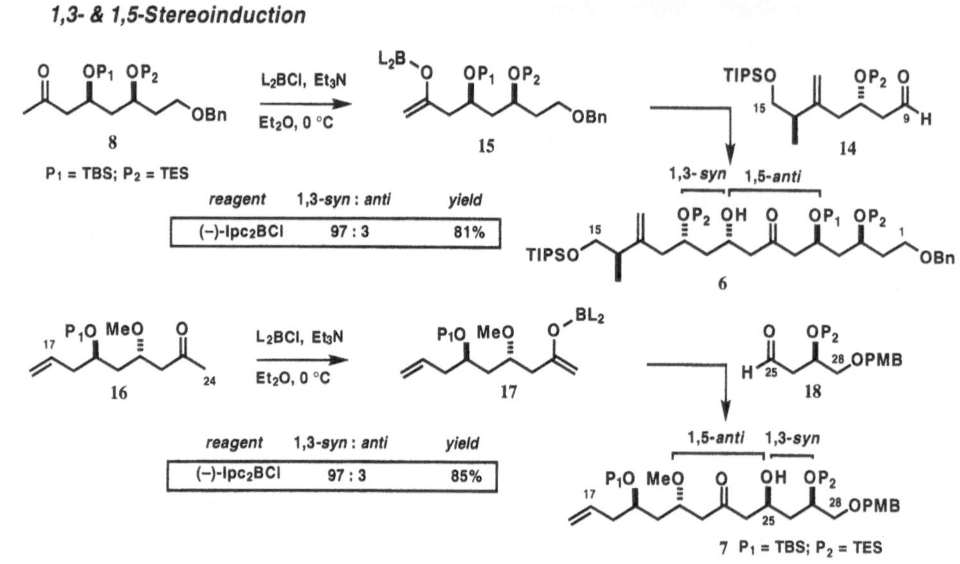

• *Fully matched 1,3-SYN and 1,5-ANTI induction from aldehyde and enolate with Ipc ligand*

Scheme 3: *Synthesis of the C_1–C_{15} and C_{17}–C_{28} acyclic precursors using boron aldol reactions*

THE AB-SPIROACETAL SUBUNIT

The AB-spiroacetal subunit of the spongistatins, *cf.* Scheme 4,[5d] possesses a thermodynamically favourable, double anomeric effect and has the C_3 and C_{11} side-chains arranged equatorially on each tetrahydropyran ring. Mild acid treatment (PPTS, MeOH) of the open-chain precursor **6** led to clean removal of both TES groups and *in situ* acetalisation to produce a single spiroacetal **19** in 88% yield. Dess-Martin oxidation and equatorial addition of MeMgBr to the resulting C_9 ketone, followed by a further 2 steps, then gave the axial, tertiary alcohol **20**. Removal of the TIPS ether in **20** and careful Dess-Martin oxidation gave the sensitive aldehyde **21**, corresponding to the AB-spiroacetal subunit having a benzyl ether at C_1. Note that appropriate manipulation of the spiroacetal **19**, including oxidation at C_1, allows access[5f] to the esters **3** (R = Me or CH_2CCl_3) shown in Scheme 1.

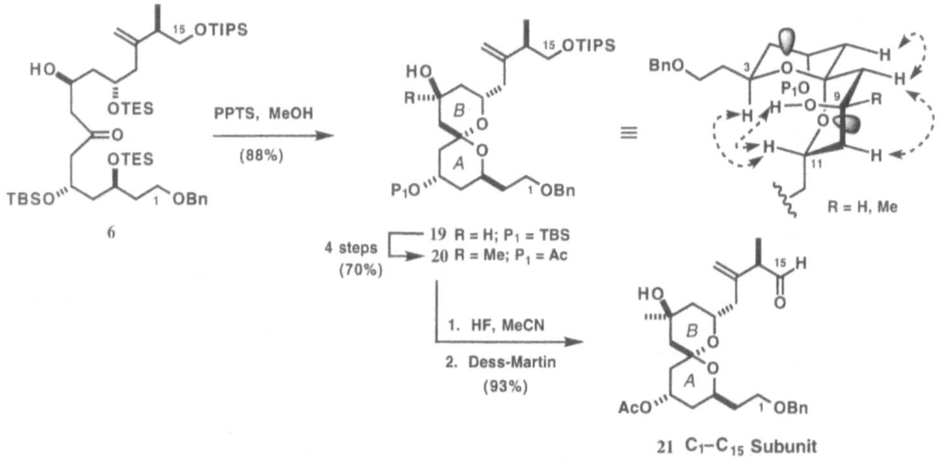

• *Single thermodynamic spiroacetal formed under mild acid treatment*

Scheme 4: *Synthesis of the AB-spiroacetal subunit*

THE CD-SPIROACETAL SUBUNIT

An important consideration here is that the CD-spiroacetal in the spongistatins does not benefit from a *double* anomeric effect. The desired CD-spiroacetal subunit 22, as shown in Scheme 5,[5e] has the C_{19}, C_{21} and C_{27} substituents arranged equatorially with only a single anomeric effect operating (for $P_1 = H$, the axial C_{25} hydroxyl can hydrogen bond to an acetal oxygen). Hence, internal acetalisation of ketone 7 may favour the isomeric (23-*epi*) spiroacetal 23, which has access to double anomeric stabilisation, rather than 22.

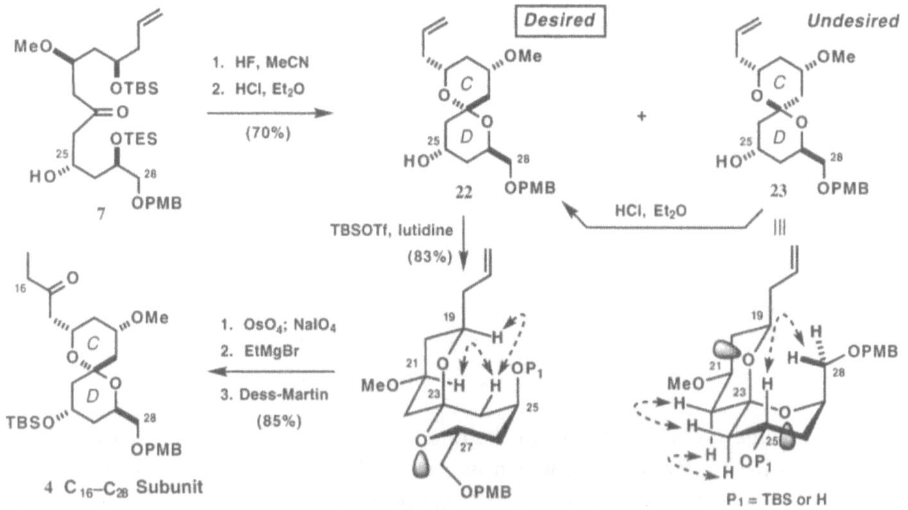

Desired

Undesired

1. HF, MeCN
2. HCl, Et$_2$O

(70%)

+

HCl, Et$_2$O

TBSOTf, lutidine
(83%)

1. OsO$_4$; NaIO$_4$
2. EtMgBr
3. Dess-Martin
(85%)

4 C$_{16}$–C$_{28}$ Subunit

P_1 = TBS or H

• *Major kinetic spiroacetal is the WRONG one*
• *Acid equilibration gives ca 1 : 1 mixture of spiroacetals*

Scheme 5: *Synthesis of the CD-spiroacetal subunit*

In practice, brief treatment of 7 with HF in MeCN led to clean desilylation and *in situ* acetalisation to give an 80% yield of the two spiroacetals 22 and 23, formed in a ratio of 1 : 5. Extensive NOE studies unambiguously showed that the major compound was the *undesired* spiroacetal 23. Fortunately, under acid treatment with anhydrous HCl in Et$_2$O (*i.e.* thermodynamic conditions), acetal equilibration occurred to give an improved, equimolar mixture of 22 and 23. Chromatographic separation and re-equilibration of 23 allowed for good conversion into the *desired* spiroacetal 22. Subsequent TBS ether formation at the C_{25}-OH and manipulation of the terminal alkene then gave the ethyl ketone 4, corresponding to the spongistatin C_{16}–C_{28} subunit required for aldol coupling with the AB-spiroacetal subunit.

ALDOL COUPLING BETWEEN THE AB- AND CD-SPIROACETAL SUBUNITS

As indicated in the retrosynthetic analysis in Scheme 1, we were attracted by the possibility of realising an *anti*-aldol coupling between the ethyl ketone 4 and the aldehyde 3 (R = Me or CH$_2$CCl$_3$). If successful, this would enable a highly convergent synthesis by directly installing the fully functionalised, bridging chain linking the AB- and CD-spiroacetal units. However, major concerns were the potential ease of enolisation of the α-methyl-β-methylene aldehyde 3 and whether, or not, the (14R)-stereocentre could be relied on to induce the required (15S,16S)-configuration.

In model aldol coupling reactions using various boron enolates of defined (E)-geometry, we demonstrated[5d] that high levels of the required Felkin-Anh control from the aldehyde component could be achieved. However, for the real coupling situation,[12] low yields were obtained on a small scale. This prompted us to examine the corresponding

303

lithium aldol reaction in Scheme 6,[5f] where now high levels of conversion could be obtained together with useful selectivity towards the required *anti* isomer 24.

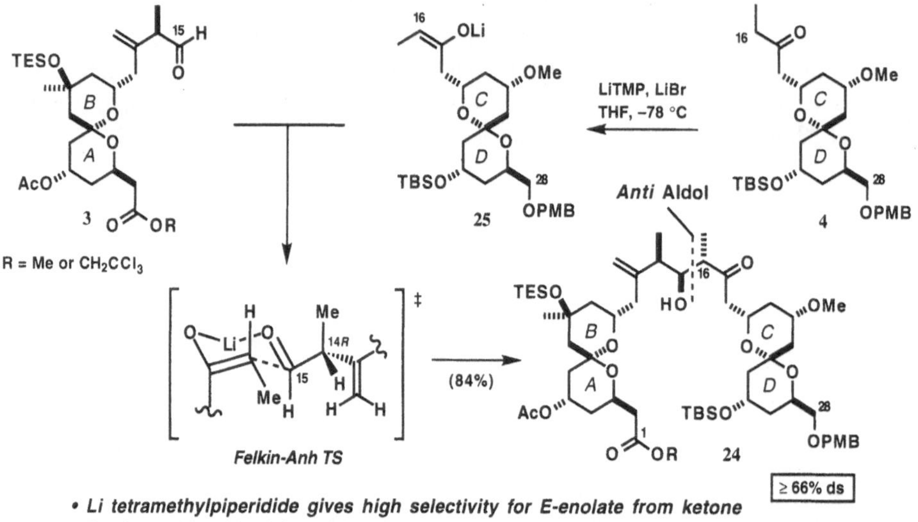

- **Li tetramethylpiperidide gives high selectivity for E-enolate from ketone**
- **Rapid addition (2 min) to aldehyde proceeds in high yield with good π-facial selection**

Scheme 6: *Key C_{15}–C_{16} aldol coupling step*

Regioselective enolisation of ketone 4 by the hindered base, LiTMP, gave predominantly the (*E*)-enolate 25, which was reacted with the aldehyde 3 in THF, under conditions of kinetic control (-78 °C, 2 min). In this way, the aldol adduct 24 was isolated as the major isomer in good yield, where (as with the boron reaction) the Felkin-Anh TS is presumably favoured. Acetylation of 24 gave the corresponding C_{15} acetate, which had ^{1}H NMR spectral data in accord with that for the corresponding region of spongistatin 1.[1a]

TOWARDS THE SYNTHESIS OF THE REMAINING C_{29}–C_{51} SUBUNIT

At this stage, we have successfully prepared the complete C_1–C_{28} subunit 24 which incorporates all of the functionality and stereochemistry of this portion of the target molecule. A suitable C_{29}–C_{51} subunit containing the E and F rings, *e.g.* the phosphonium salt 5 for use in a Wittig coupling with a C_{28} aldehyde derived from 24, is still needed to complete the total synthesis of spongistatin 1.

We have developed a synthesis of the C_{36}–C_{46} subunit 26, incorporating the fully substituted F ring, as outlined in Scheme 7.[5c] By using enolisation with $(^cC_6H_{11})_2BCl/Et_3N$ in Et_2O, the *anti* adduct 27 was obtained with >97% diastereoselectivity from the chiral ketone 28 and acetaldehyde. This substrate-controlled, boron aldol reaction [10a,13] proceeds through a highly ordered chair TS. Hydroxyl-directed reduction and acetonide formation then gave 29, which was chain extended to give the unsaturated ester 30. This alkene proved to be a good substrate for Sharpless asymmetric dihydroxylation, leading to the glycol 31 in 98% yield with high ds. A HWE chain extension was used to give 32, which on mild acid treatment underwent a hetero-Michael cyclisation to give the tetrahydropyran 26, together with its epimer at C_{43}. Equilibration of this mixture was possible using Triton methoxide in THF to give the ketone 26, which has all the substituents on the F ring equatorially disposed. We are currently exploring the elaboration of 26 into the phosphonium salt 5, which requires aldol-type chain extensions at C_{46} and C_{36} (after oxidation to the ketone at C_{35}) and introduction of the E ring.

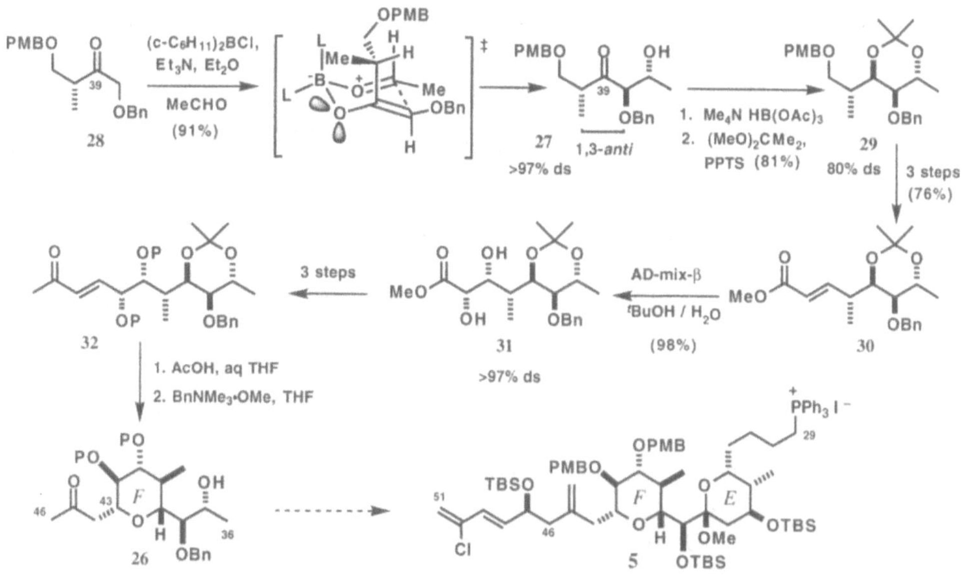

Scheme 7: *Synthesis of the C$_{36}$–C$_{46}$ subunit containing the F ring*

CONCLUDING REMARKS

Such complex, bioactive targets as the spongistatins (altohyrtins) provide an important impetus for the development of new synthetic methods and strategies. However, if sufficient synthetic material is to be made available to enable their full evaluation as anti-cancer drugs, practical routes to these rare marine macrolides[9] need to be developed. As well as completing our total synthesis, we are aware that the availability of the spongistatins (and related compounds) still remains a key challenge for the future.

ACKNOWLEDGEMENTS

This work was supported by the EPSRC (GR/L41646), NSERC (Postdoctoral Fellowship to RMO), Churchill College, Cambridge, and Merck Sharp & Dohme.

REFERENCES AND NOTES

1. (a) Pettit, G. R.; Cichacz, Z. A.; Gao, F.; Herald, C. L.; Boyd, M. R.; Schmidt, J. M.; Hooper, J. N. A. *J. Org. Chem.* **1993**, *58*, 1302. (b) Pettit, G. R.; Cichacz, Z. A.; Gao, F.; Herald, C. L.; Boyd, M. R. *J. Chem. Soc., Chem. Commun.* **1993**, 1166. (c) Pettit, G. R.; Herald, C. L.; Cichacz, Z. A.; Gao, F.; Schmidt, J. M.; Boyd, M. R.; Christie, N. D.; Boettner, F. E. *J. Chem. Soc., Chem. Commun.* **1993**, 1805. (d) Pettit, G. R.; Cichacz, Z. A.; Herald, C. L.; Gao, F.; Boyd, M. R.; Schmidt, J. M.; Hamel, E.; Bai, R. *J. Chem. Soc., Chem. Commun.* **1994**, 1605. (e) Pettit, G. R. *Pure Appl. Chem.* **1994**, *66*, 2271.

2. (a) Bai, R.; Cichacz, Z. A.; Herald, C. L.; Pettit, G. R.; Hamel, E. *Mol. Pharmacol.* **1993**, *44*, 757. (b) Bai, R.; Taylor, G. F.; Cichacz, Z. A.; Herald, C. L.; Kepler, J. A.; Pettit, G. R.; Hamel, E. *Biochemistry* **1995**, *34*, 9714.

3. (a) Kobayashi, M.; Aoki, S.; Gato, K.; Kitagawa, I. *Chem. Pharm. Bull.* **1996**, *44*, 2142. (b) Kobayashi, M.; Aoki, S.; Kitagawa, I. *Tetrahedron Lett.* **1994**, *35*, 1243. (c) Kobayashi, M.; Aoki, S.; Sakai, H.; Kawazoe, K.; Kihara, N.; Sasaki, T.; Kitagawa, I.

Tetrahedron Lett. **1993**, *34*, 2795. (d) Kobayashi, M.; Aoki, S.; Sakai, H.; Kihara, N.; Sasaki, T.; Kitagawa, I. *Chem. Pharm. Bull.* **1993**, *41*, 989.

4. For cinachyrolide A, isolated from a sponge of the genus *Cinachyra*, see: Fusetani, N.; Shinoda, K.; Matsunaga, S. *J. Am. Chem. Soc.* **1993**, *115*, 3977.

5. (a) Paterson, I.; Oballa, R. M.; Norcross, R. D. *Tetrahedron Lett.* **1996**, *37*, 8581. (b) Paterson, I.; Gibson. K. R.; Oballa, R. M. *Tetrahedron Lett.* **1996**, *37*, 8585. (c) Paterson, I. Keown, L. E. *Tetrahedron Lett.* **1997**, *38*, 5727. (d) Paterson, I.; Oballa, R. M. *Tetrahedron Lett.* **1997**, *38*, 8241. (e) Paterson, I.; Wallace, D. J.; Gibson, K. R. *Tetrahedron Lett.* **1997**, *38*, 8911. (f) Paterson, I.; Wallace, D. J.; Oballa, R. M. manuscript submitted.

6. (a) Guo, J.; Duffy, K. J.; Stevens, K. L.; Dalko, P. I.; Roth, R. M.; Hayward, M. M.; Kishi, Y. *Angew. Chem. Int. Ed. Engl.* **1998**, *37*, 187. (b) Hayward, M. M.; Roth, R. M.; Duffy, K. J.; Dalko, P. I.; Stevens, K. L.; Guo, J.; Kishi, Y. *Angew. Chem. Int. Ed. Engl.* **1998**, *37*, 192.

7. (a) Evans, D. A.; Coleman, P. J.; Dias, L. C. *Angew. Chem. Int. Ed. Engl.* **1997**, *36*, 2738. (b) Evans, D. A.; Trotter, B. W.; Côté, B.; Coleman, P. J. *Angew. Chem. Int. Ed. Engl.* **1997**, *36*, 2741. (c) Evans, D. A.; Trotter, B. W.; Côté, B.; Coleman, P. J.; Dias, L. C.; Tyler, A. N. *Angew. Chem. Int. Ed. Engl.* **1997**, *36*, 2744.

8. For other synthetic efforts, see: (a) Claffey, M. M.; Heathcock, C. H. *J. Org. Chem.* **1996**, *61*, 7646. (b) Hayes, C. J.; Heathcock, C. H. *J. Org. Chem.* **1997**, *62*, 2678. (c) Paquette, L. A.; Zuev, D. *Tetrahedron Lett.* **1997**, *38*, 5115. (d) Paquette, L. A.; Braun, A. *Tetrahedron Lett.* **1997**, *38*, 5119. (e) Smith, A. B., III; Zhuang, L.; Brook, C. S.; Boldi, A. M.; McBriar, M. D.; Moser, W. H.; Murase, N.; Nakayama, K.; Verhoest, P. R.; Lin, Q. *Tetrahedron Lett.* **1997**, *38*, 8667. (f) Smith, A. B., III; Zhuang, L.; Brook, C. S.; Lin, Q.; Moser, W. H.; Trout, R. E. L.; Boldi, A. M.;*Tetrahedron Lett.* **1997**, *38*, 8667. (g) Smith, A. B., III; Lin, Q.; Nakayama, K.; Boldi, A. M.; Brook, C. S.; McBriar, M. D.; Moser, W. H.; Sobukawa, M.; Zhuang, L. *Tetrahedron Lett.* **1997**, *38*, 8667. (h) Lemaire-Audoire, S.; Vogel, P. *Tetrahedron Lett.* **1998**, *39*, 1345. (i) Hermitage, S. A.; Roberts, S. M.; Watson, D. J. *Tetrahedron Lett.* **1998**, *39*, 3567. (j) Terauchi, T.; Nakata, M. *Tetrahedron Lett.* **1998**, *39*, 3795. (k) Zemribo, R.; Mead, K. T. *Tetrahedron Lett.* **1998**, *39*, 3895.

9. For a review on marine macrolide synthesis, see: Norcross, R. D.; Paterson, I. *Chem. Rev.* **1995**, *95*, 2041.

10. (a) For a review of asymmetric aldol reactions using boron enolates, see: Cowden, C. J.; Paterson, I. *Org. React.* **1997**, *51*, 1. (b) Paterson, I.; Goodman, J. M.; Lister, M. A.; Schumann, R. C.; McClure, C. K.; Norcross, R. D. *Tetrahedron* **1990**, *46*, 4663.

11. Evans, D. A.; Coleman, P. J.; Côté, B. *J. Org. Chem.* **1997**, *62*, 788.

12. Such an aldol coupling was employed successfully in the Evans total synthesis (*cf.* ref 7).

13. (a) Paterson, I.; Goodman, J. M.; Isaka, M. *Tetrahedron Lett.* **1989**, *30*, 7121. (b) Paterson, I.; Tillyer, R. D. *J. Org. Chem.* **1993**, *58*, 4182. (c) Paterson, I.; Norcross, R. D.; Ward, R. A.; Romea, P.; Lister, M. A. *J. Am. Chem. Soc.* **1994**, *116*, 11287.

SYNTHESIS OF HETEROCYCLIC ANTITUMOUR COMPOUNDS USING ALKYNE AND ARYNE CYCLOADDITIONS

Agustín Cobas, Mª Teresa Díaz, Sonia Escudero,
Dolores Pérez, Enrique Guitián* and Luis Castedo

Departamento de Química Orgánica y UA-CSIC
Universidad de Santiago
15706 Santiago de Compostela, Spain

The Diels-Alder reaction of α-pyrones with alkenes and alkynes is well known:[1] some 40 years ago, Wittig and co-workers reported that the reaction of α-pyrone (**1a**) with benzyne (**2**) led to formation of naphthalene (**4a**).[2] In fact, the initial adduct is the bicyclic intermediate **3a**, which undergoes a retro-Diels-Alder reaction, losing CO_2 to give the aromatic compound **4a**. The intermediate **3a** was not detected because its conversion to **4a** is favoured on both enthalpic and entropic grounds. In the intervening years the synthetic potential of this transformation for construction of aromatic molecules from α-pyrones has hardly been exploited.[3] Here we describe its application in the synthesis of some interesting compounds.

Reaction of Monocyclic α-Pyrones with Benzyne.

Benzyne (**2**)[4] reacts efficiently with pyrones bearing electron-donor or electron-acceptor substituents, as shown in Scheme 1. Thus addition of a suspension of benzenediazonium 2-carboxylate[5] to a refluxing solution of pyrone **1** in 1,2-dimethoxyethane (DME) or dioxane gave the corresponding naphthalene **4** in good to excellent yields (60-90%).[6] The tricyclic intermediate **3** was never isolated. Scheme 1 also shows, by way of example, a short synthesis of binaphthyl **5**. Here, cycloaddition of benzyne (**2**) to pyrone **1d** was followed by nickel-mediated coupling of the resulting naphthalene **4d,** which gave binaphthyl **5** in 50% yield.

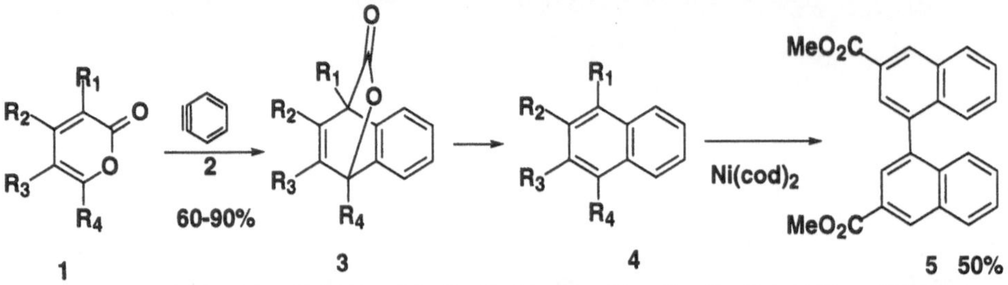

Scheme 1

a, $R_1=R_2=R_3=R_4= H$; b, $R_1= CO_2Me$, $R_2=R_3=R_4= H$; c, $R_1=R_2=R_4= H$, $R_3 = CO_2Me$;
d, $R_1= Br$, $R_2=R_4=H$, $R_3= CO_2Me$; e, $R_1=R_3= H$, $R_2= OMe$, $R_4= Me$

Scheme 2 shows that similar cycloaddition reactions can be carried out on cyclohexyne (8). This unstable cycloalkyne[7] was generated in situ by treatment of 7 with cesium fluoride,[8] compound 7 having been obtained by reducing 6 with L-Selectride and then trapping the resulting enolate with N-phenyltriflimide. The expected tetrahydronaphthalene 9 was isolated in 82% yield.[9]

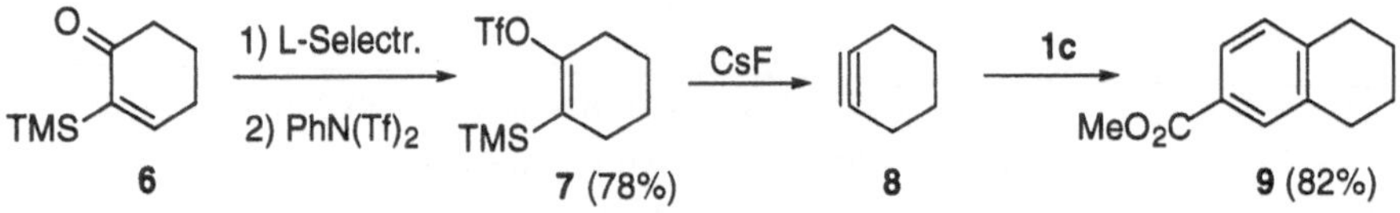

Scheme 2

Synthesis of Benzophenanthridines.

The benzophenanthridines are a group of more than 100 isoquinoline alkaloids characterized by the basic structure 10. Some of them have interesting pharmacological properties.[10] For example, fagaronine (11a) and nitidine (11b) are inhibitors of DNA topoisomerases and show antileukemic activity, though they are also toxic;[11] and both these compounds have recently been found to inhibit HIV-1 reverse transcriptase.[12]

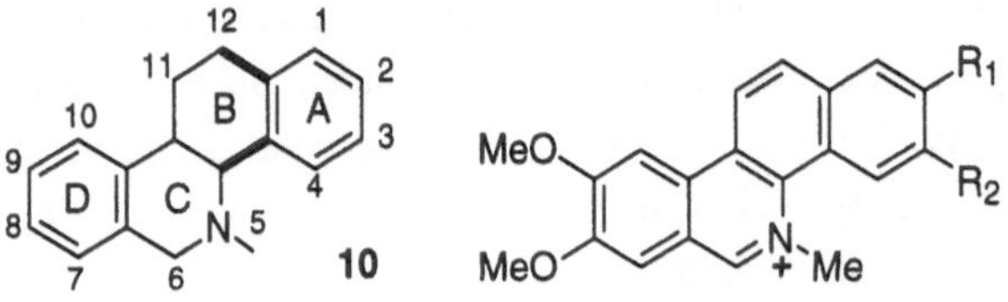

Fagaronine, 11a, , $R_1=$ OH, $R_2=$ OMe
Nitidine, 11b, $R_1+R_2=$ OCH$_2$O

Figure 1

The naphthalene moiety of benzophenanthridines can be constructed by means of a pyrone-aryne cycloaddition, as shown in Scheme 3 for nitidine (**11a**).

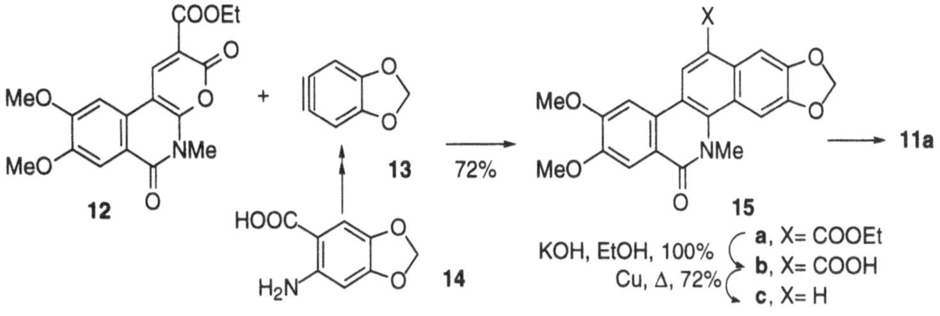

Scheme 3

Thus benzyne **13** (generated from anthranilic acid **14**[13]) was added to pyrone **12**[14] to gave adduct **15** in 72% yield -a very high yield for a cycloaddition involving a substituted benzyne such as **13**, yields of around 25% being more usual in these cases. Hydrolysis of adduct **15a** with potassium hydroxide, followed by decarboxylation with copper and quinoline, gave oxonitidine (**15c**), which can be transformed into nitidine (**11a**) by a published procedure.[15] Thus the formal synthesis of this antitumour alkaloid was completed in a short and efficient way.[16]

Synthesis of Lycorines

Several lycorines have important pharmacological properties. For example, anhydrolycorinium chloride (**17a**) has antileukemic activity, kalbretorine (**18b**) has antitumour activity, ungeremine (**17b**) is active against some types of carcinoma, and hippadine (**18a**) reversibly inhibits fertility in male rats.[17]

We have developed a new approach to lycorines that differs from that used for the synthesis of benzophenanthridines in that the dienophile is now an alkyne and the cycloaddition is intramolecular.

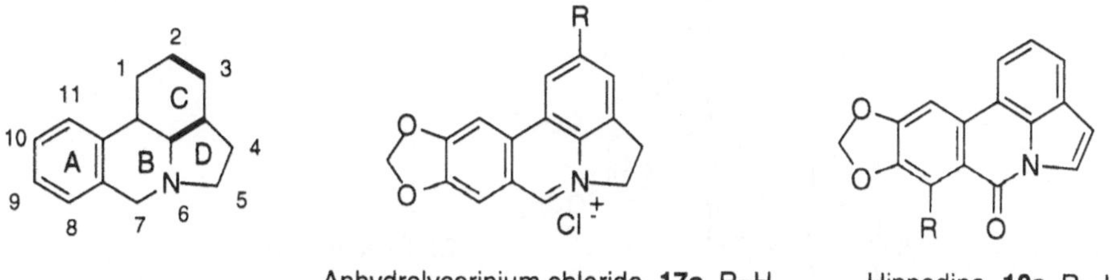

16

Anhydrolycorinium chloride, **17a**, R=H
Ungeremine, **17b**, R=OH

Hippadine, **18a**, R= H,
Kalbretorine,**18b**, R= OH,

Figure 2

Scheme 4 shows the application of this approach to the synthesis of anhydrolycorinium (**17a**) and hippadine (**18a**).

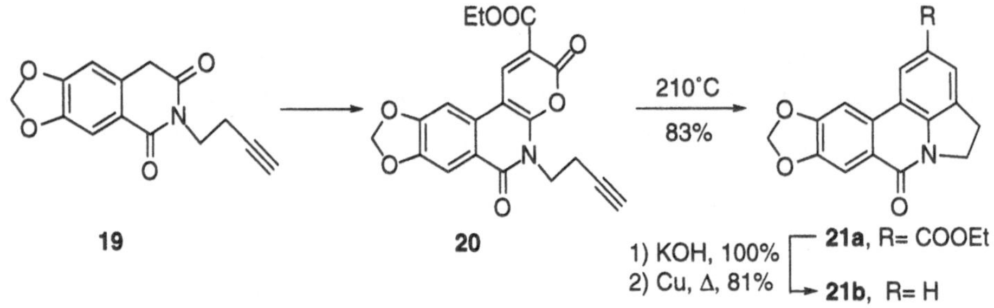

Scheme 4

Imide **19** was transformed into pyrone **20** by treating it with an ethoxyacrylate. Heating a solution of **20** in nitrobenzene at 210°C induced intramolecular cycloaddition. Ester **21a** was isolated in 83% yield. Hydrolysis of this ester, followed by decarboxylation, gave lycorine precursor **21b** whose transformation into anhydrolycorinium (**17a**)[18] and hippadine (**18a**)[19] has previously been described. Thus the formal synthesis of these pharmacologically interesting alkaloids was completed.[20]

Synthesis of Dynemicins

Dynemicin A (**22**) belongs to the group of metabolites known as enediynes, which have attracted the attention of chemists due to their unusual structure, their antibiotic and antitumour properties and their elegant mode of action.[21] The structure of dynemicin A comprises a planar fragment related to that of anthracyclinones, and an enediyne bridge, such as that found in esperamicin and calicheamicin.

We have developed our approach to dynemicin analogues that uses a pyrone cycloaddition for construction of ring C, thus allowing the intercalating capacity of the planar fragment to be varied by varying the dienophile.

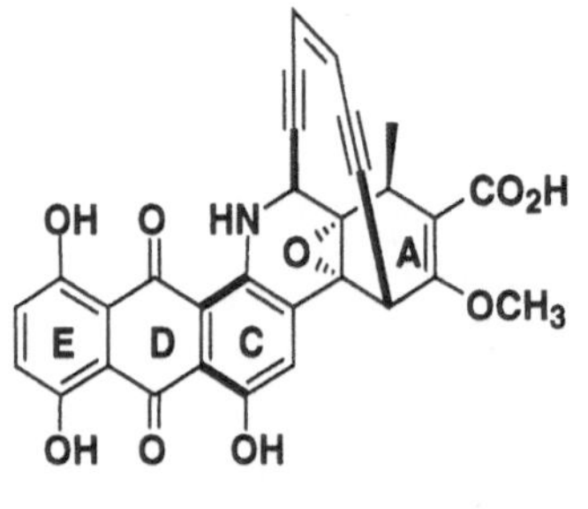

22, Dynemicin A

Figure 3

The synthesis began with the model pyrone **24**, which was prepared from imide **23** by the procedures described earlier. Cycloaddition of arynes **25a** and **25b** to **24a** and **24b** gave the expected polycyclic compounds **26** in moderate yields (approx. 70%). However reaction with 2,3-naphthalyne gave complex mixtures containing mono- and diadducts.

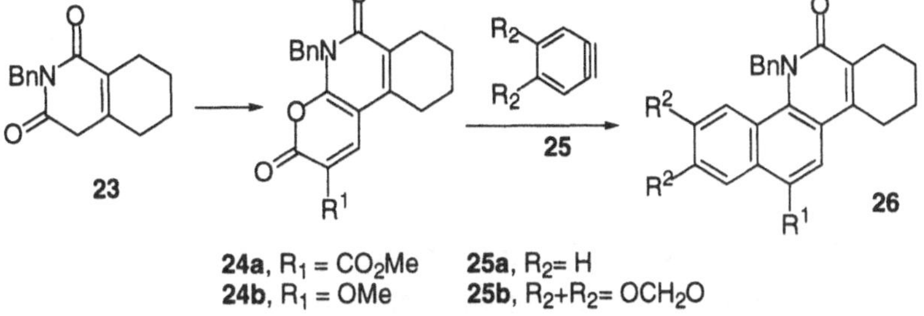
BnN
23
BnN
R1
24a, R1 = CO2Me
24b, R1 = OMe
R2
R2
25
25a, R2= H
25b, R2+R2= OCH2O
BnN
R2
R2
R1
26

Scheme 5

Cycloaddition of quinones **27** to pyrone **24** proved to be very sensitive to the nature of the R group. The best results were obtained with unsubstituted quinone **27a** and with methoxyquinone **27d**. The resulting adduct **28d** was transformed into chloro derivative **29a** by treatment with phosphorus oxychloride, and then reduced catalytically to **29b** with hydrogen.[22] We are now working on the introduction of the enediyne bridge following published procedures.[21]

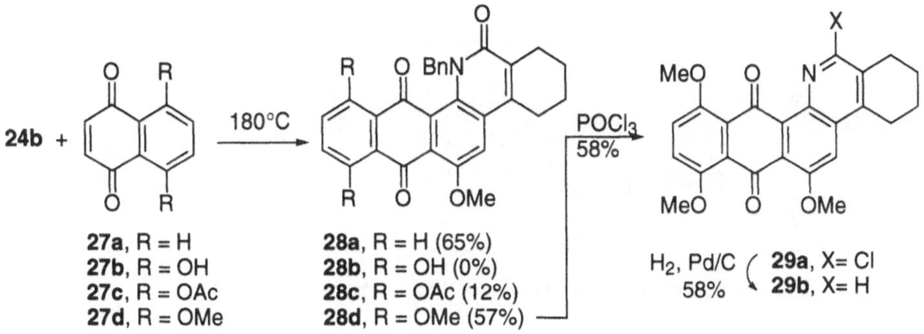
24b +
180°C
27a, R = H
27b, R = OH
27c, R = OAc
27d, R = OMe
BnN
OMe
28a, R = H (65%)
28b, R = OH (0%)
28c, R = OAc (12%)
28d, R = OMe (57%)
POCl3
58%
MeO
MeO
OMe
X
H2, Pd/C
58%
29a, X= Cl
29b, X= H

Scheme 6

Synthesis of Ellipticines.

Another group of alkaloids with interesting pharmacological properties are the ellipticines. Figure 4 shows the structure of the parent compound ellipticine (**30**) and its derivatives, elliptinium (**31a**), datelliptium (**31b**) and retelliptine (**31c**), which have all antitumour activity.[23]

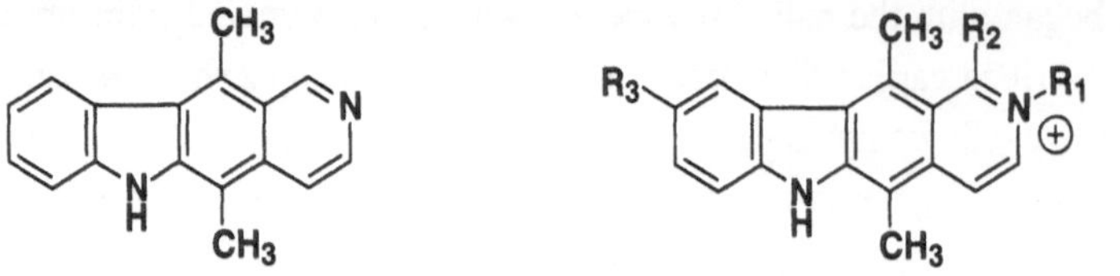

Ellipticine, **30**

Elliptinium, **31a**, R$_1$=CH$_3$, R$_2$=H, R$_3$=OH
Datelliptium, **31b**, R$_1$=CH$_2$CH$_2$N$^+$HEt$_2$, R$_2$=H, R$_3$=OH
Retelliptine, **31c**, R$_1$=H, R$_2$=NH(CH$_2$)$_3$N$^+$HEt$_2$, R$_3$=OCH⁻

Figure 4

In the last 20 years many synthetic procedures for the preparation of ellipticines have been developed. Again, we directed our attention to two approaches based on aryne cycloadditions: Moody's approach, in which 3,4-didehydropyridine (**35a**) generated by thermolysis of a triazene derivative adds to indolepyrone **36**;[24] and Gribble's approach, in which **35a** generated from 1-aminotriazolo[4,5-c]pyridine adds to furoindole **39**.[25] Both approaches, although convergent and conceptually elegant, suffer from the same limitation: the key cycloaddition step is inefficient due to its moderate yield and its lack of selectivity, generally affording 1:1 mixtures of the two possible regio-isomers in only 38-40% yield.

The regioselectivity of many nucleophilic additions and cycloadditions to arynes can be controlled by placing a suitable substituent at a position α to the triple bond, this control having been attributed to polar effects.[4,26] With this idea in mind, we devised modifications of Moody's and Gribble's approaches to ellipticines in which the pyridyne had a chlorine or bromine substituent at position 2 or 5. The required arynes were generated as shown in Scheme 7.

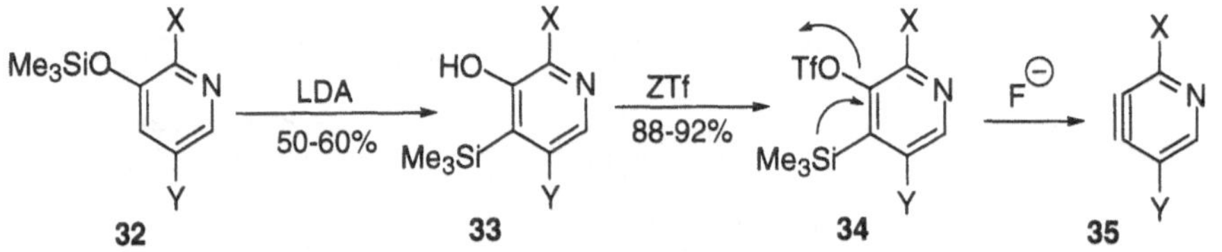

a, X=Y=H; **b**, X=Cl, Y=H; **c**, X=Br, Y=H; **d**, X=H, Y=Cl

Scheme 7

Aryne precursors **34b-d** were prepared by similar sequences in which the halopyridine **32** was firstly metallated *ortho* to the silyl ether, thus inducing migration of the trimethylsilyl group to afford pyridinols **33b-d**.[27] Then **33b,c** were reacted with triflic anhydride to obtain **34b,c**, while conversion of **33d** to **34d** required use of butyllithium and 2-[*N,N*-bis(trifluoromethyl)amino]-5-chloropyridine. Arynes **35b-d** can be generated by treating precursors **34** with cesium fluoride.

Contrary to our expectations, when aryne **35b** was generated in the presence of Moody's pyrone **36**, a 1:1 mixture of regioisomers **37b** and **38b** was obtained. Similarly,

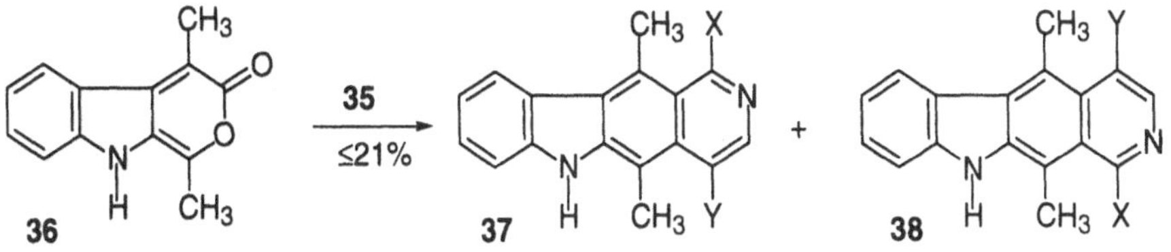

a, X=Y=H; b, X=Cl, Y=H; d, X=H, Y=Cl

Scheme 8

reaction of pyrone **36** with aryne **35d** gave a 1:1 mixture of regiosomers **37d** and **38d**.

In parallel reactions using Gribble's approach, generation of aryne **35b** in the presence of indolofuran **39** led to a 2.4:1 mixture of **40b** and **41b** in 89% yield. The regioselectivity obtained with 2-chloropyridyne (**35b**) is in keeping with the polar control hypothesis. However, reaction of 2-bromopyridyne (**35c**) and Gribble's diene **39** once again gave a 1:1 mixture of regioisomers (**40c**) and (**41c**). We then looked at the reaction of the 5-chloropyridyne **35d**, expecting the change in the substitution pattern to cause inversion of the regioselectivity. However, to our surprise a 1:1 mixture of regioisomers **40d** and **41d** was obtained.

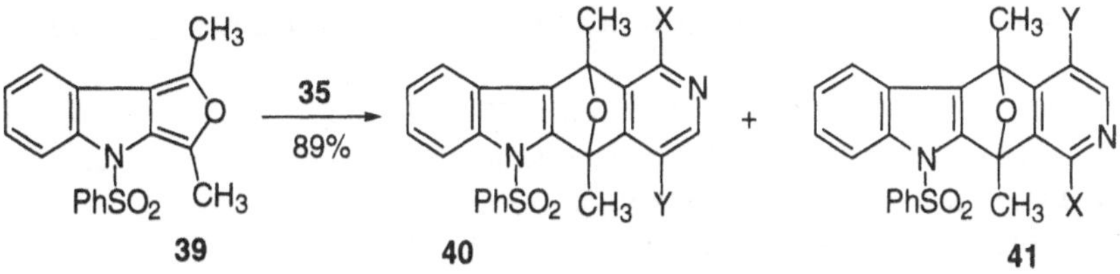

a, X=Y=H; b, X=Cl, Y=H; c, X=Br, Y=H; d, X=H, Y=Cl

Scheme 9

Finally, ellipticine (**30**) and isoellepticine were obtained from adducts **40b** and **41b**, respectively, by reductive opening of the ether bridge with NaBH$_4$ and NaOH, followed by removal of the chlorine by hydrogenolysis.[28] Thus with these modifications the yield of this approach to ellipticine was improved by a factor of 6.

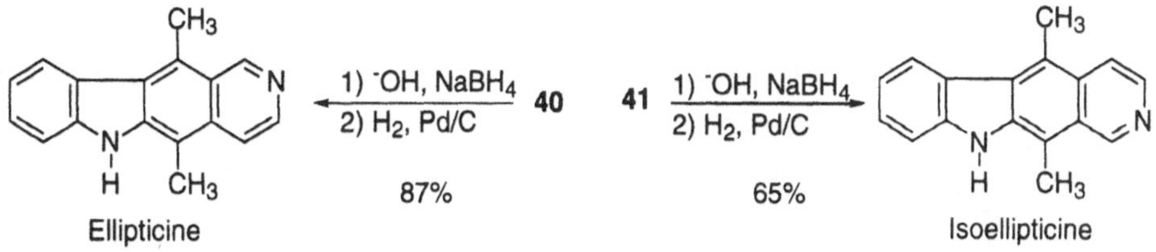

Scheme 10

ACKNOWLEDGEMENTS.

We thank the Spanish Ministry for Education and Science for financial support (Projects PB93-0533 and PB96-0764).

REFERENCES.

1. K. Afarinkia, V.Vinader, T. Nelson and G. H. Posner, *Tetrahedron 48*: 9111 (1992).
2. G. Wittig and R.W. Hoffmann, *Chem. Ber. 95*: 2718 (1962).
3. C.J. Moody, *J. Chem. Soc. Perkin Trans 1*, 2505 (1985). See also ref.24.
4. (a) R.W. Hoffmann, *Dehydrobenzene and Cycloalkynes*, Academic Press, New York, 1967. (b) H. Hart, *The Chemistry of Functional Groups, Suppl. C2: The Chemistry of Triple-Bonded Functional Groups* (Ed: S. Patai), Wiley, Chichester, 1994, p. 1017.
5. F.M. Logullo, A.H. Seitz and L. Friedman, *Org. Synth. 48*: 12 (1968).
6. S. Escudero, D. Pérez, E. Guitián and L. Castedo, *Tetrahedron Lett. 38*: 5375 (1997).
7. (a) A. Krebs and J. Wilke, *Angle Strained Cycloalkynes* in *Topics in Current Chemistry 109*: 189 (1983). (b) P. Caubere, *Chem. Rev. 93*: 2317 (1993).
8. Y. Himeshima, T. Sonoda and H. Kobayashi, *Chem. Lett.*1211 (1983). W. C. Shakespeare and R. P. Johnson, *J. Am. Chem. Soc. 112*: 8578 (1990).
9. N. Atanes, S. Escudero, D. Pérez, E. Guitián and L. Castedo, *Tetrahedron Lett. 39*: 3039 (1998).
10. V. Simànek in *The Alkaloids* (Ed. A. Brossi), Academic Press, New York, vol. 26, p.185 (1985).
11. S-D. Fang, L-K. Wang and S.M. Hecht, *J. Org. Chem. 58*: 5025 (1993).
12. G.T. Tan, J.M. Pezzuto, A.D. Kinghorn and S.H. Hughes, *J. Nat. Prod. 54*:143 (1991).
13. L. Friedman and F.M. Logullo, *J. Org. Chem. 34*: 3089 (1969).
14. O.S. Wolfbeis, I. Trummer and A. Knierzinger, *Liebigs Ann. Chem.* 811 (1981)
15. I. Ninomiya, T. Naito, H. Ishii, T. Ishida, M. Ueda and K. Harada, *J. Chem. Soc. Perkin Trans. 1* 762 (1975).
16. D. Pérez, E. Guitián and L. Castedo, *J. Org. Chem. 57*: 5911 (1992).
17. S.F. Martin in *The Alkaloids* (Ed. A. Brossi), Academic Press, New York, vol. 30, p. 252 (1987).
18. L.G. Humber et al. *J. Chem. Soc.* 4622 (1954).
19. J.W. Cook, J.D. London and P. McKloskey, *J. Chem. Soc.* 4176 (1954).
20. D. Pérez, G. Burés, E. Guitián and L. Castedo, *J. Org. Chem. 61*: 1650 (1996).
21. K.C. Nicolaou, A.L. Smith and E.W. Yue, *Proc. Natl. Acad. Sci. 90*: 5881 (1993). K.C. Nicolaou and W.-M. Dai, *Angew. Chem. Int. Ed. Engl. 30*: 1387 (1991).
22. S. Escudero, D. Pérez, E. Guitián and L. Castedo, *J. Org. Chem. 62*: 3028 (1997).
23. G.W. Gribble, *Synthesis and Antitumor Activity of Ellipticine Alkaloids and Related Compounds* in *The Alkaloids* (Ed. A. Brossi), Academic Press, New York, vol. 39, p. 239 (1990).
24. C.May, C. J. Moody, *J. Chem. Soc. Perkin Trans. 1* 247 (1988).
25. G.W. Gribble, M.G. Saulnier, M.P. Sibi and J.A. Obaza-Nutaitis, *J. Org. Chem.49*: 4518 (1984).
26. H.C. van der Plas and F. Roeterdink, in *The Chemistry of Functional Groups, Suppl. C, The Chemistry of Triple-Bonded Functional Groups* (Ed: S. Patai), Wiley, Chichester, p. 421, 1983.
27. G. Simchen and J. Pfletschinger, *Angew. Chem. Int. Ed. Engl. 15*: 428 (1976).
28. M. T. Díaz, A.Cobas, E. Guitián and L. Castedo, *Synlett* 157 (1998).

PRODIGIOSINS - A NEW FAMILY OF IMMUNOSUPPRESSANTS NOVEL AND EFFICIENT SYNTHESIS

Nicola Mongelli, Roberto D'Alessio and Ermes Vanotti

Pharmacia & Upjohn
Viale Pasteur 10, Nerviano (Mi)

Prodigiosins (Ps) are a class of naturally occurring red pigments produced by Streptomyces Genus and characterised by a peculiar 2,2'-bipyrrolyl-pyrromethene skeleton (Figure 1).

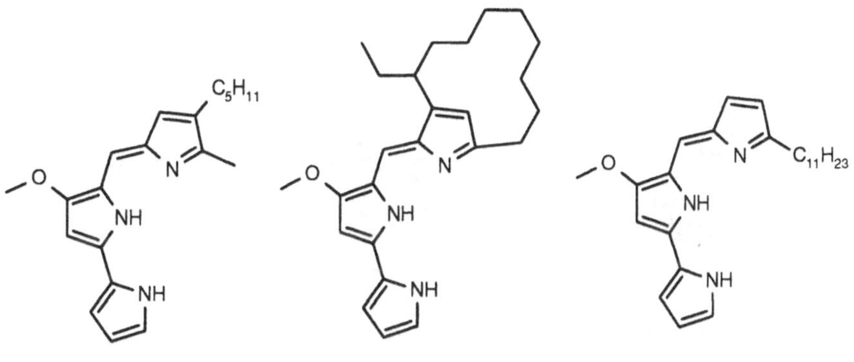

Prodigiosin Metacycloprodigiosin Undecylprodigiosin

Figure 1. Natural occurring prodigiosins

Prodigiosin, the first isolated member[1] , has been shown to possess potent antimicrobial and cytotoxic properties. It was never used in therapy because of high systemic toxicity.

Blind screening at Tokyo University demonstrated that some members of this class, and in particular undecylprodigiosin[2] (UP) and metacycloprodigiosin (MCP), were endowed with an interesting immunosuppressive activity at non cytotoxic doses[3].

Subsequent competition studies demonstrated for prodigiosins a mechanism of action different from Cyclosporin (CyA), FK506 and Rapamycin[4].

Selective inhibition of immune cell activation is actively pursued in order to control autoimmunity and organ transplant rejection. In the early fifties immunosuppression was achieved using cytotoxic drugs. The second generation of immunosoppressants was composed by antiinflammatory corticosteroids. Only the discovery of Cyclosporin A[5] revolutionised the immunosuppressive therapy.

The best treatment currently available is based on a combination of low doses of two or more agents with different mechanism of action, namely CyA (or FK 506), azathioprine (or micophenolic acid) and steroids. Nevertheless, 50% of solid organ transplantations still experience one or more acute rejection episodes and efficacy of immunosuppression in allogenic bone marrow transplantation in oncological patients is clearly unsatisfactory.

Moreover currently used immunosuppressants are associated with substantial toxicity. Thus, existing therapy could take great advantage of new immunosuppressive agents, endowed with a better pharmaco-toxicological profile and, possibly, different mechanism of action, in order to act synergistically with the presently used drugs. Since UP does not compete with CyA and FK506 for their specific binding protein cyclophilin and FKBP respectively, but it is able to block the proliferation induced by IL-2, it may act synergistically with CyA.

Natural Prodigiosins have a reduced therapeutic window because of systemic toxicity, and can not be easily used as starting substrate for semisynthetic transformations.

So we needed, for our medicinal chemistry program , a total synthesis flexible and amenable to large scale. All the described syntheses of prodigiosins[6] applied the following retrosynthetic pathway (scheme 1).

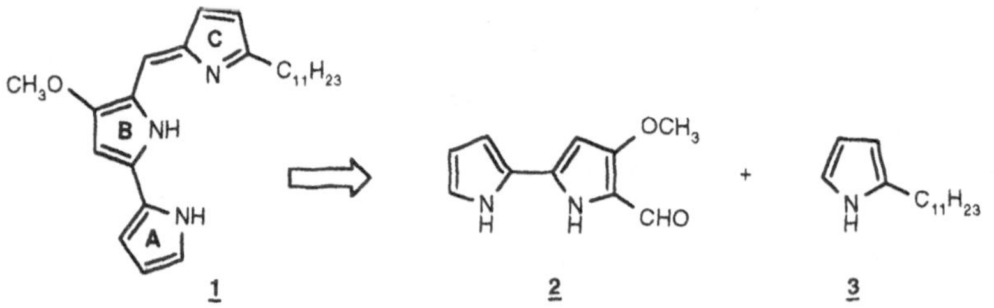

Scheme 1. Retrosynthetic pathway

The key intermediate is the 2,2'-bipyrrole aldehyde **2** that is then condensed with the third pyrrolic ring **3** in acidic conditions. All the reported methodologies , although elegant, are not suitable for scaling-up (multi-steps and low yields).

An interesting bipyrrole synthesis reported by Rapoport[7] and not utilised for prodigiosin production is the Vilsmeier-like reaction reported in Scheme 2, wherein R_2 was different from alkoxy.

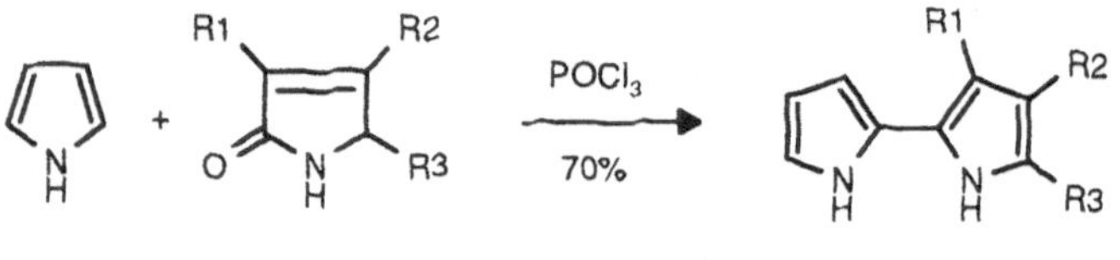

Scheme 2

316

We investigated this reaction utilising 4-methoxy-pyrrolinone **5** (scheme 3).

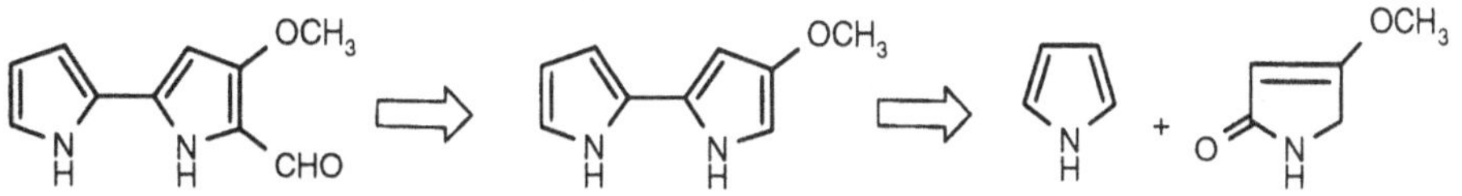

Scheme 3

This would lead to the bipyrrole **2** with the required substitution pattern.

We expected a correct regiospecific introduction of the formyl group on the most electron rich pyrrole ring. Alkoxy-pyrrolinones can be synthesised by known procedures. In analogy with the methodology reported by Rapoport, pyrrolinone **5** has been condensed with pyrrole in the presence of POCl$_3$ at room temperature, obtaining an extensive decomposition of the reagents (scheme 4).

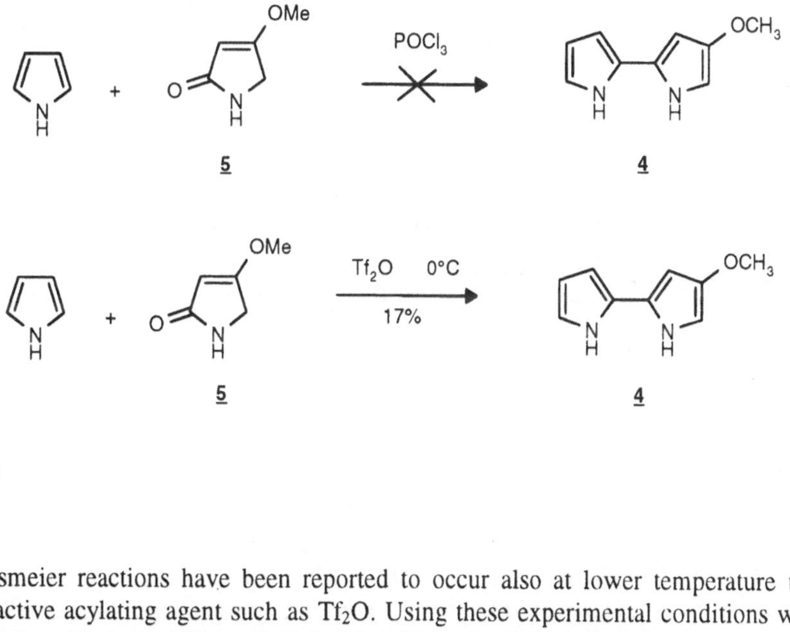

Scheme 4

Vilsmeier reactions have been reported to occur also at lower temperature using a more reactive acylating agent such as Tf$_2$O. Using these experimental conditions we were able to isolate the wanted bipyrrole derivative **4** in only 17% yield (scheme 4). Furthermore it was impossible to utilise this compound because of its low stability. Probably the presence of an electron-donating group as MeO- on a ready oxidable ring, increases the instability of the system. We thought that the insertion of a substituent able to delocalize electrons, thus reducing the electron density, would increase the stability of the oncoming bipyrrole system.

That led us to perform the Vilsmayer reaction directly on pyrrolinon **8** which has the ring C of Prodigiosin already in place (scheme 5).

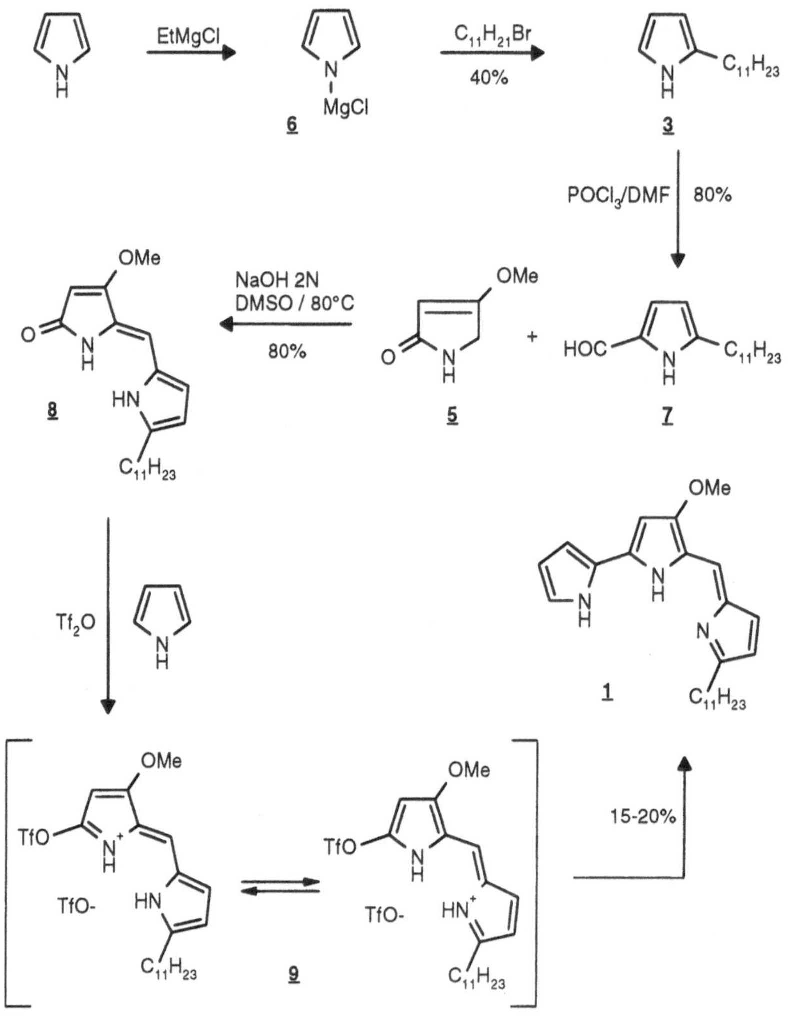

Scheme 5

The C ring of Prodigiosin can be easily obtained by known procedures and can be condensed with 4-methoxy-pyrrolinone to give the desired intermediate. The final Vilsmeier-like reaction, performed with Tf$_2$O at 0°C yielded the final prodigiosin **1** in fairly low yield. The major drawback of this approach is due to the difficult purification because of the extensive polymerisation of pyrrole (used as a solvent) in strong acidic medium. During the reaction we noticed the formation of an intermediate that turned out to be the triflate **9**, a surprisingly stable crystalline compound. This suggested us to investigate the possibility to improve the final conversion through an heteroaromatic metallorganic coupling. We tried many different metals and conditions: the one that worked better was the Suzuki reaction, that gave the coupling of triflate **9** with the pyrroleboronic acid derivative **10** in 73% yield. Scheme 6 depicts the whole synthetic pathway.

318

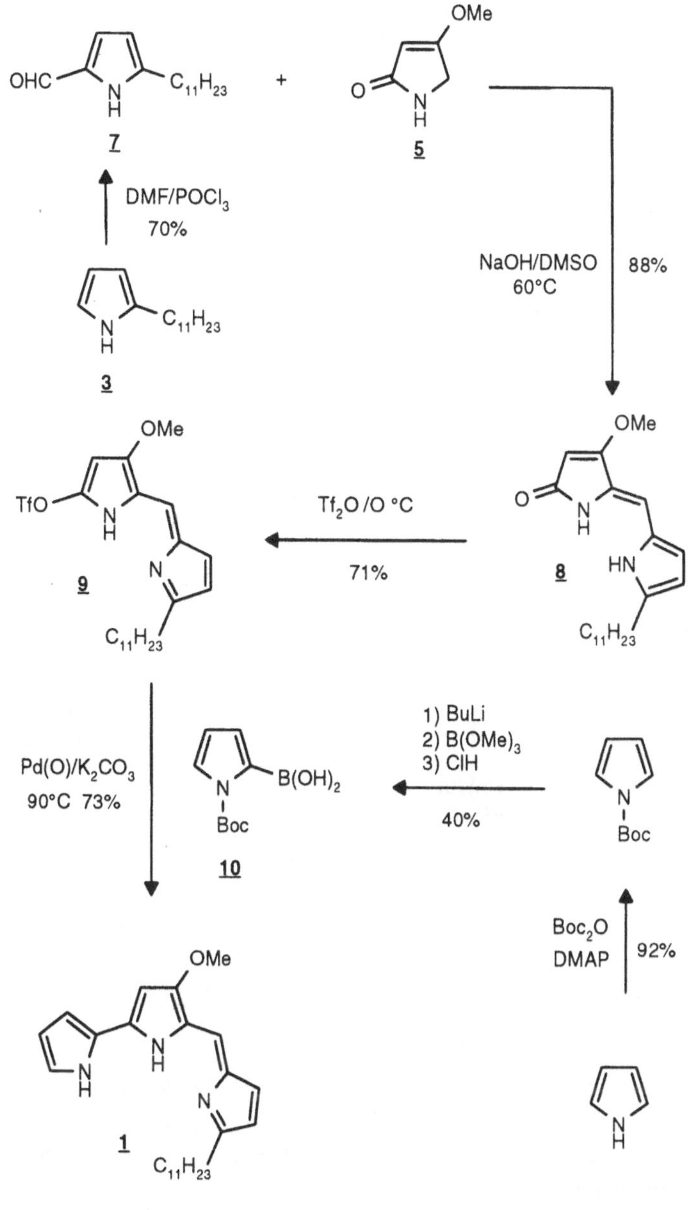

Scheme 6

Finally, we had in our hands a total synthesis efficient and suitable for scaling up. Furthermore, this synthetic approach turned out to be also quite flexible, offering the possibility to introduce many variations in different parts of the molecule. This is particularly convenient for an effective structure activity relationship study, as shown in scheme 7.

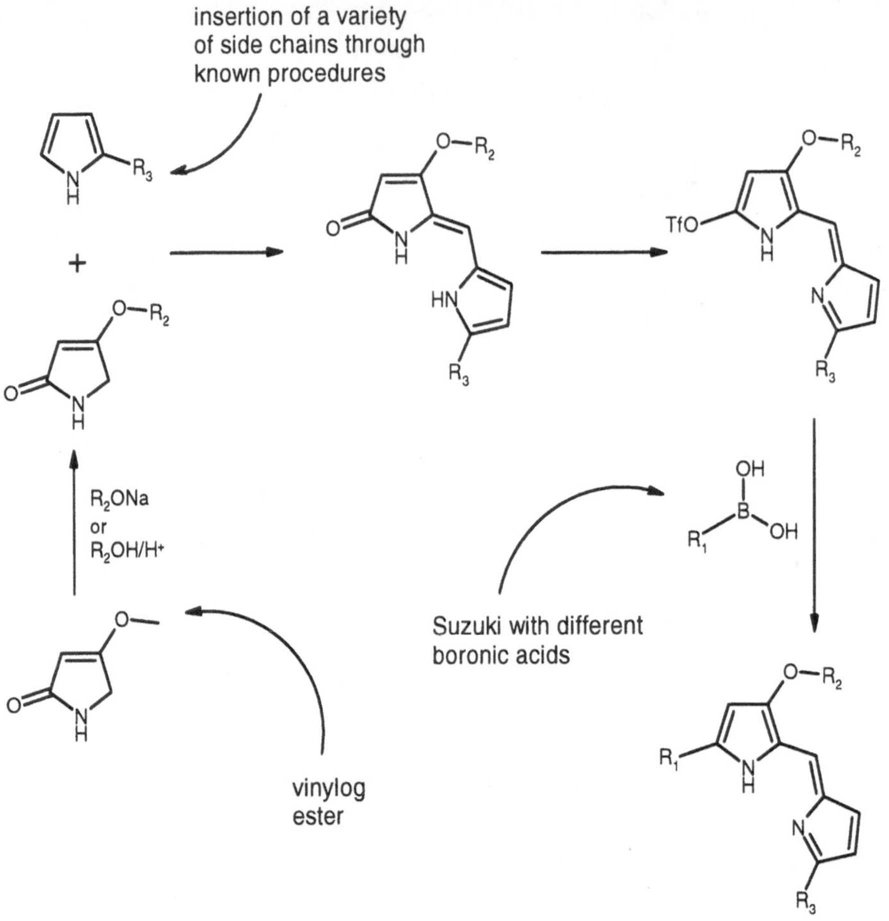

Scheme 7

In particular, the Suzuki reaction has been applied to a variety of substrates . As shown in scheme 8, the reaction is of wide applicability (pyrazole and benzoimidazole boronic acids couldn't be isolated). It should be mentioned that yields have been optimised only in case of pyrrole and indole.

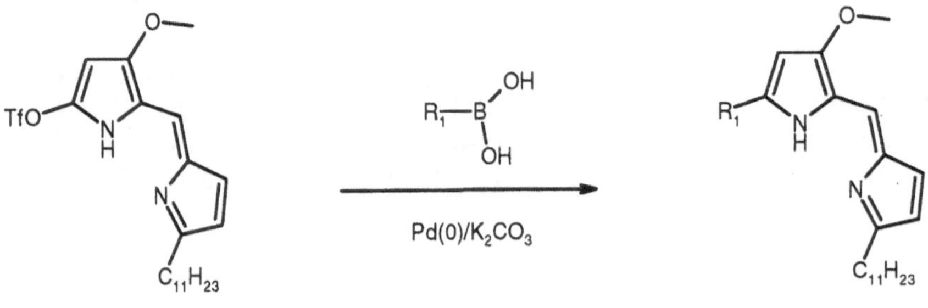

R$_1$	% Y		R$_1$	% Y
(pyrrol-2-yl)	73		(indol-2-yl)	89
(phenyl)	43		(5-methoxyindol-2-yl)	44
(thien-2-yl)	53		(5-chloroindol-2-yl)	35
(pyrazol-3-yl)	–		(benzimidazol-2-yl)	–

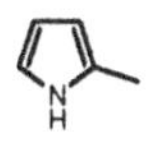

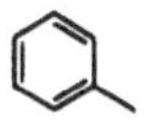

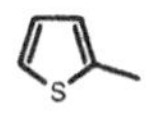

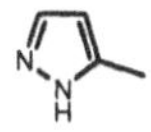

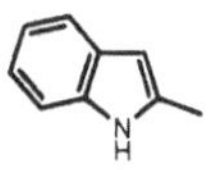

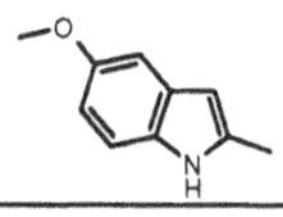

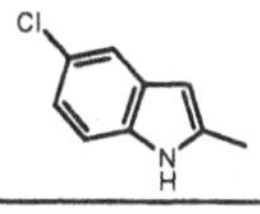

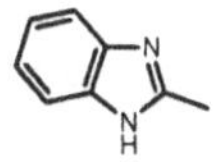

Scheme 8

Referring to the general formula shown in figure 2, the following preliminary structure activity relationships could be drawn:

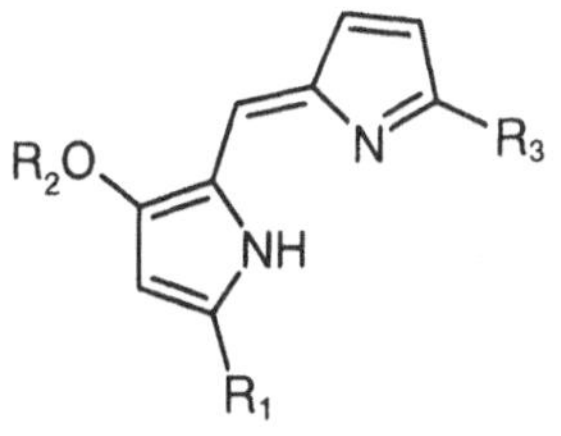

Figure 2

- The nitrogen and the electron conjugation in the ring R1 are essential for the biological activity.
- The alkoxy substituent OR$_2$ is important. Usually heavier alkoxy groups lead to a more favourable therapeutic index in vitro (expressed as the ratio between cytotoxicity and immunosuppressant activity).
- R3 side chain is not crucial for the activity and can be used to modulate the overall lipophilicity of the molecule.

Acknowledgements

We wish to acknowledge Arsenia Rossi, Carlini Orlando, Marcellino Tibolla, Mario Rossi, Pietro Motta, Antonella Ermoli and Alberto Bargiotti for their helpful contribution to this work.

1. Wrede F., Rothhass A., *Z. Physiol. Chem.* **95**, 226 1934); Thompson P.E., McCarthy D.A., Bayles A., Reinertson J.W. and Cook A.R., *Antibiotics and Cemotherapy* **6**, 337 (1956)

2. Wasserman H.H., Rodgers G. and Keith D.D., *Chem. Comm.* 825 (1966); Harashima K., Tsuchida N., Tanaka T. and Nagatsu J., *Agr. Biol. Chem.* **31**, 481 (1967)

3. Nakamura A., Nagai K., Ando K. and Tamura G. *J., Antibiotics* **39**, 1155 (1989); Tsuji R.F., Yamamoto M., Nakamura A., Kataoka T., Magae J., Nagai K. and Jamasaky M., *Antibiotics* **43**, 1293 (1990)

4. Lin J., *Immunology Today* **14**, 290 (1993)

5. Sigal H.N. and Dumont F.J., *Immunosuppression in Fundamental Immunology*, (Paul W.E. ed.), 903, Raven Press (N.Y.), (1993)

6. Rapoport H., Holden K.J., *J. Am. Chem. Soc.* **84**, 635 (1962); Boger D.L., Patel M., *J. Org. Chem.* **53**, 1405 (1988); Wasserman H.H., Lombardo L. J., *Tetrahedron Lett.* **30**, 1725 (1989); Doria G. et al., International Application WO 95/17381

7. J. Bordner, H. Rapoport, *J. Org. Chem.* **30**, 3824 (1965)

SYNTHESIS AND APPLICATIONS OF CONFORMATIONALLY CONSTRAINED PHENYLALANINE ANALOGUES

Susan E. Gibson (née Thomas),[a] Nathalie Guillo,[a] S. Barret Kalindjian,[b] Matthew J. Tozer,[b] and Nicole J. Whitcombe[a]

[a]Department of Chemistry, Imperial College of Science, Technology and Medicine, South Kensington, London SW7 2AY, U.K.
[b]James Black Foundation, 68 Half Moon Lane, London SE24 9JE, U.K.

INTRODUCTION

The use of conformationally constrained amino acids to probe how bioactive molecules bind to their receptors is a commonly used approach to the design of highly selective and active compounds.[1] Reducing the conformational freedom of a ligand may alter a) the binding affinity of the ligand at a given receptor, b) the selectivity of the ligand between different receptors, and c) the stability of the ligand with respect to enzymatic degradation. Examination of the effect of restricting the conformational freedom of a given ligand on these properties may lead to increased insight into the bioactive conformation of the ligand and hence ultimately to the generation of more potent and selective molecules. Herein we describe how we have used Heck cyclisations and radical cyclisations to generate new conformationally constrained analogues of the amino acid phenylalanine together with the first biological study based on these amino acids.

Many conformationally constrained analogues of amino acids have been synthesised and some have been used in biological studies. Focussing on phenylalanine, typical examples of conformationally constrained analogues include 3-phenylproline,[2] β,β–diphenylalanine,[3] α,β-dimethylphenylalanine[4] and β-methylphenylalanine.[5] Of particular relevance to the work described here is 1,2,3,4-tetrahydroisoquinoline-3-carboxylic acid (Tic) (Figure 1).

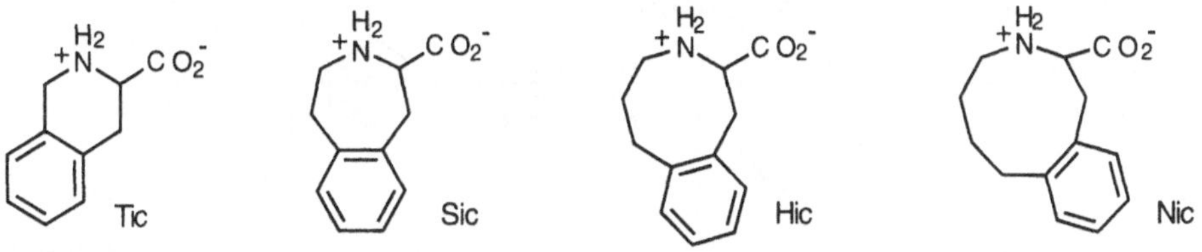

Figure 1. The conformationally constrained phenylalanine analogues, Tic, Sic, Hic and Nic.

Tic is a commercially available compound that has been used to considerable effect in many biological studies to date. In one example, replacement of Phe[2] in the dermorphin-derived δ-opioid receptor antagonist Tyr-Phe-Phe-Phe-NH$_2$ led to a dramatic improvement in its activity profile and stability (Figure 2).[6] Further modification has led to dipeptides which display some of the best δ-opioid receptor selectivities known to date.[7]

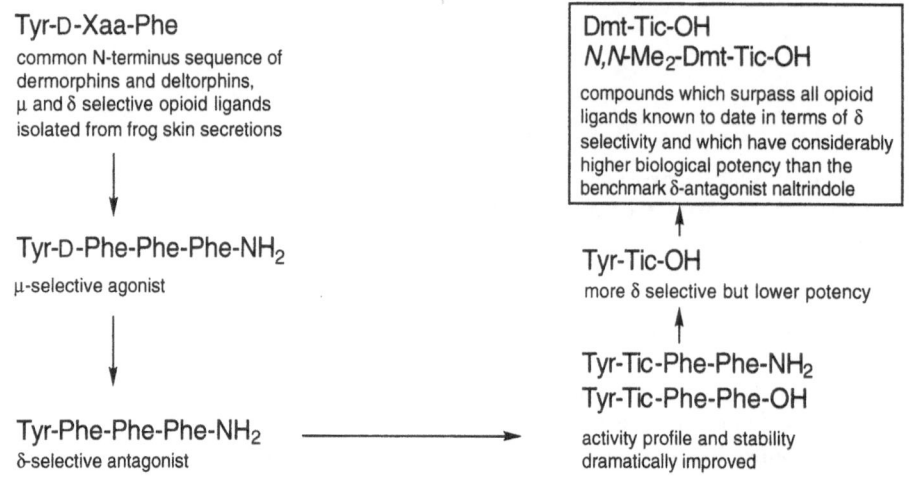

Tyr-D-Xaa-Phe

common N-terminus sequence of dermorphins and deltorphins, μ and δ selective opioid ligands isolated from frog skin secretions

↓

Tyr-D-Phe-Phe-Phe-NH₂

μ-selective agonist

↓

Tyr-Phe-Phe-Phe-NH₂
δ-selective antagonist

⟶

Dmt-Tic-OH
N,N-Me₂-Dmt-Tic-OH

compounds which surpass all opioid ligands known to date in terms of δ selectivity and which have considerably higher biological potency than the benchmark δ-antagonist naltrindole

↑

Tyr-Tic-OH
more δ selective but lower potency

↑

Tyr-Tic-Phe-Phe-NH₂
Tyr-Tic-Phe-Phe-OH

activity profile and stability dramatically improved

Figure 2. Development of highly selective and potent δ-opioid antagonists.

The current level of interest in Tic is reflected by the fact that several analogues of Tic, such as α-methyl Tic,[8] α,β-dimethyl-Tic,[4] β-phenyl-Tic,[9] benzo[f]Tic, benzo[g]Tic, and benzo[h]Tic[10] have been the targets of synthetic studies in recent years.

We thus decided to synthesise the novel 7-, 8- and 9-membered analogues of Tic *i.e.* Sic, Hic and Nic (Figure 1) in the belief that the incorporation of a series of compounds with varying degrees of conformational constraint into biologically active peptides or non-peptides would lead to greater insight into the conformational preferences of the ligand under investigation and the nature of its interaction with the active site than the incorporation of a single compound.

SYNTHESIS OF SIC, HIC AND NIC

Our initial retrosynthetic analysis of Sic, Hic and Nic, illustrated in Figure 3, depends on the intramolecular Heck reaction to create the medium-sized rings. Thus our initial targets were a series of iodoaldehydes and these were prepared using the standard chemistry illustrated in Figure 4.

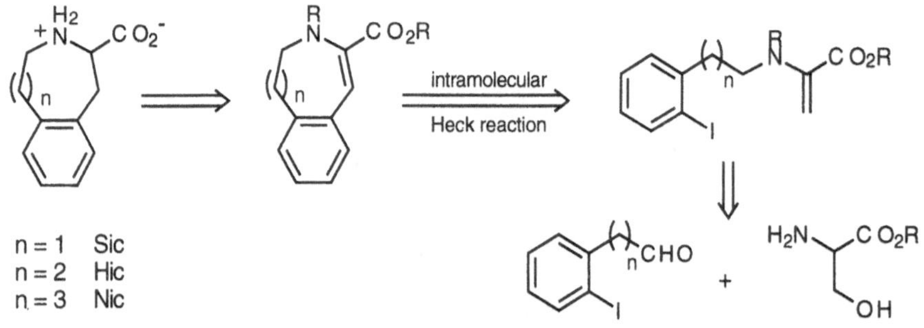

n = 1   Sic
n = 2   Hic
n = 3   Nic

intramolecular
Heck reaction

Figure 3. Retrosynthetic analysis of Sic, Hic and Nic.

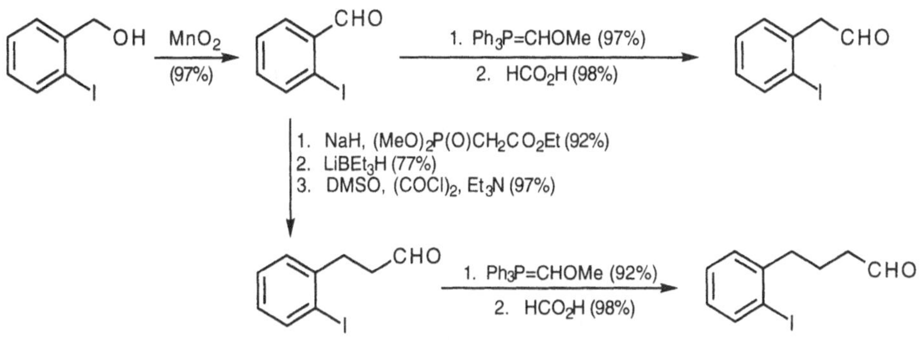

Figure 4. Synthesis of a series of iodoaldehydes.

Reductive amination of the iodoaldehydes with (±)-serine methyl ester followed by *N*-protection and introduction of a carbon-carbon double bond gave a set of potential Heck substrates (Figure 5).

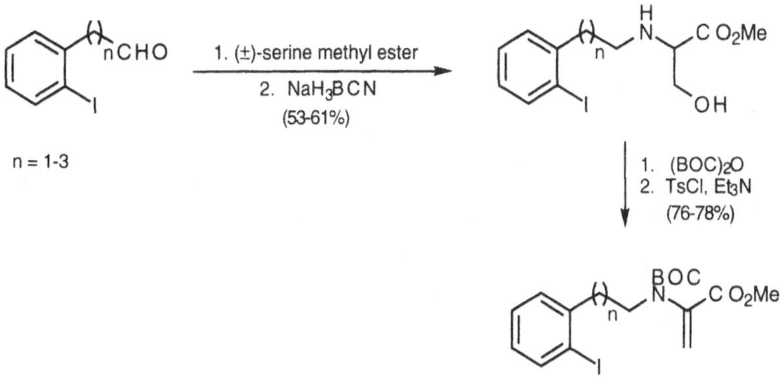

Figure 5. Formation of potential Heck substrates.

In theory, cyclisation of the Heck substrates could proceed *via* two routes: the desired *endo* pathway or an undesired *exo* pathway (Figure 6). Inspection of the literature provided little guidance to the outcome.[11] Although there are many examples of intramolecular Heck reactions being used to form 5- and 6-membered rings (almost all of which proceed *via* an *exo* pathway), and several examples of intramolecular Heck reactions being used to form large rings (most of which proceed *via* an *endo* pathway), there are few examples of the intramolecular Heck reaction being used to form medium-sized rings. Those there are gave no indication as to whether the *exo* or the *endo* mode of cyclisation would be preferred with our substrates.

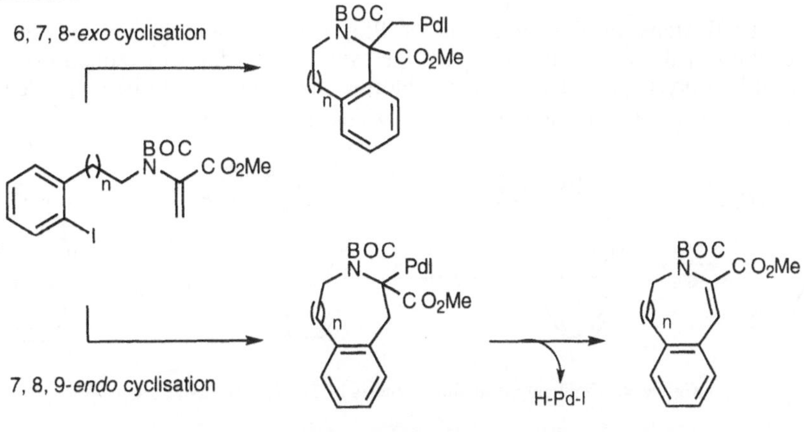

Figure 6. Possible modes of cyclisation of Heck substrates.

Nevertheless, after considerable experimentation, we identified conditions which gave the desired *endo* products in acceptable to good yields (Figure 7).[12] Interestingly, the amount of palladium catalyst required decreases as the ring size increases, probably reflecting the increasing ease of *endo* carbopalladation over *exo* carbopalladation.

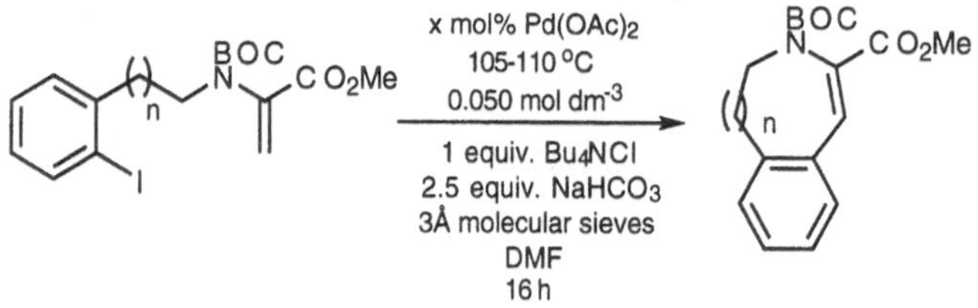

n	x	yield (%)
1	10	55
2	5	75
3	2.5	86

Figure 7. Creation of seven-, eight- and nine-membered rings *via* Heck cyclisation.

We subsequently sought to determine whether or not the desired rings could be formed *via* radical cyclisations. Again literature precedent for such cyclisations was scarce. Whilst there are several hundreds of examples of radical cyclisations leading to five- and six-membered rings, cyclisations to give seven-membered rings are relatively rare, often proceeding in moderate yield, and examples of cyclisations to eight- and nine-membered rings are limited to a mere handful.[13] We were thus pleased to discover that slow addition of Bu$_3$SnH in the presence of AIBN to the unsaturated aryl iodides gave cyclised products in good yield (Figure 8).[14] In contrast to the Heck cyclisations, the radical cyclisations gave better yields for the smaller ring sizes, and thus the two methods provide us with complementary approaches to the desired medium-sized rings.

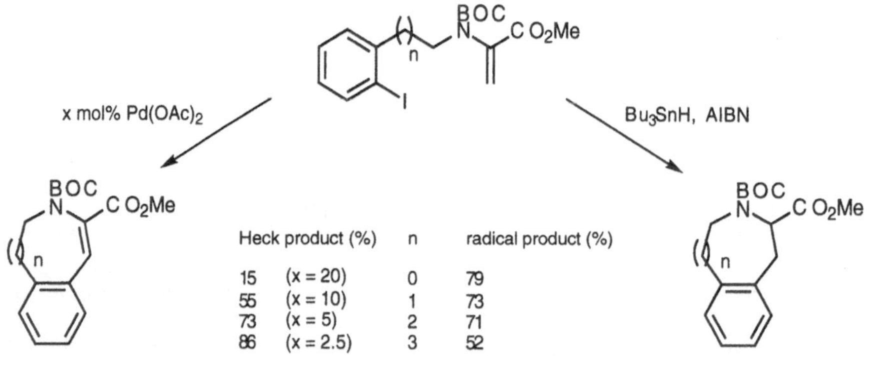

Heck product (%)		n	radical product (%)
15	(x = 20)	0	79
55	(x = 10)	1	73
73	(x = 5)	2	71
86	(x = 2.5)	3	52

Figure 8. Comparision of Heck and radical cyclisation reactions.

The final steps in the synthesis of Sic, Hic and Nic proceeded smoothly. Hydrogenation of the Heck products gave the saturated products of radical cyclisation. Subsequent hydrolysis provided the hydrochloride salts which could be converted to the neutral amino acids by treatment with propene oxide (Figure 9).[12b]

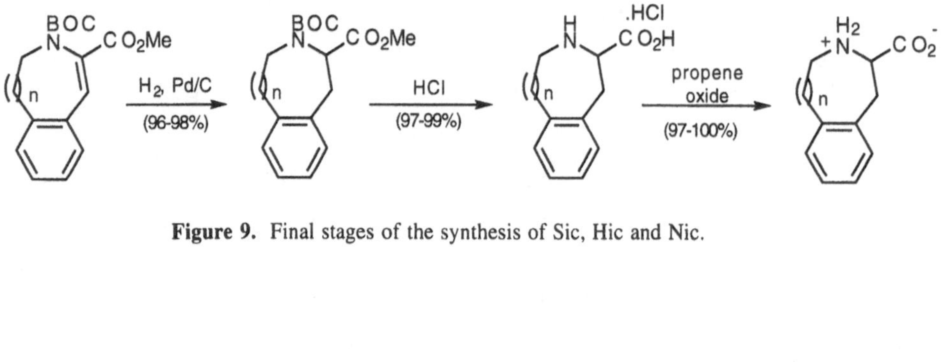

Figure 9. Final stages of the synthesis of Sic, Hic and Nic.

INCORPORATION OF SIC, HIC AND NIC INTO CCK-B/GASTRIN RECEPTOR ANTAGONISTS

Our first test of the hypothesis that incorporation of a series of compounds of varying degrees of conformational constraint into biologically active peptides or non-peptides would lead to greater insight into the conformational preferences of the ligand under investigation than the incorporation of a single compound, was carried out on a non-peptide cholecystokinin-B/gastrin receptor antagonist (Figure 10).[15]

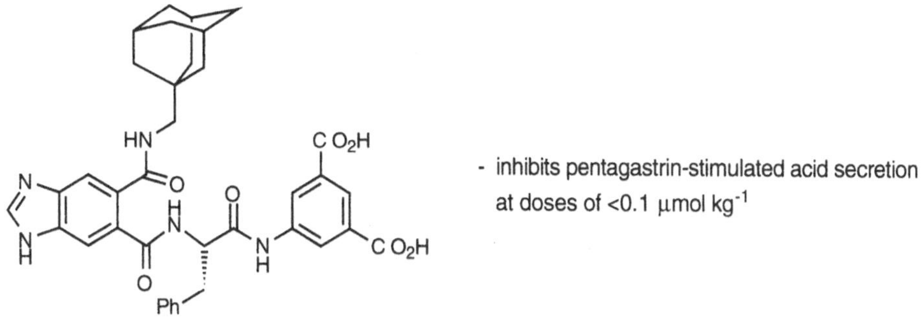

- inhibits pentagastrin-stimulated acid secretion at doses of <0.1 μmol kg^{-1}

Figure 10. The non-peptide cholecystokinin-B/gastrin receptor antagonist chosen for study.

The Tic, Sic, Hic and Nic analogues of the CCK$_B$/gastrin receptor antagonist were synthesised by appropriate modification of the synthetic route developed for the parent Phe-containing compound (Figure 11).[15]

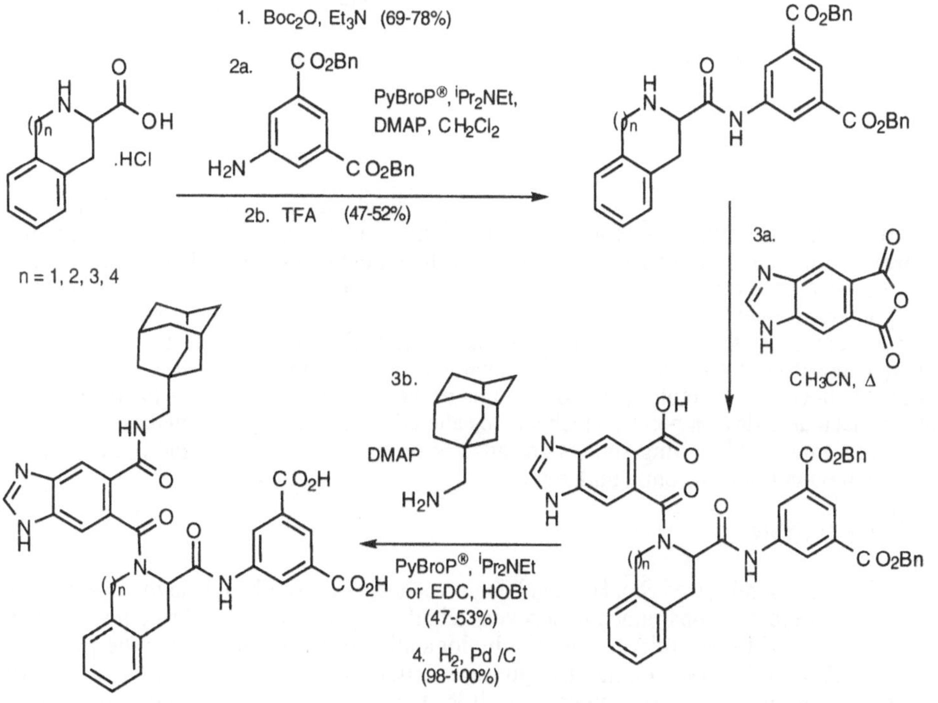

Figure 11. Synthesis of Tic-, Sic-, Hic- and Nic-containing CCK$_B$/gastrin receptor antagonists.

Affinity estimates for these compounds at CCK_B/gastrin receptors were determined in two assays: the isolated immature rat stomach, a functional bioassay; and mouse cortical homogenate, a radioligand binding assay (Figure 12),[16] and compared with a standard ligand L-365,260.

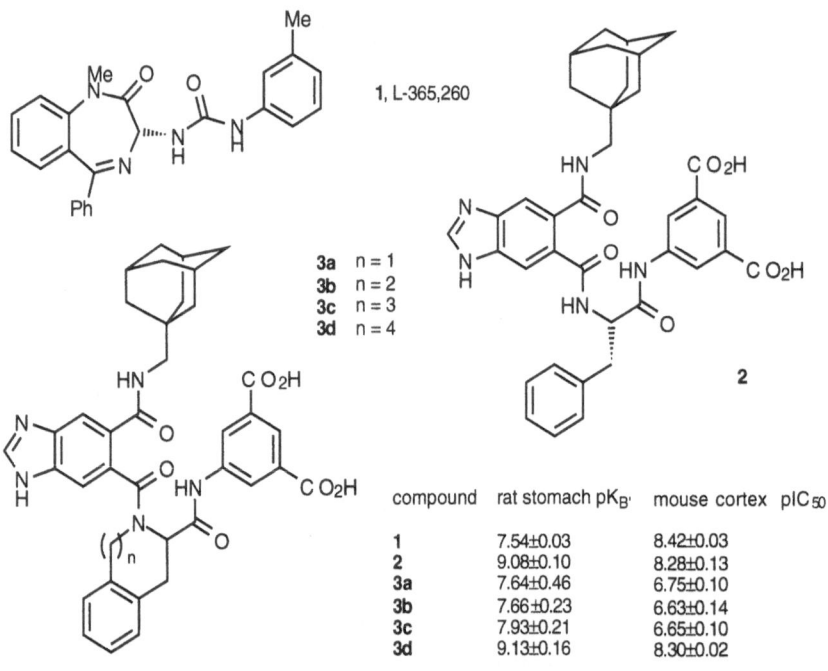

compound	rat stomach $pK_{B'}$	mouse cortex pIC_{50}
1	7.54±0.03	8.42±0.03
2	9.08±0.10	8.28±0.13
3a	7.64±0.46	6.75±0.10
3b	7.66±0.23	6.63±0.14
3c	7.93±0.21	6.65±0.10
3d	9.13±0.16	8.30±0.02

Figure 12. Receptor affinity values for CCK_B/gastrin antagonists.

Relative to the parent Phe-containing compound, the 6-, 7- and 8-membered ring analogues lost affinity in both assays. The affinities of the 9-membered ring analogue, however, were indistinguishable from those of the parent compound. Thus it can be seen that the size of the restricting ring is of importance in the biological activity of these compounds. It may be that the Nic analogue confers a favourable disposition of the aromatic ring comparable to that of the aromatic ring of phenylalanine. It is recognised, however, that the transposition of Tic, Sic, Hic and Nic analogues for phenylalanine will affect a number of factors beyond the simple spatial orientation of the aromatic ring, such as the potential for intramolecular hydrogen bonds, which may be altered by replacing the secondary amide with a tertiary amide. Modelling studies designed to probe the conformational preferences of these compounds are currently underway.

CONCLUSION

We have synthesised Sic, Hic and Nic, novel analogues of phenylalanine in which the phenyl substituent is constrained by a seven-, eight- and nine-membered ring respectively. Incorporation of these amino acids into biologically active molecules for the first time revealed that the Nic-containing analogue was better able to interact with CCK_B/gastrin receptors than the smaller ring analogues. It is evident that these constrained amino acids have a role to play, complementary to established phenylalanine analogues, in providing structural information about ligands and their corresponding sites of action.

REFERENCES

1. a) C. Toniolo, *Int. J. Peptide Protein Res.* 1990, **35**, 287; b) A. Giannis and T. Kolter, *Angew. Chem., Int. Ed. Engl.*, 1993, **32**, 1244; c) V.J. Hruby, G. Li, C. Haskell-Luevano and M. Shenderovich, *Biopoly.*, 1997, **43**, 219.
2. J.Y.L. Chung, J.T. Wasicak, W.A. Arnold, C.S. May, A.M. Nadzan, and M.W. Holladay, *J. Org. Chem.*, 1990, **55**, 270.
3. K. Hsieh, T.R. LaHann and R.C. Speth, *J. Med. Chem.*, 1989, **32**, 898.
4. W.M. Kazmierski, Z. Urbanczyk-Likowska and V.J. Hruby, *J. Org. Chem.*, 1994, **59**, 1789.
5. B.Y. Azizeh, M.D. Shenderovich, D. Trivedi, G. Li, N.S. Strum, V.J. Hruby, *J. Med. Chem.*, 1996, **39**, 2449.
6. P.W. Schiller, T.M.-D. Nguyen, G. Weltrowska, B.C. Wilkes, B.J. Marsden, C. Lemieux, and N.N. Chung, *Proc. Natl. Acad. Sci. U.S. A.*, 1992, **89**, 11871.
7. S. Salvadori, G. Balboni, R. Guerrini, R. Tomatis, C. Bianchi, S.D. Bryant, P.S. Cooper and L.H. Lazarus, *J. Med. Chem.*, 1997, **40**, 3100.
8. a) J.W. Skiles, J.T. Suh, B.E. Williams, P.R. Menard, J.N. Barton, B. Loev, H. Hones, E.S. Neiss, A. Schwah, W.S. Mann, A. Khandwala, P.S. Wolf and I Weinryb, *J. Med. Chem.*, 1986, **29**, 784; b) U. Schöllkopf, R. Hinrichs, and R. Lonsky, *Angew. Chem., Int. Ed. Engl.*, 1987, **26**, 143.
9. H.G. Chen and O.P. Goel, *Synth. Commun.*, 1995, **25**, 49.
10. C. Wang and H.I. Mosberg, *Tetrahedron Lett.*, 1995, **36**, 3623.
11. a) S.E. Gibson (née Thomas) and R.J. Middleton, *Contemp. Org. Synth.*, 1996, **3**, 447; b) E. Negishi, C. Copéret, S. Ma, S-Y. Liou and F. Liu, *Chem. Rev.*, 1996, **96**, 365.
12. a) S.E. Gibson (née Thomas) and R.J. Middleton, *J. Chem. Soc., Chem. Commun.*, 1995, 1743; b) S.E. Gibson (née Thomas), N. Guillo, R.J. Middleton, A. Thuilliez and M.J. Tozer, *J. Chem. Soc., Perkin Trans. 1*, 1997, 447.
13. B. Giese, B. Kopping, T. Göbel, J. Dickhaut, G. Thoma, K.J. Kulicka and F. Trach, *Org. React. (N.Y.)*, 1996, **48**, 315.
14. S.E. Gibson (née Thomas), N. Guillo and M.J. Tozer, *Chem. Commun.*, 1997, 637.
15. S.B. Kalindjian, I.M. Buck, J.M.R. Davies, D.J. Dunstone, M.L. Hudson, C.M.R. Low, I.M. McDonald, M.J. Pether, K.I.M. Steel, M.J. Tozer and J.G. Vinter, *J. Med. Chem.*, 1996, **39**, 1806.
16. S.E. Gibson (née Thomas), N. Guillo, S.B. Kalindjian and M.J. Tozer, *Bioorg. Med. Chem. Lett.*, 1997, **7**, 1289.

SYNTHESIS AND PROPERTIES
OF NOVEL FUNCTIONAL FULLERENE DERIVATIVES

Maurizio Prato,[1] Tatiana Da Ros,[1] Michele Maggini,[2] Dirk M. Guldi,[3] and Luigi Pasimeni[4]

[1] Dipartimento di Scienze Farmaceutiche, Piazzale Europa 1, 34127 Trieste, Italy
[2] C.M.R.O.-C.N.R., Dipartimento di Chimica Organica, Università di Padova, Italy
[3] Radiation Laboratory, University of Notre Dame, Notre Dame, IN 46656, U.S.A
[4] Dipartimento di Chimica Fisica, Università di Padova, Italy

INTRODUCTION

Among the many reasons that have made fullerenes popular in science, one is certainly geometry. [60]Fullerene, or C_{60},[1,2] the most abundant representative of the fullerene family, has the roundest shape one can imagine for a molecule. But many other properties of fullerenes and fullerene derivatives have attracted the attention of physicists and chemists, such as superconductivity,[3-6] ferromagnetism,[7] spectacular electrochemical[8,9] and photophysical behavior.[10,11]

[60]Fullerene

More interestingly from the organic chemistry point of view, C_{60} (as well as the higher fullerenes) is also a reactive chemical species. The "organic functionalization" of fullerenes has provided, in recent years, quite a number of new functional derivatives, which combine the original properties of the carbon cage with those of other interesting classes of compounds.[12-20]

SYNTHESIS AND PROPERTIES

With the aim of producing stable and characterizable fullerene derivatives, we have devised a general synthetic methodology, based on 1,3-dipolar cycloaddition of azomethine ylides.[21] This reaction provides access to a wide range of compounds in which a pyrrolidine ring is fused to a 6,6 ring junction of C_{60}, commonly termed fulleropyrrolidines or pyrrolidinofullerenes (Scheme 1).[22-30]

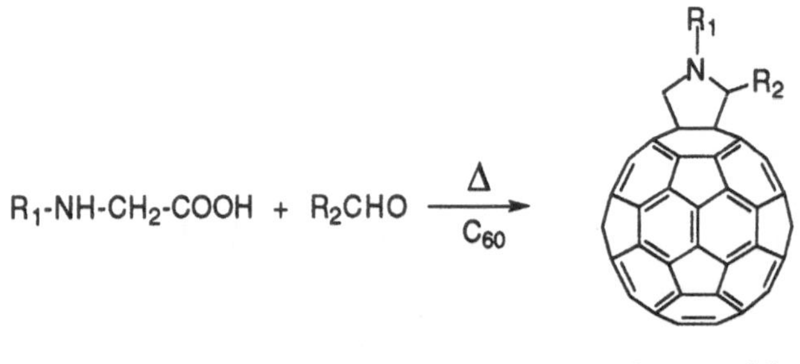

$$R_1\text{-NH-CH}_2\text{-COOH} + R_2\text{CHO} \xrightarrow[C_{60}]{\Delta}$$

fulleropyrrolidine

Scheme 1

The condensation of an N-functionalized α-amino acid with virtually any aldehyde offers the great opportunity to introduce simultaneously two functional groups, R_1 and R_2. Thus, for instance, one group might be a solubilizing appendage whereas the other is an electroactive or a photoactive unit.

Azomethine ylides can also be successfully generated via thermal ring-opening of aziridines. Thermolysis of an aziridine activated by an ester group in the presence of C_{60} generates the corresponding fulleropyrrolidine in very good yields.

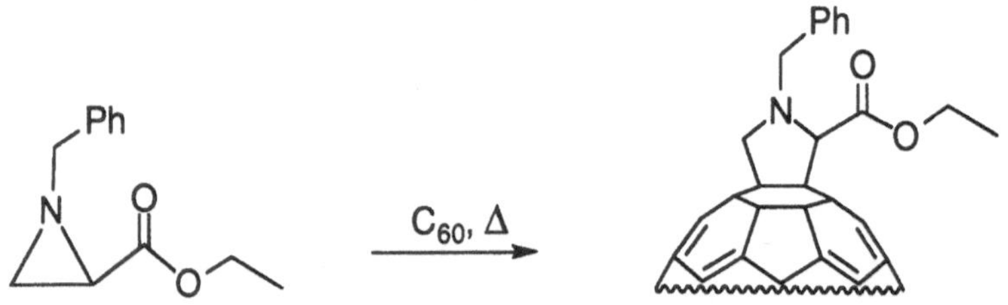

$$\xrightarrow{C_{60}, \Delta}$$

A fundamental issue to be addressed when working with fullerenes is whether or not the new derivatives retain the remarkable properties of the carbon cage. Among others, the electrochemical and photophysical properties are usually taken into consideration. The electrochemical properties of fulleropyrrolidines can be studied in terms of cyclic voltammetry.[24,26]

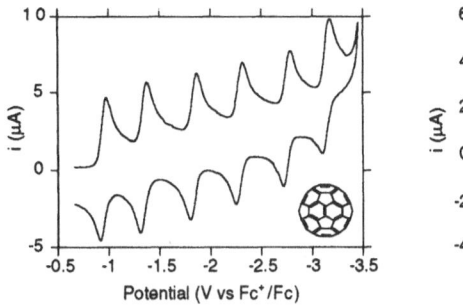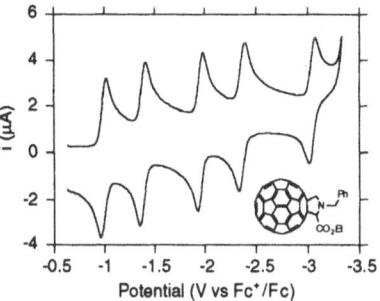

Figure 1

Fig. 1 shows the cyclic voltammograms of C_{60} and a representative fulleropyrrolidine in toluene/acetonitrile $3:1$.[26] It is easily seen that the basic electroactive behavior of C_{60} is maintained in the derivative. In the pyrrolidine derivative, however, the reduction potentials are shifted to more negative values when compared to those of C_{60}. This can be expected on the basis of the saturation of a double bond in C_{60} resulting in a decreased conjugation and an increased perturbation. Due to this effect, only five reduction peaks for the C_{60} moiety in fulleropyrrolidines are detected in the accessible potential range.

Based on the partially broken symmetry, the ground state absorption properties of the fulleropyrrolidines are even richer than those of pristine C_{60}. The absorption range in these derivatives extends throughout the entire UV-Vis region, up to 720 nm. As a direct consequence, the excited states of the fulleropyrrolidines (singlet and triplet) are very easy to access. Additionally, due to the almost quantitative generation of singlet oxygen from the singlet state of fullerene derivatives, these latter compounds have also been proposed as efficient photosensitizers for photodynamic therapy.[31,32]

The combination of the absorption properties and the electron-accepting properties of fullerenes and fullerene derivatives have found wide application in the field of photonics. C_{60} and its organic derivatives, in fact, are excellent electron acceptors both in the ground and excited states. Their involvement for conversion of light into usable electric current is an argument of high interest.[33-35] For this reason, C_{60} has been used unmodified in heterogeneous blends with conducting polymers[36,37] or strong donors like phthalocyanines.[38] Alternatively, C_{60} has been covalently linked to a donor, in order to obtain dyads or triads of various types.[34,35]

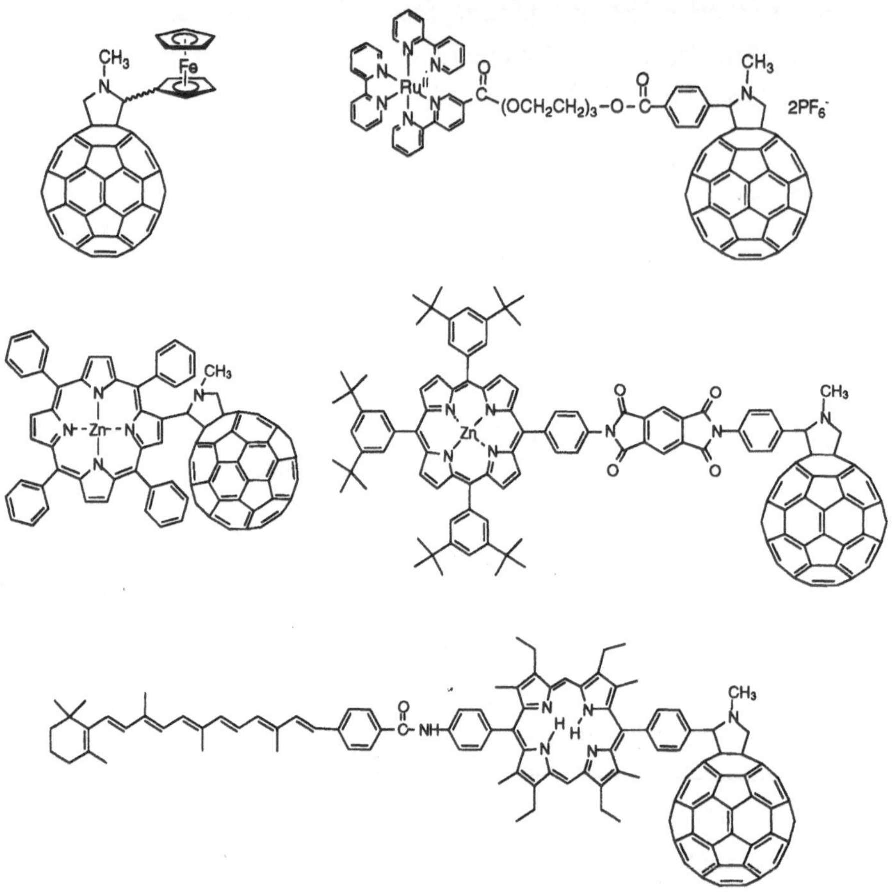

Scheme 2

Scheme 2 outlines only a few representative examples among the number of dyads and triads that have been synthesized in recent years for exploring the use of C_{60} in photoinduced processes[21,25,39-43]

The common objective of these electron and energy transfer studies is the generation of charge-separated species, with lifetimes long enough to allow their ultimate utilization, for instance, in photovoltaic solar cells.

To accelerate forward electron-transfer and slow down back electron-transfer, we have designed a system in which C_{60}, acting as an electron-acceptor unit, is noncovalently attached to a donor (Scheme 3).

In this system, a pyridine-functionalized fullerene (Py-C_{60}) coordinates, reversibly, to the zinc metal of a tetraphenyl porphyrin (ZnTPP). The equilibrium can be monitored by UV-Vis spectroscopy and is heavily shifted towards the associated complex (right side of the equation).

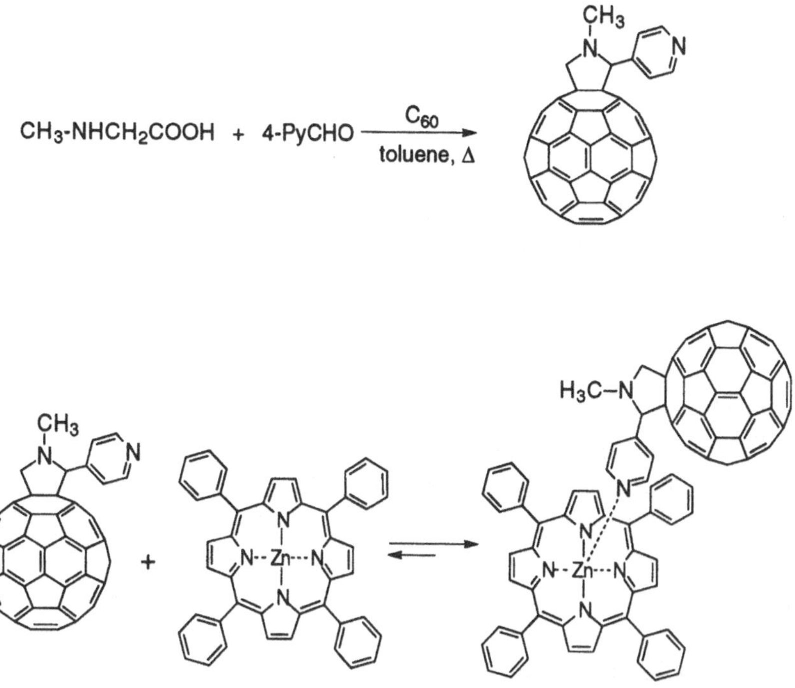

Scheme 3

Based on the spatial donor-acceptor distance, light induced electron-transfer from the porphyrin to C_{60} should proceed very fast intramolecularly. Ideally, once generated, the charge-separated species should dissociate, so that charge recombination becomes less probable (Scheme 4).

$$\text{Py-C}_{60}\text{—ZnTPP} \xrightarrow{h\nu} \text{Py-C}_{60}\text{—ZnTPP}^*$$

$$\text{Py-C}_{60}\text{—ZnTPP}^* \xrightarrow{e^- \text{ transfer}} \text{Py-C}_{60}^{\ominus}\text{—ZnTPP}^{\oplus}$$

$$\text{Py-C}_{60}^{\ominus}\text{—ZnTPP}^{\oplus} \xrightarrow[\text{dissociation}]{\text{complex}} \text{Py-C}_{60}^{\ominus} + \text{ZnTPP}^{\oplus}$$

Scheme 4

Photophysical experiments, together with time-resolved EPR spectroscopy give results that basically agree with an intramolecular electron transfer mechanism as shown in Scheme 4. Both techniques show that in polar solvents (CH_2Cl_2, THF, benzonitrile) the charge-separated state is sufficiently long-lived to let envision practical use of the energy generated. In CH_2Cl_2 and THF lifetimes of the order of 10 μs were detected, whereas in benzonitrile the ionic couple lived for several hundreds of μs. Furthermore, EPR results enabled us to detect spin exchange within the generated radical pair. Electron-transfer occurring in the dissociated complex (intermolecular electron-transfer) was also detected, giving rise to a different type of EPR signal.

CONCLUSIONS

The supramolecular dyad obtained combining a Zn-TPP together with a Py-C_{60} offers great chances for photoinduced electron-transfer processes. The similarity to the processes occurring in natural photosynthesis makes the Zn-TPP/Py-C_{60} system very promising in terms of photoconductivity and energy storage.

ACKNOWLEDGEMENTS

Financial support from MURST and CNR (legge 95/95) is gratefully acknowledged. This is contribution No. NDRL- 4079 from the Notre Dame Radiation Laboratory.

REFERENCES

1. H.W. Kroto, J.R. Heath, S.C. O'Brien, R.F. Curl and R.E. Smalley, *Nature* 318:162 (1985).
2. W. Krätschmer, L.D. Lamb, K. Fostiropoulos and D.R. Huffman, *Nature* 347:354 (1990).
3. A.F. Hebard, M.J. Rosseinski, R.C. Haddon, D.W. Murphy, S.H. Glarum, T.T.M. Palstra, A.P. Ramirez and A.R. Kortan, *Nature* 350:600 (1991).
4. K. Holczer, O. Klein, S.-M. Huang, R.B. Kaner, K.-J. Fu, R.L. Whetten and F. Diederich, *Science* 252:1154 (1991).
5. R.C. Haddon, *Acc. Chem. Res.* 25:127 (1992).
6. M.J. Rosseinsky, *J. Mater. Chem.* 5:1497 (1995).
7. P.M. Allemand, K.C. Khemani, A. Koch, F. Wudl, K. Holczer, S. Donovan, G. Gruner and J.D. Thompson, *Science* 253:301 (1991).
8. Q. Xie, E. Pérez-Cordero and L. Echegoyen, *J. Am. Chem. Soc.* 114:3978 (1992).
9. Y. Ohsawa and T. Saji, *J. Chem. Soc, Chem. Commun.* 781 (1992).
10. J.W. Arbogast, A.P. Darmanian, C.S. Foote, Y. Rubin, F.N. Diederich, M.M. Alvarez, S.J. Anz and R.L. Whetten, *J. Phys. Chem.* 95:11 (1991).
11. C.S. Foote, *Top. Curr. Chem.* 169:347 (1994).
12. F. Wudl, *Acc. Chem. Res.* 25:157 (1992).
13. R. Taylor and D.R.M. Walton, *Nature* 363:685 (1993).
14. A. Hirsch, *Angew. Chem., Int. Ed. Engl.* 32:1138 (1993).
15. A. Hirsch, *The Chemistry of the Fullerenes*, Thieme, Stuttgart (1994).
16. F. Diederich, L. Isaacs and D. Philp, *Chem. Soc. Rev.* 23:243 (1994).
17. A. Hirsch, *Synthesis* 895 (1995).
18. R. Taylor, *The Chemistry of Fullerenes*, World Scientific, Singapore (1995).
19. *Tetrahedron Symposia-in-Print Number 60, Fullerene Chemistry*, Smith, A. B., Guest Ed., 4925 (1996).
20. F. Diederich and C. Thilgen, *Science* 271:317 (1996).
21. M. Maggini, G. Scorrano and M. Prato, *J. Am. Chem. Soc.* 115:9798 (1993).
22. M. Maggini, G. Scorrano, A. Bianco, C. Toniolo, R.P. Sijbesma, F. Wudl and M. Prato, *J. Chem. Soc., Chem. Commun.* 305 (1994).
23. M. Maggini, A. Karlsson, L. Pasimeni, G. Scorrano, M. Prato and L. Valli, *Tetrahedron Lett.* 35:2985 (1994).

24. M. Maggini, A. Karlsson, G. Scorrano, G. Sandonà, G. Farnia and M. Prato, *J. Chem. Soc., Chem. Commun.* 589 (1994).

25. M. Maggini, A. Donò, G. Scorrano and M. Prato, *J. Chem. Soc., Chem. Commun.* 845 (1995).

26. M. Prato, M. Maggini, C. Giacometti, G. Scorrano, G. Sandonà and G. Farnia, *Tetrahedron* 52:5221 (1996).

27. A. Bianco, M. Maggini, G. Scorrano, C. Toniolo, G. Marconi, C. Villani and M. Prato, *J. Am. Chem. Soc.* 118:4072 (1996).

28. F. Novello, M. Prato, T. Da Ros, M. De Amici, A. Bianco, C. Toniolo and M. Maggini, *Chem. Commun.* 903 (1996).

29. T. Da Ros, M. Prato, F. Novello, M. Maggini and E. Banfi, *J. Org. Chem.* 61:9070 (1996).

30. A. Bianco, F. Gasparrini, M. Maggini, D. Misiti, A. Polese, M. Prato, G. Scorrano, C. Toniolo and C. Villani, *J. Am. Chem. Soc.* 119:7550 (1997).

31. J.L. Anderson, Y.-Z. An, Y. Rubin and C.S. Foote, *J. Am. Chem. Soc.* 116:9763 (1994).

32. H. Tokuyama, S. Yamago, E. Nakamura, T. Shiraki and Y. Sugiura, *J. Am. Chem. Soc.* 115:7918 (1993).

33. Y. Sakata, H. Imahori, H. Tsue, S. Higashida, T. Akiyama, E. Yoshizawa, M. Aoki, K. Yamada, K. Hagiwara, S. Taniguchi and T. Okada, *Pure Appl. Chem.* 69:1951 (1997).

34. H. Imahori and Y. Sakata, *Adv. Mater.* 9:537 (1997).

35. M. Prato, *J. Mater. Chem.* 7:1097 (1997).

36. Y. Wang, *Nature* 356:585 (1992).

37. N.S. Sariciftci, L. Smilowitz, A.J. Heeger and F. Wudl, *Science* 258:1474 (1992).

38. C. Schlebusch, B. Kessler, S. Cramm and W. Eberhardt, *Synth. Met.* 77:151 (1996).

39. N.S. Sariciftci, F. Wudl, A.J. Heeger, M. Maggini, G. Scorrano, M. Prato, J. Bourassa and P.C. Ford, *Chem. Phys. Lett.* 247:210 (1995).

40. T. Drovetskaya, C.A. Reed and P. Boyd, *Tetrahedron Lett.* 36:7971 (1995).

41. P.A. Liddell, D. Kuciauskas, J.P. Sumida, B. Nash, D. Nguyen, A.L. Moore, T.A. Moore and D. Gust, *J. Am. Chem. Soc.* 119:1400 (1997).

42. H. Imahori, K. Yamada, M. Hasegawa, S. Taniguchi, T. Okada and Y. Sakata, *Angew. Chem., Int. Ed. Engl.* 36:2626 (1997).

43. D.M. Guldi, M. Maggini, G. Scorrano and M. Prato, *J. Am. Chem. Soc.* 119:974 (1997).

COBALT CATALYSED SYNTHESIS OF β-TURN MIMICS AS MODEL FOR THE BIOACTIVE CONFORMATION OF HIV-1 PROTEASE INHIBITORS

Javed Iqbal, Jyoti Prokash Nandy, E. N. Prabhakaran, S. Rajesh and Shyam Krishnan.

Department Of Chemistry, Indian Institute Of Technology, Kanpur 208 016, India.

INTRODUCTION

Proteins and peptides are known to interact with macromolecular receptors to trigger biological processes. In order to interrupt these processes, medicinal chemists have developed antagonists which block the native ligand from binding to its receptor. A rational design of such antagonists is facilitated by our understanding based on the spectacular advances made by molecular biology. This has helped us in identifying the region of protein and its conformation that interacts with the native ligand. The remarkable progress in the field of molecular biology, peptide synthesis, and molecular modelling has dramatically increased understanding of the relationship between protein and peptide structure and their biological function. The recent progress in the synthesis and screening of huge peptide libraries has focused attention on small peptides as important lead structures for the development of potential therapeutic agents. The linear peptide fragments are flexible and exhibit numerous conformations in solution and even in the solid state. However, if one can restrict the conformational freedom of these linear peptides by introducing some constraints in the structure, their solution conformations can be determined by usual physical methods. The conformational restriction can help render a biologically active peptide more potent, more specific and orally active and this may give rise to species which are therapeutically useful. Such constrained structures will also shed useful information on the receptor bound conformation of the ligand. This conformational knowledge about ligands which retain affinity for receptors is then used to develop a model for the biologically active conformation of peptide antagonist, i.e., an antagonist pharmacophore. A number of X-ray structures of proteases with their corresponding peptide inhibitors show that local regions of peptides bound to active site adopt an extended conformation very similar to a protein beta sheet or beta strand. The antibody-antigen complexes also show binding of the peptide antigens in a beta turn conformation. The importance of this information for drug design is that the design of the antagonist pharmacophore may be based on a particular protein structural motif. An important structural feature of many biologically active peptides and proteins is the beta turn motif. These turns are often situated at the protein surface and usually consist of polar residues that offer the opportunity of intermolecular interactions with other protein surfaces and hence provide site for intermolecular recognition. The importance of beta turns in peptides and proteins may well be crucial in receptor interactions that ultimately lead to biological activity.

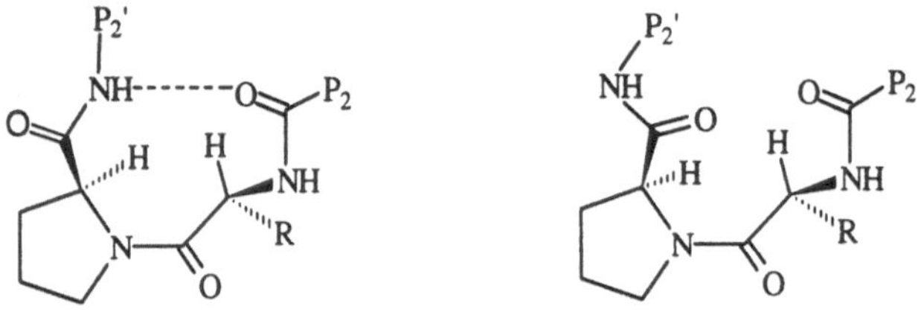

Type VIa β turn   Type VIb β turn

Figure 1

In view of the importance of constrained conformations, there have been several attempts to lock peptides into beta turn configurations and to synthesise molecules that might mimic a beta turn in an otherwise normal peptide. A beta turn[1] is a tetrapeptide sequence in which the $\alpha C_{(1)}$-$\alpha C_{(4)}$ distance is ~ 7 Å and which occurs in a non-helical region of protein. Several types of beta turns are found in proteins and the type VI turn is a unique member of the beta turn family because it is the only turn that involves an s-cis peptide bond (Figure 1). Type VI beta turns always contain a proline residue at the $i+2$ position, since peptides incorporating this amino acid are the only ones that can exist substantially in the s-cis configuration. The VI beta turn is often found[2] in peptides and proteins containing the sequence ArProAr, where Ar represents an amino acid with aryl side chain. Type VI turns are subdivided into type VIa and VIb turns. In the type VIa structures, an intramolecular hydrogen bond is formed between the i carbonyl oxygen and the $i+3$ amide hydrogen, whereas in the type VIb structures the C-terminus of the turn is away from the N-terminus which precludes the formation of hydrogen bonding seen in the former type of beta turn. The understanding of the conformation of type VI beta turn is very crucial to the development of inhibitors for HIV protease. This is mainly due to the specificity shown by the HIV protease for the selective cleavage of proline-phenyl-alanine/tyrosine amide bonds in the Matrix-Capsid domain of the gag-pol polyproteins. This kind of specificity is not exhibited by mammalian cellular proteases which are not known to efficiently hydrolyze peptide bonds involving the proline nitrogen. This unique behaviour of HIV protease makes it an attractive target for inhibition. Wong[3] and coworkers have developed pyrrolidine-containing α-hydroxy and α-ketoamide core structures **1** and **2** (Figure 2) as mechanism based inhibitors of HIV protease. Their studies have also indicated that α-hydroxy amide core structure **3** containing dipeptide isosteres exhibit high potency which is associated with the ability of the hydroxy group to hydrogen bond more effectively with the catalytic carboxyl group of aspartates present in HIV protease. The α-ketoamide was shown to undergo hydration in the active site where the hydrated bound inhibitor provided the hydrogen bonding to the aspartate carboxyls while the phenylalanine and proline side chains occupy proper S_1 and S_1' pockets. It is a well known fact that the peptides are not very efficient inhibitors as they exhibit metabolic instability and/or poor oral bioavailability. This realization has paved way for the development of potent small molecules, which retain some characteristics of the original peptide, as inhibitors for various proteases. In view of these advances, we have developed a general synthetic protocol for potent small molecule inhibitors that span P_2- P_2' subsites in HIV protease. It has been observed by Wong[3] and co-workers that hydrophobic interactions at the C- and N-terminus of the proline containing core structures play a significant role in increasing the potency of such molecules. Their studies revealed that Cbz- and t-Bu- protecting groups in compounds **1-3** were responsible for its enhanced binding with HIV protease.

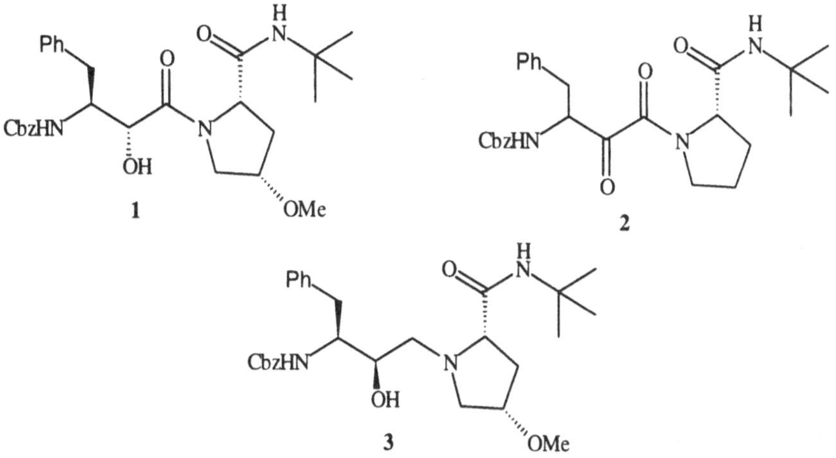

Figure 2

RESULTS AND DISCUSSION

We have designed small molecules as potent HIV protease inhibitors by incorporating features responsible for enhanced binding of these molecules with the protease. It is known that two most important interactions at the active site of the HIV protease are a) between the structural water molecule, P_1-P_1' residue carbonyls and isoleucine 50 and 150 amide hydrogens and b) between the aspartates 25 and 125 and the hydroxy group of the inhibitor. Based on these considerations, we have conceptualised a core structure possesing these salient features essential for an effective binding with the HIV protease. This core structure has the aromatic ring at the N-terminal for hydrophobic contact as well as the phenolic hydroxy group for an effective hydrogen bonding interaction with the structural water molecule (Figure 3). The α-hydroxy amide present in the core structure is suitably placed for interaction with the aspartates 25 and 125. The core structures also possess the requisite feature (i.e., a proline residue) to exhibit a type VI beta turn usually adopted by the bioactive conformation of the inhibitor on interaction with the receptor. Interestingly, if these core structures can be cyclised to the corresponding pseudopeptides then such a structure may behave as a β-turn mimic where the structural water molecule is not required as the ester carbonyl of the cyclic mimic may have the appropriate hydrogen bonding interaction with isoluecine 50 and 150 (Figure 4). We have identified the β-phenylisoserine derivative-proline derived dipeptide as the core structure which can be modified to give potent small molecule HIV protease inhibitors. These core structures can be synthesised from cinnamoyl amides (derived from amino acids) using our cobalt catalysed[4] protocol. We have recently developed a polyaniline supported cobalt complex (Co-PANI) which efficiently catalyses the conversion of cinnamoyl amides to the corresponding β-phenylisoserine derived dipeptides in good yields.

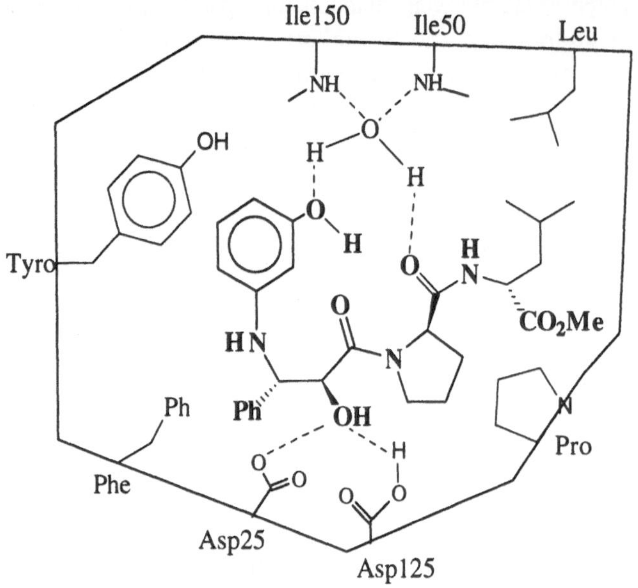

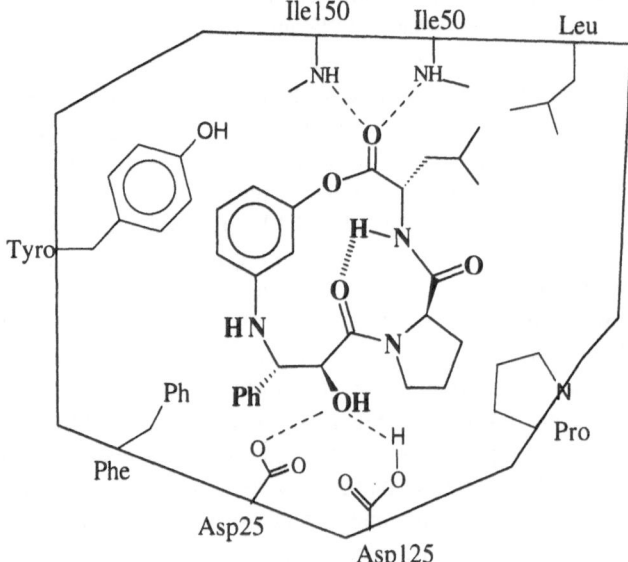

Figure 3

The polyaniline supported cobalt complex (Co-PANI) can be prepared by reacting cobalt(II) salen with polyaniline in acetic acid. Thus cinnamoyl amides can be converted to the corresponding epoxides on treatment with 2-methylpropanal in the presence of oxygen and catalytic amount of polyaniline supported cobalt complex (Co-PANI).The resulting epoxide is then opened with aniline and its derivatives in the presence of the same catalyst leading to the synthesis of dipeptides derived from β-phenylisoserine and α-amino acids. It is interesting to note that both the steps i.e., epoxidation and its opening with anilines, is catalaysed by Co-PANI and the entire transformation can be performed in one pot.

Figure 4

The one pot procedure did not afford dipeptides with high purity, however, the reaction performed on the isolated epoxide gave products in high yield and purity. Application of this protocol was demonstrated by synthesising a library[5,6] of dipeptides derived from β-phenylisoserine derivatives and various α-amino acids. It is noteworthy that the synthesis of dipeptide derivatives, derived from L-proline and 4-hydroxy L- proline **4**, is very useful as it leads to the structural analogs **6** (Scheme 1) of the core structures for HIV protease inhibitors as shown by Wong[3] and co-workers. The relative stereochemistry for compounds **6a-f** is shown in scheme 1 and in several cases the *anti* diastereomer was found to be the major product. *The opening of epoxides with aniline and its derivatives **5** provides structures having hydrophobic environment around the N-terminal of the dipeptides.*

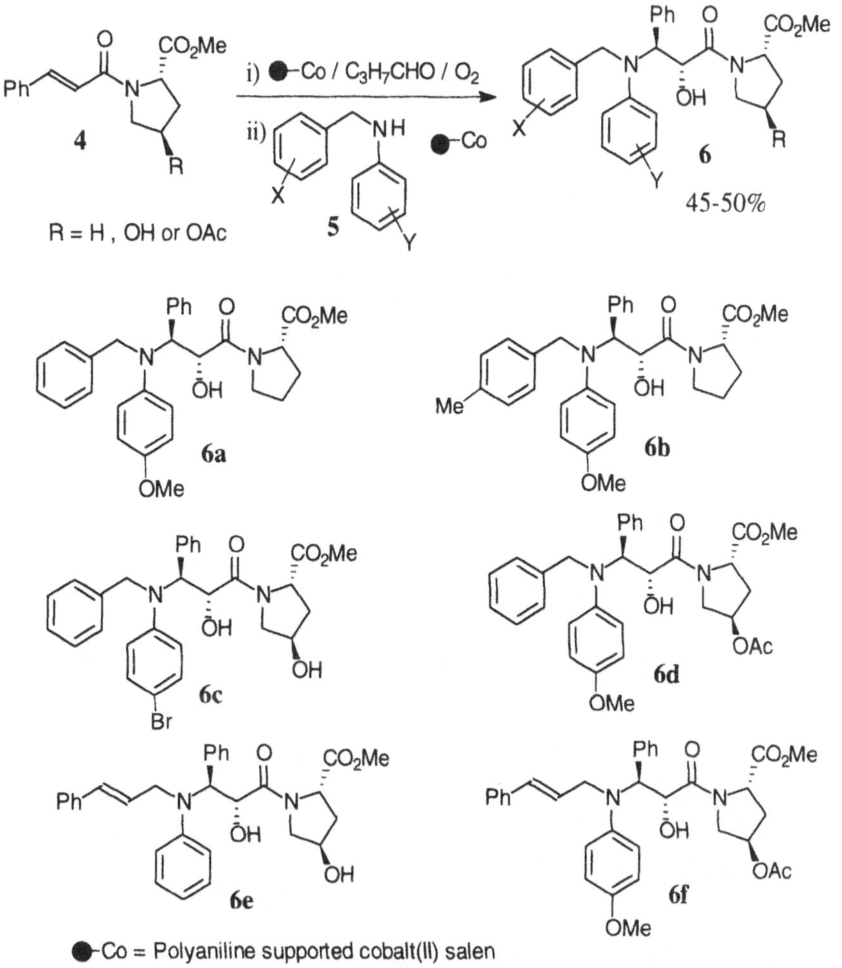

●-Co = Polyaniline supported cobalt(II) salen

Scheme 1

We have synthesised the tripeptides derived from β-phenylisoserine with the hope that such peptides will adopt a beta turn conformation. *Cyclisation of such tripeptides with an aromatic ring spacer may lock the beta turn conformation thereby rendering the resulting cyclic peptide more effective for binding with the HIV protease.* The resulting cyclic peptides are likely to exhibit the typical type VIa or VIb turn and may become potent HIV protease inhibitors. We have attempted a one pot cobalt(II) chloride opening of the dipeptide epoxide **7** and **9** with meta amino phenol (Scheme 2) to the corresponding cyclic peptide **8** and **10** respectively as a model for beta turn mimic. Our premise was based on the chemoselective opening of the epoxide with amino group of aniline and subsequent intramolecular transesterification of the C-terminal ester with the phenolic group already present on the aniline ring. Such a tandem reaction would then afford the corresponding cycilc pseudopeptide with a type VIa or VIb beta turn as envisaged in figure 4. The chemoselective opening of dipeptide epoxides **12** and **16** (obtained from **11** and **15** respectively on cobalt catalysed epoxidation) were carried out in the presence of catalytic amount of cobalt(II) chloride by meta or para hydroxy aniline leading to the formation of the tripeptide derivatives **13** and **17** respectively (Scheme 3). The phenolic group in **13** and **17** was suitably positioned for an intramolecular transesterification with the C-terminal ester group and these tripeptides therefore are useful precursors for the cyclic beta turn mimics **14** and **18** respectively. We have also carried out the synthesis of a tripeptide derivative **20** as a precursor to cyclic peptide **21** by opening the dipeptide epoxide **12** with a secondary amine **19** having a phenolic group in one of the aromatic ring (Scheme 4). Thus the alkene dipeptide **11** was converted to the corresponding epoxide **12** by polyaniline supported cobalt(II) salen catalysed reaction as described by us earlier[6]. The resulting epoxide **12** was then opened with the secondary amine **19** in the presence of catalytic amount of cobalt(II) chloride to give the corresponding tripeptide derivative **20** in good yields. The tripeptide derivatives **13**, **17** and **20** were obtained as a mixture of diastereomers, however, the *anti* diastereomers were found to be major in these cases and only the relative stereochemistry is shown for products in Scheme 3 and 4. The tripeptide **20** was subjected to cyclisation with sodium methoxide under high dilution, however, it did not afford the corresponding cyclic peptide **21** and a complex mixture of products were obtained.

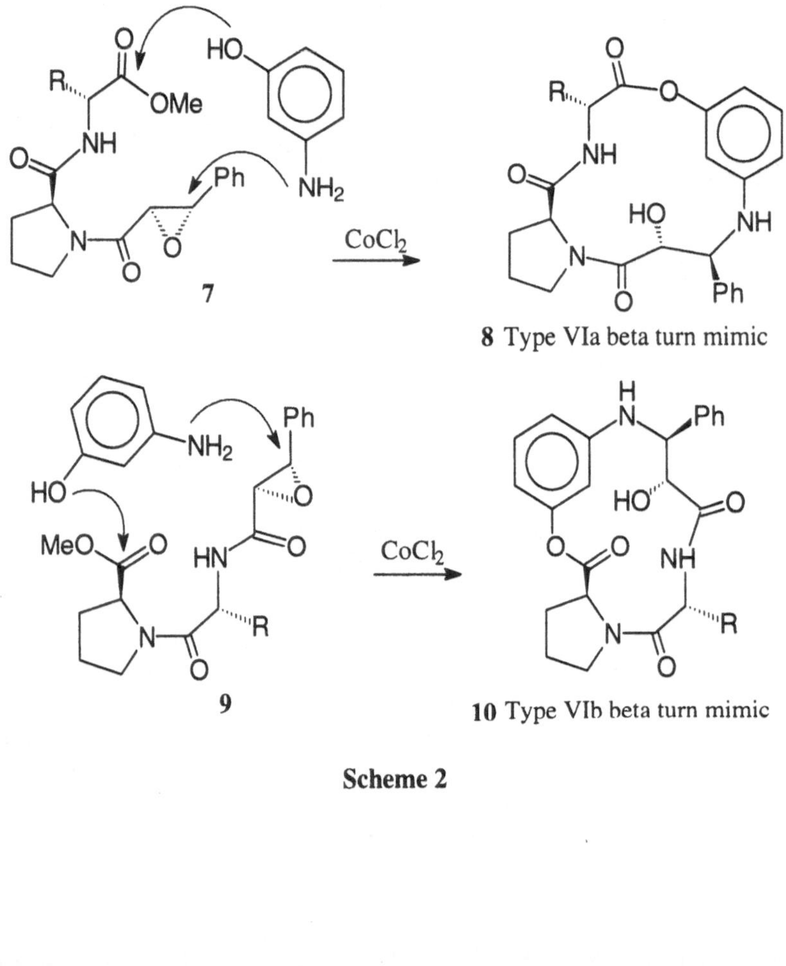

Scheme 2

344

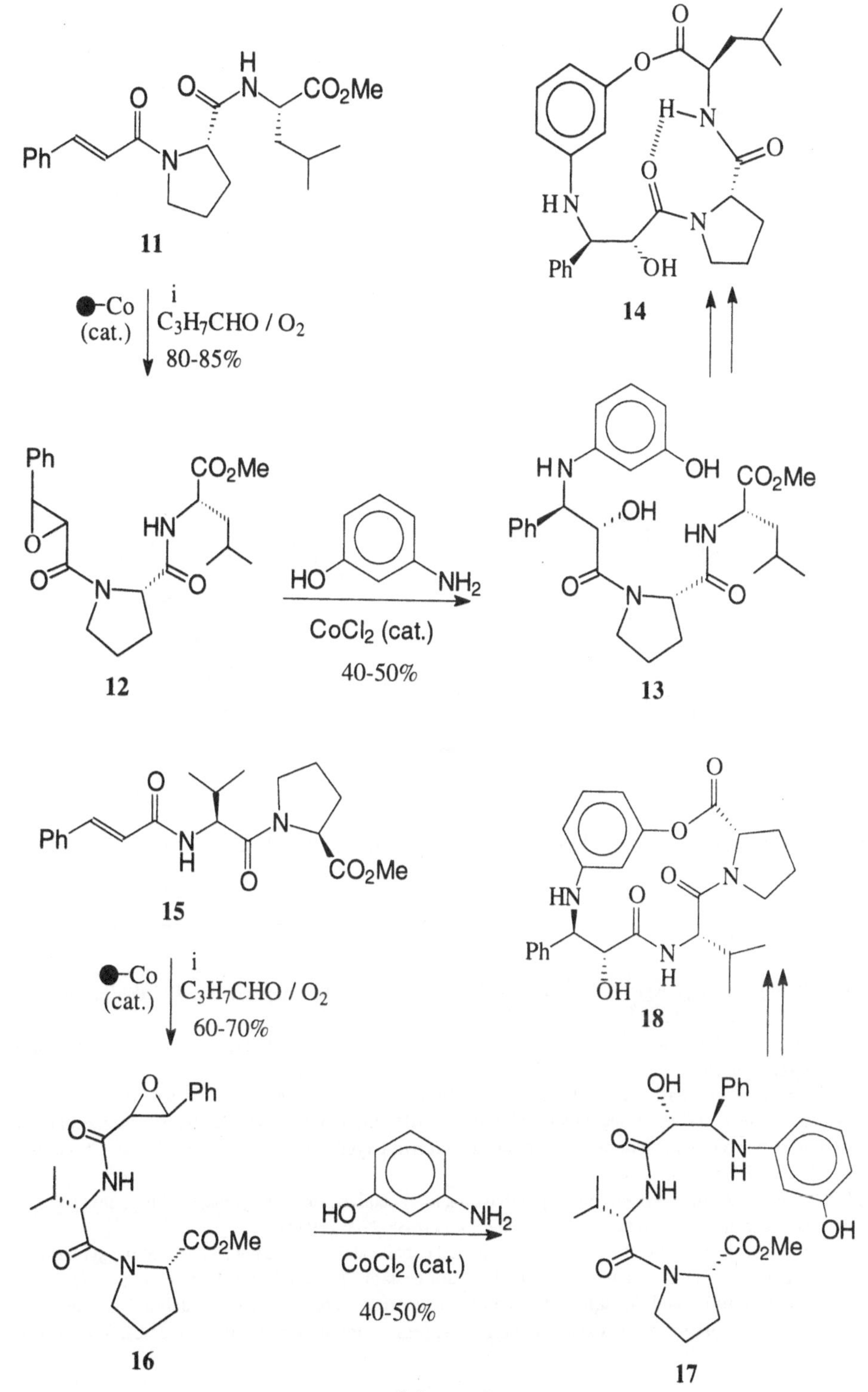

Scheme 3

In conclusion, the cobalt catalysed conversion of cinnamoyl peptides and dipeptides to the corresponding β-phenylisoserine derived dipeptides and tripeptides derivatives respectively is an efficient route for the access to proline derived core structures as potent HIV protease inhibitors. The proline derived tripeptides are potential precursors to the cyclic mimics of type VI beta turn mimics. We are currently engaged in the conversion of these tripeptides to the corresponding cyclic beta turn mimetics.

Scheme 4

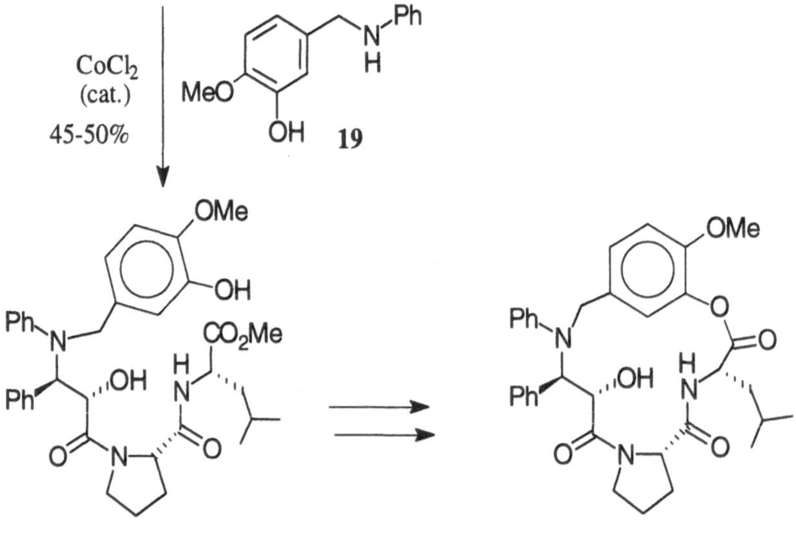

Acknowledgement: We thank DST, New Delhi for the financial support to this work.

REFERENCES

1. J. B. Ball, R. A. Hughes, P. L. Alewood and P. R. Andrews, β-Turn topography, *Tetrahedron* 49:3467 (1993) and references cited therein.
2. K. Kim and J. P. Germanas, Peptides constrained to typeVIb β-turn. 1. evidence for an exceptionally stable intermolecular hydrogen bond, *J. Org. Chem.* 62:2847 (1997).
3. D. H. Slee, K. L. Laslo, J. H. Elder, I. R. Ollmann, A. Gustchina, J. Kervinen, A. Zdanov, A. Wlodawer and C. Wong, Selectivity in inhibition of HIV and FIV protease: inhibitory and mechanistic studies of pyrrolidine-containing α-keto amide and hydroxyethylamine structures, *J. Am. Chem. Soc.* 117:11867 (1995).
4. B. C. Das and J. Iqbal, Polyaniline supported cobalt(II) catalyst: one pot synthesis of β-phenylisoserine derivatives from cinnamoyl amides, *Tetrahedron Lett.* 38:2903 (1997).
5. A. De, P. Basak and J. Iqbal, Polyaniline supported cobalt catalysed one pot stereoselective synthesis of the structural analogues of aminopeptidase inhibitor bestatin, *Tetrahedron Lett.* 38:8383 (1997).
6. T. Punniyamurthy and J. Iqbal, Polyaniline supported cobalt(II) salen catalysed synthesis of pyrrolidine-containing α-hydroxyamide core structures as inhibitors for HIV proteases, *Tetrahedron Lett.* 38:4463 (1997).

STEREOSELECTIVE SYNTHESES OF TRINEMS

Tino Rossi

Lead Discovery Department
Glaxo Wellcome Medicines Research Centre
Via Fleming 4, 37135 Verona, Italy

INTRODUCTION

Trinems (**1**, Fig. 1) can be considered the newest entry among the several classes of β-lactams discovered and published during the past 70 years since the discovery of penicillin.[1] They were first, almost contemporarily, discovered at Glaxo and Takeda in the late 80s and first appeared in the literature in 1991.[2,3] Their most important structural feature is represented by the presence of a third ring **C** (Fig. 1) fused to the bicyclic β-lactam system. This ring can be of variable size, generally from 5 to 7 member either carbocyclic or heterocyclic ring and could also bear substituents at the different positions as well as more fused rings (generally aromatic). A representation of the different subclasses of trinems reported is shown in Fig.2.

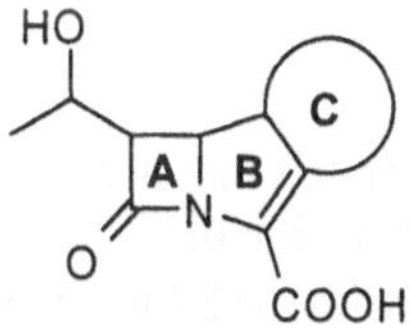

1

Figure 1. General structure of Trinems

From the biological standpoint, trinems were found to possess a broad spectrum of activity encompassing Gram positive, Gram negative and anaerobe strains, high resistance to all the most clinically relevant classes of β-lactamases and a general good stability to mammalian hydrolytic enzymes such as human renal DHP-I.[4] Tetracyclic and policyclic trinems have also been reported by Merck[5] to possess useful activity against methicillin resistant Staphylococci whose growing incidence in hospitalized patients and resistance to most of the major classes of antibacterials represents a major threat.[6]

Current Trends in Organic Synthesis
Edited by Scolastico and Nicotra, Kluwer Academic/Plenum Publishers, 1999

BIOLOGICAL PROPERTIES AND S.A.R. OF TRINEMS

Due to their particular structure trinems possess at least 4 stereogenic centres (5 or more if one or more substituents on ring **C** are present), at least 16 isomers for a single trinem structure are possible and each of them i expected to show a different biological profile.

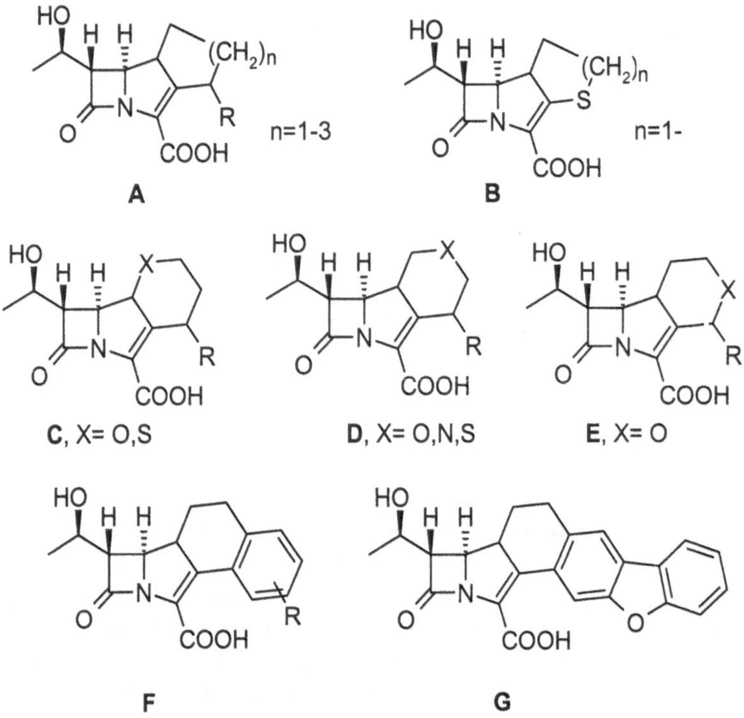

Figure 2. Classes of trinems published: structures **A-E** (Glaxo Wellcome and Takeda), **F** (Hoechst, Merck, Glaxo Wellcome) and **G** (Merck).

Early in house studies demonstrated that absolute configurations at C-9, C-10, C-12 (**2**, Fig. 3) are required in order to show the best compromise between antibacterial activity and stability to β-lactamases in line with the SAR established in the carbapenem and penem series.[7] We also found that 6-membered carbocyclic ring offers the advantage of imparting both a general very high antibacterial activity and biological stability and a good synthetic "tractability" for this class of trinems. The presence of substituents at position C-4 with particular reference to those bearing an heteroatom directly linked with C-4 is most preferred for maximizing antibacterial activity. The presence of two stereogenic centres at C-4 and C-8 indicate that 4 diastereomers are still possible. We established that isomers possessing (4*S*, 8*S*) absolute configuration possess the best compromise in terms of antibacterial potency, spectrum and stability to hydrolytic enzymes.

In particular 4-methoxy trinem, Sanfetrinem (**3**, Fig.3) both in its injectable (**3a**, sodium salt) and oral version (**3b**, sanfetrinem cilexetil) and 4-formimidoyl trinem GV129606 (**4**, Fig.3) were selected for further biological studies and promoted to the development phase.

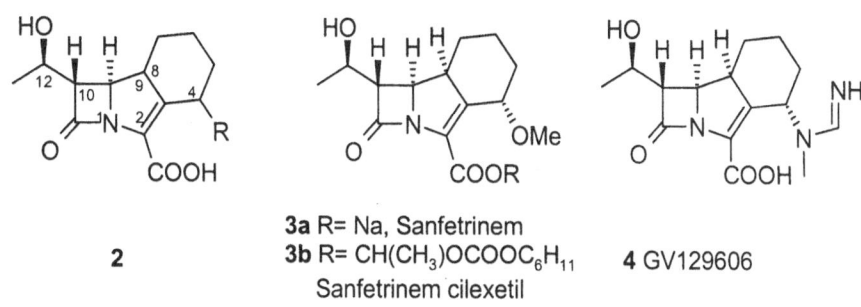

Figure 3. Structures of six membered carbocyclic trinem (**2**), Sanfetrinem and GV129606.

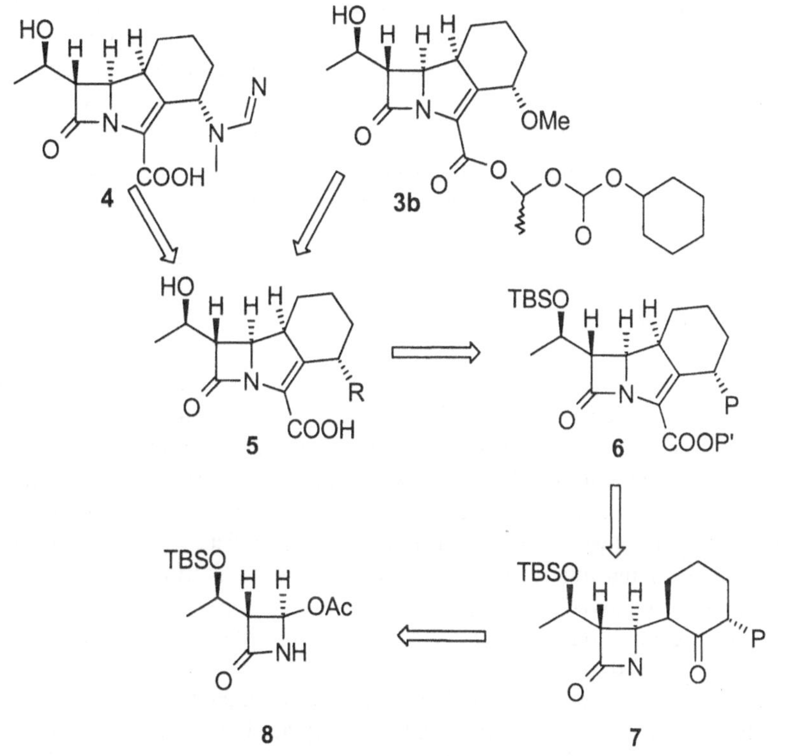

3a R= Na, Sanfetrinem
3b R= CH(CH₃)OCOOC₆H₁₁
Sanfetrinem cilexetil

4 GV129606

2

Scheme 1. Rethrosynthetic analysis of the general strategy for the synthesis of trinems

From the structural point of view a strained tricyclic structure and the presence of five stereogenic centres constitute an important challenge for the synthetic organic chemist. In this paper an account of the chemical strategies that have been pursued in order to access to all the classes of trinems will be given. Most of the examples will be taken from the most widely studied class with general structure **2** (Fig.3) but most of the methodologies can be

349

applied, and in some cases, were successfully used for accessing other subclasses of trinems.

GENERAL SYNTHETIC STRATEGY

Preliminary considerations

A retrosynthetic analysis on the synthesis of trinems is represented in Scheme 1. Commercially available acetoxy azetidinone **8** can be considered a suitable starting point; although rather expensive it is available on a large scale.

The first part of the route consists in the synthesis of ketoazetidinone **7** which contains all the stereogenic centres already in place with the appropriate configuration; moreover the substituent P already includes the appropriate protecting groups. The second stage consists in the formation of ring **B** while the final part is more related to protecting group removal and further derivatisation on the side chain at C-4. As an integrating part of the strategy the chemistry developed for this route should be flexible enough to accommodate a wide variety of rings **C** and substituents R; moreover stereoselective synthesis of key intermediates should be set up early in order to avoid tedious and complicated separation processes. Particular attention should also be paid to the synthesis of advanced intermediates that could give easily access to wide variety of derivatives minimizing the chemist's effort.

Protecting group removal

The *tert*-butyldimethylsilyl (TBS) group is the most widely used as a protecting group on the hydroxyethyl side chain. It was found generally stable under the conditions used for trinem synthesis and can be removed by fluoride ion under very mild conditions.

The standard literature conditions (n-Bu$_4$NF/AcOH in THF, room temperature)[8] was found generally effective although yields are variable and generally in the range 20-90%; removal should be always performed prior to carboxylate ester deprotection. With particularly strained and relatively unstable structures such as trinems **B** (n=1, see Fig.2), the TBS group was removed prior to ring B formation and replaced with TMS that could be removed after cyclisation without causing decomposition of the trinem structure (n-Bu$_4$NF/AcOH in THF, -30°C).[9]
synthesis of sanfetrinem and found of general utility in the trinem classes.

Highly expensive n-Bu$_4$NF (TBAF) could be replaced by the combination of solid KF (or CsF) and n-Bu$_4$NBr. The reaction temperature was also raised to 40-50°C thus compressing the reaction time from 16-40 h to 4-8 h. This procedure did not dramatically change yields and the purification protocols.

Very recently we found that commercially available Et$_3$N·3HF in N-methylpyrrolidone is a more effective and milder deprotecting agents allowing also a more rapid work up and purification. This method rapidly replaced the previous ones in the Medicinal Chemistry route to trinems.

The most widely used as protecting group for the carboxylate moiety is the allyl group both for its stability to a the most widely used reaction conditions on trinems and also because it can be removed under mild conditions by a palladium catalyzed deallylation.[10]

Particular care must be taken in the selection of the nucleophile that is acting as acceptor of the allyl moiety. In case of absence of either a basic or a positively charged moiety either on the side chain or in the ring C of the final compound, sodium

2-ethyhexanoate is the nucleophile of choice and the final trinem can be obtained as the sodium salt by precipitation from the reaction mixture.[11] In case a basic nitrogen is present, 5,5-dimethyl-1,3-dioxocyclohexane is the reagent of choice and the final compound can be obtained as internal salt by precipitation.[12] In the presence of positively charged quaternary ammonium salts n-Bu₃SnH gives the best results.[13] In Scheme 2 examples of applications of these methodologies are given.

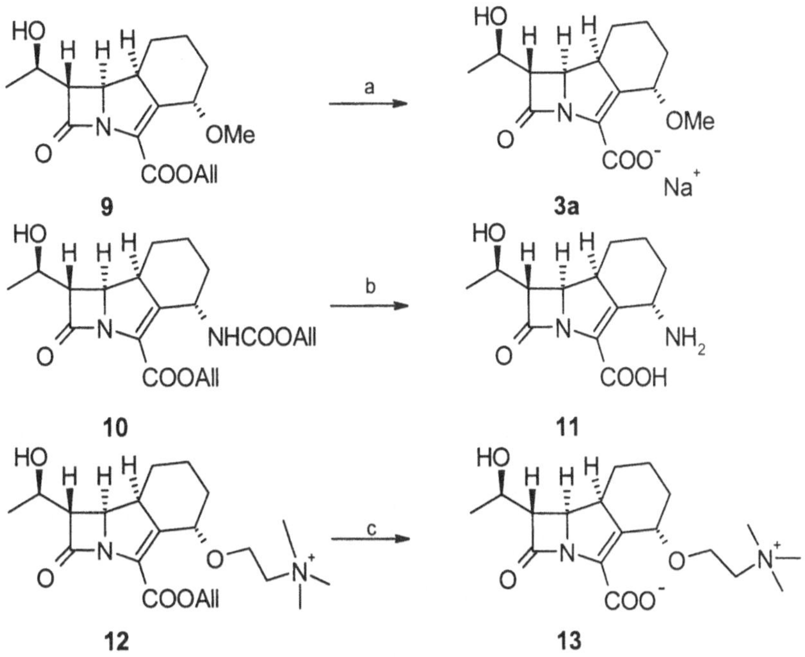

Scheme 2. a) Pd(PPh₃)₄, Na 2-ethylhexanoate, CH₂Cl₂; b) Pd(PPh₃)₄, dimedone, CH₂Cl₂; c) Pd (PPh₃)₄, n-Bu₃SnH, DMF.

A first modification in the general methodology was first introduced in the large scale
Other cleavable esters that have been employed are benzyl, substituted benzyl and diphenylmethyl that can be removed by palladium catalysed hydrogenolysis. We found that successful removal of these groups is very much substrate dependent with compounds resulting from the hydrogenation of the double bond being the most important by-products. These groups are predominantly used during scale-up studies on trinems at the development stage because they offer the advantage of giving better crystalline intermediates and minimize the risks of palladium contamination in the final products.

Cyclisation strategies

All the chemical strategies followed for formation of the trinem skeleton involved the double bond formation at C-2/C-3. From the chemistry of penems and carbapenems two routes were available and both have been successfully used (Scheme 3): the phosphorane route, first developed by Woodward,[14] and the triethylphosphite mediated oxo-oxo cyclisation (Scheme 3, Route B).[15]

Advantages shown by the phosphorane route are generally high yields and better quality of the reaction mixtures during the cyclisation step and the possibility to remove the silyl protecting group prior to cyclisation by aqueous hydrochloric acid in a two phase system. On the other hand the preparation of the phosphorane is usually long and tedious involving the transformation of hemiacetal **14** into a rather unstable chloride derivative in THF solution at low temperature followed by displacement of the chloride ion by triphenylphosphine in dry THF in the presence of lutidine. The reaction yields are variable and purification of phosphorane is complicated. We recently set up in our laboratories a more straightforward methodology in which **14** is directly converted into phosphorane **15** by portionwise addition of CBr_4 into a solution containing **14**, PPh$_3$ and lutidine at room temperature. Reaction yield were found in the 50-80% range from **7** and the purification process much simpler.

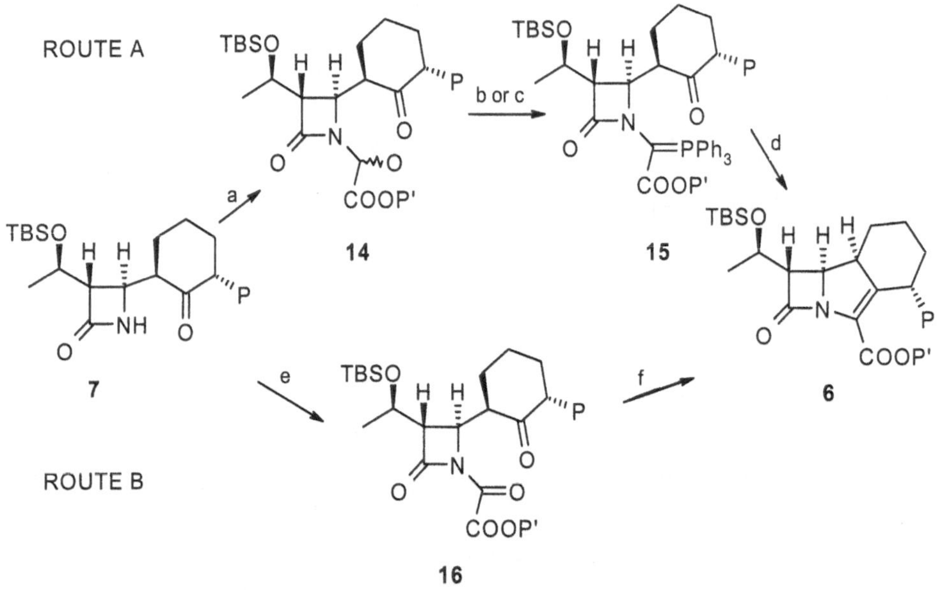

Scheme 3. a) OCHCOOAll, toluene, ref.; b) SOCl$_2$, lutidine, THF, -40°C then PPh$_3$, lutidine, 40°C; c) CBr$_4$, PPh$_3$, lutidine, DMF, 23°C; d) Δ; e) ClCOCOOAll, TEA, CH$_2$Cl$_2$; e) P(OEt)$_3$, toluene or xylene, ref.; f) MeP(OEt)$_2$, toluene, 100°C.

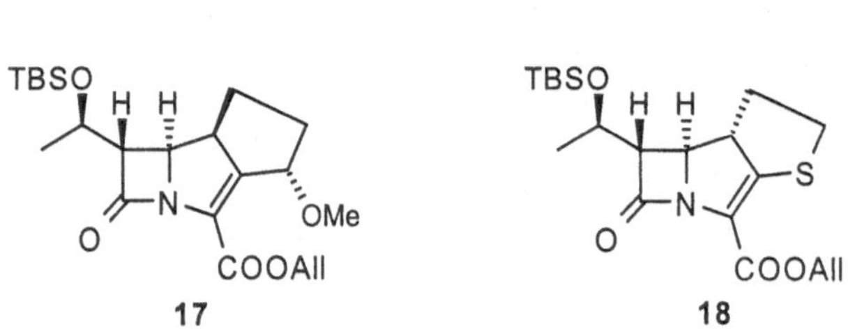

Figure 4.

The oxo-oxo cyclisation procedure involves the formation of an oxalimido derivative **16** followed by triethyl phosphite mediated ring closure at 110-140 °C depending on the substrate. These conditions are harsher compared to the phosphorane route and particularly strained trinems such as **17**[16] and **18**[9] (Fig. 4) cyclisation occurred *via* phosphorane while extensive decomposition was obtained with P(OEt)$_3$.

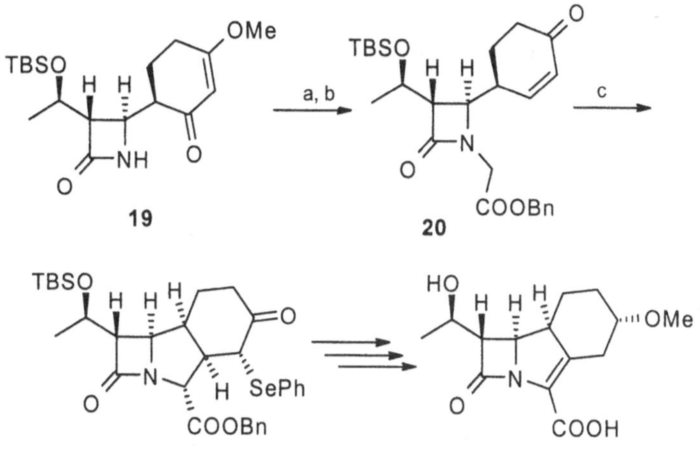

Scheme 4. a) DIBAL-H, THF, -78°C, work -up with 10% aqueous HCl; b) NaHMDS, THF then BrCH$_2$COOEt, -78 °C; c) NaHMDS, THF, -78 °C then ZnCl$_2$ the PhSeBr, -78 °C.

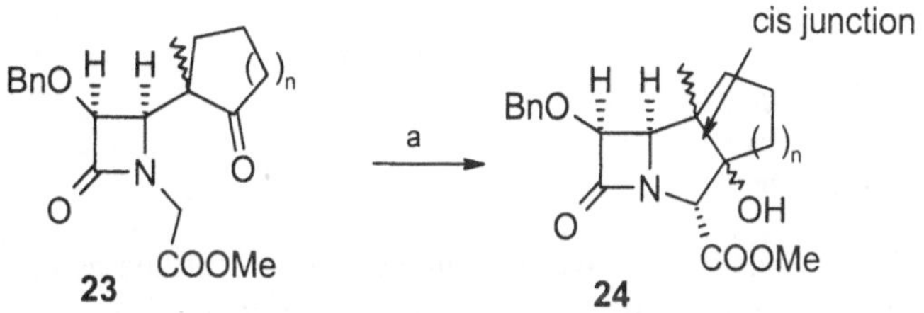

Scheme 5. a) LiHMDS, THF, -78°C.

Chemists at Hoechst[17] observed that methyl diethyl phosphite is a more effective reagent than triethyl phosphite and allowed them to run the cyclisation step at a lower temperature thus increasing yields. This has been also confirmed by us in house.

Although these proved to be methodologies of general use, other cyclisation strategies have been reported. Hanessian[18] published the synthesis of 5-methoxytrinem **22** by an elegant route involving intramolecular Michael addition on the enone moiety of compound **20** (Scheme 4). Isomerization of the double bond is another key step of the synthesis.

Alcaide[19] published a route to access the trinem skeleton *via* intramolecular aldol-type reaction (Scheme 5). A limitation of this methodology is that the stereogenic centre must be a tetrasubstituted carbon to prevent competitive enolisations that could give decomposition of the β-lactam ring *via* β-elimination. Moreover dehydration to produce the trinem moiety is required.

STEREOSELECTIVE SYNTHESIS OF KETOAZETIDINONES

Introduction

Ketoazetidinones with general structure **25** (Fig.5) can be considered key intermediates *en route* to trinems. In particular, when one or more substituents on the ketone ring, a particular care must be taken for their stereoselective production because purification of mixture of four or more isomers is usually complicated on a multigram scale. Furthermore, if a direct one step stereoselective approach is very convenient for the synthesis of large amounts of selected compounds, it is more convenient from the Medicinal Chemistry point of view to establish flexible routes to both **25** and **1** that could allow the small scale synthesis of a large number of derivatives starting from common advanced intermediates.

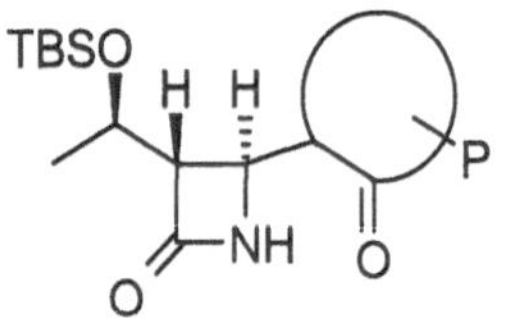

Figure 5. General structure for ketoazetidinones

Most of the examples given in the following pages will be taken from the most extensively studied series with ring **C** being a six membered carbocyclic ring but the methodologies have been successfully applied to other different ring systems.

Epoxides as key intermediates

Epoxides **28a-d** (Scheme 6) have been immediately identified as possible advanced intermediates for the production of 4-substituted trinems. Their conversion from ketoazetidinones **26a,b** could be easily achieved *via* Shapiro reaction by transformation of **26a,b** into their *p*-toluenesulphonyl hydrazone derivatives followed by treatment with excess of methyl lithium or LDA at low temperature followed by epoxidation with *m*-chloroperoxybenzoic acid. Because four epoxides are possible starting from acetoxyazetidinone **8**, a stereoselective strategy to obtain them in high yields and large amounts without needing complicated chromatographic purification stages had to be established. Reaction between **8** and trimethylsilyloxycyclohexene mediated by Lewis acids has been extensively studied (see Scheme 7) and representative results are shown in Table 1. It was found that the stereoselectivity of the reaction depends both on the Lewis acids and the substrate used. Milder Lewis acids such as $ZnCl_2$ and $BF_3 \cdot Et_2O$ influence the reaction towards a *anti* selectivity while after addition of Lewis acids no stereoselectivty is observed. when the lactam nitrogen is protected with a trimethylsilyl group the reactivity of the substrate is depressed and only stronger Lewis acids can promote the reaction and the selectivity is reversed.

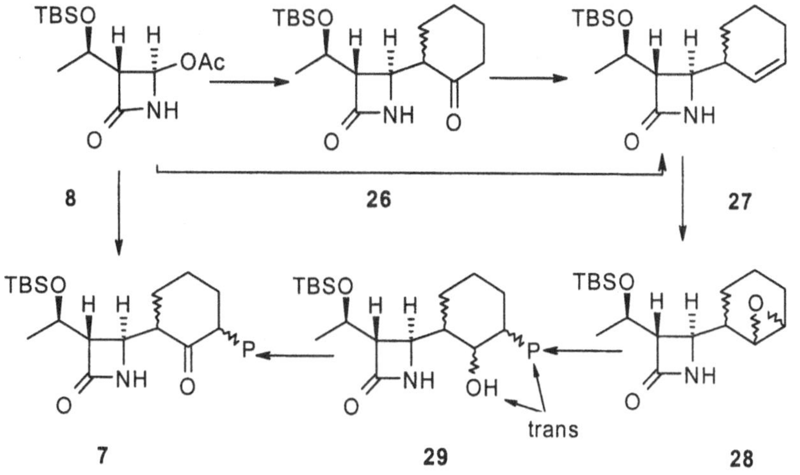

Scheme 6. General synthesis of α,α'-substituted ketoazetidinones.

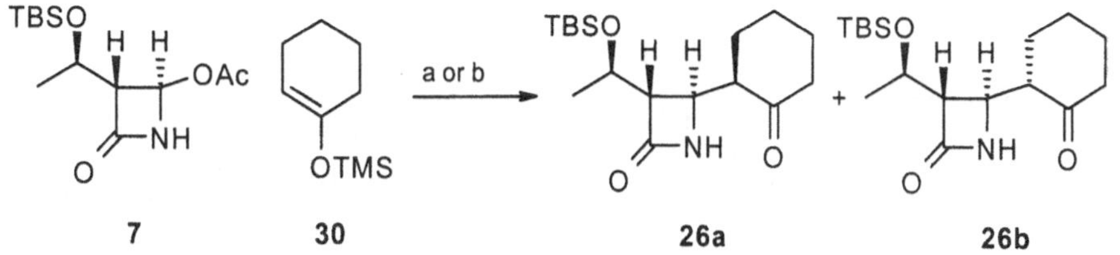

Scheme 7. a) Lewis acid, CH₂Cl₂, see Table 1; b) i: TMSCl, Et₃N, CH₃CN; ii: see Table 1; iii: KF, MeOH.

Table 1. summary of the results for the reaction between **7** and **30**.

Method	Lewis a.	Solv.	T (°C)	Combined yield	**26a**	**26b**
a	ZnCl₂	CH₂Cl₂	0 to 23	95%	3	7
a	BF₃	CH₂Cl₂	0	80%	2	8
a	TMSOTf	CH₂Cl₂	0	90%	1	1
b	SnCl₄	CH₂Cl₂	0	82%	75	25
b	TMSOTf	CH₃CN	0	97%	7	3

Epoxides **28a-d** react with nucleophiles to give α,α'-disubstituted alcohols in moderate to good yields (Scheme 9). Examples of reaction of epoxide **28b** and nucleophiles are given in Table 3; the regioselectivity observed was also complete.

Although the stereoselectivty observed is not remarkable nevertheless, for practical purposes, it enabled the production on a large scale of the desired isomers **26a** or **26b** thus avoiding long and tedious chromatographic purification.

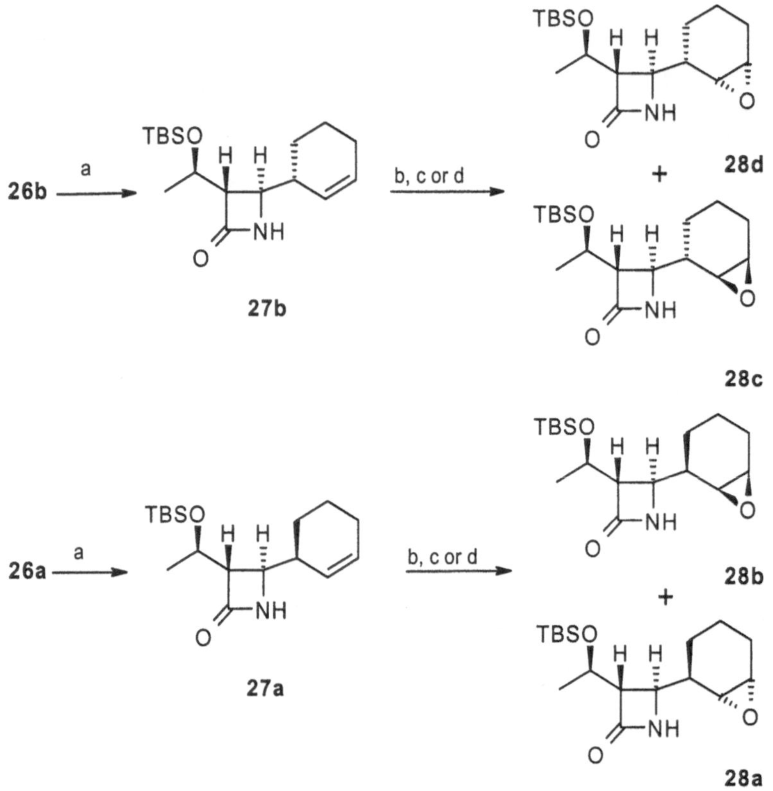

Scheme 8. a) i: TsNHNH$_2$, AcOEt; ii: LDA, THF, -30°C; b) *m*-CPBA, CH$_2$Cl$_2$; c) or Mg monoperoxyphtalate, CH$_2$Cl$_2$; d i: TBSCl, Et$_3$N, DMF; ii: *m*-CPBA, CH$_2$Cl$_2$; iii: KF, MeOH.

Table2. Epoxidation of **27a** and **27b** (see Scheme 8 for reaction conditions).

method	substrate	yield	**28d**	**28c**	substrate	yield	**28a**	**28b**
b	**27b**	90%	8	2	**27a**	76%	2	8
c					**27a**	91%	6	94
d		-	-	-	**27a**	60%	94	6

Intermediates **27a** and **27b** could be produced in large scale from **26a** and **26b** respectively. and transformed into the desired epoxides **28a-d** (as shown in Scheme 8) by simple oxidation with *meta*-chloroperoxybenzoic acid. Again the substrate is able to influence the stereoselectivity of the oxidation with *syn* isomers **28b** and **28d** preferentially formed from unprotected **27a** and **27b** and isomers *anti* being when oxidation was performed on substrates protected at the lactam nitrogen with *tert*-butyldimethylsilyl group.[20] A summary of the observed results is reported in Table 2.

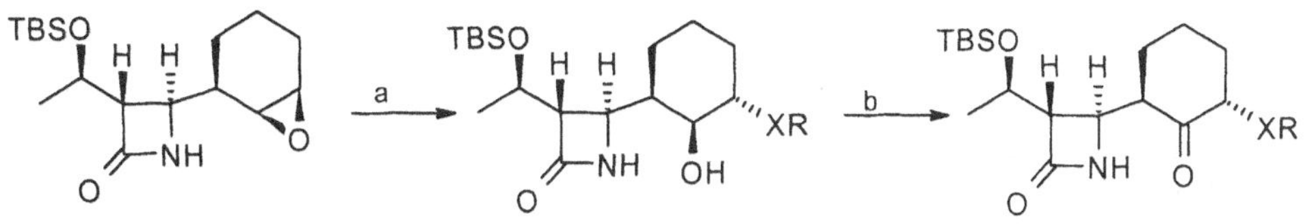

Scheme 9. a) see Table 3; b) Swern oxidation.

Table 3. Reaction of epoxide **28b** with alcohols and amines.

RXH	Solvent	Catalyst	Temperature	yield
MeOH	-	p-TSA	Reflux	80%
$HOCH_2CH_2OH$	$HOCH_2CH_2OH/CH_2Cl_2$ 20/1	p-TSA	23°C	50%
FCH_2CH_2OH	-	-	23°C	21%
$NCCH_2CH_2OH$	-	CAN	23°C	20%
NaN_3	ETOH/H2O	NH_4Cl	Reflux	73%
MeH_2	ETOH/H2O	NH_4Cl	Reflux	>95%
Me_2NH	ETOH/H2O	NH_4Cl	Reflux	55%
$CyC_5H_9NH_2$	ETOH/H2O	NH_4Cl	Reflux	45%

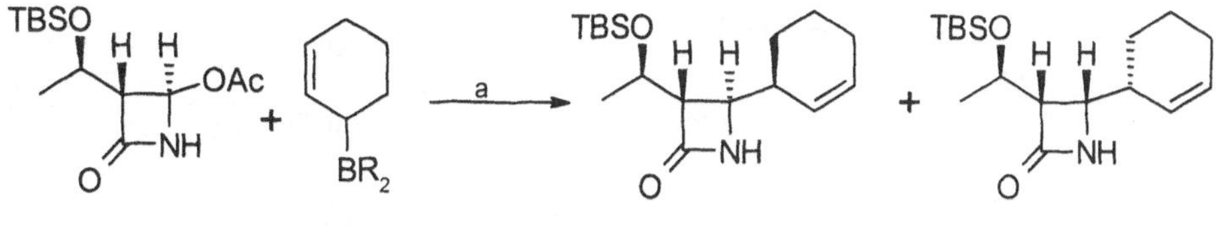

Scheme 10. a) Et_2Zn, THF, -30°C.

Stereoselective synthesis of cyclohexenyl azetidinone 27a

The finding that carbocyclic trinems with configuration (*S*) at C-9 were more biologically interesting compared to the (9*R*) isomers prompted us also to focus on the one step synthesis of **27b** that was previously obtained in 40% yield by a three steps route from 4-acetoxy azetidinone **8**..

After initial promising but not completely satisfactory results,[21] we found that dialkylcyclohexenylboranes can react under the influence of diethylzinc with **8** to give in almost quantitative yield a mixture of compounds **27a** and **33**.[22] Isomer **27b** could not be detected by HPLC and NMR analysis of the crude reaction mixture thus demonstrating the extremely high *syn* selectivity of the reaction mixture and the total reagent control on the selectivity at C-4 of the azetidinone ring. This result is unprecedented in the reactions of **8** with carbon nucleophiles with the silyloxyethyl side chain controlling the face of addition of nucleophiles

Table 4. Reaction of acetoxyazetidinone **7** with cyclohexenylboranes **33a-d**

Borane	R₂B	Solvent	yield	ratio **27a/33**
33a	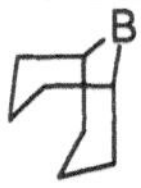	ET₂O	95%	1/1
33b	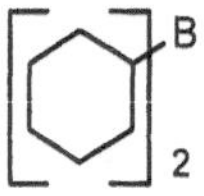	THF	98%	1/1
33c	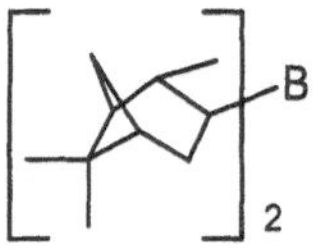	THF	95%	19/1
33d	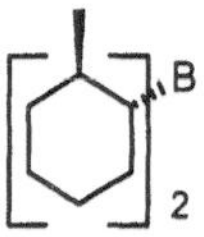	THF	96%	19/1

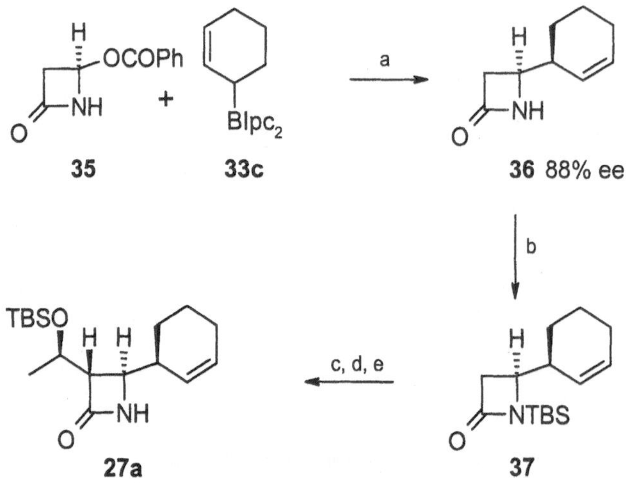

Scheme 11. a) Et₂Zn, THF, -78°C to r.t.; b) TBSCl, Et₃N, DMF; c) LDA, THF, -78°C then (CH₃)₃C(CH₃)₂SiCOCH₃; c) KOC(CH₃)₃, -78°C to r.t.; d) KF, MeOH (45% from **35**).

The use of optically pure borane **33c**, obtained from (–)-α-pinene gave preferentially the desired isomer **27a** in 90% isolated yield and 19/1 selectivity.

Further studies demonstrated that, with particularly hindered boranes, a kinetic resolution between the two enantiomeric cyclohexenyl dialkylboranes is possible: reagent (±)-**33d**, obtained form inexpensive 1-methylcylcohexene, gave similar results compared to **33c** and this allowed us to bypass the use of (–)-α-pinene.

An enantioselective access to the trinem class was also established by means of the borane chemistry (Scheme 11); racemic 4-benzoyloxyazetidinone (**35**) was transformed into the cyclohexenyl derivative **36** with 88% ee (chiral GC). The silyloxyethyl side chain was then installed by applying a well known procedure involving a silyl 1,2-migration.[23] The final compound was thus obtained after removal of the TBS protecting group on the amidic nitrogen in 45% overall yield from **35** and 90% ee by comparison with the rotation index with a pure sample of **27a**.

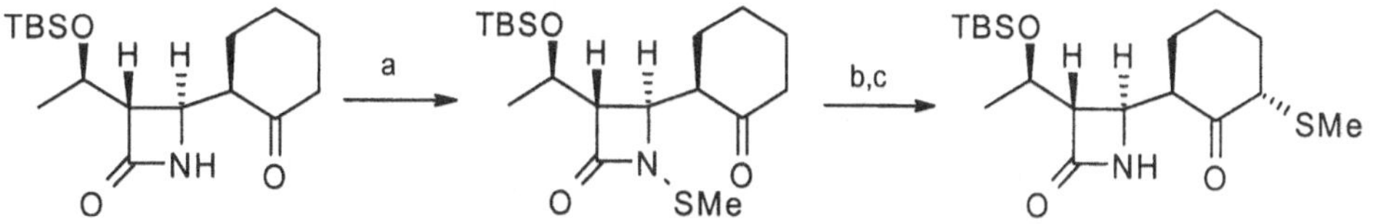

26a   **38**   **39**

Scheme12. a) LiN(SiCH$_3$)$_3$)$_2$, 0°C, THF then CH$_3$SO$_2$CH$_3$; b) LiN(SiCH$_3$)$_3$)$_2$, -78°C, THF then CH$_3$SO$_2$CH$_3$, -30°C; c) 2-PySH, CH$_2$Cl$_2$.

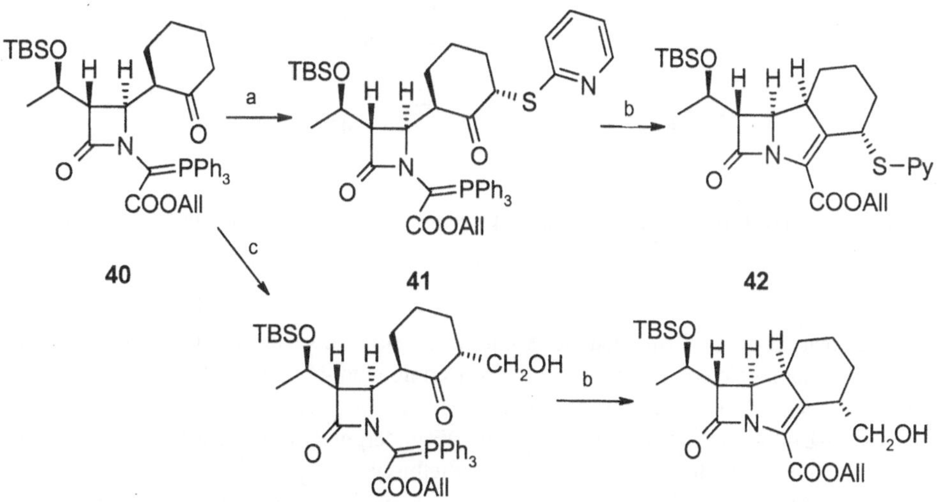

40   **41**   **42**

43   **44**

Scheme 13. a) LiN(SiCH$_3$)$_3$)$_2$, -78°C, THF then (2-PyS)$_2$; b) toluene, 100°C c) LiN(SiCH$_3$)$_3$)$_2$, -78°C, THF then HCHO.

Other routes to ketoazetidinones

Ketoazetidinones with general structure **7** can be directly obtained by treatment of **26a,b** with excess of LiN(Si(CH$_3$)$_3$)$_2$ followed by quench with an electrophile (Br$_2$, I$_2$, CH$_3$SO$_2$SCH$_3$ etc.). In this case yields are poor and mixtures of the two possible diastereomers are obtained.

The use of a suitable protecting group on the nitrogen exerts a beneficial effect on the stability of ketoazetidinones **26a,b** thus increasing chemical yields and controls the selectivity during the addition of the electrophile. Scheme 12 reports an example in which methylsulphide acts as the protecting group and intermediate **39** can be obtained as a single isomer in ca. 70% yield. A further advancement was the use of the phosphorane moiety that can function as a protective group for the lactam nitrogen and as reactant for the successive cyclisation to trinem moiety. Two representative examples are shown in Scheme 13. Phosphorane **40** undergoes enolisation with lithium bis-(trimethylsilyl)amide and reacts with electrophiles such 2,2'-dithiodipyridine or formaldehyde[24] to give intermediates **41** and **43** that can undergo cyclisation upon heating at 100°C in toluene.

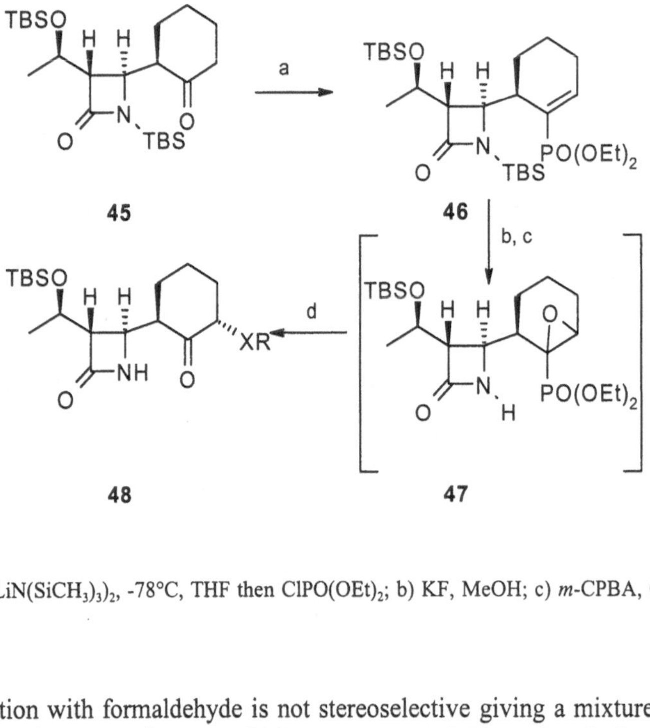

Scheme 14. a) LiN(SiCH$_3$)$_3$)$_2$, -78°C, THF then ClPO(OEt)$_2$; b) KF, MeOH; c) *m*-CPBA, CH$_2$Cl$_2$; d) RXH, 2,6-lutidine.

The reaction with formaldehyde is not stereoselective giving a mixture of epimers at the newly created stereogenic centre; however compound **44** can be obtained after heating **43a,b** in toluene while the corresponding epimer was not detected in the reaction mixture.

N-tert-Butyldimethylsilyketoazetidinone **45** undergoes lithium enolate formation and forms the enolphosphate **46** after quench with diethoxychlorophosphate.[25]

Oxidation with *m*-chloroperoxybenzoic acid yields an unstable epoxide that reacts with nucleophiles such as alcohols and amines to yield substituted ketoazetidinones in moderate yields and excellent selectivity. Primary amines require protection with an alkoxycarbonyl group after reaction with **47** before isolation. Lithium enolate derived from **45** efficiently reacts with molecular iodine (Scheme 15) to give an unstable intermediate **49** that can be reacted with soft nucleophiles such as sodium azide and both alkyl and aryl thiols under phase transfer conditions to give compound **50**, after hydrogenation to amine and protection with an allyloxycarbonyl group. It is remarkable that the absolute configuration at the newly created stereogenic centre is opposite when compared to the one obtained with the above discussed procedures thus demonstrating that an appropriate selection of the synthetic strategy can yield the products with the desired absolute configuration.[26] Another method for the stereoselective production of 6'-alkoxy - ketoazetidinones is exemplified in Scheme 16. Coupling between the silylenolether of 2-methoxycyclohex-2-enone and acetoxy azetidinone **8** produces a mixture of isomers **51a,b** in ratios that depend from the reaction conditions employed. Protection of the lactam nitrogen with a TBS group followed by hydrogenation of the double bond results in the formation of a single isomer. The same reaction performed without the use of protecting groups produces a mixture of isomers with the *syn* **52a** and **52b** prevailing.

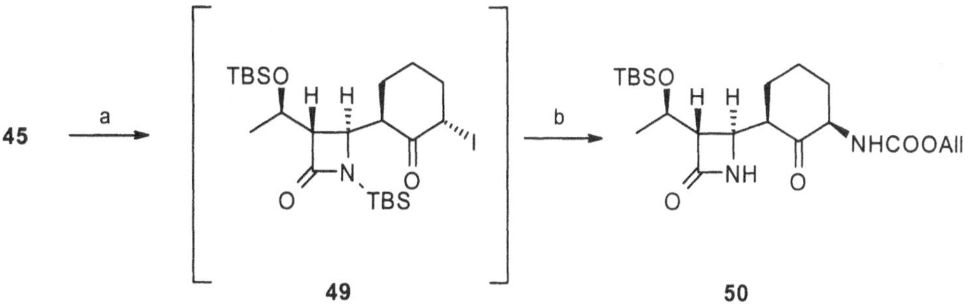

Scheme 15. a) LiN(SiCH$_3$)$_3$)$_2$, -78°C, THF then I$_2$; b) NaN$_3$; CH$_2$Cl$_2$/H$_2$O, Bu$_4$NCl; c) H$_2$, Pd/C, AcOEt; d) ClCOOAll, lutidine, AcOEt

Side chain functionalisation

During our studies we have identified a number of advanced intermediates that could be produced on a multigram scale and that could be easily transformed into a large number of trinem derivatives

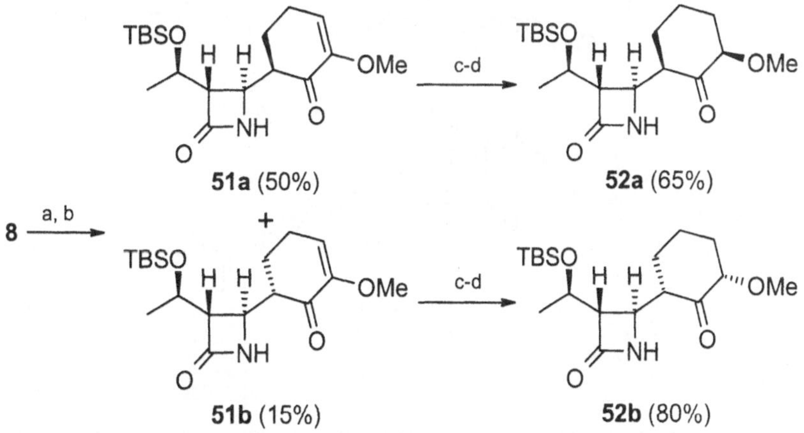

Scheme 16. a) TMSCl, TEA, CH$_3$CN; b) TMSOTf, CH$_2$Cl$_2$, 0°C, aqueous work-up then KF, MeOH; c) (CH$_3$)$_3$C(CH$_3$)$_2$SiCl, TEA, DMF; d) H$_2$, PD/C, AcOEt; d) TBAF, AcOH, THF.

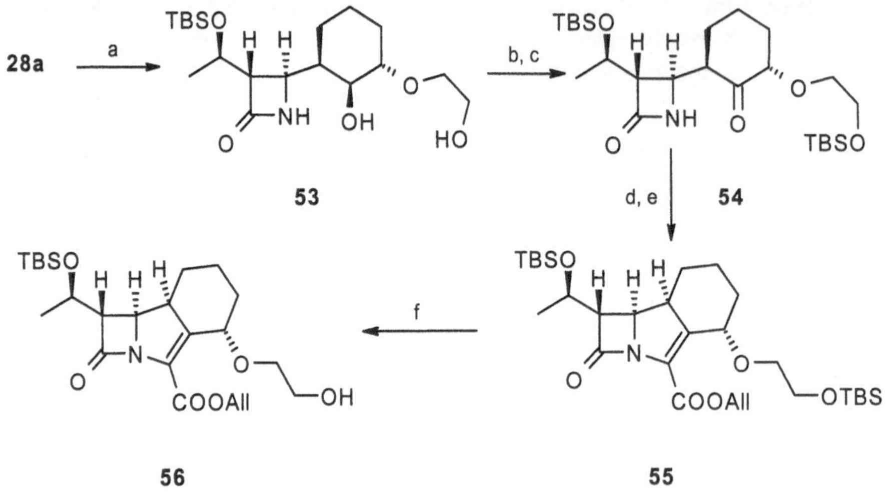

Scheme 17. a) HOCH₂CH₂OH, CH₂Cl₂, TsOH, 23°C; b) (CH₃)₃C(CH₃)₂SiCl, imidazole, DMF; c) ClCOCOCl, DMSO, Et₃N, CH₂Cl₂, -78°C; d) ClCOCOOAll, Et₃N, Ch₂Cl₂, 23°C; e) P(OEt)₃, xylene, ref.; f) TBAF, AcOH, THF, 23 °C.

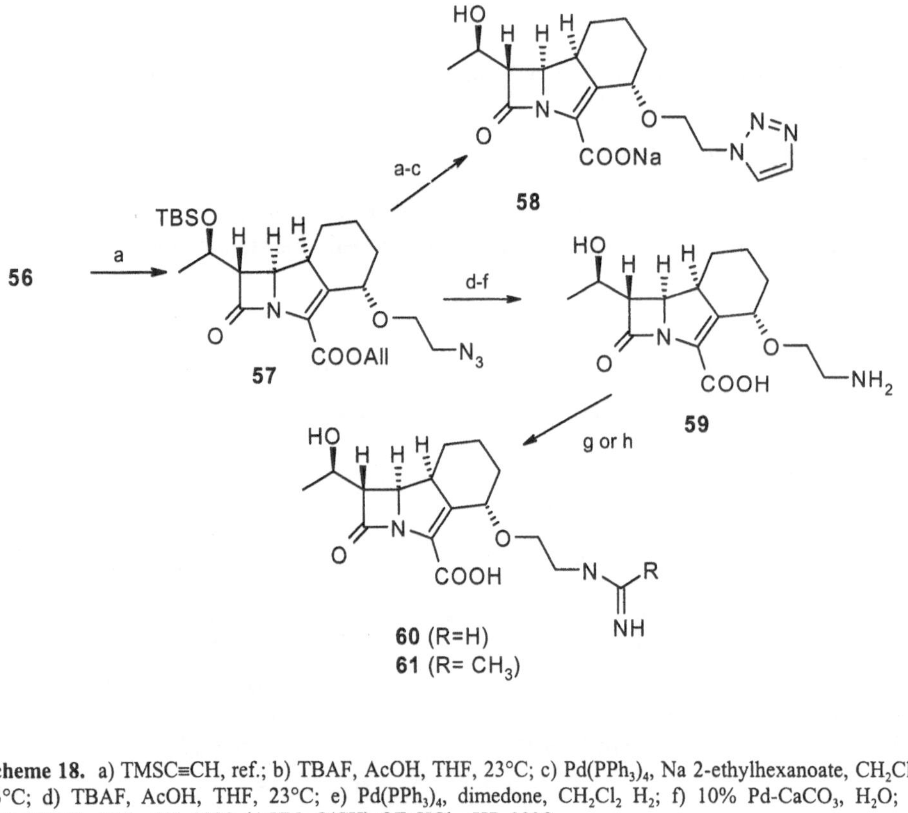

Scheme 18. a) TMSC≡CH, ref.; b) TBAF, AcOH, THF, 23°C; c) Pd(PPh₃)₄, Na 2-ethylhexanoate, CH₂Cl₂, 23°C; d) TBAF, AcOH, THF, 23°C; e) Pd(PPh₃)₄, dimedone, CH₂Cl₂ H₂; f) 10% Pd-CaCO₃, H₂O; g) HN=CHOBn·HCl, pH7, 23°C; h) HN=C(CH)₃OEt·HCl, pH7, 23°C.

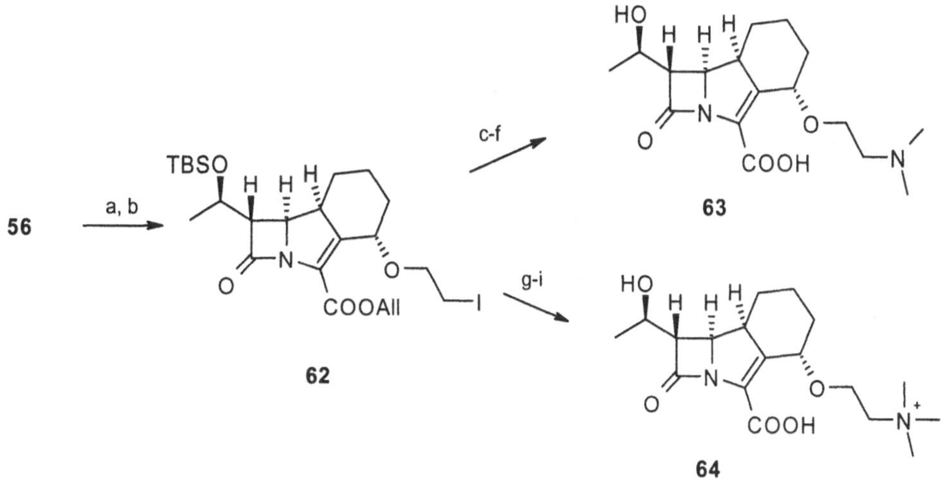

Scheme 19. a) TsCl, Et₃N, CH₂Cl₂, 23°C; b) NaI, acetone, 23°C; c) TBAF, AcOH, THF, 23°C; d) (CH₃)₃SiOTf, TEA, CH₂Cl₂, -78°C; e) Me₂NSi(CH₃)₂, 90°C; f) Pd(PPh₃)₄, Na 2-ethylhexanoate, CH₂Cl₂, 23°C; g) TBAF, AcOH, THF, 23°C; h) (CH₃)₃N, DMF, sealed tube, 90°C; i) Pd(PPh₃)₄, n-Bu₃SnH, CH₂Cl₂, 23°C.

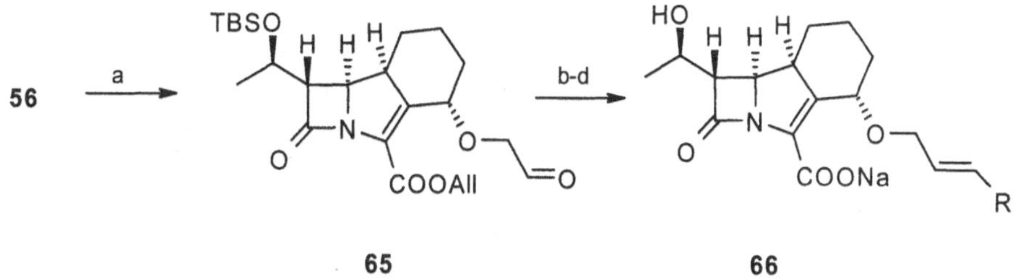

Scheme 20. a) DMSO, Et₃N, CH₂Cl₂, -78°C; b) PPh₃=CHCOR, CH₂Cl₂, 23°C; c) TBAF, AcOH, THF, 23°C; d) Pd(PPh₃)₄, Na 2-ethylhexanoate, CH₂Cl₂, 23°C.

In particular compound **56** (Scheme 17) is an interesting advanced intermediate. It can be easily produced by reaction of **28b** with ethylene glycol (Table 3) catalysed by *p*-toulenesulphonic acid followed by protection on the primary hydroxyl group, oxidation of the free secondary alcohol function to ketone, cyclisation to fully protected trinem **55** and selective removlval of the TBS group. Intermediate **56** can be transformed in azido derivative **57** by simple treatment with carbon tetrabromide, triphenyl phosphine and sodium azide in dimethylformamide[27] as reported in Scheme 18. The azide can be in turn reacted with alkynes to give triazinyl ethoxy trinems after removal of the protecting groups (In scheme the preparation of **58** is exemplified), reduced to amine **59** that can be further functionalized to trinems **60** and **61** by reaction with the appropriate imidating reagents.

Compound **56** is easily converted into the iodoethoxy derivative **62** (Scheme 19) which can react with trimethylsilyl dialkylamines to give dialkylaminoethoxy trinems (Exemplified by **63** in Scheme 19); **56** tertiary amines and pyridines to give the ammoniummethoxytrinems after removal of the protecting groups (exemplified by **64** in Scheme 19). It is worth mentioning that the trimethylsilyl group must be present on dialkylamines in order to avoid decomposition of **62** due to nucleophilic attack on the β-lactam ring.

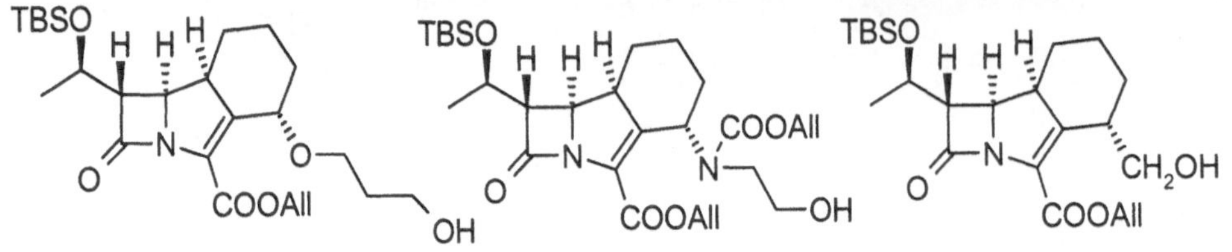

Figure 6.

Trinem **56** was also converted into the aldehyde derivative **65** (Scheme 20) that could undergo Wittig reaction with a variety of stabilized phosphoranes and transformed, after removal of the protecting groups, into allyloxytrinems **66**. The same transformations could be applied for intermediates **67, 68,** and **69**[23] (Fig. 6).

Another important group of intermediates are the fully protected 4-alkylamino trinems

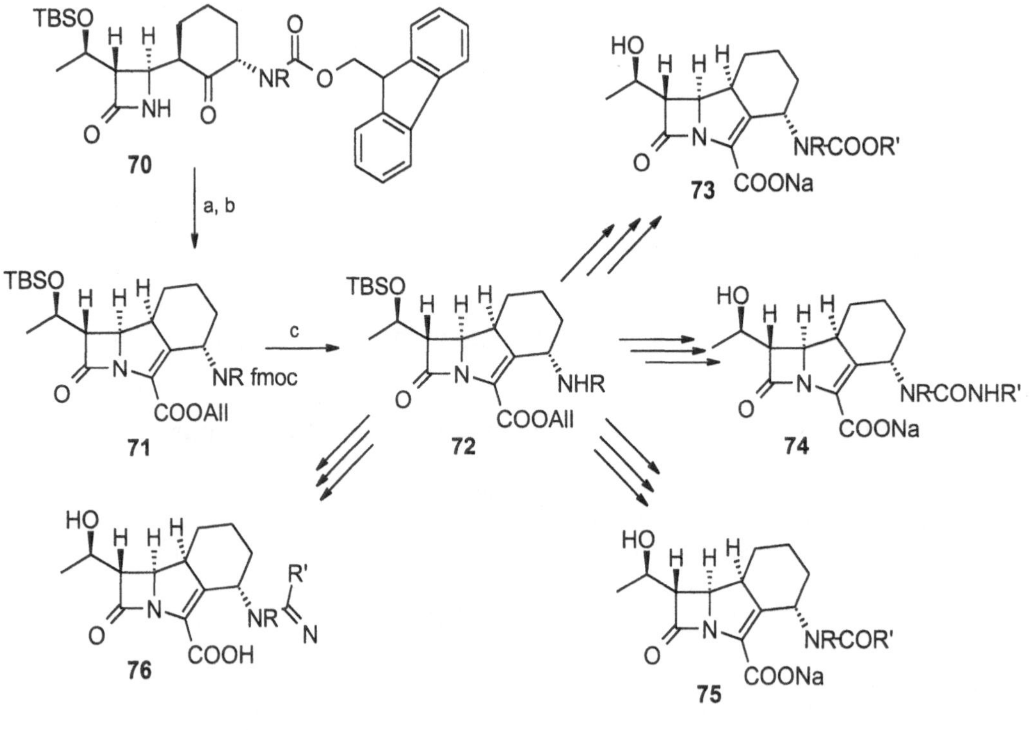

Scheme 21. a) ClCOCOOAll, Et$_3$N, CH$_2$Cl$_2$, 23°C; b) P(OEt)$_3$, xylene, ref.; c) DBU, CH$_3$CN, 80°C.

72 (Scheme 21) that can be obtained by using an appropriate protecting group on the amino moiety that is ortogonal to the protecting group on the carboxylate. It was found that the fluorenylmethoxycarbonyl moiety is suitable for this class of compounds. It can be installed during the formation of ketoazetidinones **70** (see Schemes 9 and 14) and is stable under the cyclisation conditions and is easily removed by heating **71** with a tertiary amine.

Fully protected trinem **72** react with a variety of acylating agents to give urethanes **73**, ureido trinems **74**,[28] amides **75**, and amidines **76**.

All the classes of trinems that were accessed through these advanced intermediates were found to possess a very good antibacterial activity and biological stability with the spectrum of action depending from the structure and the phisical properties of the substituents used.

CONCLUSIONS

The discovery of trinems, a novel and potent class of β- lactam antibiotics, brought important synthetic challenges both to the medicinal chemist and to the synthetic organic chemist. The task of fine tuning many structural parameters in order to maximize a variety of different biological features such as potency, spectrum of action, stability to β-lactamases and dihydopeptidases, pharmacokinetic profile was matched by the task of developing flexible, robust and stereoselective syntheses that would allow the large scale production of key intermediates.

REFERENCES

1. See for example; H. C. Neu and R. D. G. Cooper in The Chemistry of β-Lactams, M. I. Page Ed., Chapman and Hall, London, (1992).
2. B. Tamburini, A. Perboni, T. Rossi, D. Donati, D. Andreotti, G. Gaviraghi, R. Carlesso, and C. Bismara, Preparation of 10-(1-hydroxyethyl)-11-oxo-1-azatricyclo[7.2.0.0.3,8]undec-2-ene-2-carboxylic acid derivatives as antibacterials, EP 416953 A2 910313 (1991), CA 116:235337 (1991).
3. M. Sendai and T. Miwa, Preparation of tricyclic carbapenem compounds as antibiotics, EP 422596 A2 910417, (1991), CAN 115:279692.
4. a) E. Di Modugno, I. Erbetti, L. Ferrari, G. Galassi, S.M. Hammond and L. Xerri, In vitro activity of the tribactam GV104326 against gram-positive, gram-negative, and anaerobic bacteria, *Antimicrob. Agents Chemother* 38:2362 (1994); b) E. Di Modugno, R. Broggio, I. Erbetti, and J. Lowther, In vitro and in vivo antibacterial activities of GV129606, a new broad-spectrum trinem. *Antimicrob. Agents Chemother.* 41:2742-2748 (1997).
5. See for example a) K.D. Dykstra, F. DiNinno, M.L. Hammond, J.L. Huber, J.G. Sundelof, and G.G. Hammond, Synthesis and biological activity of hexacyclic 2-arylcarbapenems: Potent anti-MRS agents, Book of Abstracts, 212th ACS National Meeting, Orlando, FL, August 25-29 (1996), MEDI-036. American Chemical Society, Washington D.C. b) F. Dininno, Bridged biphenyl carbapenem antibacterial compounds, WO 9503700 A1 950209I, (1995), CA 122:265017.
6. a) D.C. Coleman, Methicillin-resistant Staphylococcus aureus: molecular epidemiology and expression of virulence determinants, *Infect. Dis. Ther.* 6:37. (1992); b) H.F. Chambers and C.J. Hackbarth, Methicillin-resistant Staphylococcus aureus: genetics and mechanisms of resistance. *Infect. Dis. Ther.* 6:21, (1992).
7. A. Andrus, F. Baker, F.A. Bouffard, L.D. Cama, B.G. Christensen, R.N. Guthikonda, J.V. Heck, D.B.R. Johnston, and W.J. Leanza, Structure-activity relationships among some totally synthetic carbapenems in Recent Adv. Chem. β-Lactam Antibiot Spec. Publ. - R. Soc. Chem. 86 (1985).
8. S. Hanessian, D. Desilets, and Y.L. Bennni, A novel ring-closure strategy for the carbapenems: the total synthesis of (+)-thienamycin. *J. Org. Chem.* 55:3098 (1990).
9. R. Di Fabio, A. Feriani, G. Gaviraghi, and T. Rossi, Synthesis and biological evaluation of 4-heterotribactams, *Bioorg. Med. Chem. Lett.* 5:1235, (1995).
10. P.D. Jeffrey, and S.V. McCombie, Homogeneous, palladium(0)-catalyzed exchange deprotection of allylic esters, carbonates and carbamates. *J. Org. Chem.* 47:587, (1982).
11. D. Andreotti, S. Biondi, R. Di Fabio, D. Donati, E. Piga, and T. Rossi, Synthesis and antibacterial activity of 4- and 8-methoxy trinems. *Bioorg Med. Chem. Lett.* 6:1683, (1996).
12. M.E. Tranquillini, G.L. Araldi, D. Donati, G. Pentassuglia, A. Pezzoli, and A. Ursini, Synthesis and antimicrobial activity of 4-amino trinems. *Bioorg. Med. Chem. Lett* 6:1683-1688, (1996).
13. S. Biondi, D. Andreotti, T. Rossi, R. Carlesso, G. Tarzia, and A. Perboni, Preparation of anellated carbapenems as antibiotics EP 502464 A1 920909. CA 117:251135.
14. a) H.R. Pflander, J. Gosteli, R.B. Woodward, and G. Rihs, Structure, reactivity, and biological activity of strained bicyclic β-lactams. *J. Am. Chem. Soc.* 103:4526, (1981); b) J. Ernst, J. Gosteli, and R.B.

Woodward, The penems, a new class of β-lactam antibiotics. 5. Total synthesis of racemic 6-α-hydroxyethylpenemcarboxylic acids. *J. Am. Chem. Soc.* 101:6310, (1979).

15. A. Afonso, F. Hon, J. Weinstein, and A.K. Ganguly, A new synthesis of penems, the oxalimide cyclization reaction, *J. Am. Chem. Soc.* 104:6138, (1982).

16. S. Biondi, G. Gaviraghi, and T. Rossi, Synthesis and biological activity of novel tricyclic β-lactams *Bioorg. Med. Chem. Lett.*6:525, (1996).

17. a) U. Gerlach, R. Hoerlein, N. Krass, R. Lattrell, T. Wollmann, M. Limbert, and A. Markus, Preparation of tetrahydronaphthocarbapenems and analogs as antibiotics, EP 92-108792 920525 (1992) CAN 118:168890; b)

18. S. Hanessian, M.J. Rozema, G.B. Reddy, and J.F. Braganza, Tricyclic β-lactams: total synthesis and antibacterial activity of 5α- and 5β-methoxy-tribactams. *Bioorg. Med. Chem. Lett.* 5:2535, (1995).

19. B. Alcaide, C. Polanco, E. Saez, and M.A. Sierra, The Intramolecular Aldol Condensation Route to Fused Bi- and Tricyclic β-Lactams, *J. Org. Chem.* 61:7125 (1996).

20. C. Marchioro, G. Pentassuglia, A. Perboni, and D. Donati, Synthesis and NMR studies of key intermediates to a new class of β-lactams: the trinems. *J. Chem. Soc. Perkin Trans. 1*, 463, (1997).

21. C. Bismara, R. Di Fabio, D. Donati, T. Rossi, and R.J. Thomas, The synthesis of a key intermediate of tricyclic beta-lactam antibiotics. *Tetrahedron Lett* 36:4283-6, (1995).

22. T. Rossi, S. Biondi, S. Contini, R.J. Thomas, and C. Marchioro, novel amidoalkylation of 4-acetoxyazetidinones with allylic boranes. A stereoselective entry into the tribactams, *J. Am. Chem. Soc* 117:9604-5, (1995).

23. F.A. Bouffard and T.N. Salzmann, A new approach to the diastereoselective synthesis of aldols. Introduction of the 6α-(1R-hydroxyethyl) side chain of the carbapenem and penem antibiotics, *Tetrahedron Lett* 26:6285-8 (1985).

24. C. Ghiron, T. Rossi, and R.J.Thomas, The stereoselective synthesis of 4-formyltrinem, a key intermediate for novel trinems. *Tetrahedron Lett* 38:3569 (1997).

25. A. Perboni, Preparation of 3-(1-hydroxyethyl)-4-[1-phosphinyloxycyclohex-1-en(oxide)-6-yl]azetidinones as antibacterial intermediates. EP 502488 A2 920909. CA 117:233706.

26. B. Tamburini, A. Perboni, T. Rossi, D. Donati, D. Andreotti, G. Gaviraghi, S. Biondi, and C. Bismara, Preparation of 4-(2-oxocyclohexyl)azetidin-2-one derivatives as intermediates for antibiotics. EP 416952 A2 910313. CA 115:279690.

27. a) R. Carlesso, S. Holman, A. Perboni, and T. Rossi, Preparation of condensed carbapenem derivatives, WO 9405666 A1 940317, CA 121:57223; b) D. Andreotti, S. Biondi, R. Di Fabio, D. Donati, E. Piga, and T. Rossi, Synthesis and biological evaluation of 4-alkoxy substituted trinems. Part I. *Bioorg. Med. Chem. Lett.* 6:2019, (1996); c) R. Di Fabio, D. Andreotti, S. Biondi, G. Gaviraghi, and T. Rossi, Synthesis and biological evaluation of 4-alkoxysubstituted trinems. Part II. *Bioorg. Med. Chem. Lett.* 6:2025 (1996).

28. A.Perboni, G. Pentassuglia, D. Andreotti, and J.A. Winders, Urea derivatives of tricyclic carbapenems as bactericides WO 9513278 A1 950518. CAN 123:198515: b) S. Gehanne, E. Piga, D. Andreotti, S. Biondi, and D. Pizzi, Synthesis and antibacterial activity of 4-ureido trinems. *Bioorg. Med. Chem. Lett.* 6:2791, (1996).

INDEX